OPTICAL
TECHNIQUES *IN*
REGENERATIVE
MEDICINE

OPTICAL TECHNIQUES *IN* REGENERATIVE MEDICINE

Edited by

Stephen P. Morgan
Felicity R. A. J. Rose
Stephen J. Matcher

CRC Press
Taylor & Francis Group
Boca Raton London New York

CRC Press is an imprint of the
Taylor & Francis Group, an **informa** business

CRC Press
Taylor & Francis Group
6000 Broken Sound Parkway NW, Suite 300
Boca Raton, FL 33487-2742

First issued in paperback 2019

© 2014 by Taylor & Francis Group, LLC
CRC Press is an imprint of Taylor & Francis Group, an Informa business

No claim to original U.S. Government works

ISBN-13: 978-1-4398-5495-2 (hbk)
ISBN-13: 978-1-138-38208-4 (pbk)

Library of Congress Cataloging-in-Publication Data

Optical techniques in regenerative medicine / editors, Stephen P. Morgan, Felicity R. Rose, Stephen J. Matcher.
    p. ; cm.
    Includes bibliographical references and index.
    ISBN 978-1-4398-5495-2 (hardcover : alk. paper)
    I. Morgan, Stephen P., editor of compilation. II. Rose, Felicity R., editor of compilation. III. Matcher, Stephen J., editor of compilation.
    [DNLM: 1. Regenerative Medicine. 2. Optical Imaging--methods. 3. Tissue Engineering. WO 515]

R857.O6
610.28--dc23   2013025790

Visit the Taylor & Francis Web site at
http://www.taylorandfrancis.com

and the CRC Press Web site at
http://www.crcpress.com

Contents

SECTION I   Introduction and Background

SECTION II   Optical Microscopy in Regenerative Medicine

SECTION III   Optical Spectroscopy in
Regenerative Medicine

Preface

Regenerative medicine is a field of research driven by the clinical need to address the shortage of organs and tissues required to restore biological function to the body following damage or disease. The term *regenerative medicine* encompasses many research themes from cellular therapy (delivering cells to the body where they form tissue or restore function *in situ*) to tissue engineering, which addresses the need to implant an already functional or near-functional tissue (cells and a supporting matrix either natural or synthetic), but also includes delivery of acellular materials that provide biological cues (either topographical or chemical) to stimulate the body's own repair mechanisms.

Tissue engineering can involve many different strategies but often involves a cell type of interest, a supportive matrix or scaffold to provide a three-dimensional structure, and a dynamic *in vitro* environment to promote generation of neo-tissue prior to implantation. The matrix needs to be porous to allow cell and nutrient ingress throughout its structure, and it needs to degrade and/or allow remodeling by the cells such that they eventually replace the scaffold with a tissue relevant matrix. In addition, it needs to guide the proliferation and differentiation of cells such that the complex cellular interactions found *in vivo* are restored and maintained resulting ultimately in tissue function. This interaction between cells and the scaffold with subsequent scaffold remodeling and tissue growth is complex and is not well understood. Efficient and information-rich methods to monitor these processes are limited, which means that multiple experiments, controlling small numbers of parameters at a time, need to be conducted to fully understand tissue formation and maturation. Only when this understanding is achieved and can be controlled will successful generation of functional tissue be a reality, leading the way to the scale-up and automation required to take full clinical advantage of the technology.

Tissue engineers largely rely on techniques that are destructive, time-consuming, and laborious to inform them about a tissue engineering strategy, and these do not allow *in situ* and spatial monitoring of tissue growth. A further major barrier to regenerative medicine is being able to monitor what is happening once the therapy is implanted into the body, as in such situations, conventional destructive techniques and/or cell labeling are not possible. This impedes translation into the clinic with researchers essentially working blind. This has led to an explosion of cross-disciplinary research with optical bioengineers to develop *in situ*, online, and nondestructive analytical techniques for both *in vitro* and *in vivo* imaging. Optical techniques can be used to image and characterize many tissue properties such as blood vessel formation and blood flow, blood oxygen saturation, extracellular matrix deposition and arrangement within the tissue, and tissue biomechanics. In addition, they can be used to fabricate tissue engineering scaffolds and characterize their properties. There is therefore a key role for optical techniques in regenerative medicine by providing tools for assessment and monitoring. In many cases, optical techniques allow for label-free imaging; however, there is also an important role for novel contrast agents such as those based on nanoparticles. These will benefit many of the techniques described, but a detailed description of novel contrast agents and reporters is beyond the scope of this book. Here we focus on the imaging techniques and instrumentation and highlight contrast agents where appropriate.

The fundamental principles of the optical techniques used and illustrated within the context of regenerative medicine are introduced in Chapter 1 by Morgan and colleagues. Fluorescence techniques are used extensively within cell biology and regenerative medicine and are described in detail here. Chapter 2 by Green and Haycock illustrates how fluorescence microscopy can be used to monitor cells and their differentiation, while Chapter 4 by Gill and Claeyssens describes using one- and two-photon polymerization for scaffold fabrication. Characterizing extracellular matrix structures such as collagen and other fibrous components can be achieved using nonlinear scattered light (second harmonic generation) as described in Chapter 3 by Matcher. Second harmonic generation microscopy is often carried out simultaneously with multiphoton microscopy and fluorescence lifetime imaging. Chapter 7 by Lloyd, Chen, and Mycek describes employing fluorescence spectroscopy to monitor biomolecules as indirect markers of various cellular processes.

Process monitoring is a mature field where image processing and machine vision can provide valuable microscopic-scale information regarding cell culture within a bioreactor, for example. Chapter 5 by Prediger and colleagues describes such optical systems and image processing techniques that can be used to provide information regarding cell number, size distribution, and morphological properties. Optical light scattering polarization spectroscopy can also be used to characterize cells, scaffolds, and tissues; Chapter 6 by Georgakoudi and Hunter describes how the properties of elastically scattered light can be related to tissue structure, function, and composition. Inelastic light scattering can be used to generate Raman spectra providing a label-free, *in situ* technique to "chemically fingerprint" cell and tissue components as described by Notingher and Pascut in Chapter 8. In addition, they discuss how the resulting anti-Stokes signal can provide contrast providing sensitivity to lipid-rich structures.

Optical coherence tomography (OCT) can achieve high-resolution 3D imaging at penetration depths of typically up to 2 mm within either a tissue or scaffold. An introduction to OCT is provided in Chapter 10 by Zhao and colleagues, and they describe using polarization-sensitive OCT imaging to monitor the elastic properties of healing tissues and using a combined multiphoton/OCT system to track stem cells. Further applications are described in Chapter 11 by Yang and colleagues including the use of OCT to analyze scaffold structure. In photoacoustic tomography, described in Chapter 12 by Steenbergen, optically absorbing structures, such as blood vessels, can be imaged by utilizing the acoustic waves generated when pulsed diffuse light is absorbed.

High-resolution 3D images of tissue constructs can be generated using optical projection tomography (OPT) and has been used extensively in embryology research and to image tissue-engineered blood vessels as described in Chapter 9 by Swoger and coauthors. Ultrasound modulated optical tomography, described in Chapter 13 by Morgan and colleagues, uses sound waves to modulate the diffuse light intensity locally within a tissue. Finally, macroscopic imaging with relatively simple instrumentation is discussed. Chapter 14 by Kirkpatrick and colleagues describe how relatively simple reflectance measurements can be taken using a white light source and image processing algorithms to monitor tissue-engineered constructs within a bioreactor. With the addition of spectral imaging, this system can be adapted to image oxygen saturation, which could be used to assess the status of an organ in the body prior to injection of stem cells or tissue regeneration. The addition of coherent illumination from a laser allows flows and biomechanics to be imaged. Bioluminescence imaging, reviewed in Chapter 15 by van Rappard and colleagues, can be used to monitor light generated within a tissue and can be used to track the population distribution of stem cells *in vivo* on a macroscopic scale.

Regenerative medicine is a dynamic and rapidly moving research area, but its success is highly dependent on achieving the correct strategy. The development of non-destructive analytical techniques, which can elucidate the stages of tissue development both *in vitro* and *in vivo* or track the fate of cells following injection, has become highly desirable. Such techniques will no doubt play a fundamental role in understanding the principles of cell–material interactions, cell behavior and differentiation, neo-tissue formation, and subsequent tissue maturation. This understanding is vital to transform regenerative medicine from academic research to a viable clinical alternative to conventional treatments.

MATLAB® is a registered trademark of The MathWorks, Inc. For product information, please contact:

The MathWorks, Inc.
3 Apple Hill Drive
Natick, MA 01760-2098 USA
Tel: 508-647-7000
Fax: 508-647-7001
E-mail: info@mathworks.com
Web: www.mathworks.com

About the Editors

Stephen P. Morgan, PhD, is a professor of biomedical engineering at the University of Nottingham. Since 1992, he has been developing novel optical techniques for imaging and spectroscopy of tissues, with a particular interest in microcirculation. The imaging devices have been used in clinical studies and have also led to a blood flow imager marketed under license to Moor Instruments. His other main research interest is developing novel methods of imaging and sensing in regenerative medicine for both *in vivo* and *in vitro* tissues. He has particular interest in imaging 3D cell culture and *in vivo* imaging of tissue regeneration.

Felicity R. A. J. Rose, PhD, is an associate professor in tissue engineering at the University of Nottingham. Felicity's research focuses on developing materials that can control stem and differentiated cell behavior during the tissue regeneration process to develop *in vivo*–like *in vitro* models for safety, toxicity, and drug screening applications, and ultimately tissues for transplantation. Knowing what is happening to a cell or tissue during *in vitro* culture is vital to any regenerative medicine process and has led Felicity to explore optical techniques for this purpose as an end user.

Stephen J. Matcher, PhD, is a reader in biomedical engineering at the University of Sheffield. Since 1992, he has conducted research on the use of optical techniques to characterize tissue structure and function. He has developed optical and near-infrared spectroscopy as a tool for the non-invasive measurement of tissue oxygenation and now concentrates on the use of optical coherence tomography and non-linear microscopy to characterize the structure of native biological tissues and tissue-engineered replacements.

Contributors

Arne Bluma
Sartorius Stedim Biotech GmbH
Göttingen, Germany

Stephen A. Boppart
Beckman Institute for Advanced
    Science and Technology
University of Illinois at
    Urbana-Champaign
Urbana, Illinois

Leng-Chun Chen
Department of Biomedical
    Engineering
University of Michigan
Ann Arbor, Michigan

Frederik Claeyssens
Department of Materials Science and
    Engineering
The Kroto Research Institute
University of Sheffield
Sheffield, United Kingdom

Andy Downes
School of Engineering
University of Edinburgh
Edinburgh, United Kingdom

Irene Georgakoudi
Biomedical Engineering Department
Tufts University
Medford, Massachusetts

Andrew A. Gill
Department of Materials Science and
    Engineering
The Kroto Research Institute
University of Sheffield
Sheffield, United Kingdom

Benedikt W. Graf
Beckman Institute for Advanced
    Science and Technology
University of Illinois at
    Urbana-Champaign
Urbana, Illinois

Nicola Green
The Kroto Research Institute
University of Sheffield
Sheffield, United Kingdom

Mark A. Haidekker
Faculty of Engineering
University of Georgia
Athens, Georgia

John W. Haycock
Department of Materials Science and
    Engineering
and
The Kroto Research Institute
University of Sheffield
Sheffield, United Kingdom

Martin Hunter
Biomedical Engineering Department
Tufts University
Medford, Massachusetts

Nam T. Huynh
Electrical Systems and Optics
    Research Division
Faculty of Engineering
University of Nottingham
Nottingham, United Kingdom

Sean J. Kirkpatrick
Department of Biomedical
    Engineering
Michigan Technological University
Houghton, Michigan

Preston Lavinghousez
Department of Radiology
and
Department of Pediatrics
Stanford University
Palo Alto, California

Patrick Lindner
Gottfried Wilhelm Leibniz University
    Hannover
Institute of Technical Chemistry
Hannover, Germany

William R. Lloyd
Department of Biomedical
    Engineering
University of Michigan
Ann Arbor, Michigan

Stephen J. Matcher
Biomedical Engineering
The Kroto Institute
University of Sheffield
Sheffield, United Kingdom

Stephen P. Morgan
Electrical Systems and Optics
    Research Division
Faculty of Engineering
University of Nottingham
Nottingham, United Kingdom

Mary-Ann Mycek
Department of Biomedical
    Engineering
University of Michigan
Ann Arbor, Michigan

Ioan Notingher
School of Physics and Astronomy
University of Nottingham
Nottingham, United Kingdom

Flavius C. Pascut
School of Physics and Astronomy
University of Nottingham
Nottingham, United Kingdom

Andreas Prediger
Gottfried Wilhelm Leibniz University
    Hannover
Institute of Technical Chemistry
Hannover, Germany

Kenneth F. Reardon
Department of Chemical and
    Biological Engineering
Colorado State University
Fort Collins, Colorado

Felicity R.A.J. Rose
School of Pharmacy
University of Nottingham
Nottingham, United Kingdom

Haowen Ruan
Electrical Systems and Optics
    Research Division
Faculty of Engineering
University of Nottingham
Nottingham, United Kingdom

Thomas Scheper
Gottfried Wilhelm Leibniz University
    Hannover
Institute of Technical Chemistry
Hannover, Germany

James Sharpe
EMBL-CRG Systems Biology Program
Centre for Genomic Regulation (CRG)
Barcelona, Spain

Wiendelt Steenbergen
MIRA Institute for Biomedical
    Technology and Technical Medicine
University of Twente
Enschede, The Netherlands

Jim Swoger
EMBL-CRG Systems Biology Program
Centre for Genomic Regulation
Barcelona, Spain

Juliaan R.M. van Rappard
Department of Radiology
Stanford University
Palo Alto, California
and
Leiden University School of Medicine
Leiden, The Netherlands

I. Alex Vitkin
Department of Medical Biophysics
University of Toronto
Toronto, Ontario, Canada

Ruikang K. Wang
Department of Bioengineering
University of Washington
Seattle, Washington

Brian C. Wilson
Department of Medical Biophysics
University of Toronto
Toronto, Ontario, Canada

Ian Wimpenny
Institute of Science and Technology in
    Medicine
Keele University
Stoke-on-Trent, United Kingdom

Joseph C. Wu
Department of Radiology
Department of Medicine
Division of Cardiology
and
Institute of Stem Cell Biology and
    Regenerative Medicine
Stanford University School of
    Medicine
Stanford, California

Ying Yang
Institute of Science and Technology in
    Medicine
Keele University
Stoke-on-Trent, United Kingdom

Youbo Zhao
Beckman Institute for Advanced
    Science and Technology
University of Illinois at
    Urbana-Champaign
Urbana, Illinois

Introduction and Background

1. The Role of Optical Techniques in Regenerative Medicine

Stephen P. Morgan, Brian C. Wilson, I. Alex Vitkin, and Felicity R.A.J. Rose

1.1 The Regenerative Medicine Process

The field of regenerative medicine is a rapidly growing area of research. This is driven by the clinical need to address the shortage of organs and tissues for replacement and to provide more suitable alternatives to current therapies where improvement in the clinical outcome is required. A broad spectrum approach to the regenerative medicine process (RMP), or tissue engineering, usually involves several key steps, as highlighted in Figure 1.1. These include

1. Selecting the most appropriate defined cell type
2. Fabricating a scaffold for cell delivery and/or to support tissue formation
3. Seeding the cells onto the scaffold for initial cell attachment

Optical Techniques in Regenerative Medicine. Edited by Stephen P. Morgan, Felicity R.A.J. Rose, and Stephen J. Matcher © 2014 CRC Press/Taylor & Francis Group, LLC. ISBN: 978-1-4398-5495-2

Chapter 1

4. A period of time in a bioreactor culture to stimulate the initial regenerative process
5. Implantation to promote maturation of the tissue, with the ultimate goal of generating a functional tissue or organ

Of course, there are variations along this theme with many different strategies employed depending upon the target tissue for regeneration, the approach taken by the research laboratory, and the final delivery mode of such therapy in the clinic. For example, the desired cell type (stem, progenitor, or differentiated) can be directly implanted into an individual and tissue regeneration occurs in the body, as in the approach taken to treat ischemic heart disease (Wu et al. 2011). Alternatively, the cells can be seeded onto a scaffold, which could be naturally occurring, for example, decellularized organ tissue (Badylak et al. 2009, 2012; Song and Ott 2011) or synthetic organ tissue (Kim et al. 2011; Williams 2009), and then implanted into the body. A period of construct growth and maturation in a bioreactor prior to implantation also may be required (Asnaghi et al. 2009; Ellis et al. 2005), which could include stimulating cell behavior, for example, by applying mechanical force (Chen and Hu 2006; Gandaglia et al. 2011). Acellular approaches (i.e., implanting a scaffold alone to allow host tissue ingress) are also being explored as such materials negate the complex regulatory and logistical issues associated with delivering cells or tissues to the body. Such materials often include growth factors to encourage and support tissue formation by cells that populate the scaffold from the surrounding tissue following implantation (Sarkanen et al. 2012).

Clearly, RMP is dynamic and highly complex, and its success relies on the appropriate interaction between all the individual components, which have to be well understood, closely monitored, and precisely controlled. Arguably, none of these requirements are currently met. In order to monitor and understand cellular interactions, proliferation, and subsequent differentiation within the microenvironment and subsequent tissue growth, it is essential to have the right tools to characterize both the initial materials and subsequent tissue development and maturation. At present it is not easy to monitor growing tissue constructs *in vitro* or the fate of cells injected straight into the body, and so tissue engineers rely heavily on destructive methods, such as biopsy followed by histology, to understand what is happening. This poses a major challenge, since these time-consuming and laborious techniques slow clinical translation. Further, biopsies may not be possible, may suffer from sampling or geographic-miss issues, and do not offer any insight into the important temporal aspect of a given process, as afforded by

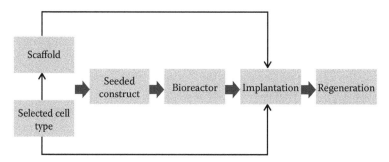

FIGURE 1.1 Overview of RMP. The purple arrows represent a conventional tissue engineering approach. The black arrows illustrate alternative approaches.

longitudinal, noninvasive tissue assessment approaches. Hence, the development of *in situ*, online, and non-destructive analytical techniques that can help understand the stages of tissue development and regeneration, both *in vitro* and *in vivo*, has become increasingly important. Biophotonic approaches are poised to play a leading role in this quest. However, the complex nature of cell differentiation and the use of largely opaque scaffolds that become more optically dense during tissue development present significant technical challenges to the biomedical optics researcher in the development of systems and techniques that align well within the RMP biological context.

1.2   Optical Techniques

The interaction of light with tissue can provide a wealth of information to the biologist. In the wavelength range considered in this book (~300–1500 nm), these interactions depend upon the molecular composition and microstructure of the tissue (Georgakoudi et al. 2008; Tuchin 2007; Wax and Backman 2009). When light illuminates tissue it can be either absorbed or scattered (Figure 1.2). If the photon energy matches the energy

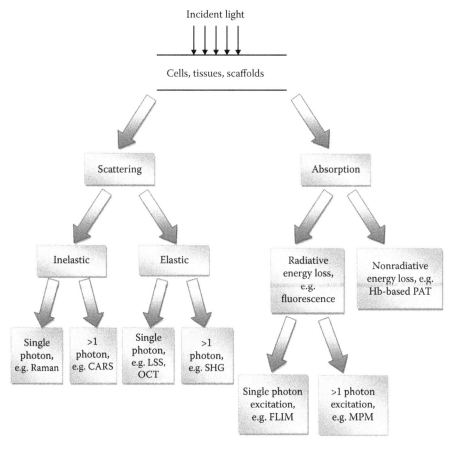

FIGURE 1.2 Interactions of light with tissue (CARS: coherent anti-Stokes Raman spectroscopy; LSS: light scattering spectroscopy; OCT: optical coherence tomography; SHG: second harmonic generation; FLIM: fluorescence lifetime imaging; MPM: multiphoton microscopy; PAT: photoacoustic tomography). Note: light can also be generated within the tissue, as in bioluminescence imaging.

Chapter 1

required to raise a molecule from the ground state to an excited state, then absorption may occur. Often the excited electron relaxes to its ground state with dissipation of energy as heat (nonradiative decay); this commonly occurs in tissue due to the presence of major chromophores such as hemoglobin in blood and melanin in skin. Measurement of the optical absorption spectrum of tissue thereby allows parameters such as blood oxygen saturation to be monitored. Alternatively, light can be re-emitted during the relaxation process (radiative decay) as fluorescence. The fluorescent light is at a longer wavelength than the excitation light, as there is some energy loss during the relaxation process ($E = hc/\lambda$, where E = energy, c = speed of light, λ = wavelength). Both the fluorescence emission spectrum and the fluorescence lifetime ($1/e$ time for excited molecules to return to the ground state) can be used to distinguish between fluorophores. Examples of naturally occurring (endogenous) fluorophores in tissue are collagen, flavin adenine dinucleotide (FAD), nicotinamide adenine dinucleotide (NADH), tyrosine, tryptophan, retinol, and elastin. Fluorescence techniques are described in more detail in Chapters 2 and 7. Usually fluorescence is due to a single photon that excites the molecule from its ground state, resulting in the emission of a single fluorescence photon when it returns to the ground state. However, it is possible for fluorescence to be obtained when two or more photons excite the molecule, resulting in the emission of a single photon when the molecule returns to its ground state. As described in Chapters 2 and 7 within this book, this property is utilized in multiphoton microscopy (MPM). MPM is particularly useful because the excitation photons are of a longer wavelength than with single-photon excitation, and thus can penetrate deeper into tissue (due to lower scattering and absorption in the near infrared spectral range) with low phototoxicity. As multiphoton excitation is a low probability event requiring simultaneous absorption of two or more photons, it usually only occurs within the focal region of a high-numerical aperture (NA) microscope objective, and so can also provide optical sectioning within a sample.

Optical scattering can also be used to characterize cells, scaffolds, and tissues (Wax and Backman 2009). Light can either be scattered elastically, in which there is no difference in energy between the incident and scattered light, or inelastically, in which there is an energy difference due to changes in the vibrational state of the molecule. As described in Chapter 6 in this book, these properties of the elastically scattered light (such as angular distribution and polarization state) can be related to tissue structure, function, and composition. When the detected light has been singly or weakly scattered (e.g., from superficial tissue), it carries information of the size, shape, and refractive index of subcellular components such as cell nuclei and membranes. Light that has been multiply scattered can be related to bulk optical properties and is useful for bulk tissue imaging, and may carry information related to, for example, collagen density, organization, and orientation. Motion of particles can change the frequency or phase content of scattered light (via the Doppler effect) and this can be used to characterize flows in tissues and scaffolds. This is utilized in many different imaging techniques, for example, Doppler optical coherence tomography (OCT) (see Chapter 11) to characterize fluid flow in scaffolds. Blood flow has been imaged directly using laser Doppler blood flowmetry, Doppler photoacoustic tomography (PAT), and indirectly in laser speckle contrast imaging (see Chapter 14).

Similar to multiphoton absorption, non-linear scattering can also occur in which two or more photons combine to form a scattered photon with precisely half the wavelength and double the energy of the excitation photon. As described in Chapter 3, this occurs in nonsymmetric structures and so is particularly useful in characterizing collagen and other fibrous structures. Second harmonic generation (SHG) microscopy is often carried out simultaneously with multiphoton absorption or fluorescence microscopy using the same instrumentation. Inelastic light scattering (see Chapter 8) occurs when part of the energy of the incident light is used to change the vibrational state of a molecule. There is a small, characteristic frequency shift in the scattered light, known as a Raman shift, which can be used to characterize the vibrational states of different molecules. The Raman effect is relatively weak, typically ~1000-fold lower than the tissue autofluorescence, but coherent anti-Stokes Raman scattering (CARS) can provide better signal-to-noise ratio. This is a label free technique that employs the two laser pulses pump and Stokes, which have wavelengths that are tuned to match the energy gap between two vibrational states of a molecule of interest. When this condition occurs, a strong anti-Stokes signal is generated, providing contrast that is predominantly due to CH_2 bands; thus, CARS is sensitive to lipid-rich structures such as myelin and adipocytes. It should be noted that, in addition to the interaction of incident light by tissue, light can also be generated within the tissue through a chemical reaction, as in bioluminescence (see Chapter 15).

Although light scattering is a rich source of information for tissue and tissue scaffold characterization, it is also the major limiting factor to achieving high resolution images at large penetration depths. As is often observed in images of cell monolayers or thin histology sections, exquisitely high spatial resolution, information-rich images can be achieved with conventional microscopes since light scattering is low. When imaging tissue, there is a trade-off between penetration depth and spatial resolution (both axial and lateral) and different optical techniques will be appropriate depending on the penetration depth required (Table 1.1). As the penetration depth increases, approaches need to be applied to overcome the effects of light scattering. Confocal microscopy uses a pinhole to reject scattered light, whereas MPM uses non-linear effects to localize the volume probed. As the non-linear emission is at a shorter wavelength to the excitation, it can be discriminated from the scattered background. Such approaches allow high spatial resolution images to be achieved up to a penetration depth of typically 300–500 μm in most tissue types. Beyond this depth, the signal-to-noise ratio becomes too low to obtain high-quality images and alternative approaches are required.

OCT (see Chapters 10 and 11) illuminates the tissue with a short coherence-length light source and interferes the scattered light from the sample with a reference beam. Only when the path lengths traveled by the light in the sample and along the reference paths are matched within the coherence length will an OCT interference signal be detected. Photon path lengths outside this coherence length window will be rejected, allowing the effects of light scattering to be reduced. This can be used to achieve imaging resolution of 3–10 μm at penetration depths of typically up to 2 mm. OCT operates at the limit of detection of weakly scattered light. To extend the penetration depth further, diffuse light needs to be used, as in diffuse optical tomography (DOT). Alternatively, in PAT (see Chapter 12), pulsed diffuse light is used to generate a small temperature rise in optically absorbing structures, such as blood vessels, whose subsequent non-radiative de-excitation and expansion results in the generation of acoustic waves that

Chapter 1

Table 1.1 Overview of Optical Techniques for Imaging and Sensing in Regenerative Medicine

Optical Technique	Measurement	Capability[a]	Role in RM Process	Comments	Chapter
Confocal/single photon fluorescence microscopy	Emission of a single photon due to single-photon absorption. Out-of-focus light eliminated by detector pinhole	Imaging exogenous fluorescence Lateral resolution 0.5 μm Axial resolution 1 μm Penetration depth up to 500 μm	Stem cell differentiation Monitoring in bioreactors *In vivo* imaging of superficial tissue	Several commercial systems available Lower cost than MPM	2 (by Green and Hayock)
MPM	Emission of a single photon due to N-photon absorption	3D imaging of endogenous fluorophores, e.g., NADH, tryptophan Lateral resolution 0.5 μm Axial resolution 1 μm Penetration depth up to 500 μm	Stem cell differentiation Monitoring in bioreactors *In vivo* imaging of superficial tissue Scaffold manufacture	Low photobleaching and phototoxicity Clinical systems available	2 (Green and Haycock); 7 (by Lloyd et al.)
SHG	Emission of a single photon due to 2-photon anisotropic scattering	No exogenous fluorophores Lateral resolution 0.5 μm Axial resolution 1 μm Penetration depth up to 1 mm	Imaging of fibrous tissue	Sensitivity to anisotropic tissue, e.g., collagen	Matcher
FLIM	Lifetime of fluorophore	Can separate multiple fluorophores within a sample Resolution depends on imaging configuration, cellular level–hundreds of micrometers	Monitoring within bioreactors Monitoring stem cell differentiation Stem cell tracking	Easily used in conjunction with other techniques, e.g, MPM	Lloyd et al.

	Principle	Capabilities	Applications	Features	Reference
Raman	Inelastic scattering from molecular vibrations (molecule in ground state)	Can probe molecular structures Resolution, cellular level - hundreds of micrometers Label-free	Stem cell differentiation Monitoring in bioreactors	*In vivo and in vitro* use Weak signals	Pascut and Notingher
CARS	Inelastic scattering from molecular vibrations (molecule in excited state)	Can probe molecular structures Resolution, cellular level Label free Lateral resolution 0.5 μm Axial resolution 1 μm Penetration depth 100 μm	Monitoring in bioreactors	Much stronger signals than Raman Low photobleaching and phototoxicity Sensitivity to lipid structures, e.g., lipid bilayer around cells Can be used in conjunction with MPM and SHG	Pascut and Notingher
LSS	Angular spectrum or polarization changes of elastically scattered light	Can obtain information about subcellular structures for superficial tissue	Monitoring in bioreactors Monitoring tissue regeneration	Polarization sensitive to collagen alignment Simple instrumentation	Georgakoudi and Hunter
OPT	Absorption or fluorescence at multiple projections	Useful for imaging thick structures 5–10 mm, which have low scattering Spatial resolution ~50–100 μm	Monitoring 3D cell culture	Relies on ideally no, or very weak, scattering Optical clearing required	Swoger et al.

(continued)

Chapter 1

Table 1.1   Overview of Optical Techniques for Imaging and Sensing in Regenerative Medicine (Continued)

Optical Technique	Measurement	Capability[a]	Role in RM Process	Comments	Chapter
OCT, DOCT, OMAG	Backscattered light that maintains coherence	Sensitive to scattering of light within coherence gate 3D imaging to depth of 2–3 mm in tissue Lateral resolution 10 μm Axial resolution 3–10 μm	Monitoring in bioreactors Identifying tissue requiring regeneration Monitoring tissue regeneration	3D imaging collagen 3D flow imaging in microvessels	Zhao et al., Yang et al.
PAT	Acoustic emission after absorption of pulsed light	Contrast due to absorption by hemoglobin Lateral resolution 1 μm Axial resolution depends on ultrasound pulse timing	Identifying tissue requiring regeneration Monitoring tissue regeneration	Imaging of single RBCs Oxygen saturation of microvessels	Steenbergen
USMOT	Ultrasound modulation of diffuse light	Contrast due to absorption, scattering, fluorescence, or potentially bioluminescence Resolution ~50 μm, scalable	Monitoring in bioreactors Identifying tissue requiring regeneration Monitoring tissue regeneration	Spatial resolution poorer than PAT for absorbing structures May be applicable for imaging necrotic tissue based on scattering changes May be applicable for imaging fluorescence and bioluminescence in 3D but signal-to-noise ratio poor	Morgan et al.

Laser speckle/laser Doppler imaging	Fluctuations in speckle pattern at detector due to motion	Contrast due to scattering of light by moving RBCs Lateral resolution hundreds of micrometers for superficial tissue	Monitoring of tissue regeneration Identify tissue requiring regeneration	Limited to 2D imaging Can image flow in microcirculation in real time Can image tissue biomechanics Simple instrumentation	Kirkpatrick et al.
BLI	Light generated within tissue due to chemical reaction	Resolution several millimeters	Stem cell tracking	Useful preclinical tool Safety concerns for clinical use	van Rappard et al.
In situ microscopy and macroscopy	Light scattered by cells, 3D constructs or fluorescence labeling	Spatial resolution scales with field of view 1 μm–1 mm	Monitoring in bioreactors of cell suspensions and 3D constructs Automation of 3D cell culture	Simple instrumentation Relies on image processing to extract useful parameters	Kirkpatrick et al., Prediger et al.
DOT	Absorption spectra of chromophores	Contrast due to oxyhemoglobin and deoxyhemoglobin absorption Or exogenous dyes or nanoparticles Spatial resolution about several millimeters	Monitoring of tissue regeneration Identify tissue requiring regeneration	2D imaging of SO_2 uses simple instrumentation 3D relies on more complex instrumentation and reconstruction algorithms	Kirkpatrick et al.

Notes: BLI, bioluminescence imaging; CARS, coherent anti-Stokes Raman scattering; (D)OCT, (Doppler) optical coherence tomography; DOT, diffuse optical tomography; FLIM, fluorescence lifetime imaging microscopy; LASCA, laser speckle contrast analysis; LSS, light scattering spectroscopy; MPM, multiphoton microscopy; OMAG, optical microangiography; OPT, optical projection tomography; PAM, photoacoustic microscopy; RBC, red blood cell; RM, regenerative medicine; SHG, second harmonic generation; USMOT, ultrasound-modulated optical tomography.

[a] Resolution and penetration depth depend upon numerical aperture, wavelength, and amount of scattering present. Typical values are presented.

Chapter 1

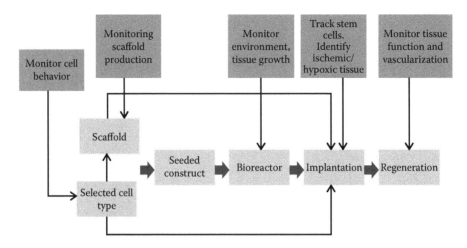

FIGURE 1.3 Potential roles for optical techniques in regenerative medicine.

are used to form an image. The major advantages of this approach are the retention of high optical absorption contrast and minimal scattering distortion due to ultrasonic readout. Ultrasound modulated optical tomography (see Chapter 13) uses sound waves to modulate the diffuse light intensity locally within tissue. Forming an image with the ultrasound modulated light can also help to overcome the effects of light scattering. In techniques such as DOT (Gibson et al. 2005) and fluorescence molecular tomography (Ntziachristos et al. 2003), heavily scattered diffuse light is detected and image reconstruction algorithms are used to form images of thick tissue (up to several centimeters) at relatively low spatial resolution (several millimeters).

Optical techniques thus offer a range of possibilities for imaging structure, composition and function of tissue, cells, and scaffolds in regenerative medicine. They can "shed light" and hence impact the RMP in a variety of ways as shown in more detail in Figure 1.3 and highlighted in the following sections.

1.3 Cell Differentiation (Selecting Cell Type in RMP)

For both safety and efficacy of treatments in regenerative medicine, it is essential to monitor cell differentiation status and phenotypic purity of stem cell populations. Transplantation of undifferentiated stem cells or cells of an unwanted phenotype can result in tissue overgrowth and even tumor formation (Reubinoff et al. 2000). Immunohistochemistry and metabolic assays can be used to monitor stem cell populations. However, this can be time consuming and often requires external labels that cannot be used for cells destined for application *in vivo*. Optical techniques, therefore, have a role to play in this aspect of the RMP by providing tools for label-free assessment and monitoring of therapeutic cells.

An excellent review (Downes et al. 2010) has highlighted Fourier transform infrared (FTIR) spectroscopy, Raman spectroscopy, and CARS as techniques with the potential to monitor stem cell differentiation. FTIR provides spectra based on the absorption of infrared light by chemical bonds within molecules. An example is the use of FTIR to monitor the differentiation of murine embryonic stem cells (Ami et al. 2008), in which, after 4 to 7 days of cell culture, specific spectral changes (enhancement of amide I band and reduction of the nucleic acid region) were used to infer the overall decrease in DNA and RNA content and increasing alpha helix protein content over time. Applications of Raman spectroscopy and CARS are described in more detail in Chapter 8 in this book. These authors have investigated the use of Raman spectroscopy in numerous RMP cases, such as differentiation of cardiomyocytes from human embryonic stem cells (Pascut et al. 2011), where glycogen was identified as an early indicator of the formation of cardiomyocytes. Murine embryonic cells have also been investigated (Notingher et al. 2004) using Raman spectroscopy: during 16 days of differentiation, a 25% decrease in an RNA-associated band (813 cm^{-1}) and a 50% decrease in a DNA-associated band (788 cm^{-1}) were observed. This is in agreement with FTIR results (Ami et al. 2008). Raman spectroscopy has advantages over IR absorption spectroscopy since its spectral features are sharper and thus more specific, and since the high water absorption of live samples limits the penetration depth in the IR. However, Raman scattering is a weak process (1 in 10^6 incident photons typically are Raman shifted), so that there is interest in improving signal-to-noise ratio using CARS. CARS has also been used to demonstrate a reduction in RNA as a result of cell differentiation (Konorov et al. 2007).

Other biophotonic techniques described in this book have also been applied to cell differentiation in the RMP context. For example, Rice et al. (2007) investigated MPM to monitor differentiation of human mesenchymal stem cells. Two-photon autofluorescence in tissue originates from intrinsic fluorophores such as nicotinamide adenine (phosphorylated) dinucleotide (NAD(P)H), flavins, retinoids porphyrins, elastin, melanin, hemoglobin, and serotonin. Using 800 nm excitation, the observed autofluorescence was due to NAD(P)H and oxidized flavoproteins. When combined with cell morphology results, these measurements could be correlated with adipogenic stem cell differentiation. Uchugonova and Konig (2008) described a system utilizing MPM, SHG, and FLIM to monitor adult stem cells in monolayers and spheroids. Within the stem cells investigated, NAD(P)H and flavoprotein fluorescence could be detected, potentially allowing discrimination between proliferating and nonproliferating cells (Zhang et al. 1997) and also metabolic changes in differentiating cells (Reyes et al. 2006). Differentiated cells produced the extracellular matrix (collagen), which could be detected using SHG microscopy.

FLIM (see Chapter 7) has recently been used to monitor stem cell differentiation in the small intestine (Stringari et al. 2012a). Analyzing the fluorescence lifetime data within an image, NADH was monitored as an indirect biomarker of cell proliferation, since proliferative cells rely on aerobic glycolysis (resulting in elevated ATP/ADP and NADH/NAD+ ratios), whereas differentiated cells rely on oxidative phosphorylation. High NADH levels are measured in highly proliferative stem cells, which can be identified within crypts in the small intestine and discriminated from differentiated cells. This approach has also been applied to monitoring the differentiation of neural cells *in vitro* (Stringari et al. 2012b). The NADH FLIM signature distinguishes noninvasively neurons from undifferentiated neural progenitor and stem cells (NPSCs) at two different developmental stages.

Chapter 1

1.4   Scaffold Production

Optical techniques can be used in both the scaffold production and quality control/ monitoring process. Chapter 4 describes the use of optical techniques to manufacture scaffolds via one- and two-photon polymerization. Optical techniques can also be used to monitor scaffold production. For example, the formation of scaffolds during a super-critical fluid process has been monitored using a simple macroscopic imaging system (Mather et al. 2009; Chapter 14). Image processing was used to track the change in size of scaffolds during the foaming process for different process parameters such as molecular weight and venting time. This provided an insight into the scaffold production process, which is beneficial for both setting the process parameters and providing real time feedback. Raman spectroscopy has been used to assess the biomaterials used in regenerated silk scaffolds (Zaharia et al. 2012).

OCT can be used to investigate the properties of pores within scaffolds (e.g., pore size and interconnectivity). However, the resulting tomographs (Yang et al. 2007; Chapter 11) may be inferior in quality to micro-CT or MR images (Mather et al. 2008; Ho and Hutmacher 2006), particularly when the scaffold itself scatters light heavily. However, OCT can provide useful information about the flows within scaffolds to investigate pore interconnectivity (Jia et al. 2009). MPM has also been applied to investigate the optical properties of scaffolds manufactured from poly(lactic acid), poly(glycolic acid), a collagen composite scaffold, Collagraft bone graft matrix, and nylon (Sun et al. 2008). The main role for optical techniques is in noninvasive monitoring of cell culture within scaffolds where micro-CT cannot be applied without damaging the sample and where MRI is not an easily accessible laboratory tool.

1.5   Monitoring in Bioreactors

Optical techniques can be used to monitor both the bioreactor environment and changes in optical properties of the tissues within it. There is a range of fiber optic sensors that can be used to monitor bioreactor properties such as oxygen, pH, and CO_2. Sensing the bioreactor environment is beyond the scope of this book, but several useful recent reviews have been published (Stich et al. 2009; Beutel and Henkel 2011; Bluma et al. 2011; Kim et al. 2012; Lourenco et al. 2012). Routine sensing and automation is in its infancy in bioreactor systems for tissue engineering, but there is a need for such approaches in order to achieve appropriately large clinical scale production (from milliliters to liters) (Archer and Williams 2005). In these aspects, much can be learned from the more mature bioprocessing field, which has implemented sensors and automation for many years.

The main focus of the techniques described in this book is on imaging and sensing the properties of tissues cultured in 3D using scaffolds. Noninvasive *in situ* monitoring is important in the development of such engineered products as 3D cell culture, though is difficult to automate simply by following a recipe as, for example, cells taken from different patients will often grow at different rates (Kinner et al. 2002). Tissue growth within a scaffold is often nonuniform and is influenced by scaffold design, and a frequently observed problem is the presence of a necrotic core due to insufficient nutrient supply throughout the scaffold (Silva et al. 2006). Noninvasive monitoring *in situ* can remove the need for offline sampling and destructive endpoint analysis.

Several optical techniques have been applied to monitoring 3D cell culture in bioreactors. The technique adopted depends upon the functional information required and also the optical penetration depth into the tissue. To minimize the effects of light scattering, some pragmatic approaches have been introduced. Thus, hollow core silica fibers have been embedded inside the scaffold (Hofman et al. 2012) and single cell imaging has been demonstrated (20–30 µm resolution). Optically transparent scaffolds have been developed to enable fundamental studies of fluid flow within them (Lagana and Raimondi 2012), although, as cells grow in 3D within such scaffolds, light scattering due to the developing tissue will limit this approach. The majority of research has been into the use of optical techniques that can be used to noninvasively monitor the scaffolds that are routinely used in 3D cell culture.

Non-linear microscopy has proven to be a useful tool for functional imaging of tissue with high spatial resolution, particularly when light scattering is relatively weak. For example, MPM has been used to monitor cell proliferation and the redox ratio (FAD/NADH) of individual cells in a 3D cell culture over a period of 6 months at depths of up to 200 µm within a silk scaffold (Quinn et al. 2012). Spatiotemporal stem cell differentiation could be tracked using the redox ratio. SHG has been used to image collagen within smooth muscle cells (SMC) located up to 120 µm deep in a poly(glycolic acid) scaffold over a period of 8 weeks in order to validate a model of growth and mechanics (Niklason et al. 2010). A combination of SHG and CARS has been used to monitor smooth muscle cells in biosynthesized cellulose scaffolds (Brackmann et al. 2011): CARS was used to monitor lipid-rich intracellular structures, while SHG was used to visualize the fibers of the cellulose scaffold (Figure 1.4). Different scaffold types (porous and compact) influenced the penetration depth of cells within the scaffold.

When optical scattering is relatively weak, optical projection tomography (OPT; Chapter 9), often aided by optical clearing agents to further reduce scattering, can be used to provide high resolution 3D images of tissue constructs. OPT has found widespread use in embryology research and has also been used to image tissue-engineered

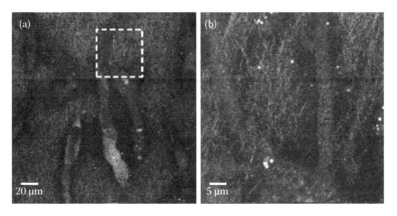

FIGURE 1.4   (See color insert.) (a) Overlay CARS (yellow) and SHG (blue) microscopy images of smooth muscle cells grown on a cellulose scaffold. Images were obtained using intrinsic contrast, with the lipids within the cells providing high contrast for CARS and the fibers of the cellulose scaffold providing SHG contrast. Cells could be tracked in real time in 3D, revealing that while cells migrate into the scaffold, they extrude filopodia on the surface. The close-up image in (b) shows the extending filopodia interspersed in the cellulose fiber network. (From Brackmann, C. et al., *Journal of Biomedical Optics*, 16:021115, 2011.)

blood vessels with the images compared to those obtained using OCT (Gladish et al. 2005). OPT has the advantage of simpler and relatively inexpensive instrumentation, since 3D images can be obtained using a charge coupled device (CCD) camera. However, OCT performs better as the light scattering increases. It is widely applied for imaging tissue constructs, as the selection of weakly scattered light from a particular depth via coherence gating can partially overcome the effects of light scattering (Mason et al. 2004). OCT has been demonstrated to be capable of monitoring cell turnover within poly(L-lactic acid) scaffolds over a period of 5 weeks. Doppler OCT has been used to monitor the fluid flow and shear stresses within chitosan scaffolds with high and low porosity (Jia et al. 2009). Polarization sensitive OCT has been used to identify collagen fiber alignment in tissue-engineered tendon (Ahearne et al. 2008). Phase-sensitive OCT can detect the micromotions of stem cells within scaffolds as a potential method of assessing cell viability (Bagnaninchi et al. 2012). OCT has also been used in combined MPM and OCT instruments for imaging 3D cell cultures (Tan et al. 2007): MPM is used to map GFP within the cells, while OCT is used to map scaffold structure and cell morphology. Metabolic states can also be monitored by spectroscopic OCT (Adie et al. 2010). More details of the application of OCT in regenerative medicine can be found in Chapters 10 and 11 in this book.

Both elastic and inelastic light scattering have been used to characterize tissue within bioreactors. Light scattering spectroscopy (see Chapter 6) has been used to monitor collagen fibril alignment in fibroblast-populated collagen gels. Using simple instrumentation, it was demonstrated that the intensity of the backscattered light was greater in a direction perpendicular to the collagen alignment due to anisotropic light scattering by collagen fibrils (Kostyuk and Brown 2004). For randomly orientated collagen, the backscattered light intensity was more uniform in all directions. This effect was used to monitor collagen alignment during gel contraction over 80 h and to monitor the effect of external tensile loading of the gel. However, since multiple-scattered light is collected, the spatial resolution is low.

The molecular signatures in Raman spectra can also be used for monitoring a 3D cell culture within bioreactors. For example, Raman microscopy has been used to study human bone marrow stromal cell proliferation and differentiation to osteoblasts over a 21-day period within a microbioreactor (Pully et al. 2010).

In terms of imaging in bioreactors, much can again be learned from the more mature bioprocessing field, in which image processing and machine vision both provide valuable information for process monitoring. Chapter 5 describes optical systems and image processing techniques that can be used to monitor cell culture at a microscopic scale within bioreactors. Properties such as cell number, cell size distribution, and morphological properties can be monitored. In regenerative medicine, this approach has been applied on a macroscopic scale to monitor tissue-engineered heart valves in a bioreactor (Ziegelmueller et al. 2010; Chapter 14).

1.6 Implantation

Prior to implantation, optical techniques could be used to identify regions of tissue in need of regeneration. Optical techniques can be used to image and characterize many tissue properties such as new blood vessel formation, blood oxygen saturation, blood flow, collagen production and alignment, and tissue biomechanics. For example, laser Doppler blood flow imaging is a useful method of predicting burn wound healing (Monstrey et al. 2011). If a region of a wound has low blood flow, then it is unlikely to heal without additional treatment such as a skin graft or tissue-engineered product. It has been suggested that oxygen saturation imaging (see Chapter 14) could be used to identify regions of ischemic tissue in cardiac applications prior to injection of stem cells into the myocardium (Wilson et al. 2009). The techniques we highlight in the following section for monitoring tissue regeneration can also be used to assess tissue that requires intervention.

Upon injection of stem cells, it is desirable to develop techniques to assess their fate and survival *in vivo*. Maintaining stem cells at the site of injection is crucial for their contribution to tissue regeneration; migration away from this site may lead to the therapy being ineffective and the possible formation of teratomas, a tumor composed of different tissue types such as hair or teeth that arises due to uncontrolled differentiation of the stem cells at distal sites. In addition, for further advancement of stem cell therapies, it is important to establish the contribution of the cells themselves to the therapeutic outcome so that the treatment can be developed further. For example, it is not clear if the beneficial effects of stem cell injection into myocardial infarct tissues in patients are due to the stem cells differentiating and forming new tissue, or due to the stimulation of tissue regeneration in response to the injection injury and introduction of foreign cells with subsequent cell signaling events (reviewed in Li et al. 2012). Nonoptical techniques such as magnetic resonance imaging (MRI) with superparamagnetic nanoparticle labeling of the stem cells, radionuclide imaging (including positron emission tomography) of stems cells that have been incubated with radioisotopes, or ultrasonography using cells labeled with microbubbles, liposomes, or nanoparticles (Nyolczas et al. 2009; Zhou et al. 2006) have been used to track stem cells within the body. These techniques are advantageous because they can be used to image patients noninvasively *in vivo* throughout the whole body, although sensitivity and resolution are often problematic. Conversely, optical techniques offer excellent intrinsic and extrinsic contrast, possess spectral dimension for added tissue specificity and differentiation, and can furnish micron-scale resolution in superficial tissues and epithelial layers of internal body cavities, for example, via microendoscopy. As such, biophotonic approaches can provide information on specific biological functions of stem cells, primarily in preclinical small animal models.

As noted in Wilson et al. (2009), there is no single technique that meets the ideal for stem cell monitoring, that is, that would allow minimally invasive tracking of individual stem cells or stem cell populations at specific locations or throughout the whole body, enable determination of their viability, provide information about the specific tissue niches to which they localize and about the particular molecular changes that occur as they interact with host tissues, enable survival to be distinguished from proliferation, and provide this information over time in a way that does not interfere with the biological processes involved.

Two techniques have been reported to tackle some aspects of the challenge of stem cell imaging *in vivo*. Intravital microscopy can be used to track individual stem cells at high resolution in superficial tissue. Contrast can be obtained using fluorescence, 2-photon fluorescence, or SHG. Bioluminescence imaging (see Chapter 15) can be used to track the population distribution of stem cells *in vivo* on a macroscopic scale. Novel work using a combined multiphoton and OCT system to track stem cells in skin is described in Chapter 10.

Some excellent examples of using intravital microscopy to track stem cells in preclinical models have been reported by Lo Celso et al. (2009a,b). A multichannel 2-photon confocal intravital microscope has been developed that can achieve video-rate imaging. Figure 1.5a shows an example of imaging the mesenchymal stem cell microenvironment (niche) of a mouse. The image shows stem cells (white), bone (blue), osteoblasts (green), and vasculature (red). Labeling of the hematopoietic stem cells (HSCs) is achieved using a lipophilic carbocyanine dye (DiD), osteoblasts can be distinguished through their expression of green fluorescent protein (GFP) in the Col2.3-GFP mouse model, vasculature is labeled using quantum dots (Qtracker655 or 800), and bone is distinguished via the second harmonic signal; note the interesting use of extrinsic labels and endogenous contrast present in this example. This system was used to monitor stem cell location and cell proliferation after gamma irradiation (Figure 1.5b) and serves to highlight how the technique could be used to monitor stem cell distribution and tissue regeneration in the RMP. CARS could further enhance intravital microscopy by providing label free imaging of, for example, lipids (Le et al. 2009; Veilleux et al. 2008). Microendoscopy (Koenig et al. 2007) can be used to provide minimally invasive imaging, allowing high resolution functional imaging of stem cell location in internal organs.

High lateral resolution implies reduction in the depth of focus of the imaging system, so tissue motion can be problematic, particularly in applications such as cardiovascular imaging. For example, intravital microscopy has been used to image fluorescently labeled fetal canine cardiomyocytes transplanted into the heart (left ventricle). Two months after cell transplantation, a fluorescent graft was visible using intravital microscopy. However, significant motion artifacts were present. Subsequent image processing using commercially available software (Axiovision, Carl Zeiss) improved the image quality, but not to the point that single cells could be observed.

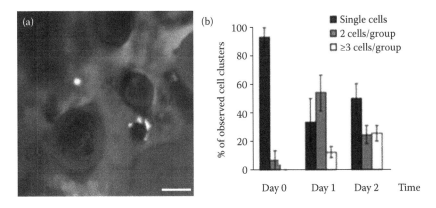

FIGURE 1.5 **(See color insert.)** (a) Image of DiD-labeled labeled HSCs (white) in the calvarium at 1 day after injection into a recipient (Col2.3 GFP mouse) showing a single cell on the left and a cluster of four cells on the right that have undergone cell division. Bone (blue), osteoblasts (green), vasculature (red), bar = 50 μm. (b) At day 0 (on the day of transplantation), most HSCs were observed as single cells, with more clusters observed in subsequent days. (From Lo Celso, C. et al., *Nature*, 457:72–76, 2009a.)

An alternative approach is to use bioluminescence imaging to track populations of stem cells that have been transfected with the luciferase gene. For example, bioluminescence imaging has been used to investigate whether it is more appropriate to administer stem cells intravenously (IV), through a coronary artery (IA), or directly into the myocardium (IM) adjacent to the infarct (Wilson et al. 2009). With IV administration, most of the cells were sequestered in the lungs. Some targeting was achieved using IA injection, but most stem cells remained in the heart by IM injection. As it involves gene modification, there are safety and ethical concerns with using bioluminescence imaging in patients, such that this approach is likely to be restricted to preclinical studies.

1.7 Regeneration

Many of the optical techniques described so far in this chapter, and in more detail within this book, can also be applied to monitor tissue regeneration events such as angiogenesis, tissue perfusion, tissue remodeling, and changes in tissue biomechanics (Wilson et al. 2009).

In vivo optical imaging can provide high resolution functional imaging of small animal models and of superficial tissue in larger animals and humans, and so is best suited for imaging the skin and eye, otherwise requiring an optical fiber-based or endoscopic approach. Clinical microscopy systems are available for skin imaging, combining multiphoton absorption, SHG, and fluorescence lifetime measurements (Koenig 2008), with similar approaches applied in microendoscopy (Koenig et al. 2007). OCT has been applied to skin imaging in numerous applications such as assessment of burn depth (Park et al. 2001) and in ophthalmology (Wotjkowski et al. 2012). It has also been implemented in endoscopes (Li et al. 2000), intravascular probes (Tearney et al. 2006), and arthroscopes (Pan et al. 2003).

For imaging *in vivo* at high penetration depths, there is an inevitable trade-off with image resolution, which limits near infrared spectroscopy and DOT to macroscopic imaging (see Chapter 14). This can be overcome to a certain extent by PAT (see Chapter 12), but even in PAT, spatial resolution scales with penetration depth.

In order for many tissues to regenerate, angiogenesis and oxygen reperfusion are essential. Given the high optical absorption contrast of oxyhemoglobin versus deoxyhemoglobin, several optical techniques can play a role in monitoring these events. As a research tool in preclinical imaging, intravital techniques, such as confocal fluorescence microscopy in which the blood stream contains a fluorophore such as a fluorescein isothiocyanate (FITC)–labeled high-molecular-weight dextran (Jiang et al. 2004), can be used be to identify new vessel formation. A second fluorescent marker can be used simultaneously to image, for example, vascular endothelial growth factor (VEGF) expression, which is a molecular promoter of new vessel formation.

Both OCT and photoacoustic tomography can be used to image new blood vessel formation without using exogenous contrast agents. An example using optical microangiography (Doppler OMAG), a variant of OCT, as shown in Figure 1.6, is how a mouse brain has been imaged through the intact skull. The green and red colors indicate flow toward and away from the direction of the incident beam, respectively. Arterioles (open-tip arrows, Figure 1.6) and venules (closed-tip arrows, Figure 1.6) could be discriminated because blood merges from the smaller vessels into the bigger vessels in venules, while the opposite is true in arterioles.

Chapter 1

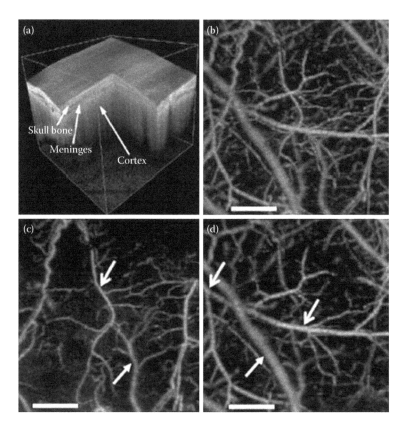

FIGURE 1.6   *In vivo* OMAG image of a mouse cortex. (a) 3D rendered microstructural image (2 × 2 × 2 mm). (b) *x–y* projection image of directional blood flow within scanned volume in (a). Directional blood flow in the meningeal layer (c) and the cortex layer (d) can be separated. Scale bar = 500 μm. (From Wang, R. K., *Optics Letters*, 33:1878–1880, 2008.)

OMAG provides high spatial resolution imaging (~10 μm isotropically) of micro-circulatory flows over a range of velocities from 4 μm/s to 23 mm/s and is described in more detail in Chapter 11 in this book. The 4 μm/s lower limit represents the best case, as it depends on the depth and orientation of the vessel. Due to its sensitivity to hemoglobin, photoacoustic microscopy can be used to map blood vessels at high spatial resolution at depths up to about 3 mm. Larger vessels can be imaged at depths up to 3 cm, with spatial resolution scaling with penetration depth. Systems incorporating two or more wavelengths have been used to image the oxygenation of blood vessels (Hu et al. 2009) and single red blood cells (Ku et al. 2010). Figure 1.7 shows images of a nude mouse ear and the oxygen saturation images obtained with a dual wavelength system. Further applications of photoacoustic tomography in regenerative medicine are described in more detail in Chapter 12 in this book.

Oxygen saturation can be imaged on a macroscopic scale using either reflectance or tomographic diffuse light techniques. As described in Chapter 14, relatively simple reflectance measurements can be taken using a white light source and multispectral or hyperspectral camera. A model of light propagation is then used to relate measurements, such as attenuation of the reflected light, to the optical properties of the tissue. The main problem with this approach is that scattered light is imaged and the path length

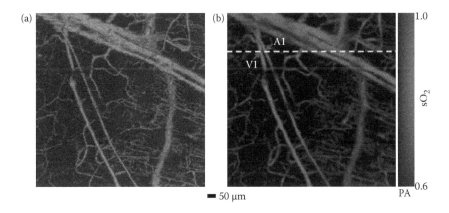

(a)   (b)   1.0

A1

V1

sO₂

■ 50 µm   PA   0.6

FIGURE 1.7   Photoacoustic microscopy image of a nude mouse ear *in vivo*. (a) Structural image at 570 nm. (b) Oxygen saturation map based on dual wavelength images (at 570 and 578 nm). A1: representative arteriole; V1: representative venule. The dotted line indicates the direction of the B-scan. (From Hu, S. et al., *Optics Express*, 17:7688–7693, 2009.)

of the emerging light is unknown, meaning that obtaining absolute quantitative data is challenging. More sophisticated DOT data can be obtained, which involves recording optical path length from time-of-flight or phase-resolved measurements (Gibson et al. 2005). Reflectance spectral imaging has been applied in imaging blood perfusion and oxygen saturation of externalized, live, and beating (porcine) heart (Nighswander-Rempel et al. 2006) and in monitoring wounds (Yudovsky et al. 2011).

Label-free optical techniques are particularly useful for monitoring tissue remodeling *in vivo*. As described previously, stem cell differentiation can be monitored by Raman spectroscopy and MPM, and these can be extended to *in vivo* monitoring techniques (Koenig 2008; Rice et al. 2007). Collagen and collagen alignment can be monitored using Raman spectroscopy, SHG, polarization sensitive OCT, and polarized light spectroscopy. Ogawa et al. (2009) have shown that NIR Raman spectra can be used to monitor type 1 collagen in cardiomyocytes in an externalized heart model. SHG microscopy combined with polarized light imaging has been used to discriminate between healthy and diseased cartilage based on collagen fibril alignment, with healthy cartilage providing a more regular structure (Mansfield et al. 2008). Polarization sensitive OCT has been used to monitor collagen alignment in healing burns *in vivo* (Park et al. 2001): changes in birefringence associated with the denaturation of collagen were used as an indicator of burn depth. Further examples of polarization sensitive OCT imaging are provided in Chapter 11 in this book.

Polarized light scattering measurements have been used to monitor collagen alignment (Ghosh et al. 2008), albeit at a lower spatial resolution than OCT. Randomly orientated collagen (e.g., due to scarring) interacts with polarized light differently than organized collagen architecture. This polarimetric approach has been used to image the remodeling of an infarction after implantation of stem cells. To take into account the effects of light scattering, a model of polarized light propagation within tissue was employed. Alternatively, polarization sensitive OCT can be used to extract weakly scattered light from a heavily scattered background (Liu et al. 2006).

Chapter 1

Useful approaches for monitoring tissue biomechanics include laser speckle imaging and optical coherence elastography (OCE) (Liang and Boppart 2010). Both approaches rely on applying an external force to the tissue and monitoring the change in the optical signal upon application or cessation. Usually, the force is applied using a mechanical actuator, but innovative approaches, such as the use of magnetic nanoparticles actuated by an external magnetic field, have also been used (Oldenburg et al. 2005). As described in Chapter 14, when coherent light is scattered by tissue, the interference of light that propagates along many random paths within the tissue results in a granular interference pattern at the detector plane. This is known as a speckle pattern. Application of a force to the tissue changes the properties of the speckle pattern (such as speckle contrast and speckle decorrelation time). These measures can be related to tissue biomechanics, since stiffer tissues result in a slowly varying speckle pattern. This approach has been used to monitor the biomechanics of excised plaques (Nadkarni et al. 2005). As the technique does not discriminate between weakly and heavily scattered light, it is usually used to image on a macroscopic scale. OCT can be used to extract weakly scattered light from a heavily scattered light background to provide high resolution 3D imaging of the elastic properties of the tissue. This approach has been notably applied in Boppart's laboratory at the University of Illinois and is described in more detail in Chapter 10 in this book.

1.8   Concluding Statements

Many different optical techniques are available with potential to be applied at different stages of the RMP. Optical techniques offer a range of possibilities for imaging the structure and function of tissue, cells, and scaffolds in regenerative medicine, as well as use in scaffold fabrication. Based on a simple model of this process (Figure 1.1), we have highlighted where the optical techniques described within this book can play roles. Light scattering is one of the major challenges associated with optical imaging of tissue. Exquisite high resolution functional images of single cells can be only obtained when scattering is relatively weak. Spatial resolution scales with penetration depth and the most appropriate optical techniques should be selected to probe different depths.

For example, tissue can be imaged at the subcellular level at penetration depths of up to about 500 µm using techniques such as MPM. OCT allows penetration to be extended to up to 2 mm with spatial resolution in the order of 10 µm. Collagen alignment, flow, and biomechanical properties can be imaged at this resolution using variants of OCT that incorporate polarization, Doppler frequency shifts, and speckle. Photoacoustic tomography and ultrasound modulated optical tomography combine optical and acoustic techniques to probe deeper tissue. Imaging and spectroscopy of diffuse light can provide useful information about tissue structure and function on the macroscopic scale. The salient features of the range of techniques described in this book are shown in Table 1.1.

References

Adie, SG, Liang, X, Kennedy, BF, John, R, Sampson, DD, Boppart, SA. 2010. Spectroscopic optical coherence elastography. *Opt. Express* 18:25519–25534.

Ahearne, M, Bagnaninchi, PO, Yang, Y, El Haj, AJ. 2008. Online monitoring of collagen fibre alignment in tissue-engineered tendon by PSOCT. *J. Tissue Eng. Regen. Med.* 2:521–524.

Ami, D, Neri, T, Natalello, A, Mereghetti, P, Doglia, SM, Zanoni, M, Zuccotti, M, Garagna, S, Redi, CA. 2008. Embryonic stem cell differentiation studied by FT-IR spectroscopy. *Biochim. Biophys. Acta* 1783:98–106.

Archer, R, Williams, DJ. 2005. Why tissue engineering needs processing engineering. *Nat. Biotech.* 23:1353–1355.

Asnaghi, MA, Jungebluth, P, Raimondi, MT, Dickinson, SC, Rees, LEN, Go, T, Cogan, TA, Dodson, A, Parnigotto, PP, Hollander, AP, Birchall, MA, Conconi, MT, Macchiarini, P, Mantero, S. 2009. A double-chamber rotating bioreactor for the development of tissue-engineered hollow organs: From concept to clinical trial. *Biomaterials* 30:5260–5269.

Badylak, SF, Freytes, DO, Gilbert, TW. 2009. Extracellular matrix as a biological scaffold material: Structure and function. *Acta Biomater.* 5:1–13.

Badylak, SF, Weiss, DJ, Caplan, A, Macchiarini, P. 2012. Engineered whole organs and complex tissues. *Lancet* 379(9819):943–952.

Bagnaninchi, PO, Holmes, C, Drummond, N, Daoud, J, Tabrizian, M. 2012. Two-dimensional and three-dimensional viability measurements of adult stem cells with optical coherence phase microscopy. *J. Biomed. Opt.* 16(8):08600.

Beutel, S, Henkel, S. 2011. *In situ* sensor techniques in modern bioprocess monitoring. *Appl. Microbiol. Biotechnol.* 91:1493–1505.

Bluma, A, Hopfner, T, Prediger, A, Glindkamp, A, Beutel, S, Scheper, T. 2011. Process analytical sensors and image-based techniques for single-use bioreactors. *Eng. Life Sci.* 11:550–553.

Brackmann, C, Esguerra, M, Olausson, D, Delbro, D, Krettek, A, Gatenholm, P, Enejder, A. 2011. Coherent anti-Stokes Raman scattering microscopy of human smooth muscle cells in bioengineered tissue scaffolds. *J. Biomed. Opt.* 16:021115.

Chen, HC, Hu, YC. 2006. Bioreactors for tissue engineering. *Biotechnol. Lett.* 28:1415–1423.

Downes, A, Mouras, R, Elfick, A. 2010. Optical spectroscopy for noninvasive monitoring of stem cell differentiation. *J. Biomed. Biotechnol.* 2010:101864.

Ellis, M, Jarman-Smith, M, Chaudhuri, JB. 2005. Bioreactor systems for tissue engineering: A four-dimensional challenge. In *Bioreactors for Tissue Engineering: Principles, Design and Operation,* 1–18, eds. Chaudhuri, J and AlRubeai, M. Springer, Netherlands.

Gandaglia, A, Bagno, A, Naso, F, Spina, M, Gerosa, G. 2011. Cells, scaffolds and bioreactors for tissue-engineered heart valves: A journey from basic concepts to contemporary developmental innovations. *Eur. J. Cardio-Thorac. Surg.* 39:523–531.

Georgakoudi, I, Rice, WL, Hronik-Tupaj, M, Kaplan, DL. 2008. Optical Spectroscopy and Imaging for the Noninvasive Evaluation of Engineered Tissues. *Tissue Eng. B* 14:321–340.

Ghosh, N, Wood, MFG, Vitkin, IA. 2008. Mueller matrix decomposition for extraction of individual polarization parameters from complex turbid media exhibiting multiple scattering, optical activity, and linear birefringence. *J. Biomed. Opt.* 13:044036.

Gibson, AP, Hebden, JC, Arridge, SR. 2005. Recent advances in diffuse optical imaging. *Phys. Med. Biol.* 50:R1–R43.

Gladish, JC, Yao, G, L' Heureux, N, Haidekker, MA. 2005. Optical transillumination tomography for imaging of tissue-engineered blood vessels. *Ann. Biomed. Eng.* 33:323–327.

Ho, ST, Hutmacher, DW. 2006. A comparison of micro CT with other techniques used in the characterization of scaffolds. *Biomaterials.* 27:1362–1376.

Hofmann, MC, Whited, BM, Mitchell, J, Vogt, WC, Criswell, T, Rylander, C, Rylander, MN, Soker, S, Wang, G, Yong, X. 2012. Scanning-fiber-based imaging method for tissue engineering. *J. Biomed. Opt.* 17:066010.

Hu, S, Maslov, K, Wang, LV. 2009. Noninvasive label-free imaging of microhemodynamics by optical-resolution photoacoustic microscopy. *Opt. Express* 17:7688–7693.

Jia, Y, Bagnaninchi, PO, Yang, Y, El Haj, A, Hinds, MT, Kirkpatrick SJ, Wang, RK. 2009. Doppler optical coherence tomography imaging of local fluid flow and shear stress within microporous scaffolds. *J. Biomed. Opt.* 14:034014.

Jiang, F, Zhang, ZG, Katakowski, M, Robin, AM, Faber, M, Zhang, F, Chopp, M. 2004. Combination therapy with antiangiogenic treatment and photodynamic therapy for the nude mouse bearing U87 glioblastoma. *Photochem. Photobiol.* 79:494–498.

Kim, BJ, Diao, J, Schuler, ML. 2012. Mini-scale bioprocessing systems for highly parallel animal cell cultures. *Biotechnol. Prog.* 28:595–607.

Chapter 1

Kim, MS, Kim, JH, Min, BH, Chun, HJ, Han, DK, Lee, HB. 2011. Polymeric scaffolds for regenerative medicine. *Polym. Rev.* 51(1):23–52.

Koenig, K. 2008. Clinical multiphoton tomography. *J. Biophotonics* 1:13–23.

Koenig, K, Ehlers, A, Riemann, I, Schenkl, S, Bueckle, R, Kaatz, M. 2007. Clinical two-photon microendoscopy. *Microsc. Res. Tech.* 70:398–402.

Konorov, SO, Glover, CH, Piret, JM, Bryan, J, Schulze, HG, Blades, MW, Turner, RF. 2007. In situ analysis of living embryonic stem cells by coherent anti-Stokes Raman microscopy. *Anal. Chem.* 79:7221–7225.

Kostyuk, O, Brown, RA. 2004. Novel spectroscopic technique for in situ monitoring of collagen fibril alignment in gels. *Biophys. J.* 87:648–655.

Ku, G, Maslov, K, Wang, LHV. 2010. Photoacoustic microscopy with 2 μm transverse resolution. *J. Biomed. Opt.* 15:021302.

Lagana, M, Raimondi, MT. 2012. A miniaturized, optically accessible bioreactor for systematic 3D tissue engineering research. *Biomed. Microdev.* 14:225–234.

Le, TT, Huff, TB, Cheng, JX. 2009. Coherent anti-Stokes Raman scattering imaging of lipids in cancer metastasis. *BMC Cancer* 9:42.

Li, SC, Acevedo, J, Wang, L, Jiang, H, Luo, JE, Pestell, RG, Loudon, WG, Chang, AC. 2012. Mechanisms for progenitor cell-mediated repair for ischemic heart injury. *Curr. Stem Cell Res. Ther.* 7:2–14.

Li, XD, Boppart, SA, Van Dam, J, Mashimo, H, Mutinga, M, Drexler, W, Klein, M, Pitris, C, Krinsky, ML, Brezinski, ME, Fujimoto, JG. 2000 Optical coherence tomography: Advanced technology for endoscopic imaging of Barrett's esophagus. *Endoscopy* 32:921–930.

Liang, X, Boppart, SA. 2010. Biomechanical properties of *in vivo* human skin from dynamic optical coherence tomography. *IEEE Trans. Biomed. Eng.* 5:953–959.

Liu, B, Harman, M, Giattina, S, Stamper, DL, Demakis, C, Chilek, M, Raby, S. 2006. Characterizing of tissue microstructure with single-detector polarization-sensitive optical coherence tomography. *Appl. Opt.* 45:4464–4479.

Lo Celso, C, Fleming, HE, Wu, JW, Zhao, CX, Miake-Lye, S, Fujisaki, J, Cote, D, Rowe, DW, Lin, CP, Scadden, DT. 2009a. Live-animal tracking of individual haematopoietic stem/progenitor cells in their niche. *Nature* 457:72–76.

Lo Celso, C, Wu, J, Lin, C. 2009b. *In vivo* imaging of HSCs and their microenvironment. *J. Biophoton.* 2:619–631.

Lourenco, ND, Lopes, JA, Almeida, CF, Sarruguca, MC, Pinheiro, HM. 2012. Bioreactor monitoring with spectroscopy and chemometrics: A review. *Ann. Bioanal. Chem.* 404:1211–1237.

Mansfield, JC, Winlove, CP, Moger, J, Matcher, SJ. 2008. Collagen fiber alignment in normal and diseased cartilage studied by polarization sensitive non-linear microscopy. *J. Biomed. Opt.* 13:044020.

Mason, C, Markusen, JF, Town, MA, Dunnill, P, Wang, RK. 2004. The potential of optical coherence tomography in the engineering of living tissue. *Phys. Med. Biol.* 49:1097–1115.

Mather, ML, Morgan, SP, White, LJ, Tai, H, Kockenberger, W, Howdle, SM, Shakesheff, KM, Crowe, JA. 2008. Image-based characterization of foamed polymeric tissue scaffolds. *Biomed. Mat.* 3:015011.

Mather, ML, Brion, M, White, LJ, Shakesheff, KM, Howdle, SM, Morgan, SP, Crowe, JA. 2009. Time-lapsed imaging for in-process evaluation of supercritical fluid processing of tissue engineering scaffolds. *Biotechnol. Prog.* 25:1176–1183.

Monstrey, SM, Hoeksema, H, Baker, RD, Jeng, J, Spence, RJ, Wilson, D, Pape, SA. 2011. Burn wound healing time assessed by laser Doppler imaging. Part 2: Validation of a dedicated colour code for image interpretation. *Burns* 37:249–256.

Nighswander-Rempel, SP, Kupriyanov, VV, Shaw, AR. 2006. Regional cardiac tissue oxygenation as a function of blood flow and pO(2): A near-infrared spectroscopic imaging study. *J. Biomed. Opt.* 11: 054004.

Niklason, LE, Yeh, AT, Calle, EA, Bai, Y, Valentin, A, Humphrey, JD. 2010. Enabling tools for engineering collagenous tissues integrating bioreactors, intravital imaging, and biomechanical modelling. *Proc. Nat. Acad. Sci. USA* 107:3335–3339.

Notingher, I, Bisson, I, Polak, JM, Hench, LL. 2004. *In situ* spectroscopic study of nucleic acids in differentiating embryonic stem cells. *Vibr. Spectrosc.* 35:199–203.

Ntziachristos, V, Bremer, C, Weissleder, R. 2003. Fluorescence imaging with near-infrared light: new technological advances that enable *in vivo* molecular imaging. *Eur. Radiol.* 13:195–208.

Nyolczas, N, Charwat, S, Posa, A, Hemetsberger, R, Pavo, N, Hemetsberger, H, Pavo, IJ, Glogar, D, Maurer, G, Gyongyosi, M. 2009. Tracking the migration of cardially delivered therapeutic stem cells *in vivo*: state of the art. *Regen. Med.* 4:407–422.

Ogawa, M, Harada, Y, Yamaoka, Fujita, K, Yaku, H, Takamutsu, T. 2009. Label-free biochemical imaging of heart tissue with high-speed spontaneous Raman microscopy. *Biochem. Biophys. Res. Commun.* 382:370–373.

Oldenburg, AL, Hansen, MN, A, Boppart, SA. 2005. Imaging magnetically labelled cells with magnetomotive optical coherence tomography. *Opt. Lett.* 30:747–749.

Pan, YT, Li, ZG, Xie, TQ, Chu, CR. 2003. Hand-held arthroscopic optical coherence tomography for *in vivo* high-resolution imaging of articular cartilage. *J. Biomed. Opt.* 8:648–654.

Park, BH, Saxer, C, Srinivas, SM, Nelson, JS, de Boer, JF. 2001. *In vivo* burn depth determination by high-speed fiber-based polarization sensitive optical coherence tomography. *J. Biomed. Opt.* 6:474–479.

Pascut, FC, Goh, HT, Welch, N, Buttery, LD, Denning, C, Notingher, I. 2011. Noninvasive detection and imaging of molecular markers in live cardiomyocytes derived from human embryonic stem cells. *Biophys. J.* 100:251–259.

Pully, VV, Lenferink, A, van Manen, HJ, Subramaniam, V, van Blitterswijk, CA, Otto, C. 2010. Microbioreactors for Raman microscopy of stromal cell differentiation. *Anal. Chem.* 82:1844–1850.

Reubinoff, BE, Pera, MF, Fong, CY, Trounson, A, Bongso, A. 2000. Embryonic stem cell lines from human blastocysts: somatic differentiation *in vitro*. *Nat. Biotechnol.* 18:399–404.

Reyes, MG, Fermanian, S, Yang, F, Zhou, SY, Herretes, S, Murphy, DB, Elisseeff, JH, Chuck, RS. 2006. Metabolic changes in mesenchymal stem cells in osteogenic medium measured by autofluorescence spectroscopy. *Stem Cells* 24:1213–1217.

Rice, WL, Kaplan, DL, Georgakoudi, I. 2007. Quantitative biomarkers of stem cell differentiation based on intrinsic two-photon excited fluorescence. *J. Biomed. Opt. Lett.* 12:060504.

Ruhparwar, A, Ghodsizad, A, Lichtenberg, A, Karck, M. 2008. Visualization of transplanted cells within the myocardium. *Int. J. Cardiol.* 124:22–26.

Sarkanen, JR, Ruusuvuori, P, Kuokkanen, H, Paavonen, T, Ylikomi, T. 2012. Bioactive acellular implant induces angiogenesis and adipogenesis and sustained soft tissue restoration *in vivo*. *Tissue Eng. Part A* 18(23–24):2568–2580.

Song, JJ, Ott, HC. 2011. Organ engineering based on decellularized matrix scaffolds. *Trends Molec. Med.* 17:424–432.

Stringari, C, Edwards, RA, Pate, KT, Waterman, ML, Donovan, PJ, Gratton, E. 2012a. Metabolic trajectory of cellular differentiation in small intestine by phasor fluorescence lifetime microscopy of NADH. *Sci. Rep.* 2:568.

Stringari C, Nourse, JL, Flanagan, LA, Gratton, E. 2012b. Phasor fluorescence lifetime microscopy of free and protein-bound NADH reveals neural stem cell differentiation potential. *PLOS One* 7(11):e48014.

Sun, Y, Tan, HY, Lin, SJ, Lee, HS, Lin, TY, Jee, SH, Young, TH, Lo, W, Chen, WL, Dong, CY. 2008. Imaging tissue engineering scaffolds using multiphoton microscopy. *Microsc. Res. Tech.* 71:140–145.

Tan, W, Vinegoni, C, Norman, JJ, Desai, TA, Boppart, SA. 2007. Imaging cellular responses to mechanical stimuli within three-dimensional tissue constructs. *Microsc. Res. Tech.* 70:361–371.

Tearney, GJ, Jang, IK, Bouma, BE. 2006. Optical coherence tomography for imaging the vulnerable plaque. *J. Biomed. Opt.* 11:021002.

Tuchin, VV. 2007. *Tissue Optics: Light Scattering Methods and Instruments for Medical Diagnosis*. SPIE Press, Bellingham.

Uchugonova, A, Konig, K. 2008. Two-photon autofluorescence and second-harmonic imaging of adult stem cells. *J. Biomed. Opt.* 13:054068.

Veilleux, I, Spencer, JA, Biss, DP, Cote, D, Lin, CP. 2008. *In vivo* cell tracking with video rate multimodality laser scanning microscopy. *IEEE J. Select. Topics Quantum Electron.* 14:10–18.

Wang, RK. 2008. Directional blood flow imaging in volumetric optical microangiography achieved by digital frequency modulation. *Opt. Lett.* 33:1878–1880.

Wax, A, Backman, V. 2009. *Biomedical Applications of Light Scattering*. McGraw-Hill, New York.

Williams, DF. 2009. On the nature of biomaterials. *Biomaterials* 30(30):5897–5909.

Wilson, BC, Vitkin, IA, Matthews, DL. 2009. The potential of biophotonic techniques in stem cell tracking and monitoring of tissue regeneration applied to cardiac stem cell therapy. *J. Biophoton.* 2:669–681.

Wotjkowski, M, Kaluzny, B, Zawadski, RJ. 2012. New directions in ophthalmic optical coherence tomography. *Optom. Vis. Sci.* 89:524–542.

Wu, KH, Mo, XM, Han, ZC, Zhou, B. 2011. Stem cell engraftment and survival in the ischemic heart. *Ann. Thor. Surg.* 92(5):1917–1925.

Yang, Y, Bagnaninchi, PO, Cuncha-Reis, C, Aydin, HM, Piskin, E, El Haj, A. 2007. Characterisation of scaffold architecture by optical coherence tomography. *Proc. SPIE*. 6439:64390G.

Chapter 1

Yudovsky, D, Nouvong, A, Schomacker, K, Pilon, L. 2011. Assessing diabetic foot ulcer development risk with hyperspectral tissue oximetry. *J. Biomed. Opt.* 16:026009.

Zaharia, C, Tudora, MR, Stanescu, PO, Vasile, E, Cincu, C. 2012. Silk fibroin films for tissue bioengineering applications. *J. Optoelectron. Adv. Mater.* 14:163–168.

Zhang, JC, Savage, HE, Sacks, PG, Delohery, T, Alfano, RR, Katz, A, Schantz, SP. 1997. Innate cellular fluorescence reflects alteration in cellular proliferation. *Lasers Surg. Med.* 20:319–331.

Zhou, R, Acton, PD, Ferrari, VA. 2006. Imaging stem cells implanted in infarcted myocardium. *J. Am. Coll. Cardiol.* 48:2094–2116.

Ziegelmueller, JA, Zaenkert, EK, Schams, R, Lackermair, S, Schmitz, C, Reichart, B, Sodian, R. 2010. Optical monitoring during bioreactor conditioning of tissue-engineered heart valves. *Am. Soc. Artificial Internal Organs J.* 56:228–231.

Optical Microscopy in Regenerative Medicine

2. Fluorescence Microscopy

Nicola Green and John W. Haycock

Chapter 2

Optical Techniques in Regenerative Medicine. Edited by Stephen P. Morgan, Felicity R.A.J. Rose, and Stephen J. Matcher © 2014 CRC Press/Taylor & Francis Group, LLC. ISBN: 978-1-4398-5495-2

2.1   Introduction

2.1.1   What Is Fluorescence Microscopy?

Fluorescence microscopy is an imaging technique that relies upon the phenomenon that when certain molecules are exposed to light of a specific wavelength, those molecules will emit light of a lower energy and increased wavelength. This emitted signal is separated from the light used to illuminate the sample using optical filters and the signal collected to generate the image. Generally, fluorescent dyes—also referred to as fluorophores—are used to stain particular components of a sample, with the location of the fluorescence indicating the location of the constituent of interest. Many biological molecules are also intrinsically fluorescent, and this property can be exploited for imaging purposes without the need for additional fluorophores.

Initially, fluorophores used for imaging were simple chemical stains, which typically were taken up by cells and incorporated into specific subcellular regions. Many of these dyes are still used by the fluorescence microscopist today, such as 4′,6-diamidino-2-phenylindole (DAPI), which binds strongly to DNA, with an increase in fluorescence once bound (Schnedl et al. 1977), and Nile Red, which can be used to fluorescently label intracellular lipids (Greenspan et al. 1985). Once it became possible to fluorescently label antibodies, the technique of immunofluorescence was developed (Coons et al. 1941). This allowed researchers to label and image the location of specific cell components, such as a particular protein or nucleic acid sequence, and fluorescence microscopy then became widespread. More recently, a number of specific fluorescent probes have been developed to monitor dynamic processes in living cells, providing the opportunity to visualize responses to changing environments (see the review by Tsien [1989] for more details).

The discovery of green fluorescent protein (GFP) in jellyfish (Shimomura et al. 1962) and the subsequent cloning of the GFP gene in the early 1990s (Prasher et al. 1992) have meant that scientists can modify cells to express proteins that include a fluorescent tag. This has dramatically enhanced studies into intermolecular interactions and subcellular localization using live cells and has seen considerable development of a range of spectral variants of this protein for such studies (reviewed in Shaner et al. [2005]).

2.1.2   Fluorescence Microscopy in Regenerative Medicine

To achieve the fundamental goals of regenerative medicine, there is a requirement for methods that enable the researcher to examine growth and behavior in a way that does not influence the physiology of the sample under study. This is particularly challenging since these samples are generally 3D tissue-engineered constructs containing live cells. Successful characterization of such specimens ensures that tissue-engineered constructs can be developed in the laboratory as experimental platforms—which are significantly more biologically relevant than a 2D cell culture (reviewed in Haycock [2011]). Alternatively, the constructs can themselves be prepared and monitored during production *in vitro* and their suitability for clinical use determined.

Standard histological protocols have previously been employed to characterize tissue-engineered constructs produced in the laboratory. These processes are invasive and time-consuming, requiring the fixing, embedding, and sectioning of samples, followed by the appropriate immunohistochemical techniques. Such extensive processing of the tissue can result in the production of artifacts and is destructive, necessitating the production of many replicates to provide material for this process, particularly if monitoring changes over time. Noninvasive techniques, especially if they can be performed on living cells without compromising their viability or behavior, are therefore highly desirable.

Fluorescence microscopy is widely used by biological scientists because it is a rapid, sensitive, and specific methodology for characterizing cells and their responses, which also permits the visualization and localization of specific proteins within the sample. Developments in the technique have allowed images to be obtained from within 3D structures, and it has become possible to image living cells, making the technique highly applicable to the field of regenerative medicine.

2.1.2.1   Advantages and Disadvantages of Fluorescent Microscopy in Regenerative Medicine

There are both advantages and disadvantages to using fluorescence microscopy to image samples. It requires only limited sample processing, compared to standard histological procedures, and this minimizes the production of artifacts within the samples. The use of low-intensity light to image samples usually has a minimal impact upon cellular integrity, and consequently, with care, imaging can be performed on live cells repeatedly, facilitating time course studies and the imaging of real-time responses.

However, as already stated, imaging generally requires the addition of contrast agents, based on chemical stains or fluorescently labeled antibodies. These produce excellent quality images permitting the localization of components and compounds within; however, they can also affect the continued viability of the sample. This has required specific methods to be developed for imaging live cells and tissues and is discussed in more detail later in this chapter.

2.1.3   Fluorescence Excitation and Emission

The theory behind fluorescence is explained in detail in *Principles of Fluorescence Spectroscopy* (Lakowicz 2006). Briefly, when fluorophores are exposed to light of a specific wavelength, some of their electrons are excited and promoted to a higher energy level. These electrons lose a small amount of energy nonradiatively, falling to the lowest vibrational level within the lowest excited state. The average time that a molecule remains in this excited state is referred to as the fluorescence lifetime, and it varies for different fluorophores but is in the nanosecond order for commonly used fluorescent dyes. The electrons then drop back to their original ground state, emitting the remaining energy as light of a lower energy and, consequently, longer wavelength. Thus, there is a difference between the excitation and emission wavelengths, called the Stokes shift. Figure 2.1 represents the energy state of an electron during the excitation and emission cycle.

The illumination light that excites the electrons can be separated from the much weaker emitted fluorescent light using appropriate filters, and it is easier to separate these when working with fluorophores that have a larger Stokes shift, since the overlap between the excitation and emission spectra is reduced.

Many fluorophores have now been developed with excitation wavelengths that cover the light spectrum from ultraviolet to near-infrared. However, it is often still

Chapter 2

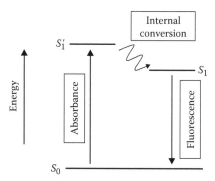

FIGURE 2.1   Jablonski diagram showing the energy state of an electron in the absorption and emission cycle of fluorescence.

difficult to predict the absorption and emission spectra of a putative fluorophore from its molecular structure, and consequently, it remains a challenge to design suitable fluorophores to match the excitation wavelengths and detection filters generally available to the researcher.

2.1.4   Fluorescence Microscope

2.1.4.1   Epifluorescence Microscope

Most fluorescence microscopes in common laboratory use are actually widefield epifluorescence microscopes, using an objective lens above the sample to both focus light upon the specimen and collect the light being emitted. These microscopes are efficient, cost effective, simple to use, readily available, and capable of imaging many fluorescent samples very effectively. The standard design of a microscope of this type is shown in Figure 2.2.

Generally, samples are illuminated with an arc lamp containing mercury vapor or, less commonly, argon vapor. An excitation filter placed between the light source and the sample is used to restrict the wavelength of light that illuminates the sample, to correspond with the excitation peak of the fluorophore. However, the light source is not sufficiently bright to allow it to be limited to a single wavelength, and light from across a waveband is required. Specific dichroic beam splitters and emission filters are then chosen to match the emission spectrum of the fluorophore and block out the excitation light ensuring that only the light emitted from the fluorophore reaches the detector. A narrow excitation waveband is consequently preferable to maximize the available detection window for the emitted light while ensuring that all excitation light is excluded from the detector. Using appropriate filters in this way, the location of a given fluorophore within the sample can be determined and an image produced.

2.1.4.1.1 Disadvantages of the Epifluorescent Microscope   A number of difficulties can be encountered when using a widefield epifluorescent microscope. Firstly, the entire specimen in the optical path is illuminated simultaneously, and the resulting fluorescence is detected from all excited fluorophores, including those that are outside the focal plane. Consequently, any sample with an axial thickness of more than about 5 μm can appear indistinct and out of focus.

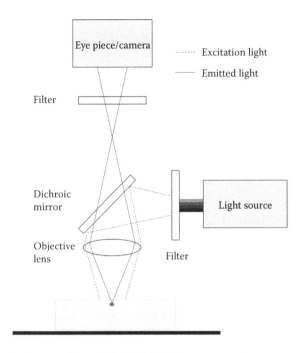

FIGURE 2.2   Schematic diagram of a typical widefield epifluorescent microscope.

Secondly, the intensity of light generated from the mercury lamp varies across the emission spectrum, with intense illumination over some bandwidths but much lower illumination at other wavelengths; there are fluorophores whose excitation maxima do not coincide with the intense bands, and excitation of these molecules is significantly reduced. One such fluorophore is the commonly used fluorescein isothiocyanate (FITC), which has an excitation maximum at 490 nm. The intensity of light from a mercury lamp is significantly reduced around this wavelength, and this decreases the relative amount of emitted light from FITC.

Finally, images from samples that contain multiple fluorescent labels are, by necessity, obtained by combining the individual images of each fluorophore. It is therefore very important to consider the excitation and emission spectra of the fluorophores used in combination to prevent the bleed through of signals caused by overlapping spectra. Although this is an issue for any fluorescence microscopy technique, it can be particularly problematic when using the standard epifluorescent microscope since the illumination of the sample by a waveband of light further increases the likelihood of exciting more than one fluorophore within the sample.

2.1.4.2   Confocal Laser Scanning Microscope

The confocal laser scanning microscope (CLSM) is a significant advance in the field of optical microscopy, developed to overcome some of the problems associated with the standard epifluorescence microscope. It permits sample visualization from within living or fixed cells, tissues, and other samples and provides the ability to collect sharply defined optical sections from differing focal planes within the sample. The standard design of a confocal microscope is shown in Figure 2.3.

Chapter 2

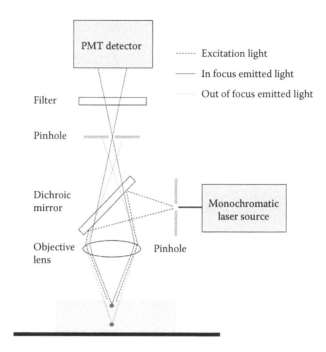

FIGURE 2.3 Schematic of a confocal microscope.

Confocal microscopes normally use a monochromatic laser to excite the sample, allowing the user to choose very specific excitation wavelengths and thus expanding the combination of fluorophores that can be used together. The laser point light source is passed through an illumination pinhole onto the sample. This ensures that only one point within a specimen is lit at any one time, eliminating stray lateral light interference and improving image contrast.

Light from the sample is collected from a single focal plane. In a standard confocal microscope, this is achieved by using a pinhole in a plane that is optically conjugated to the focal plane in front of the detector to prevent the majority of the light reflected or emitted from regions above or below the focal plane from entering the detector (Figure 2.3). The size of the detector pinhole determines the thickness of the optical slice viewed. Decreasing the pinhole reduces its axial thickness; however, this simultaneously decreases the strength of the signal reaching the detector, resulting in decreased signal-to-noise.

2.1.4.2.1 Resolution in Fluorescence Microscopy Resolution describes the capability of an imaging system to observe detail in the object being imaged. It can be defined as the minimum distance between two separate points in a field of view that can still be distinguished as two separate and distinct entities. If objects are closer together than this, they appear to blur together and become impossible to differentiate. It is necessary when using any imaging system to ensure that it has the resolving power to fulfill the experimental requirements.

The resolution achieved by fluorescence microscopy is influenced, for both confocal and epifluorescence microscopes, by the numerical aperture (NA) of the objective used, with a greater NA giving improved resolution. The wavelength of the light, both for excitation and emission, also affects the resolution, with shorter wavelengths resulting in increased resolution. The resolution achieved by CLSM is also influenced by the size of the pinhole, with a smaller pinhole leading to a higher resolution until the maximum resolution is achieved at an optimal pinhole size, determined by the excitation wavelength, objective NA, and lens magnification. Further reductions in the pinhole then lead to a decrease in signal intensity and no further increase in resolution.

In addition, since the contrast produced from the CLSM is much greater than for the epifluorescent microscope, its effective ability to resolve two closely separated features is significantly greater. There is some variability in the maximum achievable resolution quoted for CLSM, but in general, lateral resolution can reach 200 nm and axial resolution 900 nm.

2.1.4.2.2 Obtaining 3D Images and Imaging from within a Sample

By changing the focal plane from which the image is obtained, the user can obtain sequential optical slices from increasing depths within the sample. These can then be "stacked" on top of each other and combined to produce 3D images. This is called "optical sectioning" and allows the user to view inside a sample without physically disrupting it. There are, however, limits to the maximum penetration depth possible using CLSM. The working distance of the objective imposes an upper limit, since once the objective is touching the sample, it clearly cannot be focused in any further. The penetration depth is also restricted by the sample absorbing the excitation energy and scattering both excitation and emission photons. This problem is exacerbated as the imaging depth is increased, particularly if the sample is highly light-scattering or energy absorbent. Also, if there is a refractive index mismatch between the sample medium, which is normally aqueous for biological samples, and the objective immersion medium, which can often be oil, the detected signal is reduced, which reduces the depth from which a satisfactory signal can be obtained. However, the former issue can be resolved by use of long working distance objective lenses and the latter issue to some extent by the introduction of water-immersion and water-dipping objectives. If the conditions are optimized, an imaging depth of 50 to 100 μm is possible, depending on the nature of the sample.

2.1.4.2.3 Photobleaching and Phototoxicity

Photobleaching, or the irreversible destruction of a fluorophore by exposure to light, occurs when electrons are excited and while in this excited state modified by interactions with other molecules within the environment. This prevents the electrons from returning to the ground state and thereafter they are unable to emit light. The average number of excitation/emission cycles a fluorophore can go through before being irreversibly modified depends upon both the fluorophore and the local environment.

Phototoxicity is a related problem that arises when the excited electron reacts with molecular oxygen within the local environment, producing reactive oxygen species (ROS). This may cause significant damage to cellular structures, which can result in a loss of viability, and introduces particular difficulties when imaging live samples (Knight et al. 2003).

It is worth noting when using CLSM that the light detected is from a single focal plane within the sample; however, all fluorophores within the laser light path are actually excited and this has associated problems with sample photobleaching and phototoxicity. Indeed, for some of the more sensitive fluorophores, exposure to the laser even when not in the focal plane may result in molecules that are photobleached before a single image is captured from that plane.

Chapter 2

2.1.4.2.4 Disadvantages of the CLSM The successful use of CLSM requires the user to have a greater level of experience and training than is necessary for the widefield epifluorescent microscope. The equipment is also two- to threefold more expensive. However, when sample depths are greater than approximately 5 μm, the benefits of the CLSM over the epifluorescent microscope will often warrant its use, in particular, for 3D imaging and/or when there is multiple fluorophore labeling of a single sample.

2.1.5   Two-Photon Excitation Microscopy

As has already been discussed, CLSM generates fluorophore excitation above and below the focal plane, inducing phototoxicity and photobleaching outside the focal plane. Furthermore, the sample itself can absorb the short wavelength excitation beam, significantly reducing the penetration depth. The development of two-photon excitation microscopy has greatly improved the situation with regard to both of these problems. Commercial "turn-key" two-photon microscopes are also now available, providing researchers who lack specialized laser optics expertise with access to two-photon microscopy. This technique is now becoming increasingly popular.

Two-photon excitation involves the almost simultaneous (within 10^{-18} s) absorption of energy from two photons, each contributing half of the total energy required to induce fluorescence, as shown in the Jablonski diagram (Figure 2.4). The excitation wavelength required is therefore longer than the emission wavelength, providing an anti-Stokes shift.

The density of photons is high at the focal point but falls off rapidly above and below this point. Since excitation requires the simultaneous absorption of two photons, the probability that a fluorophore will be excited is highest at the focal point and, less than 1 μm above or below this, rapidly drops to near zero. As a result, only fluorophores at the focal point are excited, removing the need for a pinhole between the sample and the detector and providing one of the major advantages of the two-photon microscope. Photobleaching and phototoxicity issues are also limited to the focal plane, and the problems associated with these are therefore greatly ameliorated.

In order to achieve the required photon density at the focal point, a very high laser power is required, which would cause massive damage to the sample. This is avoided by using femtosecond pulsed lasers with an extremely high peak power but, due to the low ratio of pulse duration to gap duration, an average power similar to the standard CLSM.

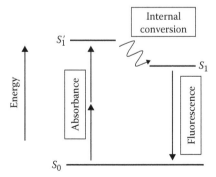

FIGURE 2.4   Jablonski diagram showing the energy state of an electron in the absorption and emission cycle of fluorescence, with two-photon excitation.

The laser and objectives required are discussed in more detail in relation to second harmonic generation (SHG) imaging in Chapter 4.

The longer excitation wavelengths used in two-photon microscopy also have reduced absorption and scattering coefficients in biological tissues. Consequently, two-photon excitation microscopy permits fluorescence imaging from much deeper within the sample (Centonze and White 1998). Furthermore, since a pinhole is no longer required in front of the detector, external (nondescanned) detectors placed as close to the collection lens as possible can be used to provide the most efficient detection system and so increase the maximum penetration depth still further.

Two-photon excitation microscopy, with its considerably improved light-gathering efficiency and reduced damage to the specimen, is therefore a very important tool for the non-destructive fluorescent imaging of live 3D constructs. However, additional costs associated with this configuration means that this approach is still perhaps not readily available to many researchers.

2.2 Fluorescence Microscopy for Clinical and Biological Measurements

2.2.1 Noninvasive Imaging for Continuous Study of Live Cells

One of the key challenges in regenerative medicine is the requirement for repeated images of the same live sample over time—without the imaging process itself resulting in damage to the sample, or indeed perturbing or compromising the behavior of the cells. It may be necessary to obtain images of live cells as they respond to the environment, grow, or develop over a time course of minutes, hours, days, or even weeks.

2.2.1.1   Short-Term Imaging of Live Cells

Sometimes, when biochemical processes or intracellular interactions are under study, the images from live cells only need to be acquired over a relatively short time. These can be readily studied by fluorescence microscopy. The key methods that can be used to study the short-term response of live cells are discussed below.

2.2.1.1.1 Fluorescence Recovery after Photobleaching   Fluorescence recovery after photobleaching (FRAP) is a technique that characterizes the movement of a specific population of fluorescent molecules within an area of the sample. For a detailed discussion of the protocol as used in 2D cultures, see the review by Reits and Neefjes (2001). Briefly, intense laser light is used to photobleach a specific area of a sample, and the subsequent movement of fluorescent molecules from adjacent areas into the bleached region is monitored. This dynamic process is the recovery of the fluorescent signal. It is also possible to observe the fluorescence loss induced by photobleaching (FLIP), which occurs outside the bleached area through an influx of photobleached molecules together with an efflux of fluorescently labeled molecules. Since the movement of molecules within the cells is usually very rapid, it is not essential to maintain cell viability for extended periods in order to perform FRAP studies.

Although FRAP has been widely used in 2D cultures, it is an emerging technique for 3D *in vitro* constructs and regenerative medicine. A small number of studies have been performed using fluorescently labeled dextran, with one group studying the diffusion of

Chapter 2

solutes in cartilage by monitoring the fluorescence of tissue-engineered cartilage constructs over time and under different culture conditions (Leddy and Guilak 2003; Leddy et al. 2004). They observed a greater diffusivity for tissue-engineered constructs compared to native cartilage, but demonstrated that the constructs decreased in diffusivity over time and attributed this to the synthesis of new cartilage matrix during culture. In a second study (Gefen et al. 2008), researchers examined the diffusion capacity of fluorescent dextran within tissue-engineered skeletal muscle when subjected to compression strain. They examined the impact of muscle compression on metabolite diffusivity and the subsequent increased risk of developing pressure-related deep tissue injury and demonstrated that compression strains comparable to local partial muscle ischemia significantly reduced the diffusivity of proteins by the order of approximately 50%.

2.2.1.1.2 Fluorescence Resonance Energy Transfer (FRET)　FRET, or fluorescence (sometimes referred to as Förster) resonance energy transfer, also requires cell viability to be maintained for only a relatively short time. This technique is used to study intermolecular or intramolecular interactions by monitoring the transfer of emitted energy from one donor fluorophore to an immediately adjacent acceptor fluorophore. The acceptor is excited by this process, and its fluorescence can then be detected (see the review by Periasamy [2001] for more details). This phenomenon can only be observed when the donor and acceptor fluorophores are around 8 to 10 nm or less apart. Donor and acceptor fluorophores must also be carefully chosen to ensure that the donor emission spectrum overlaps with the acceptor excitation spectrum, with a greater overlap generally producing a more efficient transfer of energy.

This technique has been highly developed for use with 2D cell cultures but has only recently been applied to 3D constructs. For example, Kubow et al. (2009) used FRET to study how cells reseeded into a decellularized scaffold remodeled the matrix and how chemical cross-linking of the scaffold affected this. They used fibronectin labeled with Alexa-Fluor 488 (FRET donor) and Alexa-Fluor 546 (FRET acceptor), mixed with 90% unlabeled fibronectin to prevent intermolecular FRET. This ensured that FRET would only occur if donor and acceptor fluorophores on a single fibronectin fiber were in close proximity, with donor emission resulting in excitation of the acceptor and a shift in emission from green-blue (donor emission) to yellow (acceptor emission). FRET was then used to demonstrate that scaffold decellularization eliminated cell-generated tension, allowing the fibers to partially contract and induce an increase in the FRET intensity ratio. This is seen in the original figure as a color shift from green-blue to yellow between the 4-day matrix and the scaffold with the cells removed. The figure is reproduced here in grayscale, with the dynamic ranges adjusted to demonstrate the FRET principle, such that lighter areas indicate increased FRET (Figure 2.5).

Using FRET, they then observed that seeding of fibroblasts onto the acellular scaffold resulted in scaffold remodeling as the fibers were progressively stretched, indicated by a reduction in FRET. Chemical cross-linking of the scaffold, however, was seen to prevent this remodeling from occurring.

2.2.1.2　Extended Imaging of Live Cells

Much of the work within regenerative medicine requires that live cells and tissues are characterized over extended periods and fluorescence microscopy can play a vital role

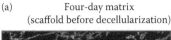

(a) Four-day matrix (scaffold before decellularization)

(b) Decellularized scaffold

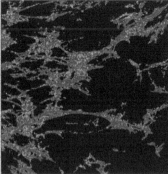

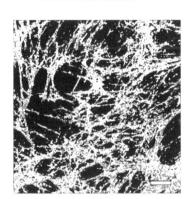

FIGURE 2.5 Use of FRET to study the impact of decellularization on fibronectin unfolding. Fibronectin labeled with Alexa-Fluor 488 (FRET donor) and Alexa-Fluor 546 (FRET acceptor) has been incorporated into the matrix. Representative FRET intensity ratio measurements of the cell-derived ECM are shown before (a) and after decellularization (b). The images are single planes from a z-stack. The original image is in color and is color-coded by ratio intensity, with a shift from green-blue (a) to yellow (b) indicating an increase in FRET. However, for the purposes of this book, it has been reproduced in grayscale and the dynamic ranges adjusted to illustrate the principle of FRET (amended from Kubow, K. E., Klotzsch, E., Smith, M. L. et al. 2009. Crosslinking of cell-derived 3D scaffolds up-regulates the stretching and unfolding of new extracellular matrix assembled by reseeded cells. *Integr. Biol. (Camb.)* 1: 635–648. Reproduced by permission of The Royal Society of Chemistry.)

in sample visualization, allowing repeated images to be obtained from a single live sample nondestructively. An obvious use of fluorescence microscopy and more specifically CLSM is therefore to obtain a time course of 3D images of a growing tissue.

Longitudinal images of live tissue constructs have been collected by a few research groups. In a study into the matrix remodeling process implicated in development and wound repair, GFP-expressing fibroblasts were grown on a scaffold that included labeled beads and the live construct imaged repeatedly over several hours (Tamariz and Grinnell 2002). The movement of the beads over time was used to monitor local matrix remodeling and required the culture to remain viable throughout the course of the experiment.

Tan et al. (2004) meanwhile obtained images from a tissue construct of GFP-expressing fibroblasts on a chitosan scaffold, for up to a week. They were able to observe changes in cell morphology, matrix secretion, and vinculin expression over time that reflected an increase in cell–cell interactions and cell–matrix adhesion throughout the culture period.

2.3 Technical Implementation of Fluorescence Microscopy

2.3.1 Imaging Living Cells

One aspect of CLSM can cause difficulties when attempting to image live cells or tissues. Most standard confocal microscopes use a raster scanning method to excite the

Chapter 2

fluorophore and collect the emitted signal on a point-by-point basis. Although the dwell time at any point is very short, generally a few microseconds, the time taken to generate the whole image can be several seconds, depending upon the total number of pixels within an image. This delay in total acquisition time prevents the monitoring of cellular responses in real time. In addition, the limited laser dwell time for each point within the image requires that the excitation energy is sufficiently high to generate a detectable response during that short dwell period. This high laser energy can cause the photobleaching and phototoxicity problems previously discussed, particularly in samples where repeated imaging is required.

In order to resolve this, the spinning disk confocal microscope has been developed. These microscopes illuminate the sample via a spread laser light projected onto the sample through a disk with a pattern of slits. This allows several thousand points to be exposed to the laser light at one time and the emission signal from each of them detected simultaneously; spinning the disk then allows all points on the sample to be illuminated. In this way, images can be collected at approximately 30 frames per second, facilitating real-time imaging. However, this increase in speed does come with a slight reduction in axial resolution compared to the CLSM. Currently, the CLSM is more versatile, with many techniques yet to be adapted for the spinning disk system; however, it is likely over time that this will change as use of the spinning disk system becomes more widespread.

2.3.1.1 Important Considerations for Live Cell Imaging

Fluorescence microscopy techniques have been established for the study of live cells in 2D cultures, but are often equally applicable to a 3D tissue or *in vitro* construct, particularly when obtained using the optical sectioning capabilities of the confocal or two-photon microscope. However, there are a number of challenges when working with live cells that the user must resolve in order to perform these types of studies, many of which relate to the timescale over which the cells or tissues are to be imaged. In all cases, the sample must remain viable and function normally for the duration of the experiment, in a configuration that also allows access for the microscope objective.

2.3.1.1.1 Short-Term Maintenance of Viability In experiments where the cells only need to remain viable for a very limited period, it is often sufficient to ensure the sample is imaged in an aqueous, pH-buffered environment such as PBS or cell culture media. A heated stage may also be included to maintain cells at 37°C.

When cells are in 2D, it can be sufficient to grow them on a glass coverslip and invert this onto a slide for imaging under an upright microscope, using vacuum grease to seal the edges and contain a small amount of culture medium between the slide and coverslip. However, this is less suitable for imaging thicker 3D structures, where the sample cannot easily be sandwiched between the slide and coverslip. Instead, images can be obtained using a microscope with an upright configuration in conjunction with water dipping objectives. In addition to a heated stage, it may be necessary to heat the objective itself through a heated collar, since the thermal gradient between the heated sample and an unheated objective can result in a loss of thermal control in the sample area in contact with the objective. This relatively basic, low-cost approach is suited to the FRAP, FLIP, and FRET techniques described previously if the data are required from only one time point. However, dipping an objective into the culture medium will clearly compromise the sterility of the sample and so is not suitable for longer-term or repeated imaging.

2.3.1.1.2 Maintenance of Sterility

For experiments in which it is necessary to repeatedly image the same sample over the course of days or even weeks, protocols are required that allow this to be done without exposure to the nonsterile environment.

A number of modified tissue culture plates have been developed for which at least part of the culture surface is of optical quality plastic or glass and is suitable for use with a microscope. There are several companies that provide slides, wells, and Petri dishes with such optically optimized, biocompatible surfaces including Ibidi (Munich, Germany), Thermo Scientific (Braunschweig, Germany), PAA (Yeovil, UK), MaTek Corp (Massachusetts, USA), and Merck Millipore (Massachusetts, USA). To ensure contact between the cells and the coverslip-equivalent surface, the optically optimized surface is on the bottom of the plate or well, and these products require an inverted microscope configuration. Lids are used to preserve the sterility of the sample, and it is not necessary to remove these during the imaging process; however, there is no environmental control and additional equipment, or the repeated transfer of the sample between the microscope and the incubator is necessary to maintain cell viability for extended periods.

Although these culture dishes have been developed for use with 2D cultures, it can be possible to use them for the growth and subsequent imaging of live organotypic 3D constructs. They are particularly suitable if the area of interest is on the underside of the construct or the sample is of limited thickness, thus allowing penetration throughout the whole sample. Unfortunately, samples in regenerative medicine can have the cells seeded on the uppermost surface of a thick construct. It is then necessary to invert the sample, under sterile conditions and prior to imaging, to ensure the surface of interest is in contact with the optical surface. Despite these complications, this method has been used successfully to image a 3D lung organotypic model, consisting of a coculture of a GFP-expressing mammary tumor cell line and primary lung cells on a synthetic foam scaffold, at four time points over a 5-day period (Martin et al. 2008). In these experiments (Figure 2.6), the researchers incorporated positional markers into the organotypic model to ensure that the same area was imaged each time to produce a number of z-stacks from multiple locations within the sample for each time point. The tissue was then returned to the incubator between each image collection step.

Heated stages that accept the optically optimized cell culture chambers are available, ensuring that the temperature of the sample is maintained during the imaging process, improving cell viability and extending the potential imaging period. Without additional environmental controls, however, oxygenation, pH, and humidity levels remain an issue, and the sample must still be returned to an incubator for long-term culture.

2.3.1.1.3 Environmental Control

For long-term, repeated imaging of live cells and tissues, it is important to consider controlling additional environmental factors. Localized regulation of pH can be achieved using small chambers that fit around the sample and provide a controlled supply of carbon dioxide. These chambers can also be used to control the oxygen and humidity levels around the culture, facilitating the manipulation of hypoxic conditions within the sample and preventing evaporation of medium from the sample.

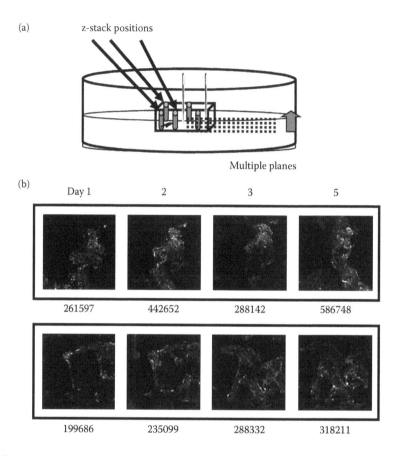

FIGURE 2.6   CLSM was performed on cocultures of a GFP-expressing mammary tumor cell line and primary lung cells on a synthetic foam scaffold, taking z-stacks at multiple locations in each culture. Positional markers were used to ensure the image was acquired from the same location at each time point (a). Images in (b) are shown for one single plane within one z-stack over 5 days for a culture containing breast and lung cells (top panel) and breast cells only (bottom panel). GFP expression in the form of total integrated GFP intensity has been used as a marker of mammary tumor cell growth, and the values are given below the micrographs for the plane shown. (With kind permission from Springer Science+Business Media: *Clin. Exp. Metastasis*, Establishment and quantitative imaging of a 3D lung organotypic model of mammary tumor outgrowth, 25, 2008, 877–885, Martin, M. D., Fingleton, B., Lynch, C. C. et al.)

The use of heated stages, particularly for longer-term cultures, can be problematic, since the direct heat source can result in localized scorching of the sample. This issue can be ameliorated with heated water jackets to provide the physical contact with the sample. It is also possible to place the microscope stage and optics within an environmental chamber and control the temperature through a blown air system. This also makes it possible to maintain humidity and gas levels. However, when sample sterility is required, these systems are still only applicable for inverted microscopes, requiring the area of interest to be located at the bottom of the sample, very close to the coverslip.

2.3.1.1.4 Bioreactors   Bioreactors, in which conditions can be closely controlled to permit or induce certain responses in living cells or tissue, have been developed for the growth of tissue-engineered samples, as described in a number of reviews (Rauh et al. 2011; Naing and Williams 2011; Ellis et al. 2005). Bioreactors can be used to control the concentration of gases and nutrients within the culture medium and to expose the specimens to given physical stimuli. They have been used to culture a number of different tissue types; for example, recent tissues grown in bioreactors include bladder (Wei et al. 2011), heart valves (Sierad et al. 2010), skin (Sun et al. 2005), cartilage (Ng et al. 2009), and nerves (T. Sun et al. 2008). Unfortunately, despite the increased prevalence of bioreactors, their design is such that it is still generally necessary to remove the tissue from the bioreactor before imaging under a microscope. For example, Bolgen et al. (2008) employed a bioreactor to examine the effects of perfusion rate upon an osteoblast-like cell line, and Srouji et al. (2008) cultured human mesenchymal stem cells (MSCs) within bioreactors on electrospun scaffolds. In both instances, imaging was only possible after terminating the culture and removing it from the bioreactor.

Attempts are now under way to develop a bioreactor that would also facilitate imaging of the developing tissue without necessitating removal from the bioreactor. Such systems have been developed for the repeated imaging of 2D bioreactor cultures (Schwartlander et al. 2007), and a bioreactor system has recently been described by Kluge et al. (2011) that allows the application of a range of mechanical loading regimes onto a 3D tissue-engineered scaffold in conjunction with repeated noninvasive imaging of the sample. The authors incorporated both a nonpermanent cell tracer dye, DiI, for short-term studies of cell growth and a GFP-expressing cell line for longer-term work and were able to demonstrate the capability to nondestructively track cells and scaffold that had undergone mechanical strain over an extended period. Figure 2.7 shows images that were obtained at three time points during the 2-week experiment, where daily mechanical loading was applied between days 5 and 14 to silk fibers cultured with GFP-expressing human fibroblasts.

In this study, the researchers were able to maintain sterility and culture conditions during both stretching and imaging modalities, isolate the sample environments to prevent cross-contamination or unintentional biological signaling between samples, and ensure that the samples were optically accessible to a microscope. However, the system still relies on an inverted microscope imaging samples from beneath, and further work is needed to allow samples to be imaged within bioreactors where the cells of interest are on the upper surface of the construct.

2.3.1.1.5 Maintaining Samples in Focus Another problem with the long term imaging of live cells arises because of the precise optical sectioning capabilities of confocal microscopes. At the start of the image capture process, an area of interest is set up within the focal plane of the microscope, but in order for useful images to be captured over time the focus needs to remain on the area of interest. However, the mobility of cells within a sample, the potential for movement of the microscope stage (which is more likely to occur with the use of heated stages), and the potential for movement of the specimen itself mean that this can be very difficult to achieve. The easiest way to resolve this is to capture a 3D image at each time point, allowing the researcher to observe cells as they exit and enter each focal plane. However, this increases both the time required to capture images and the potential for photobleaching and phototoxicity. Even when capturing a 3D image, it is still often necessary to monitor the sample during the course of the experiment to review and possibly readjust the focal plane.

Chapter 2

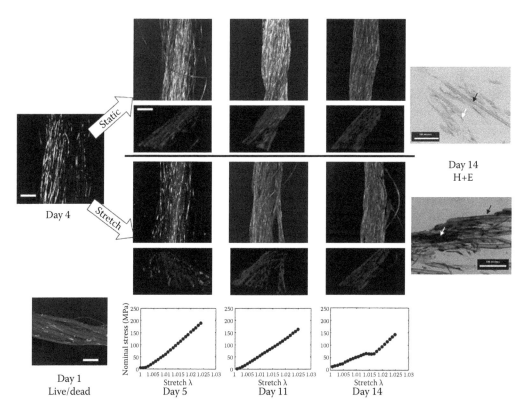

FIGURE 2.7 (See color insert.) GFP-transfected dermal fibroblasts (green) grown to near confluence over 5 days on silk fibers (red). Daily mechanical loading was then applied until day 14 and confocal scans taken daily. Images are shown for day 4, before the effect of mechanical loading and for days 5, 11, and 14 after stretching was applied. Live/dead staining is shown for day 1 after seeding. Scale bar is 300 μm. H+E staining is shown for the end point following stretched (top) or static (bottom) conditions. Scale bar is 100 μm. Black arrows indicate silk fibers; white arrows indicate cells. Mechanical test data at each time point are also shown. (With kind permission from Springer Science+Business Media: *Ann. Biomed. Eng.*, Bioreactor system using noninvasive imaging and mechanical stretch for biomaterial screening, 39, 2011, 1390–1402, Kluge, J. A., Leisk, G. G., Cardwell, R. D. et al.)

2.3.2 Labeling of Samples

2.3.2.1 Labeling Fixed Cells

When obtaining an image of fixed cells, labeling can be done by incorporating fluorescent stains or via immunofluorescent techniques, and many studies have been performed that have used such exogenous fluorophores. Fluorescent microscopy of fixed cells has been used in the field of regenerative medicine, for example, to assess cell viability, proliferation, and adhesion in 3D tissue-engineered samples (Xu et al. 2004; Martina et al. 2005; Landis et al. 2006; Mahoney and Anseth 2006; Chen et al. 2010; Tritz-Schiavi et al. 2010; Bolgen et al. 2011); to characterize stem cell and precursor cell morphology (Leong et al. 2006; Kaewkhaw et al. 2011; Wang et al. 2011); to assess novel scaffold materials and designs (Turner et al. 2004; Deshpande et al. 2010; Lin et al. 2010); or to evaluate the impact of the physical environment on cell behavior (Bolgen et al. 2008; Tigli and Gumusderelioglu 2009; Morelli et al. 2010). Figure 2.8 shows some example images of fixed and fluorescently labeled tissue-engineered constructs obtained by fluorescent microscopy. However, since the cells have been fixed, these samples have been limited to a single time point measurement, and multiple samples fixed at various time points are required to investigate any changes during the culture period.

2.3.2.2 Labeling Live Cells

Unfortunately, immunofluorescent techniques cannot be readily applied to living cells and tissues since they are impermeable to the relatively large labeled antibodies. Immunofluorescence in live cells is therefore restricted to imaging extracellular or

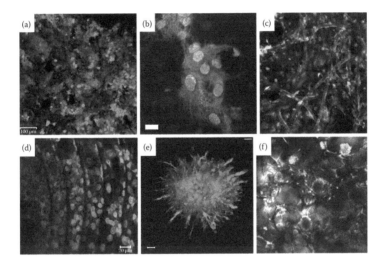

FIGURE 2.8 **(See color insert.)** Fluorescent images of fixed and stained tissue constructs. (a) Live (green) and dead (red) stained chondrocyte-seeded cryogels after 15 days culture. (Bolgen, N., Yang, Y., Korkusuz, P. et al. 3D ingrowth of bovine articular chondrocytes in biodegradable cryogel scaffolds for cartilage tissue engineering. *J. Tissue Eng Regen. Med.* 2011. 5. 770–779. Copyright Wiley-VCH Verlag GmbH & Co. KGaA. Reproduced with permission.) (b) Osteoblasts on EBPADMA-300 scaffolds after 3 days culture, showing nuclei (green), actin fibers (red), and photoscaffold (white). (From Landis, F. A. et al., *Biomacromolecules.* 7: 1751–1757, 2006.) (c) Colocalization of endothelial cells (green) and laminin matrix (red) on a hyaluronan-based scaffold. (Reprinted from *Biomaterials*, 25, Turner, N. J., Kielty, C. M., Walker, M. G., and Canfield, A. E., A novel hyaluronan-based biomaterial (Hyaff-11) as a scaffold for endothelial cells in tissue-engineered vascular grafts, 5955–5964, Copyright 2004, with permission from Elsevier.) (d) Corneal cells on an acellular fish-scale derived scaffold after 7 days culture, showing nuclei (green) and actin fibers (red). (From Lin, C. C. et al., *Eur. Cell Mater.* 19: 50–57, 2010.) (e) Projection of z-stack (1 mm thick) of live (green) or dead (red) staining of neural cells in degradable hydrogels after 14 days culture. (Reprinted from *Biomaterials*, 27, Mahoney, M. J., and Anseth, K. S., Three-dimensional growth and function of neural tissue in degradable polyethylene glycol hydrogels, 2265–2274, Copyright 2006, with permission from Elsevier) (f) Chondrogenic cells cultured on BMP-6 loaded chitosan scaffolds cultured in dynamic conditions after 4 weeks, showing actin fibers (green) and DNA (red) and background view of scaffold morphology. (Tigli, R. S., and Gumusderelioglu, M., Chondrogenesis on BMP-6 loaded chitosan scaffolds in stationary and dynamic cultures. *Biotechnol. Bioeng.* 2009. 104. 601–610. Copyright Wiley-VCH Verlag GmbH & Co. KGaA. Reproduced with permission.)

Chapter 2

endocytosed proteins. Furthermore, some stains capable of entering the cells are toxic and will cause a loss in viability over the longer term. Instead, contrast needs to be introduced into the image through endogenous fluorescence, fluorescent probes that do not compromise cell viability, or the use of cells that have been modified to express proteins such as the green, yellow, cyan, and red fluorescent proteins. These procedures are discussed in more detail below.

2.3.2.2.1 Endogenous Fluorescence   Some biological molecules naturally fluoresce when exposed to light of an appropriate wavelength. This means that certain images of tissue can be obtained without the addition of exogenous labels. The excitation and emission maxima of some intrinsically fluorescent molecules that are commonly found in biological samples are given in Table 2.1.

There are numerous studies using cells grown in 2D cultures that utilize intrinsic fluorescence to image live cells, generally focusing upon the pyridinic and flavin coenzymes. For example, Rajwa et al. (2007) used the autofluorescence of the plasma membranes of human hepatocytes to image live cells and also found that they were able to relate the presence and intensity of this fluorescence to the presence of redox agents, the metabolic state, and the membrane integrity of the cell.

Endogenous fluorescence has also been exploited very recently to nondestructively monitor cell viability within tissue-engineered constructs (Dittmar et al. 2012), taking advantage of the change in the spectral characteristics of light emitted by live and dead cells. This work used both single-photon and two-photon excitations and demonstrated that for constructs containing mostly live or dead cells, single-photon imaging was less accurate than two-photon imaging in determining cell viability. Studies have also shown that it is possible to image cells in tissue sections, snap frozen in liquid nitrogen (Rigacci et al. 2000), relying only on the endogenous fluorescence of biomolecules within the sample (Figure 2.9).

Other studies have used intrinsic fluorescence to further investigate oxidative stress, which is known to play a part in a wide variety of conditions and diseases such as diabetes, cancer, and neurodegeneration. Since nicotinamide adenine dinucleotide (NADH)

Table 2.1   Excitation and Emission Maxima for a Range of Intrinsically Fluorescent Biological Molecules

Molecule	Maximum Excitation Wavelength (λ = nm)	Maximum Emission Wavelength (λ = nm)
Collagen	270–370	305–450
Elastin	300–340	420–460
Flavins	375–450	520–550
Lipofuscin	340–390	430–490
NADH	350	450
NADPH	336	464
Phenylalanine	257	282
Retinol	325	400
Tyrosine	283	305
Tryptophan	280	350

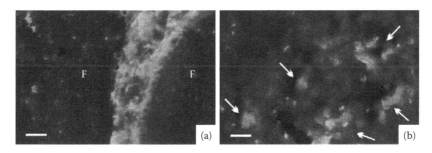

FIGURE 2.9 (a) Endogenous fluorescence image of hyperplastic lymphoid tissue (365 nm excitation) with a strongly fluorescent connective trabeculae composed mainly of collagen and elastin, separating two follicles (F). (b) Endogenous fluorescence of Hodgkin's lymphoma tissue (365 nm excitation) with small groups of large cells with intense fluorescence (arrows). Scale bar represents 25 μm. (Rigacci, L., Alterini, R., Bernabei, P. A. et al., Multispectral imaging autofluorescence microscopy for the analysis of lymphnode tissues. *Photochem. Photobiol.* 2000. 71. 737–742. Copyright Wiley-VCH Verlag GmbH & Co. KGaA. Reproduced with permission.)

reduced and flavin adenine dinucleotide (FAD) fluoresce at different wavelengths, it is possible to determine the intracellular ratio of the two molecules and, from this, the reduction–oxidation capacity or redox potential of the cell. This enables researchers to nondestructively investigate the impact of disease or metabolic change on cell and tissue behavior in living cells. A clear description of the principles and protocol for obtaining the optical redox potential, in this case with two-photon excitation, has been published by Skala and Ramanujam (2010). Although the optical redox ratio can be obtained with single-photon excitation, when continued cell viability or images from deeper within tissues are required, the method is usually performed using two-photon excitation (see Section 2.3.3). A recent example of a redox ratio study using single-photon excitation is described by Walsh et al. (2012). The ratio of NADH to FAD was acquired for a panel of breast cancer cell lines, with an increase in the redox ratio being indicative of increased metabolic activity. It was found that the ratio was highest in cancer cell lines overexpressing human epidermal growth factor receptor 2 (HER2), with HER2 positive tumors shown previously to display aggressive cancer progression.

To date, however, there is very little published work using endogenous fluorescence in the field of regenerative medicine, and more work is required to fully exploit the intrinsic fluorescent properties of these biological molecules.

2.3.2.2.2 Probes for the Study of Live Cell Function

Many fluorescent indicators are commercially available for monitoring physiological processes as they occur within live cells. These stains are capable of entering a live cell without need for permeabilization and are often reported to have no significant impact on cell viability or behavior. They allow the tracking of cell growth and proliferation within samples or the monitoring of specific intracellular processes during culture and in response to the environment, but are generally not yet suitable for inclusion in samples destined for clinical use.

There are dyes that can be loaded into cells before or after incorporation into a 3D structure. The dye diffuses readily into live cells and is then retained either through conversion by intracellular enzymes into a membrane impermanent molecule or via interactions with intracellular components. Some of these dyes can also be used as indicators of cell viability as they only become fluorescent by the action of enzymes found within live

Chapter 2

cells. Calcein AM (Molecular Probes, Life Technologies), for example, is a nonfluorescent cell-permeant dye that is converted to green fluorescent calcein by intracellular esterases.

A popular family of reagents includes the CellTrackers (Molecular Probes, Life Technologies), which contain diacetate groups conjugated to the fluorophore, facilitating passage through the cell membrane and retention following intracellular esterase removal. Thereafter, integration with glutathione permits chemical stability and an ability to be repeatedly imaged over several days. A number of different colored fluorophores can be incorporated for double- or triple-labeling cocultured samples. The dyes are retained within the cell and its daughters through several generations without being transferred to adjacent unlabeled cells. Consequently, they are ideal medium-term tracers for transplanted cells or tissues. These probes have been used, for example, to study the *in vitro* reentry of cells into an acellular heart scaffold (Ott et al. 2008), the interactions between cell types in a reconstructed skin model (Sun et al. 2006), and the potential for cell sorting during culture of neocartilage grafts (Hayes et al. 2007). Figure 2.10 shows an image obtained from a coculture of two cell types each labeled with a different colored CellTracker dye prior to incorporation into a fibrin gel, facilitating the study of the interaction of the two cell types within the skin model. In all the reports above, the samples were fixed in formalin prior to imaging; however, fixing of the sample is not necessary for imaging purposes and live samples could be imaged in this way.

Fluorescent probes are also available that allow the imaging of ROS and hydrogen peroxide signaling (Dickinson et al. 2010; Forkink et al. 2010); intracellular concentrations of specific ions such as Ca^{2+} (Devinney et al. 2005; De et al. 2008), Zn^{2+} (Devinney et al. 2005), and Mg^{2+} (Komatsu et al. 2004); and intracellular pH (Balut et al. 2008)—all in living cells. These probes have all been used in 2D cultures; however, if they can be

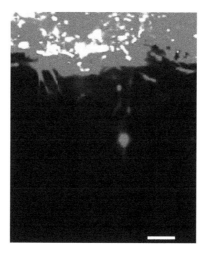

FIGURE 2.10 Coculture of human fibroblasts incorporating Red CellTracker and human keratinocytes incorporating Green CellTracker cultured in thin fibrin gels. The image has been reproduced in grayscale from the original color image. Keratinocytes appear white (yellow in the original image), due to the overlapping presence of fibroblasts (stained red) and keratinocytes (stained green) in the same area. Scale bar is 100 μm. (Reprinted from *Biomaterials*, 27, Sun, T., Haycock, J., and MacNeil, S., *In situ* image analysis of interactions between normal human keratinocytes and fibroblasts cultured in three-dimensional fibrin gels, 3459–3465, Copyright 2006, with permission from Elsevier.)

taken up by cells within a 3D sample, through the use of increased probe concentrations or extended incubation times, imaging of live tissues labeled with these probes should also be possible. However, it will be necessary to ascertain the impact of these incorporated probes on the long-term growth and viability of 3D constructs.

2.3.2.2.3 Incorporating the GFP Family of Proteins into Cells

Advances in molecular biology have enabled cells to be modified to express GFP or one of its spectral variants. The DNA encoding GFP can be linked to the DNA for virtually any protein of interest, producing a protein with a fluorescent tag. This allows localization of the protein by fluorescence microscopy without the requirement for prior fixing or permeabilization and enables responses to environmental changes to be monitored. Since fluorescence is incorporated into the living cell and all daughter cells, this also facilitates the dynamic imaging of living cells during culture. There are reviews, book chapters, and whole books that discuss the use of GFP and its spectral variants in 2D fluorescence microscopy (Goodwin 1999; Haseloff 1999; Ward and Lippincott-Schwartz 2006; Wiedenmann et al. 2009; Chudakov et al. 2010); however, its use in 3D tissue engineering is currently much less widespread.

The transfection and incorporation of DNA into the genome of immortal cell lines to produce stable GFP expressing cell lines is a routine procedure; however, it can be more challenging to introduce DNA into tissue-derived, mortal cells, and it is such cells that are frequently used for tissue engineering. These cells can only be maintained in culture for a limited number of passages, which can restrict the opportunities to obtain and verify the stable incorporation of DNA into the genome. Also, some primary cells can respond to the transfection process by terminally differentiating, removing their ability to be incorporated into 3D constructs. The solution is often to accept that DNA incorporation may be transient rather than stable, to use transgenic animals that express GFP as the source of the primary cells, or to introduce the DNA directly into the genome of human stem cells, which can be maintained in the multipotent proliferative stage prior to differentiation. It has been shown, for example, that human adipose derived stem cells can maintain a high level of GFP expression without influencing the multilineage potential of the cells (Lin et al. 2006).

Canton et al. (2007) transiently expressed a GFP-NFκB reporter construct in human fibroblasts while the cells were growing in a tissue-engineered skin construct to produce an experimental platform for the imaging of NFκB activation following exposure to various stimuli. GFP was also transiently expressed in fibroblasts used to monitor cell adhesion within a 3D construct during matrix remodeling (Tamariz and Grinnell 2002). Transgenic mice that express GFP have been used to provide GFP-labeled stem cells for the monitoring of the stem cell differentiation process via fluorescence microscopy (Grottkau et al. 2010) and GFP-labeled primary fibroblasts for incorporation into a range of 3D constructs (Tan et al. 2007). Human stem cells have been modified to express GFP and used in skin grafts to monitor their incorporation into skin defects in mice (Perng et al. 2008), or into a tissue-engineered periosteum, subsequently implanted into bone defects in mice (Schonmeyr et al. 2009).

GFP is of great benefit when using tissue-engineered samples as 3D experimental platforms; however, there are still many challenges with using such modified cells in samples that are destined for clinical use in humans.

Chapter 2

2.3.2.3　Dealing with Autofluorescence

When imaging samples of biological complexity (e.g., tissues rather than cultured cells), nonlabeled structures such as the extracellular matrix frequently create unwanted autofluorescence. Synthetic scaffolds may also be highly autofluorescent. Although this can be used to the researcher's advantage to image samples without additional fluorophores, it can also be problematic since it can interfere with the primary signal under study and can prevent the acquisition of useful or interpretable images. This has necessitated the development of methods that allow the researcher to identify and remove background fluorescence from an image.

2.3.2.3.1 Spectral Imaging　Spectral imaging relies upon the idea that each fluorescent molecule, whether it is an exogenous fluorophore or an intrinsically fluorescent biomolecule, has a characteristic emission spectrum that can be used to differentiate the fluorescent signals. Rather than detecting only the light around the emission maximum of a fluorophore, spectral imaging uses additional hardware to acquire a complete spectrum of the emitted light for each pixel within an image. Reference spectra are first acquired for each fluorophore or background signal within a sample, a multispectral image of a sample is then obtained, and a computational process is performed to analyze the signal obtained from each pixel to determine the proportion of each fluorescent molecule within that spectrum. In this way, the characteristic spectrum of the background collagen fluorescence can be subtracted from the image or differentiated from other fluorescent molecules.

This technique, discussed in more detail in the review by Dickinson et al. (2001), can also be used to discriminate between fluorophores with highly overlapping spectra, greatly expanding the combinations of fluorophores available for use within a single sample. However, this process does require the collection of a significantly greater number of photons from each pixel point to generate an image. Consequently, images obtained in this way are more susceptible to noise, and it can be a challenge to generate good images if the fluorescence level from the fluorophore of interest is weak.

2.3.2.3.2 Fluorescence Lifetime Imaging Microscopy, Long Life Fluorophores, and Time Resolved Emission Imaging Microscopy　Fluorescence lifetime imaging microscopy (FLIM) has seen most use as an imaging method to detect photosensitizer compounds in cells and tumors. It was developed in the 1980s (see review by Oida et al. 1993) and increased in popularity during the 1990s. However, its use is still largely confined to specialist laboratories due in part to the need for fast lasers and detectors. The principle of the technique relies on producing an image based on differences in the exponential decay rate of fluorescence from a fluorescent sample. Thus, the lifetime of the emitted signal, rather than its intensity, is used to create an image, and this is typically measured over a nanosecond timeframe. In 2D cell culture, FLIM has been used to study EGF receptor signaling (Wouters and Bastiaens 1999) and ErbB1 receptor trafficking (Verveer et al. 2000), and it has also been used clinically via multiphoton tomography for detecting intradermal cancer cells (Konig 2008).

Attempts have also been made to use FLIM to remove background autofluorescence by utilizing differences in the timescale of the emission of the probe versus the background. However, commercially available fluorophores usually have very short fluorescence lifetimes. For example, fluorescein (as FITC) has an emission lifetime of 3.6 ns, which is very similar to that of the extracellular matrix proteins such as collagen. Recent work has seen the development of a new class of fluorescent (or more accurately phosphorescent) platinum II (Pt(II)) labels for biological imaging (Botchway et al. 2008). These molecules are composed of a terdentate complex with a centrally located platinum ion, which is linked to two pyridine rings and a phenyl group. One of the key properties of these compounds is the emissive lifetime, which is measured in microseconds rather than nanoseconds. The extended lifetime is due to the structure of the molecule, which has a carbon–platinum bond at its center. While this is a very short bond, it is also very stable, and this property contributes directly to the rigidity of the overall complex—and in turn the extended photostability of the label.

The development of the Pt(II) labels has enabled the parallel development of a fluorescence-imaging technique called time resolved emission imaging microscopy (TREM). This exploits the extended lifetime of these labels and is very similar in principle to FLIM but utilizes a much longer timescale of microseconds rather than nanoseconds. This has a number of practical advantages, notably that longer timescales offer an improved ability to discriminate over much longer differences in lifetime, and therefore one can use TREM in time-gated experiments for distinguishing between preferentially labeled targets from short-lived autofluorescent species (de Haas et al. 1999). Also, TREM has an advantage in not requiring fast excitation or detection methods (in contrast to FLIM) and so can be undertaken using nanosecond lasers and slower gated detectors.

Pioneering studies by TREM have shown the suitability of Pt(II) complexes for live cell imaging (Botchway et al. 2008), for which the direct labeling of human fibroblasts, melanoma cells, and Chinese hamster ovary cells was achieved using micromolar concentrations of the label. Intracellular accumulation was detected within 5 min, and the compounds did not affect cellular viability over 24 h. Single- and two-photon excitation showed the labels to have an intrinsic high affinity for cell nuclei, binding preferentially to DNA and RNA. A number of applications are now possible for the combination of TREM and Pt(II) labels in tissues and 3D *in vitro* models, especially via Pt(II) conjugation to antibodies for the discrimination of autofluorescence and identification of long-lived emissive structures therein.

2.3.3   Two-Photon Excitation Imaging

All the techniques described so far in this chapter using the CLSM can also be performed using a two-photon excitation system. Furthermore, two-photon excitation microscopy can go beyond the capabilities of single-photon imaging and produce images via SHG. Using this technique, which is discussed in more detail in Chapter 4, images of collagen structure can be obtained without the requirement for exogenous labels. By combining the two-photon benefits of increased depth penetration and reduced damage to the sample with two-photon autofluorescence and SHG imaging, two-photon microscopy has been able to produce complementary information from within many live biological samples without the need for additional fluorophores. However, two-photon excitation microscopy is still a relatively specialized technique with comparatively few published uses in regenerative medicine to date.

Chapter 2

Two-photon autofluorescence in combination with SHG imaging has been used to characterize a number of commonly used tissue engineering scaffolds (Y. Sun et al. 2008); assess the impact of various heart valve decellularization protocols upon ECM structure (Zhou et al. 2010); and image hMSCs cultured on 3D silk scaffolds (Rice et al. 2008) or hMSCs seeded on polyglycolic acid-based scaffolds and cultured in chrondo-genic induction media (Lee et al. 2006).

The benefits of two-photon excitation have also been combined with other techniques discussed in this chapter to obtain images from live samples. Repeated *in situ* observation was required for the 3D culture of primary liver cells grown in a perfusion bioreactor (Powers et al. 2002). Researchers employed two-photon microscopy to excite recombinant EGFP in the hepatocytes to enable them to obtain images from the bioreactors over extended periods without damage to the tissue or loss in viability.

Optical redox images have also been obtained from live tissue biopsies. Two-photon optical redox images of normal, precancerous, and cancerous squamous epithelial tissues demonstrated the capability to image the tissues and differentiate between the tissue types (Skala et al. 2005), while redox images of live ovarian biopsies over time differentiated between normal and cancerous tissue (Kirkpatrick et al. 2007). Recently, two-photon derived optical redox images have been obtained from hASCs and HMVECs seeded on porous silk scaffolds nondestructively over a 6-month period (Quinn et al. 2012). The authors developed a method to automatically identify cells within a 3D image and calculate the optical redox status of these cells. In this way, they were able to observe metabolic changes associated with stem cell differentiation and lipogenesis over the 6-month culture period.

It is also becoming apparent that two-photon excitation has added benefits for spectral imaging, including easier access to higher excitation energies. In addition, the tunable lasers that are often used in two-photon microscopy facilitate image acquisition at incremental excitation wavelengths, allowing excitation spectra to be easily obtained and unmixed in a similar way to emission spectra. Grosberg et al. (2011) used this technique to image the intrinsic fluorescence from fresh gastrointestinal tissue samples. They employed a range of excitation wavelengths to obtain images and derived excitation spectra from four major tissue types present in the gastrointestinal tract. They were then able to use these excitation spectra to differentiate between normal and neoplastic tissue, providing a technique that could be performed on fresh, living tissue as an alternative to conventional histology of fixed tissues.

2.4   Example Applications

2.4.1   Characterization of Stem Cell Differentiation

A particular focus for research within regenerative medicine is the use of stem cells for tissue replacement and repair. Stem cells have the capacity to self-renew indefinitely and differentiate into cells with specialized functions (Sylvester and Longaker 2004; Mitalipov and Wolf 2009). It may even be possible to use autologous stem cells derived from the patient undergoing treatment, which would greatly reduce the potential for rejection. In order to advance the therapeutic potential of stem cells for the treatment of human disease, it is necessary to gain an understanding of their biological properties. The ability of fluorescence microscopy to image cells *in situ* means that this technique is ideally suited to monitoring the development and differentiation of stem cells.

For example, fluorescence microscopy was used to assess the potential of using stem cells to regenerate liver tissue following injury (Cho et al. 2009). GFP-expressing bone marrow cells were subdivided into three populations—mononuclear cells, MSCs, and hematopoietic stem cells, and then transplanted into liver-injured mice. Mice that received MSCs showed the highest levels of GFP in liver tissue, suggesting these cells have the highest potential for regenerating injured liver tissue. Human MSC differentiation to smooth muscle cells following exposure to cyclic strain has also been assessed using fluorescence microscopy (Ghazanfari et al. 2009). After undergoing strain, cells were labeled with FITC tagged anti-smooth muscle α-actin antibodies and imaged. It was observed that axial strain significantly upregulated expression of α-actin, indicating that cyclic strain alone induced differentiation to smooth muscle cells without the addition of growth factors.

The impact of the anatomical source of stem cells upon subsequent differentiation has also been investigated. Adipose-derived stem cells (ASCs) obtained from three distinct anatomical sites were induced to differentiate into Schwann-like cells and the morphological changes following differentiation assessed by S100β immunofluorescent labeling and imaging (Kaewkhaw et al. 2011). Cell morphology was quantified by determining the cell aspect ratio, with differentiated cells having a more elongated bipolar morphology and a higher aspect ratio than undifferentiated cells. Interestingly, the anatomical source of the ASC was found to impact upon the cell aspect ratio, indicating that stem cell source is an important consideration for stem cell therapies. ASCs have also been investigated as a source of interstitial cells for use in tissue-engineered heart valves (Colazzo et al. 2010). Endothelial differentiation was induced and the resulting cells characterized by immunofluorescence staining for a range of endothelial markers. The results indicated that ASCs were capable of differentiating into endothelial-like cells, suggesting that autologous stem cells could be used for the tissue engineering of heart valves.

In all these examples, the ability to image the stem cells during growth and differentiation was critical, and fluorescence microscopy was an essential tool for obtaining such images.

2.4.2 Developing Vascularized Tissue–Engineered Constructs

Without a blood supply, oxygen reaches cells by diffusion alone, limiting the thickness of a 3D tissue-engineered construct. The production of vascularized tissue is also crucial for its successful implantation, survival, and integration within the patient. Consequently, another key area of interest for regenerative medicine is the ability to produce 3D constructs that also contain functional blood vessels.

CLSM has proved very useful in monitoring microvessel formation within the constructs. Much of the research has focused upon the impact of incorporating additional cell types into constructs to promote vascularization. For example, Vo et al. (2010) used vascular smooth muscle-like cells (SMLCs), derived from human embryonic stem cells, to support *in vitro* vasculature formation by interacting with human endothelial progenitor cells (EPCs). The SMLCs and EPCs were labeled with different fluorescent membrane dyes, seeded in Matrigel, and imaged to determine the extent of tube formation. It was shown that increasing the ratio of SMLCs to EPCs produced longer, thicker tubes with less complex networks. SMLCs also stabilized and prolonged the formation of capillary-like structures on the Matrigel, suggesting that the SMLCs could be used to support engineered vascular networks *in vitro*. Fluorescent microscopy was also used to examine the effect of including fibroblasts, vascular endothelial growth factor (VEGF), and basic fibroblast growth factor (bFGF) upon the formation of vessel-like structures by human umbilical vein endothelial cells (HUVECs) cultured on a fibrin gel embedded with polymer microfilaments (Sukmana and Vermette 2010). The HUVEC cells were labeled with CellTracer, 10% of the fibrin within the gel was conjugated with Alexa Fluor 546, and the constructs were imaged. It was shown that fibroblasts significantly improved the maturation of microvessels compared to cultures that received growth factors alone. HUVECs have also been cocultured with human ASCs and grown on biodegradable 3D aqueous silk scaffolds (Kang et al. 2009). The HUVECs and hASCs were engineered to express GFP or tomato red fluorescent protein, respectively, and the constructs were imaged during the culture period. It was observed that the cells showed increased organization over time, and by day 14, there was extensive alignment of HUVECs within the adipocytes.

The fluorescence microscope has also been modified to permit intravital imaging, and this has facilitated the imaging of constructs following implantation in animal models to monitor the subsequent vascularization of these constructs. For example, Schumann et al. (2009) used intravital fluorescence microscopy to investigate the effect of seeding a collagen-coated poly(L-lactide-co-glycolic acid) (PLGA) scaffold with bone marrow MSCs, osteoblast-like cells, or a combination of the two, upon subsequent vascularization of the construct following implantation into mice. Intravenous injections of FITC-dextran provided contrast enhancement staining of the blood plasma and rhodamine 6G directly stained white blood cells. They observed that vascularization was accelerated in all scaffolds that had been preseeded with cells, independent of the cell type.

Finally, *in vivo* prevascularization has been investigated by intravital fluorescence microscopy (Laschke et al. 2010). Nanosized hydroxyapatite particles/poly(ester-urethane) scaffolds were implanted into GFP mice to allow the ingrowth of granulation tissue with GFP-positive blood vessels. The constructs were then transferred into recipient mice to study the connection of host to graft blood vessels using intravital fluorescence microscopy. In some instances, the constructs were implanted immediately, while others were first embedded in Matrigel and implanted immediately or after 3 days *in vitro* cultivation. Vascularization of the constructs cultivated in Matrigel gave elevated functional microvessel density and an increase in GFP-positive blood vessels growing into the surrounding host tissue, indicating that precultivation in an angiogenic extracellular matrix can accelerate the formation of a blood supply to prevascularized tissue.

The production of 3D images of constructs has proved particularly useful in studies of vascularization, and as a consequence, fluorescence microscopy, and more specifically CLSM, has been very useful for this area of research.

2.5 Conclusions

Fluorescence microscopy is an essential tool for use in regenerative medicine. Samples can be imaged nondestructively and repeatedly with a minimum of prior processing. The CLSM, with its additional optical sectioning capabilities, can obtain clear images from within 3D structures and render 3D representations of the samples, making it simple to obtain images from within tissues and scaffolds. Two-photon excitation takes this one step further with increased depth penetration and an additional reduction in damage to the sample, making it possible to image live tissue samples without any significant impact upon the viability, proliferation, or physiological behavior of the cells.

However, the techniques have been developed for use with 2D cultures, and it is only recently that many of these techniques have begun to be incorporated into the characterization of the more complex 3D samples.

Recent developments have also allowed fluorescence imaging to occur *in vivo*. Skin graft regeneration has been studied in mice using GFP-labeled human bone marrow stem cells, incorporated into a construct and placed on a skin defect created on the back of nude mice. It was then possible to observe the incorporation of the labeled stem cells into the regenerating skin using fluorescence microscopy and a standard stereo microscope setup (Perng et al. 2008). In situations where the organ or tissue of interest is within the body, laser scanning microscopes are now commercially available for intravital imaging. These microscopes have ultraslim objectives and flexible observation angles for less invasive *in vitro* and *in situ* fluorescence imaging. However, the procedure is only suitable for small animals and still necessitates some surgical intervention to allow the objective to access the organ or tissue of interest.

Finally, to take full advantage of fluorescence microscopy, there are still issues that need to be addressed. Bioreactors suitable for growing 3D constructs that also enable images to be captured in any orientation and from any point within the bioreactor need to be developed. Methods for dealing with scaffold autofluorescence, such as techniques like TREM and FLIM that use fluorophores with longer fluorescence lifetimes or the avoidance of autofluorescence by using fluorophores with longer excitation wavelengths, need further investigation. There is also a need to develop a larger range of dyes with the required fluorescence characteristics to exploit the full potential of these techniques. Techniques to incorporate contrast into live 3D constructs also need further attention. Studies into the incorporation of fluorescent dyes into live 3D constructs and their subsequent impact upon the growth and development are required, and if necessary, new stains need to be developed. Indicator dyes that respond to changes in protein conformation, phosphorylation, and dephosphorylation, or specific molecular interactions by altering their emission spectra could also be developed, which are suitable for live 3D constructs. This would provide the researcher with the tools to observe specific intracellular events through the monitoring of fluorescence emission spectra. The advantages of increased sample penetration and reduced damage to cell viability conferred by two-photon excitation microscopy also need to be exploited more fully through both endogenous and exogenous fluorescence imaging.

However, even without these developments, fluorescence microscopy is a tool that is already successfully employed in regenerative medicine and has the potential to be even more widely applied in the future.

Chapter 2

References

Balut, C., vandeVen, M., Despa, S. et al. 2008. Measurement of cytosolic and mitochondrial pH in living cells during reversible metabolic inhibition. *Kidney Int.* 73: 226–232.

Bolgen, N., Yang, Y., Korkusuz, P. et al. 2008. Three-dimensional ingrowth of bone cells within biodegradable cryogel scaffolds in bioreactors at different regimes. *Tissue Eng Part A* 14: 1743–1750.

Bolgen, N., Yang, Y., Korkusuz, P. et al. 2011. 3D ingrowth of bovine articular chondrocytes in biodegradable cryogel scaffolds for cartilage tissue engineering. *J. Tissue Eng Regen. Med.* 5: 770–779.

Botchway, S. W., Charnley, M., Haycock, J. W. et al. 2008. Time-resolved and two-photon emission imaging microscopy of live cells with inert platinum complexes. *Proc. Natl. Acad. Sci. U. S. A.* 105: 16071–16076.

Canton, I., Sarwar, U., Kemp, E. H. et al. 2007. Real-time detection of stress in 3D tissue-engineered constructs using NF-kappaB activation in transiently transfected human dermal fibroblast cells. *Tissue Eng* 13: 1013–1024.

Centonze, V. E., and White, J. G. 1998. Multiphoton excitation provides optical sections from deeper within scattering specimens than confocal imaging. *Biophys. J.* 75: 2015–2024.

Chen, J. P., Li, S. F., and Chiang, Y. P. 2010. Bioactive collagen-grafted poly-L-lactic acid nanofibrous membrane for cartilage tissue engineering. *J. Nanosci. Nanotechnol.* 10: 5393–5398.

Cho, K. A., Ju, S. Y., Cho, S. J. et al. 2009. Mesenchymal stem cells showed the highest potential for the regeneration of injured liver tissue compared with other subpopulations of the bone marrow. *Cell Biol. Int.* 33: 772–777.

Chudakov, D. M., Matz, M. V., Lukyanov, S., and Lukyanov, K. A. 2010. Fluorescent proteins and their applications in imaging living cells and tissues. *Physiol Rev.* 90: 1103–1163.

Colazzo, F., Chester, A. H., Taylor, P. M., and Yacoub, M. H. 2010. Induction of mesenchymal to endothelial transformation of adipose-derived stem cells. *J. Heart Valve Dis.* 19: 736–744.

Coons, A. H., Creech, H. J., and Jones, R. N. 1941. Immunological properties of an antibody containing a fluorescent group. *Proc. Soc. Exp. Biol. Med.* 47: 200–202.

De, P., I., Pintelon, I., Brouns, I. et al. 2008. Functional live cell imaging of the pulmonary neuroepithelial body microenvironment. *Am. J. Respir. Cell Mol. Biol.* 39: 180–189.

de Haas, R. R., van Gijlswijk, R. P., van der Tol, E. B. et al. 1999. Phosphorescent platinum/palladium coproporphyrins for time-resolved luminescence microscopy. *J. Histochem. Cytochem.* 47: 183–196.

Deshpande, P., McKean, R., Blackwood, K. A. et al. 2010. Using poly(lactide-co-glycolide) electrospun scaffolds to deliver cultured epithelial cells to the cornea. *Regen. Med.* 5: 395–401

Devinney, M. J., Reynolds, I. J., and Dineley, K. E. 2005. Simultaneous detection of intracellular free calcium and zinc using fura-2FF and FluoZin-3. *Cell Calcium* 37: 225–232.

Dickinson, M. E., Bearman, G., Tille, S., Lansford, R., and Fraser, S. E. 2001. Multi-spectral imaging and linear unmixing add a whole new dimension to laser scanning fluorescence microscopy. *Biotechniques* 31: 1272–1278.

Dickinson, B. C., Huynh, C., and Chang, C. J. 2010. A palette of fluorescent probes with varying emission colors for imaging hydrogen peroxide signaling in living cells. *J. Am. Chem. Soc.* 132: 5906–5915.

Dittmar, R., Potier, E., van, Z. M., and Ito, K. 2012. Assessment of cell viability in three-dimensional scaffolds using cellular auto-fluorescence. *Tissue Eng Part C. Methods* 18: 198–204.

Ellis, M., Jarman-Smith, M., and Chaudhuri, J. B. 2005. Bioreactor systems for tissue engineering: A four dimensional challenge. In *Bioreactors for Tissue Engineering: Principles, Design and Operation*, eds. J. B. Chaudhuri, and M. AlRubeai, 1–18. Dordrecht, Netherlands: Springer.

Forkink, M., Smeitink, J. A., Brock, R., Willems, P. H., and Koopman, W. J. 2010. Detection and manipulation of mitochondrial reactive oxygen species in mammalian cells. *Biochim. Biophys. Acta* 1797: 1034–1044.

Gefen, A., Cornelissen, L. H., Gawlitta, D., Bader, D. L., and Oomens, C. W. J. 2008. The free diffusion of macromolecules in tissue-engineered skeletal muscle subjected to large compression strains. *J. Biomech.* 41: 845–853.

Ghazanfari, S., Tafazzoli-Shadpour, M., and Shokrgozar, M. A. 2009. Effects of cyclic stretch on proliferation of mesenchymal stem cells and their differentiation to smooth muscle cells. *Biochem. Biophys. Res. Commun.* 388: 601–605.

Goodwin, P. C. 1999. GFP biofluorescence: imaging gene expression and protein dynamics in living cells. Design considerations for a fluorescence imaging laboratory. *Methods Cell Biol.* 58: 343–367.

Greenspan, P., Mayer, E. P., and Fowler, S. D. 1985. Nile red: a selective fluorescent stain for intracellular lipid droplets. *J. Cell Biol.* 100: 965–973.

Grosberg, L. E., Radosevich, A. J., Asfaha, S., Wang, T. C., and Hillman, E. M. 2011. Spectral characterization and unmixing of intrinsic contrast in intact normal and diseased gastric tissues using hyperspectral two-photon microscopy. *PLoS One* 6: e19925.

Grottkau, B. E., Purudappa, P. P., and Lin, Y. F. 2010. Multilineage differentiation of dental pulp stem cells from green fluorescent protein transgenic mice. *Int. J. Oral Sci.* 2: 21–27.

Haseloff, J. 1999. GFP variants for multispectral imaging of living cells. *Methods Cell Biol.* 58: 139–151.

Haycock, J. W. 2011. 3D cell culture: a review of current approaches and techniques. *Methods Mol. Biol.* 695: 1–15.

Hayes, A. J., Hall, A., Brown, L., Tubo, R., and Caterson, B. 2007. Macromolecular organization and *in vitro* growth characteristics of scaffold-free neocartilage grafts. *J. Histochem. Cytochem.* 55: 853–866.

Kaewkhaw, R., Scutt, A. M., and Haycock, J. W. 2011. Anatomical site influences the differentiation of adipose-derived stem cells for Schwann-cell phenotype and function. *Glia* 59: 734–749.

Kang, J. H., Gimble, J. M., and Kaplan, D. L. 2009. *In vitro* 3D model for human vascularized adipose tissue. *Tissue Eng Part A* 15: 2227–2236.

Kirkpatrick, N. D., Brewer, M. A., and Utzinger, U. 2007. Endogenous optical biomarkers of ovarian cancer evaluated with multiphoton microscopy. *Cancer Epidemiol. Biomarkers Prev.* 16: 2048–2057.

Kluge, J. A., Leisk, G. G., Cardwell, R. D. et al. 2011. Bioreactor system using noninvasive imaging and mechanical stretch for biomaterial screening. *Ann. Biomed. Eng* 39: 1390–1402.

Knight, M. M., Roberts, S. R., Lee, D. A., and Bader, D. L. 2003. Live cell imaging using confocal microscopy induces intracellular calcium transients and cell death. *Am. J. Physiol. Cell Physiol.* 284: C1083–C1089.

Komatsu, H., Iwasawa, N., Citterio, D. et al. 2004. Design and synthesis of highly sensitive and selective fluorescein-derived magnesium fluorescent probes and application to intracellular 3D Mg2+ imaging. *J. Am. Chem. Soc.* 126: 16353–16360.

Konig, K. 2008. Clinical multiphoton tomography. *J. Biophotonics.* 1: 13–23.

Kubow, K. E., Klotzsch, E., Smith, M. L. et al. 2009. Crosslinking of cell-derived 3D scaffolds up-regulates the stretching and unfolding of new extracellular matrix assembled by reseeded cells. *Integr. Biol. (Camb.)* 1: 635–648.

Lakowicz, J. R. 2006. *Principles of Fluorescence Spectroscopy*. New York: Springer.

Landis, F. A., Stephens, J. S., Cooper, J. A., Cicerone, M. T., and Lin-Gibson, S. 2006. Tissue engineering scaffolds based on photocured dimethacrylate polymers for *in vitro* optical imaging. *Biomacromolecules.* 7: 1751–1757.

Laschke, M. W., Mussawy, H., Schuler, S. et al. 2010. Promoting external inosculation of prevascularised tissue constructs by pre-cultivation in an angiogenic extracellular matrix. *Eur. Cell Mater.* 20: 356–366.

Leddy, H. A., Awad, H. A., and Guilak, F. 2004. Molecular diffusion in tissue-engineered cartilage constructs: Effects of scaffold material, time, and culture conditions. *J. Biomed. Mater. Res. B Appl. Biomater.* 70B: 397–406.

Leddy, H. A., and Guilak, F. 2003. Site-specific molecular diffusion in articular cartilage measured using fluorescence recovery after photobleaching. *Ann. Biomed. Eng* 31: 753–760.

Lee, H. S., Teng, S. W., Chen, H. C. et al. 2006. Imaging human bone marrow stem cell morphogenesis in polyglycolic acid scaffold by multiphoton microscopy. *Tissue Eng* 12: 2835–2841.

Leong, D. T., Khor, W. M., Chew, F. T., Lim, T. C., and Hutmacher, D. W. 2006. Characterization of osteogenically induced adipose tissue-derived precursor cells in 2-dimensional and 3-dimensional environments. *Cells Tissues. Organs* 182: 1–11.

Lin, C. C., Ritch, R., Lin, S. M. et al. 2010. A new fish scale-derived scaffold for corneal regeneration. *Eur. Cell Mater.* 19: 50–57.

Lin, Y., Liu, L., Li, Z. et al. 2006. Pluripotency potential of human adipose-derived stem cells marked with exogenous green fluorescent protein. *Mol. Cell Biochem.* 291: 1–10.

Mahoney, M. J., and Anseth, K. S. 2006. Three-dimensional growth and function of neural tissue in degradable polyethylene glycol hydrogels. *Biomaterials* 27: 2265–2274.

Martin, M. D., Fingleton, B., Lynch, C. C. et al. 2008. Establishment and quantitative imaging of a 3D lung organotypic model of mammary tumor outgrowth. *Clin. Exp. Metastasis* 25: 877–885.

Martina, M., Subramanyam, G., Weaver, J. C. et al. 2005. Developing macroporous bicontinuous materials as scaffolds for tissue engineering. *Biomaterials* 26: 5609–5616.

Mitalipov, S., and Wolf, D. 2009. Totipotency, pluripotency and nuclear reprogramming. *Adv. Biochem. Eng Biotechnol.* 114: 185–199.

Morelli, S., Salerno, S., Piscioneri, A. et al. 2010. Influence of micro-patterned PLLA membranes on outgrowth and orientation of hippocampal neurites. *Biomaterials* 31: 7000–7011.

Chapter 2

Naing, M. W., and Williams, D. J. 2011. Three-dimensional culture and bioreactors for cellular therapies. *Cytotherapy.* 13: 391–399.

Ng, K. W., Mauck, R. L., Wang, C. C. et al. 2009. Duty cycle of deformational loading influences the growth of engineered articular cartilage. *Cell Mol. Bioeng.* 2: 386–394.

Oida, T., Sako, Y., and Kusumi, A. 1993. Fluorescence lifetime imaging microscopy (flimscopy). Methodology development and application to studies of endosome fusion in single cells. *Biophys. J.* 64: 676–685.

Ott, H. C., Matthiesen, T. S., Goh, S. K. et al. 2008. Perfusion-decellularized matrix: using nature's platform to engineer a bioartificial heart. *Nat. Med.* 14: 213–221.

Periasamy, A. 2001. Fluorescence resonance energy transfer microscopy: a mini review. *J. Biomed. Opt.* 6: 287–291.

Perng, C. K., Kao, C. L., Yang, Y. P. et al. 2008. Culturing adult human bone marrow stem cells on gelatin scaffold with pNIPAAm as transplanted grafts for skin regeneration. *J. Biomed. Mater. Res. A* 84: 622–630.

Powers, M. J., Domansky, K., Kaazempur-Mofrad, M. R. et al. 2002. A microfabricated array bioreactor for perfused 3D liver culture. *Biotechnol. Bioeng.* 78: 257–269.

Prasher, D. C., Eckenrode, V. K., Ward, W. W., Prendergast, F. G., and Cormier, M. J. 1992. Primary structure of the Aequorea-Victoria green-fluorescent protein. *Gene* 111: 229–233.

Quinn, K. P., Bellas, E., Fourligas, N. et al. 2012. Characterization of metabolic changes associated with the functional development of 3D engineered tissues by non-invasive, dynamic measurement of individual cell redox ratios. *Biomaterials* 33: 5341–5348.

Rajwa, B., Bernas, T., Acker, H., Dobrucki, J., and Robinson, J. P. 2007. Single- and two-photon spectral imaging of intrinsic fluorescence of transformed human hepatocytes. *Microsc. Res. Tech.* 70: 869–879.

Rauh, J., Milan, F., Gunther, K. P., and Stiehler, M. 2011. Bioreactor systems for bone tissue engineering. *Tissue Eng Part B Rev.* 17: 263–280.

Reits, E. A., and Neefjes, J. J. 2001. From fixed to FRAP: measuring protein mobility and activity in living cells. *Nat. Cell Biol.* 3: E145–E147.

Rice, W. L., Firdous, S., Gupta, S. et al. 2008. Non-invasive characterization of structure and morphology of silk fibroin biomaterials using non-linear microscopy. *Biomaterials* 29: 2015–2024.

Rigacci, L., Alterini, R., Bernabei, P. A. et al. 2000. Multispectral imaging autofluorescence microscopy for the analysis of lymph-node tissues. *Photochem. Photobiol.* 71: 737–742.

Schnedl, W., Mikelsaar, A. V., Breitenbach, M., and Dann, O. 1977. DIPI and DAPI: fluorescence banding with only negligible fading. *Hum. Genet.* 36: 167–172.

Schonmeyr, B., Clavin, N., Avraham, T., Longo, V., and Mehrara, B. J. 2009. Synthesis of a tissue-engineered periosteum with acellular dermal matrix and cultured mesenchymal stem cells. *Tissue Eng Part A* 15: 1833–1841.

Schumann, P., Tavassol, F., Lindhorst, D. et al. 2009. Consequences of seeded cell type on vascularization of tissue engineering constructs *in vivo*. *Microvasc. Res.* 78: 180–190.

Schwartlander, R., Schmid, J., Brandenburg, B. et al. 2007. Continuously microscopically observed and process-controlled cell culture within the SlideReactor: proof of a new concept for cell characterization. *Tissue Eng* 13: 187–196.

Shaner, N. C., Steinbach, P. A., and Tsien, R. Y. 2005. A guide to choosing fluorescent proteins. *Nat. Methods* 2: 905–909.

Shimomura, O., Johnson, F. H., and Saiga, Y. 1962. Extraction, purification and properties of aequorin, a bioluminescent protein from the luminous hydromedusan, Aequorea. *J. Cell Comp. Physiol.* 59: 223–239.

Sierad, L. N., Simionescu, A., Albers, C. et al. 2010. Design and testing of a pulsatile conditioning system for dynamic endothelialization of polyphenol-stabilized tissue engineered heart valves. *Cardiovasc. Eng Technol.* 1: 138–153.

Skala, M., and Ramanujam, N. 2010. Multiphoton redox ratio imaging for metabolic monitoring *in vivo*. *Methods Mol. Biol.* 594: 155–162.

Skala, M. C., Squirrell, J. M., Vrotsos, K. M. et al. 2005. Multiphoton microscopy of endogenous fluorescence differentiates normal, precancerous, and cancerous squamous epithelial tissues. *Cancer Res.* 65: 1180–1186.

Srouji, S., Kizhner, T., Suss-Tobi, E., Livne, E., and Zussman, E. 2008. 3-D Nanofibrous electrospun multilayered construct is an alternative ECM mimicking scaffold. *J. Mater. Sci. Mater. Med.* 19: 1249–1255.

Sukmana, I., and Vermette, P. 2010. The effects of co-culture with fibroblasts and angiogenic growth factors on microvascular maturation and multi-cellular lumen formation in HUVEC-oriented polymer fibre constructs. *Biomaterials* 31: 5091–5099.

Sun, T., Haycock, J., and MacNeil, S. 2006. In situ image analysis of interactions between normal human keratinocytes and fibroblasts cultured in three-dimensional fibrin gels. *Biomaterials* 27: 3459–3465.

Sun, T., Norton, D., Haycock, J. W., Ryan, A. J., and MacNeil, S. 2005. Development of a closed bioreactor system for culture of tissue-engineered skin at an air-liquid interface. *Tissue Eng* 11: 1824–1831.

Sun, T., Norton, D., Vickers, N. et al. 2008. Development of a bioreactor for evaluating novel nerve conduits. *Biotechnol. Bioeng.* 99: 1250–1260.

Sun, Y., Tan, H. Y., Lin, S. J. et al. 2008. Imaging tissue engineering scaffolds using multiphoton microscopy. *Microsc. Res. Tech.* 71: 140–145.

Sylvester, K. G., and Longaker, M. T. 2004. Stem cells: review and update. *Arch. Surg.* 139: 93–99.

Tamariz, E., and Grinnell, F. 2002. Modulation of fibroblast morphology and adhesion during collagen matrix remodeling. *Mol. Biol. Cell* 13: 3915–3929.

Tan, W., Sendemir-Urkmez, A., Fahrner, L. J. et al. 2004. Structural and functional optical imaging of three-dimensional engineered tissue development. *Tissue Eng* 10: 1747–1756.

Tan, W., Vinegoni, C., Norman, J. J., Desai, T. A., and Boppart, S. A. 2007. Imaging cellular responses to mechanical stimuli within three-dimensional tissue constructs. *Microsc. Res. Tech.* 70: 361–371.

Tigli, R. S., and Gumusderelioglu, M. 2009. Chondrogenesis on BMP-6 loaded chitosan scaffolds in stationary and dynamic cultures. *Biotechnol. Bioeng.* 104: 601–610.

Tritz-Schiavi, J., Charif, N., Henrionnet, C. et al. 2010. Original approach for cartilage tissue engineering with mesenchymal stem cells. *Biomed. Mater. Eng* 20: 167–174.

Tsien, R. Y. 1989. Fluorescent probes of cell signaling. *Annu. Rev. Neurosci.* 12: 227–253.

Turner, N. J., Kielty, C. M., Walker, M. G., and Canfield, A. E. 2004. A novel hyaluronan-based biomaterial (Hyaff-11) as a scaffold for endothelial cells in tissue engineered vascular grafts. *Biomaterials* 25: 5955–5964.

Verveer, P. J., Wouters, F. S., Reynolds, A. R., and Bastiaens, P. I. 2000. Quantitative imaging of lateral ErbB1 receptor signal propagation in the plasma membrane. *Science* 290: 1567–1570.

Vo, E., Hanjaya-Putra, D., Zha, Y., Kusuma, S., and Gerecht, S. 2010. Smooth-muscle-like cells derived from human embryonic stem cells support and augment cord-like structures *in vitro*. *Stem Cell Rev.* 6: 237–247.

Walsh, A., Cook, R. S., Rexer, B., Arteaga, C. L., and Skala, M. C. 2012. Optical imaging of metabolism in HER2 overexpressing breast cancer cells. *Biomed. Opt. Express* 3: 75–85.

Wang, G., Zheng, L., Zhao, H. et al. 2011. Construction of A fluorescent nanostructured chitosan-hydroxyapatite scaffold by nanocrystallon induced biomimetic mineralization and its cell biocompatibility. *ACS Appl. Mater. Interfaces.* 3: 1692–1701.

Ward, T. H., and Lippincott-Schwartz, J. 2006. The uses of green fluorescent protein in mammalian cells. *Methods Biochem. Anal.* 47: 305–337.

Wei, X., Li, D. B., Xu, F. et al. 2011. A novel bioreactor to simulate urinary bladder mechanical properties and compliance for bladder functional tissue engineering. *Chin Med. J. (Engl.)* 124: 568–573.

Wiedenmann, J., Oswald, F., and Nienhaus, G. U. 2009. Fluorescent proteins for live cell imaging: opportunities, limitations, and challenges. *IUBMB. Life* 61: 1029–1042.

Wouters, F. S., and Bastiaens, P. I. 1999. Fluorescence lifetime imaging of receptor tyrosine kinase activity in cells. *Curr. Biol.* 9: 1127–1130.

Xu, C., Inai, R., Kotaki, M., and Ramakrishna, S. 2004. Electrospun nanofiber fabrication as synthetic extracellular matrix and its potential for vascular tissue engineering. *Tissue Eng* 10: 1160–1168.

Zhou, J., Fritze, O., Schleicher, M. et al. 2010. Impact of heart valve decellularization on 3-D ultrastructure, immunogenicity and thrombogenicity. *Biomaterials* 31: 2549–2554.

Chapter 2

3. Second–Harmonic Generation

Stephen J. Matcher

Chapter 3

Optical Techniques in Regenerative Medicine. Edited by Stephen P. Morgan, Felicity R.A.J. Rose, and Stephen J. Matcher © 2014 CRC Press/Taylor & Francis Group, LLC. ISBN: 978-1-4398-5495-2

3.1   Introduction

Tissue engineering involves an attempt to replace natural tissues and organs of the body with synthetic replacements that closely resemble the original tissue. Often it is hoped that the body will eventually remodel the implanted construct into something that is functionally identical to native tissue, and this means successfully regenerating both the cellular and extracellular components of the tissue. The dominant component of the extracellular matrix (ECM) is the protein collagen. There are over 20 isoforms of this structural protein, of which types I, II, and IV are most abundant in mammalian tissues. Collagen gives tissues their tensile strength, and the exceptional mechanical properties are closely associated with the ability of this protein to form ordered, semi-crystalline domains. Much effort in tissue engineering is thus devoted to inducing the production of various collagen types by cells such as fibroblasts. Chemical stimuli such as ascorbate (Franceschi et al. 1994) and mechanical stimuli such as applied strain and/or scaffold fiber alignment (Lee et al. 2005) are known to upregulate collagen production by fibroblasts and osteoblasts. To assess the extent to which the cells are manufacturing structurally organized collagen of the correct type, one would ideally like a noninvasive imaging tool that is sensitive to collagen levels and also to the degree of crystallinity of the molecules. Over the past decade, second harmonic generation (SHG) microscopy has seen a resurgence in interest as a tool to study collagen (and some other) ECM molecules. The chief defining features of the technique that account for this are as follows:

1. The technique is based on optical microscopy and so has a spatial resolution measured in hundreds of nanometers.
2. No access to complicated radiation sources is needed, unlike x-ray diffraction techniques.
3. 3D volumetric images of collagen levels can be produced, due to the intrinsic depth-sectioning properties of SHG imaging.
4. The technique is very closely related to two-photon emission fluorescence (TPEF) microscopy, so that a lab equipped with the TPEF technique is also set up to perform SHG microscopy.
5. Within the usual access requirements of optical microscopy, the technique is noninvasive and nondestructive.
6. Nonlinear photodamage to tissues is inherently lower than TPEF.
7. The signal is strongly related not just to the abundance of collagen molecules (as with fluorescence) but also on the degree of their spatial ordering.
8. The technique exploits purely endogenous contrast, so that no dyes or other labels need be added.
9. Some discrimination between different collagen types (especially types I and II) may be possible.
10. The local orientation of the fibers, potentially in 3D, can be deduced even when the fibers themselves cannot be resolved, by combining SHG with polarimetry.
11. Advances in laser technology mean that the technique can potentially be implemented considerably more cheaply than TPEF microscopy (although of course TPEF data are highly desirable to have also).

It is the goal of the remainder of this chapter to explain the physical basis of the technique in a nonmathematical way and hence explain why these advantages arise, to describe the measurements that can be made by SHG, and finally, to survey some of the uses to which the technique has been put by the tissue engineering community to date.

3.2 Basic Physical Concepts Behind SHG Microscopy

When a light wave interacts with matter, a scattered light wave is generally produced. This scattered wave can have many different properties depending on the wavelength of the light and the properties of the material. It might be of the same wavelength as the incident light (elastic scattering), or it may have a longer wavelength (Raman scattering, fluorescence emission). Some processes can generate light of a shorter wavelength, which means an increase in photon energy. From the photon point of view, energy conservation requires that two or more photons from the incident light beam combine together to produce a single photon in the scattered beam. Processes in this category include coherent anti-Stokes Raman scattering (CARS—as discussed in Chapter 8 in this book) and harmonic generation, of which SHG is our focus in this chapter. The wavelength shift between the incident and scattered photon in SHG is very precisely defined: the scattered photon is of precisely half the wavelength (hence twice the energy) of the incident photon.

Regardless of the type of interaction that occurs, it turns out that all the various types of scattered waves can be related to the incident wave by a single physical parameter, the susceptibility. This is a material property, just like the elastic modulus or the heat capacity of a material. Just as these properties can vary (elastic modulus varies in general with the amount a material has been stretched, by its temperature, etc.), so the susceptibility of a material depends on many factors. Chief among these is the wavelength of the incident EM radiation, but, as we shall see, the intensity of the light is also important under certain circumstances. Atomic and molecular conditions within the material are also important, and the susceptibility varies, for example, with the atomic and molecular structure of the material (diamond and carbon-black have very different optical properties while both being formed purely of packed carbon atoms, the difference lying solely in the crystal lattice structures being different). The susceptibility of a material describes in detail how an incident light wave will be scattered by the material. Knowing how the susceptibility varies with light wavelength, intensity, and polarization and also how it varies with the intrinsic and environmental conditions of the material thus form the foundation of optics: the science of manipulating light–matter interactions.

Any biological tissue has a susceptibility and its variation with frequency/wavelength explains why our bodies are transparent to x-rays and also to the radio waves used to make an MRI but are largely opaque to visible light. The susceptibility of most tissues dictates that ultraviolet, visible, and infrared photons are mainly elastically scattered, that is, there is no wavelength change and the scattered light has exactly the same wavelength as the incident light. However, other processes can also occur albeit much less efficiently. Biological tissues can generate low levels of endogenous fluorescence and even lower levels of Raman scattered light. These signals carry a great deal of information about the chemical composition and chemical environment within the tissue. These signals, although weak, scale in intensity in direct proportion to the intensity of the incident (excitation) light beam. Emission that scales in direct proportion to excitation is said to be a linear form of light–tissue interaction. It can be shown that such processes involve the conversion of one photon into another photon, possibly of different energy. Hence, such processes are also known as one-photon processes.

3.2.1 Linear and Nonlinear Susceptibility

Under certain circumstances, however, something rather different can occur. Recall that light is a wave the frequency and wavelength of which are linked by the requirement that their product always equals a constant: the speed of light. Then light of wavelength 500 nm has a frequency of 6×10^{14} Hz. Sound is also a wave, albeit one of much lower frequency: middle "C" on the piano is around 262 Hz. Imagine listening to piano music on the radio. With the volume turned down, the notes are pleasant and melodious, whereas if the volume is turned to maximum, then the music not only becomes unpleasantly noisy but the purity of the notes themselves becomes severely degraded by distortion. Because the radio circuits and loudspeaker cannot reproduce arbitrarily loud sounds, turning the volume too high eventually causes the output (the loudspeaker sound wave amplitude) to cease to be proportionally related to the input (which ultimately is the amplitude of the piano sound as it was recorded in the studio). This phenomenon is often referred to as harmonic distortion because if one analyzes carefully the distorted sound of a pure middle C tone, for example, then one finds that the distorted note sounds bad because it now contains a mixture of the fundamental frequency 262 Hz plus, in varying proportions, the harmonic overtones of 262 Hz, that is, the frequencies 524 Hz (= 2 × 262 Hz; known in music as the "first overtone" but in physics as the "second harmonic"), 786 Hz (3 × 262 Hz, i.e., the "second overtone" or "third harmonic"), and so on. The amounts of these harmonic overtones that are present depend directly on the degree of distortion, and this in turn depends directly on to what extent the radio circuitry is being overdriven by having the volume turned too high.

Importantly, biological tissues exposed to incident light waves behave in an analogous manner. When a biological tissue elastically scatters an incident light beam, a direct analogy can be made with the radio converting an input signal (the recorded piano sound) into an output signal (the generated loudspeaker sound). Tissue converts the input light wave into an output scattered wave. If the amplitude of this incident signal is increased, then the system can start to generate a distorted output that contains harmonic overtones of the fundamental. If the incident light wave has a wavelength of 500 nm and so a frequency of 6×10^{14} Hz, then the overtones have frequencies of 12×10^{14} Hz (wavelength 250 nm), 18×10^{14} Hz (wavelength 166 nm), and so on (250 nm is the second harmonic, 166 nm is the third harmonic, etc.). Whereas a small radio can be made to generate harmonic distortion rather easily, biological tissues must be subjected to input signals of a magnitude not typically found in nature in order to generate measurable levels of even the second harmonic. Indeed, in a typical experiment to generate second harmonic signals from collagen, the instantaneous light intensity can reach levels over 1 trillion (1×10^{12}) times higher than that of direct tropical sunlight. It is a remarkable testament both to the ingenuity of optical scientists and the robustness of nature that it is comparatively simple these days to generate such conditions harmlessly in the lab and also that live biological cells generally seem unperturbed by being exposed to these conditions. Some of the reasons for this advantageous state of affairs will be discussed in the following sections as well as elsewhere in this book (see especially Chapter 2).

From a photon perspective, the generation of, for example, the second harmonic overtone involves the conversion of two photons of equal energy into a single photon of twice the energy. For this to occur, both input photons must in a sense be scattered by a given molecule at the same instant of time, or at least they must both interact with this molecule within a very short (e.g., 1 trillionth of a second) time window. The intensity of a light beam measures the probability that a photon in the beam will interact with a molecule irradiated by that beam in any given interval of time. Hence, doubling the light intensity doubles this probability. Recalling a basic law of probabilities, namely, that the probability of two independent events occurring together is given by the product of the probabilities that each event occurs in isolation, then we see that doubling the light intensity would increase the probability of SHG by a factor of $2 \times 2 = 4$. This rule turns out to be quite general, and the intensity of third harmonic generation (THG) is found to increase as the cube of the input light intensity. This can be very powerful because increasing the laser power by ×10 in an SHG experiment can boost signal levels by × 100 (or × 1000 in a THG experiment). The same argument applies to TPEF.

An important difference between SHG and TPEF can be seen directly by considering the processes at a photon level. SHG involves the conversion of two photons of equal energy into a single scattered photon of exactly twice the energy, whereas TPEF involves the generation of a scattered fluorescence photon of which the energy is somewhat smaller (hence longer wavelength) than the combined energy of the excitation photons. This energy difference (the "Stokes shift") is deposited into the tissue where it can cause local heating or bleaching of the fluorophore. There is no Stokes shift for SHG, that is, there is zero energy available to be deposited into the tissue by nonlinear absorption. Thus, in SHG imaging, the only tissue heating or fluorophore bleaching comes from one-photon absorption of the excitation light or of the SHG emission. Given the wavelengths involved, this occurs to a much lesser extent than heating via nonlinear absorption; hence, tissues can generally tolerate higher excitation power levels in an SHG experiment than in a TPEF experiment.

3.2.2 Nonsymmetric Structures

We have noted that SHG microscopy requires that a biological sample be illuminated with a very intense light wave; however, SHG also makes important restrictions on the microscopic structure of the sample. To understand why this is so, let us recall that a light beam is an electromagnetic wave and that consequently the electric field experienced by a molecule varies in time sinusoidally as the wave travels past it. The same graph also describes the sound pressure emitted by the radio we introduced in the earlier discussion of distortion and harmonic generation, so let us briefly return to that example and consider in slightly more detail precisely how the waveform is distorted within a single oscillation cycle. At the top of Figure 3.1, we see an undistorted pure sinusoidal waveform and below it two types of distorted waveforms. The first is symmetric between the positive and negative parts of the waveform, whereas the other is distorted in an asymmetric way; specifically, the waveform has saturated (reached a maximum value) on the negative excursion but not the positive. In the left and right columns below these, we have plotted a single cycle of each distorted waveform and below these the fundamental and the first two harmonic overtones, namely, the second and third harmonics. During a single cycle of the fundamental, the second harmonic undergoes two cycles and the third harmonic three. The distorted waveform is the summation of the undistorted (a.k.a. the fundamental) waveform and appropriately weighted harmonic overtones (in principle, an indefinite number; however, in practice, the weight given to ever higher harmonics falls rapidly, rapidly becoming negligible). Let us pay

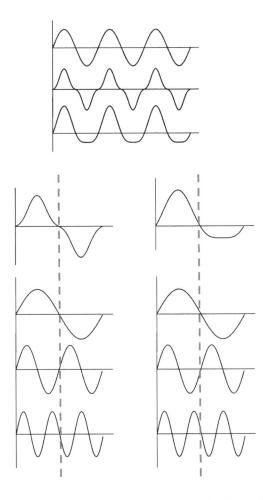

FIGURE 3.1   Harmonic content of symmetrically and asymmetrically distorted waveforms differs fundamentally. The symmetry between the positive and negative phases of a symmetrically distorted waveform (left column) means it cannot contain any even harmonic components. This implies that SHG requires an asymmetrically distorted waveform (right column), which in turn requires a physical asymmetry to be present in the material.

particular attention to the sign of the harmonics in the positive part (i.e., the first half) and negative part (i.e., the second half) of the waveform. The form of the fundamental in the second half of the excursion is inverted compared with its form in the first half of the cycle, and this property is shared by the third harmonic. The second harmonic, however, has the same form in both halves. This pattern repeats for all harmonics: the even harmonics (second, fourth, sixth, etc.) have the identical form in both halves, while the odd harmonics (third, fifth, seventh, etc.) have the inverted form in the second half relative to the first. Now a distorted wave such as that shown on the left, which is distorted equally in both halves of the fundamental, has this same characteristic in common with the fundamental and all the odd harmonics.

Since the distorted waveform is the sum of the fundamental and harmonics, it follows immediately that the distorted waveform must only contain the fundamental and odd harmonics. It cannot contain the second harmonic (or any even harmonic) because this would upset the precise inversion symmetry between the first and second half of the cycle. To summarize this very important point, a waveform that is distorted but remains symmetric with respect to the positive and negative phases of a single cycle of the waveform cannot contain any even harmonic overtones. In order for a distorted waveform to contain the second (etc.) harmonic, it must display an asymmetry between the positive and negative phases, as in the rightmost waveform. This means that whatever mechanism is responsible for distorting the waveform (e.g., the electronic amplifier circuitry in the radio) must act differently on the positive phase of the input waveform as on the negative phase of the cycle. For SHG from tissue, the mechanism responsible for scattering the input wave arises from the electrons within the specific molecules within the tissue. Consider the two hypothetical molecules shown in Figure 3.2, in which we imagine that all atomic centers and bonds are identical.

Further imagine a light wave propagating into the plane of the page. Positive and "negative" phases of the wave physically correspond to the electric field of the light wave

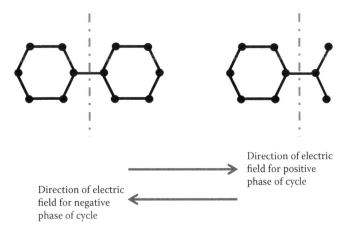

Direction of electric field for positive phase of cycle

Direction of electric field for negative phase of cycle

FIGURE 3.2 Hypothetical molecule on the left possesses inversion symmetry about the dashed vertical mirror plane while the one on the right does not. For a light wave plane polarized in the direction of the arrows, only the molecule on the right can generate SHG.

pointing to the right and left, respectively (right and left being defined as direction in space along which the incident light wave is "plane polarized"; see Section 3.4.2). The hypothetical molecule on the left has the property of inversion symmetry with respect to left and right (i.e., it looks identical to its mirror image reflected in a plane denoted by the dashed vertical line: a so-called "mirror plane"). The molecule on the right lacks such inversion symmetry because its mirror image appears to be different when reflected in that plane. Now recall that it is a fundamental requirement for SHG that the response of the system to the incident driving wave must be different for the positive and negative phases of one oscillation cycle. For our molecule, these phases correspond to light wave electric field pointing to the right and left, respectively (if appropriately plane-polarized). The molecule that displays inversion symmetry cannot respond differently when the electric field is pointing left rather than when it is pointing right because these two situations are perfect mirror images of each other. Only the molecule that lacks the mirror symmetry might be expected to display a different response: there being a clear physical difference between having the electric field pointing from the hexagonal ring structure to the y-shaped structure and *vice versa*. Finally, we reach our very important conclusion. Only molecules that lack the property of inversion symmetry are capable of generating second (and indeed any even) harmonic. The mere presence of an SHG response can therefore tell us something about the structure of a molecule even if that molecule itself is too small to be directly imaged.

3.3   Practical Implementation

This previous discussion leads us to define some of the basic requirements (and features) of SHG and other forms of nonlinear microscopy. Most SHG experiments are performed on systems designed for two-photon microscopy, which in turn are generally modified one-photon laser scanning confocal microscopes.

3.3.1   Basic Requirements

The chief requirement is that the tissue be illuminated by a sufficiently intense light beam that measurable amounts of SHG are produced. This requirement is essentially identical to that in TPEF microscopy and is met in the same way. The intensity that can be produced by any thermal light source such as a light bulb is inadequate; however, laser light has much greater directionality and so can be focused down to a diffraction-limited spot less than a micron in diameter. The laser of choice for TPEF is the tunable titanium–sapphire laser because its output wavelength can be tuned in the region 680–1000 nm, which is ideal for exciting most biological fluorophores. It can generate 100 fs pulses with output power exceeding 20 mW across this range, which is generally adequate for biological samples. These same attributes make it ideal for SHG also. SHG offers more flexibility over the laser source, however, and in particular, much of the early work in the 1980s was performed using the Q-switched NdYAG laser emitting at 1064 nm, a wavelength too long for TPEF work. This is potentially interesting as extremely low-cost NdYAG microchip lasers are now commercially available, which could make SHG microscopes considerably cheaper to construct than TPEF microscopes.

To focus the beam down to a submicron spot, high-performance microscope lenses must be used; in particular, they must possess a high numerical aperture (NA). The NA describes the cone angle of light rays that are brought to a focus, with a value 1.0 indicating that the cone angle reaches the theoretical maximum of 180° (in the case that the sample immersion medium is air; for water and oil, the values are 1.3 and 1.5, respectively). Diffraction theory tells us that the focus spot size of the laser has a diameter of about λ/NA (λ denotes the wavelength of the laser) so that a NdYAG laser with λ = 1 μm wavelength focused by a NA 0.5 objective produces a spot of diameter around 2 μm. A second requirement that is placed on the objective is that the glass from which it is made and the coatings that are applied should be chosen so as to maximize transmission both of the excitation light and also the emission light. In conventional bright-field microscopy, these wavelengths are of course identical; however, for SHG, the excitation light might be at 800 nm, in which case the SHG is emitted at 400 nm. These wavelengths are sufficiently different that they present specific challenges to the lens designer, and so it should not be assumed that the standard objectives supplied with a microscope will all perform equally well in an SHG experiment. Manufacturers such as Zeiss helpfully provide a multiphoton star rating for many of their objectives and so, for example, we can note that whereas the c-apo-chromat 63×/1.2 water-immersion objective has a five-star multiphoton rating, the N-achroplan 63×/0.9 water-immersion objective has only one star. The difference lies in the higher NA, the better throughput especially for longer excitation wavelengths, and also optimized correction of aberrations in the near infrared (NIR). The SHG neophyte is strongly encouraged to discuss their experiment with an expert in order to ensure that all elements are optimized.

3.3.2 Lab-Based Systems

A number of manufacturers supply complete multiphoton microscopy platforms. The multiphoton option is generally supplied as an additional feature to a single-photon confocal fluorescence microscope, since the microscope requirements are very similar. There are however some major differences between one-photon and two-photon systems:

1. The excitation wavelength is in the NIR for multiphoton systems rather than the UV-visible, and hence it is advantageous to choose an objective with good transmission in this range, especially near the edge of the tuning curve for the laser, where the output power can be low (typically beyond 900 nm).
2. Filters must be specified to offer excellent blocking in the NIR as well as in the UV-visible because elastically backscattered excitation light is orders of magnitude stronger than the multiphoton signal.
3. The detector requirements are different because of the relaxation of the requirement for a confocal pinhole in multiphoton microscopy. The most efficient detection scheme for multiphoton work is simply that which collects the most light and hence the so-called "non-descanned" detectors are placed as close to the collection lens (microscope objective or condenser lens depending on whether epidetection or transmission detection is performed) as possible. This is particularly important for epidetection of nonlinearly produced harmonic signals such

as SHG (and also THG, CARS, etc.), because the emitted radiation is mostly forward directed. Epidetection of these signals therefore relies heavily on the detection of elastically backscattered, forward directed light (Evans et al. 2005). Since this light is very diffuse after multiple scatterings, its efficient detection requires non-descanned detectors.

4. For SHG detection, the filter requirements are different because the emission linewidth is very narrow (typically 10 nm for a 100 fs pulse laser). The filter must be exactly matched to equal half the emission wavelength. A convenient and flexible alternative is to use a wavelength-resolved detector such as a spectrograph (Zoumi et al. 2002; Theodossiou et al. 2006). This (1) removes the need to change the filter when the excitation wavelength is changed and (2) allows the simultaneous collection of SHG and TPEF. The Zeiss LSM 510 META is a 32-channel multispectral detector that can collect up to 32 images, each centered on a user-specified wavelength and with a user-specified bandwidth. This makes the separation of SHG and fluorescence emission much easier and facilitates the generation of coregistered SHG/TPEF images.

5. An advantage can be gained by using a "pulse picker" rather than a neutral density filter to attenuate the excitation beam down to the levels required in a biological experiment. The pulse picker is basically a fast optical shutter that can block selected pulses from the laser. For example, if the laser power output is 2 W and this must be attenuated to 20 mW, let us say, to avoid burning the sample, then that the pulse picker would be programmed to only transmit 1 pulse in every 100 from the laser. Compared with the alternative of attenuating all pulses by 100× using a neutral density filter, the approach of using the pulse picker gives 100× more two-photon or SHG signal. This is because of the square-law dependency between the excitation energy per pulse and the emission signal per pulse.

3.3.3 *In Vivo* Systems

At the time of writing, there is one commercially available multiphoton imaging system that is certified for use in humans. The DermaInspect system from JenLab GmbH is a portable multiphoton tomograph suitable for imaging human skin and other accessible tissues *in vivo* (Konig and Riemann 2003). A titanium–sapphire laser, imaging optics, and data processing system are integrated together onto a trolley that can be brought to the patient. The beam delivery system is a rigid, articulated arm with a scan unit at the distal end. The scan unit contains the objective, galvo scanners, filters, and detectors. The system offers TPEF and SHG detection options and has recently been upgraded to offer CARS. Images of the superficial layers of human skin can generally be acquired using a few milliwatts of excitation power. Multimodal images revealing cellular organelles (via NADH/FAD autofluorescence), ECM collagen (SHG), and cellular membranes (CARS of CH_2 stretching modes in phospholipids) can be obtained in the clinic. Figure 3.3 shows illustrative results taken with the system. Note especially the signal from the dermis under 820 nm excitation, which is predominantly SHG from dermal collagen.

Stratum corneum   Stratum spinosum   Dermis

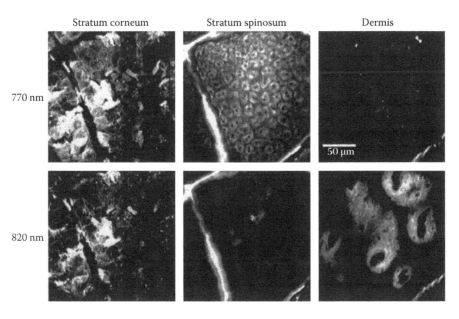

770 nm

50 µm

820 nm

FIGURE 3.3 Two-photon images of human skin *in vivo* taken with the DermaInspect system at various depths and using two different excitation wavelengths. Note the strong signal from the dermis at 820 nm, which is mainly SHG generated by dermal collagen. (Reproduced from Konig, K., and I. Riemann, *J Biomed Opt* 8(3): 432–439, 2003. With permission.)

3.3.4 Endoscopic Approaches

Multiphoton microscopy is generally performed using a benchtop microscope and is therefore restricted to situations in which a reasonable degree of physical access can be gained. The high NA objective lenses used generally have small working distances; a figure of 0.2 mm not being uncommon. The tissue must therefore be brought fairly close to the front of the objective, which clearly limits applications that require tissue constructs to be studied in a bioreactor or even postimplantation in patients. Standard high-NA objectives are available with working distances up to 3 mm (e.g., Olympus XLUMPLANFL, 20×, N.A. 0.95, W.D. 2 mm), and more recently, the company introduced MicroProbe "stick" objectives that use a gradient-index (GRIN) focusing lens to allow access through a much narrower channel (<1.5 mm) than is needed to accommodate a standard objective. These objectives have recently been combined with multiphoton microscopy to allow so-called "endoscopic" multiphoton microscopy (Heider et al. 2010). Indeed this approach has recently been taken even further, with fiber-optic based two-photon fluorescence and SHG microscopes demonstrated by several groups (Fu et al. 2006; Liu et al. 2009). The challenge, especially for SHG detection in the epidirection, is that the radiation that must be collected is extremely diffuse and so generally requires a large area light detector placed close to the tissue. In order to relay the excitation light to the tissue, however, it must be delivered by a narrow-core fiber (to allow tight focusing of the beam). The solution to this problem is the double-clad fiber optic, in which light is relayed back up a central narrow channel, and backscattered signal is relayed back down a larger concentric cladding layer that surrounds the central core. More recently, Fu and

Gu (2008) demonstrated that polarized-SHG measurements (see the next section) can also be performed using the double-clad fiber system. Such technology opens up the possibility to monitor SHG signals much deeper inside constructs than is possible with external optics, since light scattering limits the practical imaging depth to a few hundred microns beneath the tissue surface typically.

3.4   SHG Measurement Types

The most basic measurement that can be made on the SHG signal is, of course, its intensity. This relates directly to the magnitude of the second-order susceptibility and thus to the abundance of SHG-active molecules (i.e., "harmonophores"). Because of the requirement that the system lack inversion symmetry, it is also the case that the signal intensity relates to the degree of spatial order among these harmonophores (since a completely random distribution will possess inversion symmetry even if the individual harmonophores do not). Basic measurements of intensity thus provide a powerful insight into the degree of microstructural order or crystallinity in the tissue. However, one of the powerful attributes of SHG imaging is that more elaborate measurements that reveal more subtle features of the sample are also possible.

3.4.1   Forward/Backward Intensity Ratio

A major difference between SHG and TPEF concerns the directionality of the emitted radiation. Fluorescence (be it single or multiphoton) is fundamentally a random process in which the emitted photon has essentially lost all memory of the directional properties of the excitation photon(s) that created it and so the fluorescence is emitted isotropically in biological tissues. In contrast, SHG emission is not random but retains a strong imprint of the direction of the excitation photon(s). Hence, it is found that the SHG emission is much stronger in the "forward" direction (i.e., in the transmitted light detection) than in the backward (i.e., the epidirection) direction. This behavior has two fundamental causes. Firstly, SHG is said to be "coherent," which is a technical term describing the relationship between SHG emissions from different harmonophores. With coherent emission such as SHG, the wave amplitudes from the harmonophores can combine constructively or destructively, and it turns out that the backward propagating component is weak because backward propagating waves from different harmonophores have a tendency to combine destructively. In contrast, TPEF from different fluorescent molecules (i.e., "fluorophores") does not combine coherently and there is no constructive or destructive interference between waves, but rather the total fluorescent signal is just the additive sum of the fluorescent intensity from each fluorophore. Secondly, it turns out that the directionality of the emitted radiation is closely related to the overall size of the emitting structure (i.e., the spatial extent of the population of contributing harmonophores) or more precisely the size relative to the wavelength of the radiation. Objects that are similar in size or larger than the wavelength emit highly directional radiation, whereas objects much smaller than the wavelength emit radiation more isotropically. A good example is to compare the nondirectional properties of a mobile phone antenna with the highly directional properties of a satellite TV dish. The mobile phone antenna is small compared to the wavelength of mobile phone radio waves (about 15 cm), whereas satellite dishes are large compared to the wavelength of satellite TV broadcasts (about 3 cm). Hence, the satellite TV dish requires accurate alignment with the satellite, whereas the mobile phone can be used in more or less any orientation relative to the mobile phone mast.

The practical application of this in SHG imaging is that the measurement of the SHG intensity collected in the transmitted light direction can be compared with that collected in the epidirection and used to estimate the size of the fibrillar structures that generate the SHG, even if the fibrils themselves cannot be directly visualized. Zipfel et al. (2003), for example, compare images of collagen gels in which the fibrils are either parallel to the incident beam or at right angles to it. The backward directed signal from a 10 μm thick gel is much weaker when the fibrils are aligned in the beam propagation direction. This can be roughly understood in the terms framed above: the size of the SHG structure appears greater (when measured along the propagation direction) in the first case, and hence the signal is more highly directed into the forward direction. A more sophisticated analysis can be performed using the so-called "phase matching" considerations; however, the essential idea that isotropic emission implies small structures is found still to be true.

Practically speaking, such measurements place some limitations on the sample, which must generally be thin (a few tens of microns); otherwise, strong elastic backscatter will start to distort the forward/backward ratio (decreasing it). However, progress has been made in this area by using Monte Carlo modeling techniques to account for elastic scatter, thus enabling an important practical application in the study of ECM remodeling during cancer progression (Campagnola 2011). Figure 3.4 illustrates typical data and model results.

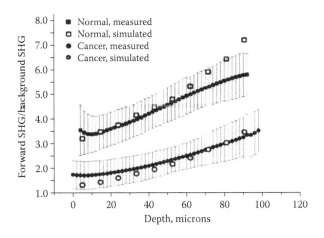

FIGURE 3.4   Forward/backward intensity ratio of SHG can carry important diagnostic information. The upper and lower lines compare the ratio for normal (upper) and cancerous (lower) ovarian tissue samples. The difference can be explained very well in terms of differences in the scattering properties of the tissues, as shown by Monte Carlo simulations (solid lines). (Reproduced from Campagnola, P., *Anal Chem* 83: 3224–3231, 2011. With permission.)

Chapter 3

3.4.2 Polarized Light Measurements

One of the most interesting features of SHG emission is the strong polarimetric information that is carried. Recall that electromagnetic radiation is a transverse wave (like a wave on a stretched rope or string; the oscillatory movement of the rope is orthogonal to the direction along the rope, i.e., the direction of wave propagation). Since for a rope that is strung horizontally, the plane in which the rope moves during its oscillatory wave motion could be vertical or horizontal (i.e., confined to a plane parallel to the ground), we speak of vertically and horizontally polarized waves on the string. Of course, the rope could vibrate in a plane at 45° to the floor as well (or indeed an infinite number of choices for the angle); however, it can be shown that all these motions of the rope can be considered to be the sum of motion purely in the horizontal plane and motion purely in the vertical plane. Thus, two fundamental polarization states for the wave on the rope can be defined: horizontal and vertical linear polarizations. Light (and other electromagnetic radiation, e.g., radio waves) shares this feature. The observant reader may, for instance, have noticed that some of the TV antennae fitted to properties in his/her locale are oriented so that the protruding rods—forming the "herringbone" pattern—are horizontal, whereas others are oriented so that these rods are vertical. This is because the former antennae are receiving signals from a TV transmitter emitting horizontally plane polarized radio waves, while the latter are pointing toward a different transmitter: one emitting vertically plane polarized radio waves. The antenna is maximally sensitive when the rods are parallel to the plane of polarization. Lasers such as the Coherent Chameleon and the SpectraPhysics Tsunami adopt the convention of emitting horizontally polarized light so that when mounted on the bench, a beam of light is emitted horizontally and the plane of wave vibration is a plane parallel to the surface of the bench. After being deflected by various mirrors, if an upright microscope is used, then the beam travels vertically downward onto the sample. The plane of polarization must always remain orthogonal to the beam direction, and so it is again parallel to the surface of the bench. However, in which particular direction in this plane the wave oscillation now occurs is determined by factors such as the precise arrangement of beam-turning mirrors used. The microscope installer is best placed to advise on this or the direction may be determined experimentally using a linear polarizer and a power meter (see later). In any case, it will be oriented at some defined angle relative to a reference direction, for example, the lateral movement direction of the microscope sample stage.

The significance of this discussion to SHG from biological tissues is that the polarization direction corresponds to the direction in which the electrical field associated with the light wave points. By changing the polarization direction, we can therefore rotate the electric field direction relative to the SHG-active molecules in the tissue. Recall that SHG emission fundamentally requires the molecule to lack inversion symmetry about a mirror plane. In Figure 3.2, we showed a molecule that lacked inversion symmetry about a mirror plane (the vertical dashed line). Hence, it may emit SHG if the electric field is oriented horizontally, that is, if the incident wave is horizontally polarized. However, note that this particular molecule does possess mirror symmetry about a mirror plane that is horizontal because the top and bottom halves are indeed mirror images of each other. Hence, SHG cannot be emitted if the incident wave is vertically polarized. Two very important consequences thus follow.

Firstly, a medium will emit no SHG, regardless of the polarization direction, if no mirror planes can be found about which the molecule appears to lack inversion symmetry; in other words, if the molecule is completely centrosymmetric, it will not produce SHG under any circumstances. An immediate practical consequence of this is that only the fibrillar collagens such as type I and type II are efficient SHG generators. Collagen type IV, which predominantly forms sheet-like structures especially in the epithelial basement membrane and the lens capsule, generates essentially no SHG. Cox et al. (2003) used the phenomenon to differentiate between collagen type I and type III by noting that type I is by far the more SHG active, whereas both types display similar contrast under Sirius-red staining. Figure 3.5 shows illustrative results.

Secondly, for noncentrosymmetric molecules, information can be gleaned about the local orientation of the molecule by studying how the SHG intensity varies as the polarization direction is rotated relative to the molecule. For our hypothetical molecule in Figure 3.2, the dashed axis will be aligned at 90° to the polarization direction when the SHG intensity falls to zero. This structural information can be obtained even when the molecules themselves are below the resolution limit of the microscope. In the next section, we will study these ideas in somewhat more detail.

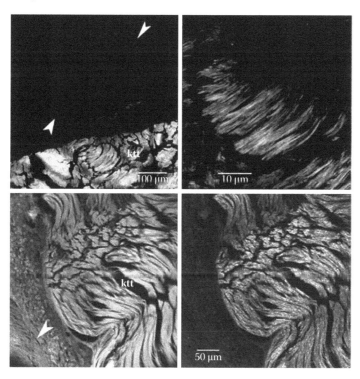

FIGURE 3.5 **(See color insert.)** SHG and two-photon fluorescence give complementary information as shown by these SHG (cyan) and fluorescence (red) micrographs from picrosirius-stained sections of kangaroo tail tendon (ktt) embedded in rat paravertebral muscle (arrowed). The tendon is predominantly type I collagen, which generates intense SHG. The paravertebral tissue is dominated by type III collagen, which shows much weaker SHG activity, but which binds strongly to the picrosirius red. Note also the much greater contrast between these collagen types when imaging in the forward direction (bottom-right) compared with the backward direction (bottom-left). This is because forward directed emission from type I collagen is dominated by intense SHG. (Reproduced from Cox, G. et al., *J Struc Biol* 141: 53–62, 2003. With permission.)

Chapter 3

3.4.2.1 Behavior Predicted by C∞ and Kleinman Symmetry

An early theoretical and experimental study into the effects of light polarization state on SHG production by rat tail tendon collagen was carried out by Freund et al. (1986). They focused on SHG in tendon by building a simplified model of this tissue as an array of parallel-aligned, infinitely long cylinders of material, with the cylinders having the same nonlinear susceptibility but a variety of diameters. Technically each fiber is said to possess "C∞" symmetry, because a cylinder cross section is a circle: a regular polygon with infinitely many sides. A further type of symmetry can also be applied, the so-called Kleinman symmetry. This is more abstract in nature, but essentially it is assumed to be a valid assumption whenever the excitation photon energy is significantly lower than that needed to produce direct autofluorescence from collagen. By making these simplifications, Freund et al. (1986) derived particularly simple equations to describe how the SHG emission intensity varies when linearly polarized light strikes a collagen fiber. When the plane of linear polarization makes an angle θ to the cylinder long axis and the excitation beam travels orthogonal to the fiber, then the forward-emitted SHG has a total unpolarized intensity (i.e., the detected light intensity, assuming that no polarizer is placed between the sample and the light detector) given by the equation

$$I(\theta) \propto [\rho \cos^2 \theta + \sin^2 \theta]^2 + [\sin 2\theta]^2 \tag{3.1}$$

The quantity ρ is significant because under the assumptions of C∞ and Kleinman symmetry (and assuming that only relative SHG intensity measurements are available), it is a single number that characterizes the second-order susceptibility of the material. Two important consequences follow:

1. The intensity varies considerably depending on how the fibers are oriented with respect to the plane of polarization of the laser. For tendon collagen ($\rho = 1.8$), the intensity versus angle reaches its minimum value when the laser polarization direction is perpendicular to the local cylinder long axis. Hence, the fiber orientation can be determined provided the polarization direction of the excitation beam at the sample is known.
2. The quantity ρ can be determined by considering the detailed form of the intensity versus angle. This leads to the possibility of discriminating different types of tissue within an SHG image if these tissues have differing values of ρ.

This measurement of intensity versus angle can be obtained, for example, by rotating the specimen underneath the microscope objective. This method has the disadvantage, however, that in practice, it is difficult to mechanically rotate a specimen without also translating it, especially given the small field of view in such experiments (typically 0.1 × 0.1 mm). A more reliable method is to rotate the plane of linear polarization of the laser. Most lasers used for multiphoton microscopy produce a linearly polarized output, and the system installer will generally arrange for this polarization state to be preserved after the beam has propagated through the microscope optics and emerges from the objective lens. An optical half-wave plate inserted into a linearly polarized beam has the effect of rotating the plane of linear polarization (at an angular rate equal to twice the rate of rotation of the wave plate itself). The best position for the half-wave plate is directly behind the main objective. Microscopes such as the Zeiss Axiovert 200M inverted microscope have a filter slider position conveniently placed beneath the objective. They also supply a cassette that locates into this slot and that can accommodate a wave plate into a holder. The holder in turn is connected by a rubber drive belt to a user-operable graduated rotation control. When the user rotates the control, the wave plate rotates beneath the objective and thus effects the desired rotation of the polarization direction. In practice, care must be taken (1) to verify that the polarization state at the sample is indeed linear and (2) to ascertain the precise orientation of the polarization direction relative to the sample. A simple polarizer, which can itself be rotated and which can be placed between the objective and a power meter, can resolve both questions. The polarizer is rotated whilst carefully monitoring the power meter reading. The plane of linear polarization and the polarizer transmission axis are aligned when the power meter reading is maximized and they are perpendicular when the reading is minimized. The polarization state is essentially linear if the ratio of the maximum to minimum power meter readings is comparable with the polarizer extinction ratio (value available from the manufacturer and generally at least 100). Ratios significantly less than this indicate that the polarization state is not linear and the microscope manufacturer should be consulted for advice.

As an example of what may be observed, Figure 3.6 shows results obtained by Tiaho et al. (2007) on adult xenopus gastrocnemius muscle. Note that for tendon, the SHG is minimized when the polarization vector is orthogonal to the fibers, whereas for muscle, it is minimized when these directions are aligned. This is because polarized SHG is sensitive to the differing orientation of SHG harmonophores relative to the macroscopically visible fiber direction.

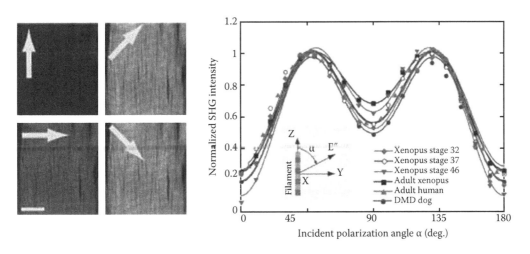

FIGURE 3.6 Different SHG emitters can be differentiated based on their polarization response. By measuring the SHG intensity as a function of the angle, as the plane of polarization onto the sample is varied, curves such as these can be generated. Myosin shows a minimum SHG intensity when the polarization is aligned with the macroscopic fiber axis, whereas for collagen the minimum is at 90°. The difference arises because of the different microstructural arrangement of molecular harmonophores relative to the macroscopic fiber direction. (Reproduced from Tiaho, F. et al., *Opt Express* 15(19): 12286, 2007. With permission.)

Chapter 3

Stoller et al. (2002) designed a rotating polarization SHG microscope with no moving parts using a voltage controllable wave plate, which removes the tedious need for manual operation. By rotating the polarization orientation through 360° at high frequencies (4 kHz), an electronic lock-in detector can be used to automatically determine both the fiber orientation direction and the susceptibility constant ρ as functions of spatial position in the image, that is, orientation and susceptibility images can be produced in addition to the usual SHG intensity images. Stoller et al. (2002) have demonstrated fiber orientation images in rat tail tendon and bovine tendon fascia as well as in porcine cornea and human intervertebral disk. The effects of fascicle disruption via water-induced swelling on both SHG intensity and polarization sensitivity were also studied, as were the effects of interfibrillar cross-linking, the latter being studied on *in vitro* preparations of precipitated collagen. Isolated domains of fibrillar ordering were noted in both porcine cornea and bovine tendon fascia, although the spatial scale of the domains in the former tissue was approximately an order of magnitude larger than in the latter. As expected, rat tail tendon showed a highly regular orientation image and intervertebral disk showed two domains in which the average orientation differed by 60°. This presumably relates to the collagen structure in the annulus fibrosus, where the type I fibers are oriented at ±55° relative to the main compression axis (Klein and Hukins 1982). Interestingly, studies on precipitated collagen fibrils from both acid-soluble and neutral salt-soluble collagen before and after spontaneous cross-linking induced by native aldehydes suggest that neither the SHG intensity nor the polarization sensitivity is greatly influenced by the degree of cross-linking. This confirms the assumption that the SHG signal arises from the collagen alpha-helix, and this stands in contrast to the autofluorescence signal from collagen, which appears to be strongly dependent on the abundance of GTA-type and pyridinium-type cross-linkages between fibrils (Raub et al. 2010). Combined autofluorescence and SHG therefore have the potential to provide detailed information on the microstructural parameters that are likely to determine the ultimate mechanical properties of collagenous connective tissues.

Other biological tissues that have been found to display polarization-sensitive SHG include muscle tissue (Plotnikov et al. 2006; Tiaho et al. 2007) and articular cartilage (Yeh et al. 2005; Mansfield et al. 2008).

3.4.2.2 α-Helix Pitch Angle

The previous discussion shows how the 2D orientation of the fibers can be inferred from polarization-resolved SHG measurements. By "2D orientation," we refer to a situation in which it is implicitly assumed that the fibers are oriented in the en-face plane, that is, lie normal to the direction of the incident light beam. Of course, in general, we cannot assume that this is so; even if the macroscopically observable fibrils are oriented in this way, the structure at the molecular level will be more complicated. Plotnikov et al. (2006) used a general theory of SHG to relate the signal from an individual molecule (a.k.a. a "harmonophore"), with an arbitrary 3D orientation in space, to the macroscopically observable SHG signal generated from an image voxel when imaging muscle myosin. Since myosin has an alpha-helix structure, they calculated the SHG signal from a distribution of individual harmonophores whose polar angle relative to the helix axis is constant but whose azimuth angles are uniformly distributed in the range 0° to 360°.

The microscope image voxel contains many hundreds of harmonophores so that an average is taken over all the harmonophore orientations. Tiaho et al. (2007) showed that this analysis leads to a simple relationship between ρ and the mean polar tilt angle θ_e of the harmonophores:

$$\cos^2\theta_e = \frac{\rho}{2+\rho} \tag{3.2}$$

Given the imaging geometry and the helical structure of the myosin, this angle is seen to relate directly to the pitch angle of the helix. Myosin is found to have a mean polar tilt angle of 62° and collagen type I 49°. Both angles are in good agreement with the helix angles inferred from x-ray diffraction data (68° and 45°, respectively).

A tantalizing view of what might be possible with carefully controlled polarimetric measurements is the ability to use this approach to separate collagen types I and II without using exogenous labels (Su et al. 2010). These two collagen types are both fibrillar but differ in the nature of the three alpha-helices forming a fibril. Careful measurements comparing rat-tail tendon (predominantly type I collagen) with rat trachea cartilage (predominantly type II) under 780 nm excitation suggest that the alpha-helix pitch angles of these two isoforms are sufficiently different that they can be differentiated in polarization-resolved SHG images (see Figure 3.7). Interestingly, when human mesenchymal stem cells were seeded onto a chitosan scaffold, and induced to differentiate toward a chondrocyte lineage using added TGF-β3, the polarized SHG images tended to agree with the results of direct immunohistochemistry labeling of collagen I and II. A tendency for both types to be produced by the hMSCs was noted, as was a tendency for the type II collagen to localize around the edges of type I fibril bundles. Such observations are potentially significant in cartilage tissue engineering as the goal is to recreate normal hyaline cartilage, which contains a majority of type II collagen and little type I while avoiding the production of abnormal fibrocartilage (Hunziker 2001).

3.4.3 Wavelength–Resolved Measurements

Fluorescence emission is characterized by strong dependence on the excitation wavelength. Beyond a certain wavelength, the photon energy is insufficient to promote transitions between the real electronic energy levels of the fluorophore and so fluorescence cannot be excited, whereas if the photon energy is too high, then it is not resonant with the energy band gap and so fluorescence is also weak. Within this range, however, the excitation–emission curve gives much information about the species that are present. A useful example is that of the respiratory chain enzymes NADH and FAD, the autofluorescence spectra of which show peak two-photon excitation efficiency at 800 and 890 nm, respectively, so that by combining images obtained at each of these wavelengths, their relative concentrations (which correlate with the mitochondrial redox potential) can thus be determined without labeling (Skala et al. 2007).

The situation with SHG is different. The radiation itself has a narrow linewidth–$1/\sqrt{2}$ that of the laser (Zipfel et al. 2003)—and so the wavelength-dependent spectrum of SHG is defined in terms of the variation in SHG intensity with respect to excitation wavelength, keeping the power onto the sample constant. Ultimately, the SHG intensity

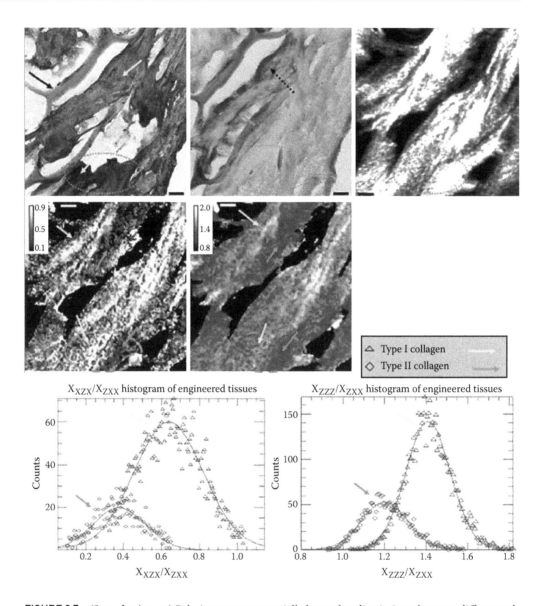

FIGURE 3.7   (See color insert.) Polarimetry can potentially be used to discriminate between different collagen types, as shown by these results. By fitting the polarization response on a pixel-by-pixel basis and extracting estimates of rho by fitting (3.1 to the data, histograms of rho values can be assembled for collagens of known type. It is found that type II collagen produces significantly different values to type I. Color-coded images may then be produced, which map these collagen subtypes without the need for immunohistochemical labeling. (Reproduced from Su, P. et al., *Biomaterials* 31: 9415–9421, 2010. With permission.)

is determined by the nonlinear susceptibility χ as well as by the polarization state of the light, the degree of phase-matching, etc. None of these parameters varies especially strongly with wavelength; indeed the often-assumed Kleinman symmetry is an assumption that the susceptibility is wavelength-independent (Plotnikov et al. 2006), and so it is not expected that the wavelength dependence of SHG is as strong an information carrier as that of fluorescence or CARS. Nonetheless, some workers have investigated the issue, although the information is not exhaustive and sometimes appears contradictory. The measurements must be made with care as many microscopes show significant wavelength-dependent changes in optical efficiency, both for excitation and emission wavelengths, which must be properly accounted for.

Zoumi et al. (2002) performed experiments on collagen-gel–based RAFT skin models (polymerized rat-tail collagen gel plus human neonatal foreskin fibroblasts), finding distinct wavelength dependence to the epidetected SHG excitation efficiency. The SHG efficiency peaked at 800 nm, falling to either side of this before rising again at 740 nm. This increase at short wavelengths potentially relates to a phenomenon called "resonance enhancement." Many light–tissue interactions, including Raman and harmonic generation, increase in efficiency when the excitation wavelength is closely matched to an electronic resonance (i.e., where strong absorption or fluorescence occurs). Since the peak one-photon excitation wavelength for collagen autofluorescence is at 330 nm (Richards-Kortum and Sevick-Muraca 1996), the rise in SHG intensity at 740 nm could be explained by this. The peak at 800 nm was interpreted in terms of an optimization of the local phase matching within the excitation focal volume. Because SHG is a coherent process, the signals generated at different locations within this volume combined by constructive or destructive interference. The condition for constructive interference between two points separated by a fixed distance is that an integer number of wavelengths should fit into this distance, and hence, the choice of excitation wavelength has an influence. The situation is complicated because also one must consider how well the excitation and emission waves remain in phase as they propagate through the focal volume; however, Zoumi et al. (2002) present a simple analysis that can explain the 800 nm peak in SHG intensity. A key point is that this mainly reflects sensitivity to the local optical properties of the material (e.g., the refractive indices at the fundamental and second harmonic wavelengths) rather than to the properties of the collagen molecules themselves. Unfortunately, the literature is not wholly consistent on this point. Zipfel et al. (2003) present spectra of normalized SHG emission using excitation wavelengths from 730 to 908 nm. Both rat tail tendon and purified collagen I/III gels produce identical results, with SHG intensity steadily falling with increasing wavelength, that is, no 800 nm peak was observed. Theodossiou et al. (2006) compared the wavelength dependence of SHG from rat hind limb tendon in the forward and backward directions using excitation from 830 to 910 nm. In the forward direction, the SHG maximum is found to be at a longer wavelength than found by Zoumi et al. (2002)—880 nm—and to be very low at 800 nm. They interpret this as a consequence of the near-perfect phase-matching at 880 nm due to the specific wavelength dependence of the refractive index; however, this theory is not fully developed. In the backscatter direction, multiple peaks of comparable intensity occur at this and other wavelengths. The results appear to be broadly similar comparing rat tendon with bovine Achilles tendon.

In summary, the wavelength dependence of SHG has been measured by a few groups but with significantly differing results. Theoretically, it seems likely that the results are more strongly determined by the light transport properties of the medium than by specific microstructural properties of the collagen; however, this is not established and could be an interesting area for future studies. From a practical point of view, there is one important consequence of these studies. In highly crystalline collagenous tissues such as tendon, the SHG efficiency does not fall precipitously with increasing excitation wavelength; however, beyond 800 nm, the majority of autofluorescent compounds

Chapter 3

(especially collagen itself) show very large drops in fluorescence yield. Hence, increasing the excitation wavelength well beyond 800 nm is a potential strategy to suppress unwanted autofluorescence and maximize sensitivity to SHG. The increasing availability of very compact and low-cost pulsed laser sources at 1064 nm and longer wavelengths makes this attractive from the practical standpoint also.

3.5   Applications of SHG Microscopy in Tissue Engineering

Having described the key features of SHG microscopy, we will now consider specific applications that have been reported by the tissue engineering and ECM biology communities.

3.5.1   Abnormal Collagen Structure in Disease

An application with strong clinical potential is in the assessment of collagen structural changes that are associated with invasive cancer. That malignant cells are capable of altering their local ECM scaffold is well known. For example, excessive fibrosis is partly responsible for the pronounced change in tissue mechanical properties, especially the increased stiffness of tumors relative to healthy tissue (which allows for tumor detection by mechanical palpation). Also it is established that tumor invasion through healthy stroma is facilitated by the secretion of matrix metalloproteinases (MMPs), which cleave the fibrils in normal collagenous tissues (Stamenkovic 2003). The collagen content of solid tumors also appears to be predictive of the permeability of the tumor mass to administered drugs (Netti et al. 2000). Interestingly, the collagen content, rather than the glycosaminoglycan content, seems to be the more important parameter. Brown et al. (2003) used SHG imaging in the epidirection to visualize the type I collagen content of human melanoma, adenocarcinoma, and soft-tissue sarcoma tumors grown in a dorsal skin chamber in immunodeficient mice. The fibrillar type I collagen showed a strong SHG signal in both *in vitro* gel preparations and in the tumor model, in contrast to preparations containing other ECM components such as laminin and type IV collagen. Exogenous collagenase was found to significantly reduce the SHG signal from individual fibers while preserving their length, suggesting that the collagenase "dissolves" the fibers rather than cleaves them. In comparison, exogenous relaxin (a nontoxic hormone that is secreted during pregnancy) caused significant shortening of the collagen fibers but preserved their intensity, suggesting that the hormone cleaves the fibers. This was found to correlate with significantly increased permeability of the tumor mass to IgG and dextran, with potential applications in improving drug delivery efficiencies. There is an increasing interest in the use of tissue-engineered models of tumor invasion both to assess imaging technologies and to assess response to therapies (Smith et al. 2010). SHG is therefore ideally placed to help in the assessment of these models.

SHG is also a powerful means to monitor changes in the ECM structure that are brought about by tumor invasion. Campagnola (2011) describes measurements of the ratio of transmitted to backscattered SHG in ovarian cancer. The measurement is made on histological slices of tissue using a dual transmitted and epidetection SHG microscope. It is found that the malignant ovarian tissue is more efficient at generating backscattered SHG, that is, the ratio forward SHG/backward SHG is lower for malignant tissues. This appears to be the result of changes in the fibrillar density and ordering, which are both increased as a result of tumor remodeling.

3.5.2 Scaffold Characterization

Tissue engineering involves the attempt to replicate living tissues by seeding some form of scaffold with cells in order to produce a construct that can then be implanted into the body. Although a recent paradigm involves using decellularized allografts or xenografts to obtain a collagenous scaffold with an architecture that is virtually identical to the native tissue, the use of artificial biodegradable fibrous scaffolds remains a popular approach (Badylak 2002). The hope is that both prior to implantation and after, the implanted cells within the scaffold will begin to synthesize new collagen, which will eventually form an interconnected network with the necessary mechanical properties to constitute a new ECM. At the same time, the artificial polymer scaffold is designed to degrade by hydrolysis. The ability of SHG to image collagen without the need for labels clearly makes it an excellent candidate for noninvasive monitoring of construct development. However, since SHG has very little chemical specificity, it cannot be used to directly separate signal produced by collagen from signal produced by other noncentrosymmetric structures, and so it is important to understand the type of SHG signal that can be produced by various types of biodegradable polymer scaffold.

Many natural and synthetic polymers have been evaluated for use as tissue-engineering scaffolds. A material famous since antiquity and yet still finding new applications is silk, which is continuing to attract interest as a potential tissue-engineering scaffold due to its good biocompatibility, strength, and biodegradability (Shao and Vollrath 2002). Rice et al. (2008) evaluated the appearance of various types of silk scaffold using two-photon and second harmonic microscopy, comparing native silk fibers with methanol-treated aqueous film, compressed/stretched films, a hydrogel, and a synthetic scaffold fabricated from silk fibroin aqueous solution using a salt leaching method. This type of silk contains, in its natural state, two primary domains: namely, type I silk (with random coil and randomly aligned alpha-helix structures) and type II silk (primarily aligned beta-sheet structures). It was found that whereas the fluorescence signal (proportional to beta-sheet content) was significant among all these materials, the SHG varied greatly. Indeed, fibroin aqueous films and hydrogels showed no SHG at all, whereas natural fibers showed strong SHG. The processing of silk to form films and scaffolds largely destroys the alignment of the beta-sheet structures; hence the SHG signal is lost. This illustrates again how SHG carries important information about the microstructure of a material and not just the molecular composition. Since β-sheet content and alignment are determinants of the mechanical properties and degradation kinetics of the scaffold and can be modified by suitable processing, SHG could have important applications in monitoring this type of scaffold.

Other biopolymer scaffolds have also been characterized using TPEF and SHG. Sun et al. (2008) characterized five different types of polymer scaffold: open-cell polylactic acid, polyglycolic acid, collagen composite scaffold, nylon, and Collagraft: a bone graft

Chapter 3

substitute comprising hydroxyapatite/tricalcium-phosphate beads and fibrillar collagen. All five scaffolds were found to generate measurable SHG in the epidirection as well as distinctive autofluorescence (the spectral properties of which can differentiate between polymer types). Importantly, it was demonstrated that for blended materials such as Collagraft, the SHG and autofluorescence are not colocalized but provide complementary information about the spatial distributions of the different components in the blend (see Figure 3.8).

The effect of scaffold structure on the ability of smooth muscle cells to migrate into deeper areas of the scaffold has been studied Brackmann et al. (2011) using dual modality CARS and SHG microscopy. SHG proved effective at defining the morphology of the biosynthesized cellulose scaffold, whereas CARS imaging using the 2845 cm^{-1} CH$_2$ mode of lipid offered excellent label-free visualization of the cell membrane of human smooth muscle cells isolated from the tunica media and seeded onto the scaffold. Using this approach, it was demonstrated that compact cellulose scaffolds severely impede cell migration from the surface, with invasion depths being less than 10 μm. In contrast,

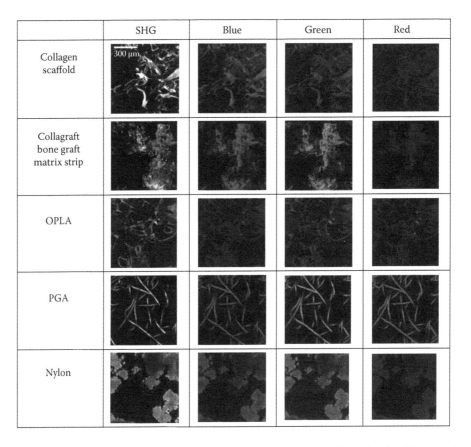

FIGURE 3.8 (See color insert.) Scaffold materials can also generate significant levels of SHG, as shown by the images in the left column. These signals must be carefully characterized as they will potentially interfere with weak signals generated by low levels of collagen production. Note also that the SHG and autofluorescence from composite Collagraft material are not colocalized and hence are complementary. (Reproduced from Sun, Y. et al., *Microsc Res Tech* 71: 140–145, 2008. With permission.)

cells seeded onto the surface of a more open porous cellulose scaffold of presumably identical chemical composition readily migrated to depths exceeding 40 μm in the same amount of time.

3.5.3   Collagen Production via Scaffold Composition

A potentially important modulator of collagen production in a tissue-engineered composite is the chemical composition of the scaffold. Deng et al. (2003) investigated the ability of rabbit articular cartilage chondrocytes to generate type II collagen after being seeded onto artificial polymer scaffolds. The scaffolds were a copolymer blend of poly-hydroxybutyrate (PHB) and poly(hydroxybutyrate-co-hydroxyhexanoate) (PHBHHx) with varying relative fractions of the two, fabricated into a porous scaffold structure using a salt leaching process. Rabbit chondrocytes cultured on these scaffolds showed markedly elevated levels of both GAG and collagen II production especially when a copolymer blend of 1:2 PHB/PHBHHx was used. SHG imaging in the forward direction showed a strong signal from the polymer scaffold itself and revealed that the optimum copolymer blend appeared to produce an increased number of micropores in the 5 to 10 μm size range, which in turn appeared to produce better anchoring points for type II collagen fibrils.

Filová et al. (2010) used TPEF and SHG to study collagen production by rabbit chondrocytes seeded onto (1) cross-linked gelatin scaffold and (2) polycaprolactone foam scaffolds fabricated by a salt leaching method. Cells were incubated in Iscove's Modified Dulbecco's Medium supplemented by 10% fetal bovine serum for 14 days but were otherwise unstimulated. Exogenous labeling for type II collagen using Cy3-conjugated antibodies was combined with SHG imaging. The PCL scaffold was observable by autofluorescence but not by SHG. Clear differences in the localization between labeled collagen II and SHG were found. Cy3 fluorescence was generally located close to the cells, suggesting a preponderance of immature collagen in this region (Figure 3.9). For the gelatin scaffolds, little SHG was found, suggesting that this scaffold does not promote the formation of aligned, crystalline collagen. SHG was markedly higher for chondrocytes seeded onto PCL foam, especially in areas of higher cell concentrations.

A major goal of tissue engineering is to produce a collagenous matrix with reasonable mechanical properties. Raub et al. (2010) used SHG to study the relationship between collagen gel density, cross-linking, cellularity, and ultimate mechanical stiffness. It was found that SHG intensity and image morphology (e.g., speckle texture, brightness histogram skew, etc.) are predictive of the mechanical properties of the collagen gels (Raub et al. 2010). If made quantitative, this could prove to be a useful way of indirectly probing the local mechanical properties of constructs at different stages of development.

Chen et al. (2010) used TPEF and SHG microscopy to study the production of collagen by human mesenchymal stem cells seeded onto a chitosan scaffold. Using 780 nm excitation, they noted that the scaffold displayed strong autofluorescence but negligible SHG. hMSCs were stimulated to differentiate toward a chrondrogenic lineage by the addition of the induction factor TGF-β_3, and SHG imaging at 390 nm showed evidence of collagen production from as early as day 7. Interestingly, mRNA expression suggested that this was type I collagen since type II mRNA was not detected until day 21. Analysis of the SHG intensity versus time suggests that collagen production increased

Chapter 3

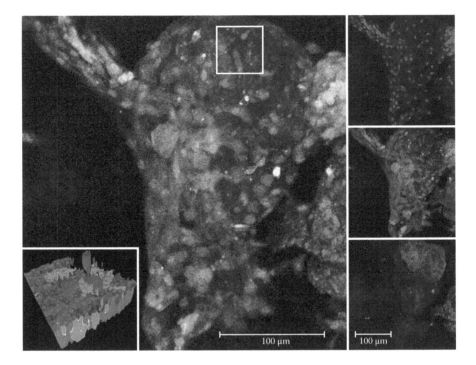

FIGURE 3.9   (See color insert.) Simultaneous 3D mapping of procollagen and collagen deposition in a porous polycaprolactone scaffold can be achieved using a combination of immunolabeled TPEF and SHG imaging. The three images on the right show, from the top down respectively, cell nuclei labeled using propidium iodide, procollagen labeled using antibody-conjugated Cy3, and mature collagen generating epidetected SHG. (Reproduced from Filová, E. et al., *J Biomed Opt* 15: 066011, 2010. With permission.)

exponentially with time for the first 11 days before reaching a plateau. Taken together, these observations could suggest that type II collagen represents only a small fraction of the total collagen produced, which is significant because one of the main motivations to use chondrogenic hMSCs is to produce synthetic cartilage, which is predominantly type II. Alternatively, it could be that type II collagen possesses a small second-order susceptibility compared with type I collagen so that its production at later stages does not strongly affect the overall SHG signal. Other observations of note from this study were that SHG production was largely confined to the surface of the scaffold, consistent with the finding that hMSCs seeded onto the surface of the scaffold do not migrate easily into deeper regions. The SHG that is produced also shows little attachment to the scaffold but instead forms extended regions within individual scaffold pores, with some interconnections between neighboring regions. Such observations can potentially help to determine whether the collagen scaffold is likely to possess mechanical properties that are similar to those of the native tissue.

3.5.4   Collagen Production via Applied Mechanical Stimulation

It is a well-established paradigm in tissue engineering that cells can upregulate their production of ECM proteins if they experience mechanical stress. Ultimately, this is presumably how the body responds to increased loading in order to lay down increased soft tissue and bone mass. SHG is an ideal tool to study this process longitudinally. Hu et al. (2009) studied fiber alignment in fibroblast-seeded collagen gels by constructing a biaxial mechanical stretcher that can be placed on the stage of a multiphoton microscope. The device allows differential strains to be applied to the long and short arms of cruciform-shaped gels such that SHG imaging can be performed at various locations and time points and also with the constructs under tension and relaxed. Fiber alignment is determined by power spectral analysis of SHG intensity images and an "alignment index" calculated. It is found, for example, that anisotropic stretching of the gels along the long arm produces increasingly aligned fibers but that this alignment is largely reversible upon deloading of the construct up until day 6, whereupon the directional alignment becomes essentially a permanent feature.

3.5.5 Collagenous Scaffold Production

Another exciting application of SHG imaging is in the characterization of the scaffold production process itself. Many materials have been proposed as synthetic scaffolds and electrospun polymers such as PLGA are very popular. However, the body's natural biopolymer collagen would be the material of choice if it could be fabricated into similar structures as electrospun polymers. Huang et al. (2001) first demonstrated the feasibility of electrospinning collagen by first dissolving it in a fluoroalcohol and then extruding from a needle at a high electric potential. However, further work on the resulting materials demonstrated a number of poor material properties in the final material, most noticeably water solubility reminiscent of denatured collagen (gelatin).

Zeugolis et al. (2008) thoroughly studied the properties of electrospun nanofibers of collagen using a variety of techniques including transmission SHG imaging. In contrast to native tendon, self-assembled (in phosphate buffer + polyethylene glycol medium) collagen fibrils, and extruded (into the same medium) micron-scale collagen fibers, the electrospun fibers were found to generate essentially no SHG signal at all (see Figure 3.10). This, along with TEM imaging, differential scanning calorimetry, and gel electrophoresis, provides compelling evidence that the combination of aggressive fluoroalcohols and exposure to high voltages essentially destroys the crystalline triple-helix structure of the collagen molecules and converts it to gelatin. Improved fabrication techniques are therefore required if nanoscale fibers of collagen are required for scaffold fabrication.

In a more recent development, Caves et al. (2010) successfully fabricated continuous lengths of 50 μm collagen fiber by simultaneously extruding acid-soluble collagen monomer and a PEG-based wet spinning buffer such that the collagen coagulated in the buffer to form a filament that was then passed through an ethanol drying bath before collection onto a reel at rates of up to 60 m/h. An important step prior to cross-linking in glutaraldehyde vapor is incubation for up to 48 h in a phosphate buffer, which promotes self-assembly of the collagen fibrils into axially ordered fiber structures. SHG imaging was used to confirm the molecular ordering, which is a key determinant of the mechanical strength and viscoelastic properties. The wet fiber ultimate tensile strength of the fiber, 94 MPa, and modulus of 0.7 GPa, is significantly better than other wet-spun collagen fibers and is comparable with that of native tendon.

Chapter 3

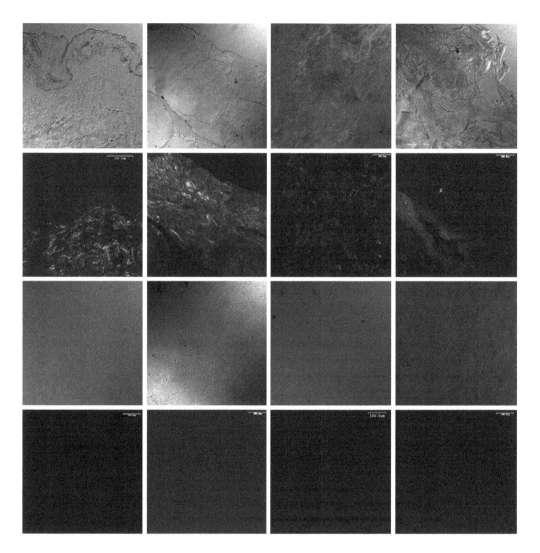

FIGURE 3.10   SHG can give insights into the nanostructure of synthetic scaffolds such as electrospun collagen. The first and third rows show light micrographs of various tissues and below them are corresponding SHG micrographs. Note the strong SHG from native collagen in human skin (row 2, column 1) and rat tail tendon (row 2, column 2), whereas it is absent from electrospun collagen nanofibers (row 4, columns 2–4). This has been interpreted as evidence that the process of electrospinning effectively degrades the collagen and converts it into gelatin. (Adapted from Zeugolis, D. I. S. et al., *Biomaterials* 29: 2293–2305, 2008.)

3.5.6   Collagen Alignment via Aligned Scaffolds

Lee et al. (2006) used SHG and two-photon autofluorescence microscopy to study collagen production by cultured human mesenchymal stem cells seeded onto a polyglycolic acid nonwoven fiber scaffold. The PGA itself is highly autofluorescent as well as weakly SHG-active. The seeded stem cells were stimulated to produce collagen via the addition of l-ascorbic acid and recombinant human transforming growth factor TGF-β_3. SHG demonstrates that as early as week 1, SHG-active material (presumably collagen) is deposited into the scaffold, with a large increase at week 3. Significantly, the alignment of the SHG-emitting filaments appears to be generally aligned with the scaffold fiber direction initially; however, this alignment tends to reduce for filaments that are located further from the scaffold fiber. mRNA analysis confirms a strong increase in type II collagen by week 3.

3.6 Conclusions

Over the past decade, SHG microscopy has moved from being an esoteric research tool into one that many biological labs are now able to implement. This has been driven by a number of factors, including reductions in the associated equipment costs and also an increased awareness of the closely related technique of two-photon microscopy, which shares most of the equipment needs of SHG microscopy. Also of relevance, however, has been the growing awareness of the close interaction between cells and their extracellular environment and the ability of these cells to modulate their phenotype and their protein expression levels in response to cues from the ECM. Given the importance of collagen in forming the ECM and also its ability to generate very high levels of SHG, the use of SHG in biology in general and tissue engineering in particular seems likely to continue to grow in importance.

Of course, SHG imaging is not a universal panacea. Whilst providing excellent label-free contrast for fibrillar collagens, it is not yet established that the signal can match the specificity of immunohistochemistry in terms of, say, differentiating between different isoforms of collagen. The interpretation of the images is also nontrivial as the signal brightness can be influenced by many factors in addition to the local density of collagen, including the degree of spatial ordering, the 3D fibrillar orientation relative to the light polarization direction, the degree of phase matching, and, when working in the epidirection, the degree of elastic light scattering at the emission wavelength.

The numerous unique features of SHG microscopy must be set against these potential limitations. SHG is the only microscopy modality that can noninvasively and nondestructively map the production of collagen in a 3D environment, with a subcellular spatial resolution and with sensitivity not just to the abundance of collagen molecules but also to their microscopic structural arrangement. The technique is also potentially applicable *in vivo* (instrumentation already exists commercially to allow this). An emerging paradigm in tissue engineering is the realization that the body itself is an excellent bioreactor in which to develop artificial constructs such as bone (Stevens et al. 2005), and so monitoring techniques that provide measurements *in vivo* could help to quantify the success of this strategy.

Looking to the near-term future of the technique, systems with significantly reduced cost and more flexible access requirements are not difficult to envisage. While traditionally SHG microscopy is performed on bulky and expensive research microscopes using expensive titanium–sapphire lasers, SHG can also be successfully implemented using lower-cost fiber lasers (Liu et al. 2009). There is an acknowledged need in the tissue engineering community for online monitoring tools, and so systems that are cheap and compact enough to dedicate to a bioreactor facility could offer great benefits in terms of continuous longitudinal studies. *In situ* two-photon microscopy has been demonstrated in a perfusion bioreactor designed for perfused 3D liver culture (Powers et al. 2002), and the extension to SHG imaging should be straightforward.

Chapter 3

In summary, SHG is a tool with a great many benefits to offer to the tissue engineering community. It is hoped that this chapter has helped to demystify the technique for the nonspecialist and will encourage an even greater uptake of the method within the tissue engineering community.

References

Badylak, S.F. 2002. The extracellular matrix as a scaffold for tissue reconstruction. *Semin Cell Dev Biol* 13: 377–383.

Brackmann, C., M. Esguerra, D. Olausson et al. 2011. Coherent anti-Stokes Raman scattering microscopy of human smooth muscle cells in bioengineered tissue scaffolds. *J Biomed Opt* 16(2): 021115.

Brown, E., T. McKee, E. diTomaso et al. 2003. Dynamic imaging of collagen and its modulation in tumors *in vivo* using second-harmonic generation. *Nat Med* 9: 796–800.

Campagnola, P. 2011. Second harmonic generation imaging microscopy: applications to diseases diagnostics. *Anal Chem* 83: 3224–3231.

Caves, J.M., V.A. Kumar, J. Wen et al. 2010. Fibrillogenesis in continuously spun synthetic collagen fiber. *J Biomed Mater Res Part B: Appl Biomater* 93B: 24–38.

Chen, W., C. Huang, L. Chiou et al. 2010. Multiphoton imaging and quantitative analysis of collagen production by chondrogenic human mesenchymal stem cells cultured in chitosan scaffold. *Tissue Eng* 16: 913–920.

Cox, G., E. Kable, A. Jones et al. 2003. 3-Dimensional imaging of collagen using second harmonic generation *J Struc Bio* 141: 53–62.

Deng, Y., X. Lin, Z. Zheng et al. 2003. Poly(hydroxybutyrate-co-hydroxyhexanoate) promoted production of extracellular matrix of articular cartilage chondrocytes *in vitro*. *Biomaterials* 24: 4273–4281.

Evans, C.L., E.O. Potma, M. Puoris'haag et al. 2005. Chemical imaging of tissue *in vivo* with video-rate coherent anti-Stokes Raman scattering microscopy. *PNAS* 102(46): 16807–16812.

Filová, E., Z. Burdíková, M. Rampichová et al. 2010. Analysis and three-dimensional visualization of collagen in artificial scaffolds using nonlinear microscopy techniques. *J Biomed Opt* 15: 066011.

Franceschi, R.T., B.S. Iyer, and Y.Q. Cui. 1994. Effects of ascorbic-acid on collagen matrix formation and osteoblast differentiation in murine MC3T3-E1 cells. *J Bone Miner Res* 9(6): 843–854.

Freund, I., M. Deutsch, and A. Sprecher. 1986. Connective tissue polarity: Optical second-harmonic microscopy, crossed-beam summation, and small-angle scattering in rat-tail tendon. *Biophys J* 50: 693–712.

Fu, L., and M. Gu. 2008. Polarization anisotropy in fiber-optic second harmonic generation microscopy. *Opt Express* 16(7): 5000–5006.

Fu, L., A. Jain, H. Xie et al. 2006. Nonlinear optical endoscopy based on a double-clad photonic crystal fiber and a MEMS mirror. *Opt Express* 14: 1027–1032.

Heider, B., J.L. Nathanson, E.Y. Isacoff et al. 2010. Two-photon imaging of calcium in virally transfected striate cortical neurons of behaving monkey. *PLoS ONE* 5(11): e13829.

Hu, J., J.D. Humphrey, and A.T. Yeh. 2009. Characterization of engineered tissue development under biaxial stretch using nonlinear optical microscopy. *Tissue Eng* 15: 1553–1564.

Huang, L., K. Nagapudi, R. Apkarian et al. 2001. Engineered collagen–PEO nanofibers and fabrics. *J Biomater Sci Polymer Ed* 12(9): 979–993.

Hunziker, E.B. 2001. Articular cartilage repair: basic science and clinical progress. A review of the current status and prospects. *Osteoarthritis Cartilage* 10: 432–463.

Klein, J.A. and D.W.L. Hukins. 1982. Collagen fibre orientation in the annulus fibrosus of intervertebral disc during bending and torsion measured by x-ray diffraction. *Biochim Biophys Acta* 719: 98–101.

Konig, K., and I. Riemann. 2003. High-resolution multiphoton tomography of human skin with subcellular spatial resolution and picosecond time resolution. *J Biomed Opt* 8(3): 432–439.

Lee, C.H., H.J. Shin, I.H. Cho et al. 2005. Nanofiber alignment and direction of mechanical strain affect the ECM production of human ACL fibroblast. *Biomaterials* 26(11): 1261–1270.

Lee, H., S. Teng, H. Chen et al. 2006. Imaging human bone marrow stem cell morphogenesis in polyglycolic acid scaffold by multiphoton microscopy. *Tissue Eng* 12: 2835–2841.

Liu, G., T. Xie, I.V. Tomov et al. 2009. Rotational multiphoton endoscopy with a 1 micron fiber laser system. *Opt Lett* 34(15): 2249–2251.

Mansfield, J.C., C.P. Winlove, J.M. Moger et al. 2008. Collagen fibre arrangement in normal and diseased cartilage studied by polarization sensitive non-linear microscopy. *J Biomed Opt* 13(4): 044020.

Netti, P.A., D.A. Berk, M.A. Swartz et al. 2000. Role of extracellular matrix assembly in interstitial transport in solid tumors. *Cancer Res* 60: 2497–2503.

Plotnikov, S.V., A.C. Millard, P.J. Campagnola et al. 2006. Characterization of the myosin-based source for second-harmonic generation from muscle sarcomeres. *Biophys J* 90: 693–703.

Powers, M.J., K. Domansky, and M.R. Kaazempur-Mofrad. 2002. A microfabricated array bioreactor for perfused 3D liver culture. *Biotechnol Bioeng* 78(3): 257–269.

Raub, C.B., A.J. Putnam, B.J. Tromberg et al. 2010. Predicting bulk mechanical properties of cellularized collagen gels using multiphoton microscopy. *Acta Biomater* 6: 4657–4665.

Rice, W.L., S. Firdous, S. Gupta et al. 2008. Non-invasive characterization of structure and morphology of silk fibroin biomaterials using non-linear microscopy. *Biomaterials* 29(13): 2015–2024.

Richards-Kortum, R., E. Sevick-Muraca. 1996. Quantitative optical spectroscopy for tissue diagnosis. *Annu Rev Phys Chem* 47: 555–606.

Shao, Z., and F. Vollrath. 2002. Surprising strength of silkworm silk. *Nature* 418: 741.

Skala, M.C., K.M. Riching, A. Gendron-Fitzpatrick et al. 2007. *In vivo* multiphoton microscopy of NADH and FAD redox states, fluorescence lifetimes, and cellular morphology in precancerous epithelia. *PNAS* 104: 19494–19499.

Smith, L.E., R. Smallwood and S. MacNeil. 2010. A comparison of imaging methodologies for 3D tissue engineering. *Microsc Res Tech* 73: 1123–1133.

Stamenkovic, I. 2003. Extracellular matrix remodelling: the role of matrix metalloproteinases. *J Pathol* 200: 448–464.

Stevens, M., R.P. Marini, D. Schaefer et al. 2005. *In vivo* engineering of organs: The bone bioreactor. *PNAS* 102(32): 11450–11455.

Stoller, P., K.M. Reiser, P.M. Celliers et al. 2002. Polarization-modulated second harmonic generation in collagen. *Biophys J* 82: 3330–3342.

Su, P., W. Chen, T. Li et al. 2010. The discrimination of type I and type II collagen and the label-free imaging of engineered cartilage tissue. *Biomaterials* 31: 9415–9421.

Sun, Y., H. Tan, S. Lin et al. 2008. Imaging tissue engineering scaffolds using multiphoton microscopy. *Microsc Res Tech* 71: 140–145.

Theodossiou, T.A., C. Thrasivoulou, C. Ekwobi et al. 2006. Second harmonic generation confocal microscopy of collagen type I from rat tendon cryosections. *Biophys J* 91: 4665–4677.

Tiaho, F., G. Recher, and D. Rouède. 2007. Estimation of helical angles of myosin and collagen by second harmonic generation imaging microscopy. *Opt Express* 15(19): 12286.

Yeh, A.T., M.J. Hammer-Wilson, D.C. Van Sickle et al. 2005. Nonlinear optical microscopy of articular cartilage. *Osteoarthritis Cartilage* 13: 345–352.

Zeugolis, D.I.S., T. Khew, E.S.Y. Yew et al. 2008. Electro-spinning of pure collagen nano-fibres—Just an expensive way to make gelatin? *Biomaterials* 29: 2293–2305.

Zipfel, W.R., R.M. Williams, R. Christie et al. 2003. Live tissue intrinsic emission microscopy using multi-photon-excited native fluorescence and second harmonic generation. *PNAS* 100(12): 7075–7080.

Zoumi, A., A. Yeh, and B.J. Tromberg. 2002. Imaging cells and extracellular matrix *in vivo* by using second-harmonic generation and two-photon excited fluorescence. *PNAS* 99(17): 11014–11019.

Chapter 3

4. Two-Photon Polymerization for Tissue-Engineered Scaffold Fabrication

Andrew A. Gill and Frederik Claeyssens

Chapter 4

Optical Techniques in Regenerative Medicine. Edited by Stephen P. Morgan, Felicity R.A.J. Rose, and Stephen J. Matcher © 2014 CRC Press/Taylor & Francis Group, LLC. ISBN: 978-1-4398-5495-2

The production of user-defined 3D microstructures from biocompatible and bio-degradable materials via free-form fabrication opens up a wide range of possibilities for the creation of a new generation of structurally optimized and chemically functionalized tissue scaffolds, as well as other implantable medical devices such as microneedle arrays and patient specific prostheses. In this review we will highlight the production of user-defined tissue engineering scaffolds via laser direct write. This technique utilizes photopolymerization processes (both 1- and 2-photon) to produce (sub)micrometer feature size 3D objects. We will review both (i) the main fabrication processes used and (ii) high impact applications studied by researchers in this emerging field. Common fabrication systems will be discussed as well as the expanding range of materials that may be used. The design of efficient photoinitiators facilitating the use of relatively inexpensive microlaser systems will be highlighted as well as future perspectives for the technology.

4.1 Introduction

4.1.1 Microstereolithography

Microstereolithography refers to a class of rapid prototyping technologies first explored in the 1980s (Doraiswamy et al. 2006). The technique relies on the photocuring of a liquid prepolymer or resin according to a predetermined design to create a solid polymerized object. Typically either a pattern of light is formed by a static or dynamic masking technique and projected into a thin layer of photosensitized prepolymer (Ha et al. 2008) (projection microstereolithography), or a focused laser beam is scanned over the photosensitized polymer according to a computer design, writing out the desired image (scanning microstereolithography) (Gandhi and Deshmukh 2010). When curing has occurred, the remaining liquid prepolymer is washed out by a solvent (developer) leaving the cured pattern intact. To create multilayered objects, consecutive layers of prepolymer are added and cured before the unsolidified polymer or resin is washed out to give a three-dimensional structure.

4.1.2 Microstereolithography in Comparison with Other Techniques

The use of microstereolithography for the fabrication of tissue scaffolds and other medical devices has evolved into a significant research area since its conception as a rapid prototyping technology in the 1980s. One of the main advantages of microstereolithography, as in other direct write techniques for scaffold fabrication, is the ability to create scaffolds with a specific pore geometry and distribution simply by creating a 3D model in computer aided design and manufacturing (CAD-CAM) software. Preexisting technologies for the creation of bulk porous structures such as particulate leaching, gas foaming, freeze drying, and phase separation are all capable of creating bulk structures with a high degree of porosity, but give little control over pore distribution, interconnectivity, and pore geometry, as well as the effect these features have on the bulk scaffold mechanical properties (Coutu et al. 2009).

Layer-by-layer solid free-form fabrication (SFF) techniques such as laser sintering, bioprinting, or bioplotting using nozzle ejection systems provide a method of creating structures with designer features (Hollister 2005). The mechanical performance of scaffolds produced using SFF scaffolds is far superior to that of bulk processed ones. The maximum compressive modulus of scaffolds made by traditional processing techniques such as porogen leaching is around 0.4 MPa (Ma and Choi 2001), as porosity is inversely related to mechanical strength. The mechanical modulus of hard tissue (e.g., bone) is between 10 and 1500 MPa (Goulet et al. 1994; Hollister 2005), and clearly these scaffolds do not possess sufficient strength to match these properties. Fused deposition modeling of polycaprolactone (PCL) was used to produce scaffolds with a high porosity (48%–77%) with compressive moduli ranging from 4 to 77 MPa and yield strength from 2.58 to 3.32 MPa (Hutmacher et al. 2001).

Computational models can be used to determine the optimum scaffold macro to nanostructure with respect to nutrient and metabolite diffusion, cell spreading and attachment, and also scaffold mechanical strength (Chua et al. 2003a,b). The benefit of a computer-designed scaffold architecture that facilitates diffusion within a highly connected network of pores is demonstrated in Figure 4.1, which shows that cells within the microstereolithography produced scaffold are evenly distributed throughout the structure, whereas cells in the comparison scaffold produced by salt leaching are limited to the very surface of the structure.

Another possibility with microstereolithography is the fabrication of patient-specific scaffolds and other medical devices. In one example, a scaled down model of a human

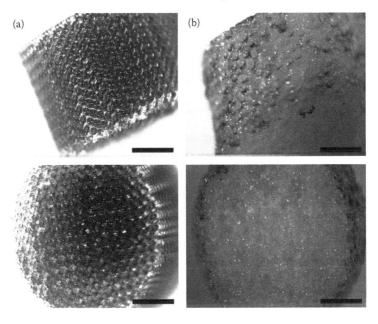

(a)   (b)

FIGURE 4.1   Scaffolds prepared by microstereolithography (a) with gyroid architecture and scaffolds prepared by salt leaching (b) seeded with cells in static culture. Stereomicroscope images were taken of cells stained with methylene blue (appearing here as black). Scale bars are 2 mm. (Adapted from Melchels, F. P. W. et al., *Acta Biomaterialia* 6: 4208–4217, 2010. With permission.)

Chapter 4

kidney was produced using deconstructed medical CT data as a proof of concept (Choi et al. 2009a). The fabrication of scaffolds with custom architecture allows the balancing of mass transport and mechanical support within the structure, which then allows an even distribution of cells to be achieved (Hollister 2005). Rapid prototyping can be viewed as an economic method of producing scaffolds with high reproducibility in terms of microstructure and mechanical properties (Hutmacher et al. 2004).

4.1.3 Two-Photon Polymerization

Microstereolithography systems based on one-photon processes rely on the linear absorption of light by a photoinitiator within the volume of the prepolymer, as described by the Beer–Lambert law. Drawbacks with these systems include overpenetration of light within the photosensitive resin leading to overcuring of layers, light scattering within the resin reducing resolution, and the accumulation of partially cured oligomers within the internal features of the fabricated structures (Choi et al. 2009a). Furthermore, the need to fabricate structures in a layer-by-layer process can lead to "stepping effects," resulting in small ridges on the external features of the structure. Oxygen quenching of radicals on the surface can further complicate the curing process, requiring materials to be purged with nitrogen or argon prior to curing and nitrogen blanketing during the cure step.

To overcome these drawbacks, careful calibration of the cure depth, exposure time, and light intensity is required (see Figure 4.2). Light-absorbing dyes have been used to successfully improve resolution by reducing penetration and scattering (Choi et al. 2009b) however, when preparing implantable devices for regenerative medicine, the effects of these additives on the biocompatibility of the fabricated structures require further investigation. Cure times are also increased due to the absorption of light by the dye, which reduces the curing rate. Purging individual layers between exposures with an inert liquid such as perfluorohexane has also been shown to improve feature resolution by washing away partially cured oligomers (Han et al. 2008), improving internal feature definition but further complicating the fabrication process.

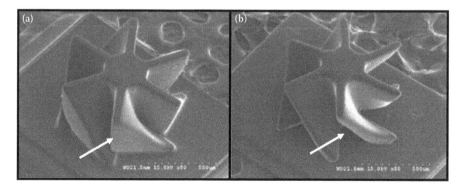

FIGURE 4.2 Effect of cure depth control on resolution in the *z*-plane. The microrotor (a) presents overcuring of the vertical features (indicated by arrow); the structure (b) was fabricated using careful control of cure time to prevent "curing through" of features. (Adapted from Choi, J. W. et al., *Rapid Prototyping Journal* 15: 59–70, 2009. With permission.)

Nonlinear optical processes have been utilized to overcome the problems of typical one-photon lithography methods and were first applied to lithography in 1992 by Wu et al. (1992) with the aim of improving the resolution of UV lithography for micro-electronics applications. These multiphoton processes were first predicted by Maria Goppert-Mayer (1931) and demonstrated by Kaiser and Garrett (1961).

Similar to multiphoton microscopy, two-photon polymerization (TPP) relies on the nonlinear absorption of two photons simultaneously by a photoinitiator to initiate curing processes. Due to the short lifetimes of the intermediate states involved (around 10^{-15} s for true three-dimensional TPP (Lee et al. 2006), as discussed later), initiation only occurs at very high photon density, as predicted by the theory in 1931.

Polymerization is usually achieved by focusing femtosecond pulsed radiation from a titanium:sapphire laser through a high numerical aperture (NA) lens, with initiation occurring at the focal point of the objective. Since polymerization is only initiated at the focal point of the objective, by scanning the focal point within the volume of a photosensitive resin, three-dimensional structures can be written out according to a computer model in a CAD-CAM fashion (Serbin and Chichkov 2003; Stute et al. 2004). The majority of materials used for TPP are the same as those used in UV photolithography due to their availability. Organically modified ceramic resists (ORMOCERS), for example, give excellent structuring results in both UV curing and TPP (Lee et al. 2006). Initially explored for more typical microfabrication applications such as photonics and micro-optics (Ha et al. 2008; Gandhi and Deshmukh 2010; Choi et al. 2009), research using these materials has attracted the attention of tissue engineers. The high resolution of TPP allows the creation of designer tissue scaffolds with defined pore size (Claeyssens et al. 2009), microneedles with reduced penetration cross section (Doraiswamy et al. 2006), microvalves with defined internal structure (Schizas et al. 2010), and a wide range of other exciting possibilities. These will be discussed in detail in Section 4.4.

The choice of photoinitiators has also, until recently, relied on commercially available initiators for one-photon processes such as Michler's ketone (4,4'-bis(N,N-dimethyl-amino)benzophenone), a commonly used UV photoinitiator. Efforts are now under way to create more suitable photoinitiators with an improved two-photon absorption cross section (Lemercier et al. 2006). Although the majority of published work relies on femtosecond-pulsed Ti:sapphire lasers emitting at around 800 nm, impressive results have also been achieved with less powerful laser systems. Q-switched Nd:YAG micro-lasers emitting at 1064 and 532 nm with peak powers between 0.1 and 1 kW, subnanosecond/picosecond pulse durations and repetition rates on the kilohertz scale have also been shown to be capable of initiating TPP (Wang et al. 2002).

The efficiency with which TPP can be initiated is determined by not only the two-photon absorption cross section of the photoinitiator (a measure of the ability of the initiator to absorb two photons simultaneously) but also the ability of the initiator to generate radicals in the excited state. A good initiator will require a low power to efficiently generate polymerization. Additionally, the energy gap between the initiation threshold and the power threshold at which material damage occurs should be as high as possible for reproducible structuring. The design of efficient photoinitiators is highlighted in Section 4.3.

Chapter 4

4.2 Clinical and Biological Measurement Challenges

4.2.1 Introduction

TPP trades fabrication speed for resolution. Whereas typical one-photon scanning systems can have a write speed of up to 500 mm s^{-1}, TPP is approximately a thousand-fold slower at around 1 mm s^{-1} (Stampfl et al. 2008). The benefit of this trade-off is that resolutions of 200–300 nm can be achieved routinely (Serbin et al. 2003), and resolutions as low as 100 nm have been achieved, for example, by using quenchers to prevent polymerization from spreading beyond the focal point (Takada et al. 2005).

Due to its precise nature, the technique is ideally suited to the fabrication of small-scale prototypes and high-resolution microdevices with niche applications. Additionally, the true three-dimensional writing ability of TPP makes it ideal for creating seamless microdevices with complex internal parts, such as microactuation valves (Schizas et al. 2010) and microphotonics (Ostendorf and Chichkov 2006).

As the field of scaffold fabrication by TPP grows, the capabilities of the technique are increasing, due to advances in photoinitiators and lasers (Wang et al. 2002), computer model optimization (Park et al. 2005), parallel processing (Kato et al. 2005), optics and instrument design (Hsieh et al. 2010), and materials (Melchels et al. 2009). The field is now nearing a point where it can grow from a small-scale prototyping technique limited to a narrow range of materials to a state-of-the-art technology that can be exploited for the fabrication of scaffolds with unparalleled resolution fabricated from a diverse range of polymers (Gill and Claeyssens 2011), functionalized biological materials (Ovsianikov et al. 2011b), and hydrogels (Jhaveri et al. 2009).

TPP presents a method for the production of patient-specific devices with unparalleled resolution and is ideal in situations in which cells require contact guidance or a precise local environment in order to promote functional recovery. One example of this is in nerve entubulation devices or nerve guidance conduits (NGCs), where microstructure has been demonstrated to enhance functional recovery of transected nerves (Hsu and Ni 2009). Microstereolithography and TPP may be used to optimize and produce NGCs with performance superior to simple hollow tube devices (Melissinaki et al. 2011).

4.2.2 Tissue Scaffolds and Extracellular Matrix

A particular challenge that TPP is suited to address is that of tissue scaffold fabrication (Hsieh et al. 2010). Tissue scaffolds generally provide an artificial extracellular matrix (ECM) in which cells can proliferate and grow, forming an engineered tissue (Langer and Vacanti 1993). Additionally, these scaffolds may be made from a biodegradable material that breaks down as the cells form their own ECM, eventually disappearing completely (Atala et al. 2006). The purpose of the scaffold is to provide mechanical protection for the growing tissue, as well as to provide a general structure for cell attachment (Bellucci et al. 2011). The three-dimensional microenvironment of the tissue scaffold is also believed to affect the behavior and proliferation of the cells within, with cells cultured in a 2D monolayer fashion not behaving as closely to those in normal living tissue compared to their 3D scaffold cultured counterparts (Abbott 2003).

It has been demonstrated that different cell types prefer different internal pore geometries, and so in creating custom scaffolds, the pore shape must be tailored to the desired tissue type. Furthermore, different cell types have different optimum pore sizes; for example, bone cells (osteoblasts) function optimally in scaffolds with a pore size of around 350 µm, and liver cells (hepatocytes) prefer smaller pores of around 20 µm and fibroblasts 5–15 µm (Whang et al. 1999).

Typical bulk scaffold fabrication methods consist of techniques such as particulate leaching or gas blowing to create a porous foam. Scaffolds fabricated in this way tend to perform poorly, with cell ingrowth being confined to the first 200 µm of the scaffold below the surface (Awwad et al. 1986; Colton 1995). Additionally, these scaffolds typically do not give sufficient mechanical strength for load-bearing applications such as bone scaffolds, or even soft tissues (Hollister 2005). Mass transport is also poor due to low pore interconnectivity (Melchels et al. 2010).

Stereolithography (including TPP) is an ideal technique for addressing these issues. Pore geometry and the overall scaffold porosity can be changed simply by changing the computer model, and furthermore, medical imaging data such as computerized tomography (CT) and magnetic resonance imaging (MRI) may be converted directly to CAD models in order to produce patient- and injury-specific scaffolds (Seol et al. 2012).

4.2.3   Angiogenesis and Vascularization

A key problem facing the tissue engineering community is scaffold vascularization, the formation of a vascular network within scaffolds to distribute cells and nutrients (Wang et al. 2012). The scaffold porosity must be high enough to allow sufficient fluid flow for nutrient transport and waste metabolite removal (Bettahalli et al. 2011); however, if the porosity is too high, the scaffold mechanical properties break down. Using microfabrication techniques, a network of vessels may be written into produced scaffolds allowing for the formation of a vascular network postimplantation (Kaully et al. 2009).

4.3   Experimental Implementation of the Method

4.3.1   Introduction

TPP can achieve higher resolution three-dimensional structuring than any commercially competing technology [however, experimental techniques such as two-color lithography can demonstrate better resolution (Scott et al. 2009)]. According to Abbe's diffraction limit, the resolution of a focused laser is limited by the wavelength of light used and the NA of the focusing objective, preventing one-photon–based stereolithography techniques from achieving a submicrometer resolution. Other high-resolution techniques such as e-beam lithography are limited to surface effects, unlike TPP, which is an in-volume technique (Farsari et al. 2010). Although there are exceptions such as carbon vapor deposition (Van Dorp and Hagen 2008), these techniques are not well suited to scaffold fabrication due to limitations, for example, on the materials that can be used. Sub-100 nm three-dimensional structuring with TPP has been achieved (Haske et al. 2007).

4.3.2 Initiation of Polymerization Processes

Two-photon absorption occurs by two mechanisms: sequential and simultaneous absorption of two photons. In sequential absorption, one photon is absorbed by the excited species promoting the absorber to a real excited state. A second photon is then absorbed within the lifetime of this excited species (10^{-4} to 10^{-9} s) (Lee et al. 2006). The presence of the real intermediate state would require the material to absorb at this wavelength, and hence, absorption would be governed by the Beer–Lambert law (Farsari et al. 2010). TPP, on the other hand, is driven by the simultaneous absorption of two photons of light by the photoinitiator. As there is no real intermediate state, the material is transparent to light of the wavelength of the exciting radiation; two photons must arrive within the lifetime of the virtual excited state (10^{-15} s) to initiate polymerization. The short lifetime of the virtual state means that high intensities of light are required. This is typically achieved with a femtosecond-pulsed Ti:sapphire laser. Ti:sapphire lasers are favored because of their short pulse length, which limits thermal damage or burning of the sample, and because these lasers typically emit light at around 800 nm, which is appropriate for the two-photon excitation of a wide range of UV-sensitive photoinitiators. Most UV initiators and curable materials are also transparent at this wavelength, allowing light to be focused within the volume of the material and interact only with the desired initiator via two-photon mechanisms.

The two-photon absorption cross section (σTPA) of an initiator determines the suitability of an initiator for TPP and describes the ability of the initiator to absorb two photons simultaneously. Molecular engineering of photoinitiators seeks to maximize the two-photon absorption cross section by conjugating electron donating (D) or electron accepting (A) groups to a π charge transfer system in a symmetric sequence (D–π–A–π–D), allowing stabilization of the intermediate state by charge delocalization. Unsymmetrical "push–pull" molecules, for example, a D–π–A system, also enhance the two-photon susceptibility of the molecule, increasing the two-photon absorption cross section (Baldeck et al. 2010; Gan et al. 2009). The development of better photoinitiators is an important step in making TPP systems more available and reducing the dependence on expensive femtosecond-pulsed laser systems.

Another important factor for efficient photoinitiators is their ability to generate radicals in the excited state. Initiator–coinitiator systems that facilitate radical transfer from initiator to polymer have been explored (Belfield et al. 2000) as well as molecules that (1) have improved absorption cross sections and (2) contain amine groups that facilitate radical generation (Kuebler et al. 2001). Two-photon sensitive initiators for Ti:sapphire lasers emitting at around 800 nm have seen extensive optimization (Lemercier et al. 2006). Additionally, initiators tuned to the operating wavelengths of Q-switched Nd:YAG microlasers (532 and 1064 nm) have been synthesized to facilitate efficient structuring with lasers of much lower specification and cost, and these lasers are expected to become a routine alternative to more powerful femtosecond Ti:Sapphire systems (Baldeck et al. 2010).

4.3.3   Experimental Setups for TPP

The CAD-CAM nature of TPP makes it an attractive technique for tissue scaffold fabrication, allowing a diverse range of scaffold architectures to be explored simply by changing the computer model. Typical setups combine a femtosecond-pulsed Ti:sapphire laser emitting in the region of 800 nm with a Galvano-scanner controlled by CAD-CAM software, which moves the focus of the beam within the horizontal plane of a high NA objective. Vertical stepping is most commonly achieved using a piezoelectric stage. Online monitoring may be achieved by focusing a CCD camera through the focusing system using a dichroic mirror to image the curing process within the volume of the resin. Since most materials undergo a change of refractive index upon curing, the polymerization process can be visualized by the appearance of the structure within the polymer. A schematic of a typical TPP system is shown in Figure 4.3.

A three-dimensional computer model of the desired structure is generated with modeling software. This model is then cut into a series of horizontal slices, which are written out in a layer-by-layer fashion by scanning the beam within the volume of the resin, with the height of the substrate being adjusted by a high-precision piezoelectric stage. Either the entire volume of the structure is cured in the writing step (raster scanning) or only the outline is cured in the writing step (contour scanning) (Lee et al. 2006), and the inner volume of the structure is cured in a postprocessing step using one-photon curing (Sun and Kawata 2003).

4.3.4   Materials

Early research into the use of TPP in areas such as photonics (Ostendorf and Chichkov 2006) required materials that gave good structuring results without placing too many requirements on the specific chemical nature of the resulting structure beyond physical robustness. Well-characterized photoresist materials used in conventional lithographic

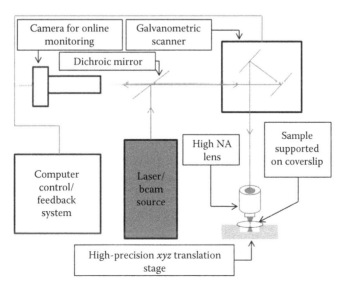

FIGURE 4.3   Simplified schematic of a typical TPP system showing key components of the device.

Chapter 4

techniques, such as Ormocer and SU8, fulfilled these demands and have been used in a wide range of applications. The demonstrated biocompatibility (Doraiswamy et al. 2005) of Ormocer saw it used in a range of medical studies as discussed in Section 4.4. Additionally, these materials exhibit low shrinkage (Winfield and O'Brien 2011), a problem often encountered with small molecule resins (Farsari et al. 2006), which can reduce final structure quality.

The drawback with these materials for the creation of tissue scaffolds is that they are not bioresorbable (biodegradable) and, once implanted into the body, will remain there permanently or until removed, requiring follow-up surgery.

Bioresorbable, photocurable polymers based on well-characterized polymeric biomaterials such as polylactic acid (PLA), PCL, polyglycolic acid (PGA), and trimethylene carbonate (TMC), among others, have been developed (Kwon and Matsuda 2005; Mizutani and Matsuda 2002a,b; Matsuda et al. 2004; Bat et al. 2011; Melchels et al. 2010, 2011; Melchels et al. 2009; Pego et al. 2003; Seck et al. 2010; Claeyssens et al. 2009). The common feature of these polymers is the inclusion of hydrolyzable ester bonds between monomers, which allow them to degrade following implantation. Typically a low molecular weight oligomer is prepared from a multiarmed initiator, which serves as the core for ring opening polymerization with the selected monomer using stannous octoate catalyst (Gill and Claeyssens 2011). Photo cross-linking end groups such as methacrylates or coumarin are then added to make the oligomers photocurable. Microwave-assisted synthesis has recently been employed in the preparation of polyethylene glycol (PEG)–based resorbable hydrogels incorporating poly(lactic acid) (PLA) segments and methacrylate end groups, presenting a rapid and easily accessible method of preparing photocurable materials with a diverse range of monomers (Seck et al. 2010). Structuring of photocurable gelatin has also recently been reported using methacrylate functionalization (Ovsianikov et al. 2011b).

Another important class of materials commonly used for tissue engineering purposes are hydrogels (Lee and Mooney 2001; Shin et al. 2012). These soft polymers are attractive scaffold materials due to their biocompatibility and tunable cell adhesiveness (Halstenberg et al. 2002; Luo and Shoichet 2004). Photocurable hydrogel materials such as PEG diacrylate (PEGDA) are amenable to direct structuring in aqueous media (Jhaveri et al. 2009) or photo cross-linking via one-photon curing to create a cross-linked network followed by three-dimensional patterning of photolabile chemical cues, which influence the behavior of cells contained within the gel (Luo and Shoichet 2004).

A range of techniques have been investigated for patterning hydrogels such as one-photon lithography (Yu and Ober 2003), microfluidic patterning (Tan and Desai 2003), and 3D printing (Mironov et al. 2003), among other additive layer technologies (Mironov et al. 2008). These layer-by-layer processing techniques give the desired control over features of individual layers; however, in the vertical or "Z" plane, the coherence between individual layers is poor (Luo and Shoichet 2004). In this respect, there is an clear advantage of using "in-volume" two-photon direct-write processes over these competing technologies.

Hydrogels allow the encapsulation of live cells into scaffolds. For example, in the study by Lee et al. (2008), a photocurable and degradable acrylate–PEG–(peptide-PEG)$_n$–acrylate hydrogel was prepared and used to encapsulate human dermal fibroblasts by cross-linking via one-photon curing. The encapsulated cells were then soaked with a solution containing acrylate conjugated PEG-RGDSK (Arg-Gly-Asp-Ser-Lys, a cell adhesive ligand), and the two-photon laser scanning technique was used to selectively conjugate the cell adhesive ligand throughout the matrix according to a predetermined design. This technique was demonstrated to guide cell migration throughout the hydrogel. Similar techniques have been applied in order to pattern growth factors and other proteins within hydrogel constructs (Wylie et al. 2011; Wylie and Shoichet 2011) and influence stem cell differentiation.

4.3.5   Improving the Efficiency and Cost of TPP

Barriers to the more mainstream adoption of TPP as a routine research technique include the costs involved, mainly resulting from the high-power laser systems required, and additionally the small scale of the structures that can be fabricated due to the focusing range of the types of optics used and the low scan speed required for curing with appropriate voxel dimensions.

Many TPP systems use oil immersion lenses due to their high NA and precise focusing. Moving from oil lenses to air lenses overcomes the height limitations imposed by the need to dip the lens within a small drop of oil, as demonstrated in the study by Hsieh et al. (2010). Structures fabricated with oil lenses are limited to a height of about 1 mm (the focal length of the objective, which cannot be lifted from the oil droplet); however, using an air lens, the authors claim a scan height of 30 mm. A scan speed of 30 mm s^{-1} is also reported along with a resolution of around 100 nm. Finally, the authors produced a cubic porous scaffold applied to hepatocyte culture with dimensions of 2.5 mm^3, which was produced in around 2 h, a feasible timescale for routine scaffold manufacture.

A range of optical components and techniques have also been investigated for improving the processing time required for TPP. Parallel processing using refractive and diffractive optics to split the beam into an array of smaller beams has been demonstrated (Kato et al. 2005; Winfield et al. 2007a), and up to 227 structures have been fabricated simultaneously in one fabrication cycle (Bhuian et al. 2007). Axicon lenses have also been explored as a way of generating three-dimensional shapes in a single exposure (Winfield et al. 2007b; Bhuian et al. 2007).

A common method for creating structures from a 3D computer model via TPP is the single-dimensional scanning method (SSM) (Park et al. 2005). A 3D digital model of the object to be fabricated is converted to a polyhedral one and then sliced into a series of parallel layers along a plane defined by the user. These layers are then converted into a scanning path for the laser. The slowest fabrication method is to write out the entire volume of the structure using this map (raster scanning). A more efficient way is to only write out the solid outer shell or "contours" of the structure [contour scanning method (CSM)], also known as vector scanning (Sun and Kawata 2003), solidifying the internal volume of the structure postdevelopment by exposure, for example, to UV light. This method reduces the number of individual voxels (volume pixels), which make up the structure. As demonstrated by Sun and Kawata (2003), a "microbull" structure composed of 2×10^6 voxels was fabricated by raster scanning in a time of 3 h. The outer shell of the bull could be recreated accurately using only 5% of these voxels, and using the CSM, the bull was recreated in only 13 min.

Chapter 4

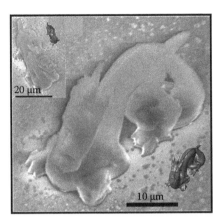

FIGURE 4.4　Microdragon fabricated using TSM technique, showing 3D model (inset). (Adapted from Liao, C. Y. et al., *Applied Physics Letters* 91: 033108, 2007. With permission.)

Contour scanning has been further improved to compensate for difficulties in creating near-flat surfaces within polymerized objects. If an object created by contour scanning is not sealed sufficiently before development, the outer shell can tear and release the inner unpolymerized resin, leading to collapse of the structure. As the fabricated surface approaches horizontal, careful stepping of the layers is required to create sufficient overlap, preventing breakage. This requires a large number of voxels and leads to very long fabrication times. A two-dimensional slicing method (TSM) has been developed to simplify the scan paths of structures with near-flat surfaces (Liao et al. 2007). In one example, a polyhedral model of a microdragon sliced by SSM consists of 871,414 polygons, but by TSM, it is simplified down to 16,620 polygons. Using TSM, the fabrication took 19 min, while it is estimated that using SSM, it would have taken 12.66 h. The fabricated dragon is shown in Figure 4.4.

4.4　Example Applications

4.4.1　Introduction

As discussed in earlier sections, microstereolithography is well suited to the production of tissue scaffolds with optimum pore geometry with respect to nutrient and metabolite diffusion. User-defined scaffold structuring via TPP allows the investigation and production of scaffolds, which can be tailored to the ideal pore size and geometry of a given tissue simply by adjusting the 3D model in a CAD-CAM software suite. It has been demonstrated that microscale features can have a profound effect on cell behavior (Chen et al. 1997; Dike et al. 1999), as well as cell migration and scaffold mechanical properties. Additionally, features such as vasculature for nutrient/metabolite transport and features that can give mechanical strength to the structure can be written directly into the scaffold.

Beyond the development of tissue scaffolds, the technique is finding applications in patient-specific prostheses and geometrically optimized implantable devices such as microneedles and cell delivery vehicles, which will be discussed below.

4.4.2   Implantable Medical Devices Created by TPP

Initial work into the fabrication of permanent scaffolds from off-the-shelf materials such as ORMOCER® and SU8 assessed the biocompatibility of these materials (Schlie et al. 2007; Ovsianikov et al. 2007d) and demonstrated cell growth on small-scale scaffold-like structures. The suitability of these and similar materials for two-photon–based microstructuring and their demonstrated biocompatibility and cell adhesiveness led to a number of publications regarding their use in small-scale medical devices (Schizas et al. 2010; Doraiswamy et al. 2006; Ovsianikov et al. 2007c; Ovsianikov et al. 2007b).

4.4.2.1   Prosthesis

Ovsianikov et al. (2007a) used TPP of ORMOCER for the fabrication of ossicular replacement prostheses, implantable devices that are intended to improve hearing by reconstructing ossicles, which are structures located within the inner ear that may be damaged by disease (Albu et al. 1998). This application requires structures that exhibit good stability and mechanical properties, retaining their structure throughout the lifetime of the implant. The material must be nontoxic and cell adhesive, and the device must have good acoustic transmission. Preexisting surgical options include reshaping of autologous inner ear bone tissue (Colletti and Fiorino 1999), which is not always available due to the extent of the damage, or cadaveric donor tissue, now disfavored due to the risk of infection (Glasscock et al. 1988) or degradation of stiffness. Mass-produced artificial implants do not take into account the variability of individual patient anatomy. The authors demonstrated that TPP may be used to generate easily implanted devices, which can easily be tailored to the needs of a particular patient.

4.4.2.2   Microneedles

ORMOCER Microneedle arrays fabricated by TPP have been explored for applications such as transdermal drug delivery (Ovsianikov et al. 2007c). The CAD-CAM nature of TPP processing allows for total control over features such as surface area, penetration cross section, mechanical strength, and diffusion within the needles simply by changing the computer model (Doraiswamy et al. 2006; Ovsianikov et al. 2007c; Ovsianikov et al. 2007b). Comparable microcone arrays fabricated using one-photon based layer-by-layer microstereolithography fabricated by Kwon and Matsuda (2005) show visible ridges due to the curing of individual layers consecutively, an effect that the high resolution of two-photon direct write eliminates.

Microneedle arrays fabricated by TPP have also been used for the delivery of fluorescent quantum dots to porcine skin, facilitating imaging by multiphoton microscopy (Gittard et al. 2011). Material biocompatibility was assessed by the proliferation of neonatal human epidermal keratinocytes and human dermal fibroblasts. The authors indicate that TPP microneedle fabrication allows the delivery of theranostic agents (agents that can indicate the correct therapeutic process for individual patients) to epidermal, dermal, or subdermal tissues depending on the computer design in a patient-specific manner.

One concern with implantable microneedle arrays is the risk of infection. Microneedle arrays prepared by TPP and replicated by PDMS stamping were prepared from a mixture

Chapter 4

of PEGDA and 2 mg/mL of the antimicrobial agent gentamicin sulfate. The efficacy of this agent in inhibiting the growth of the pathogen *Staphylococcus aureus* was demonstrated by agar plating assay (Gittard et al. 2010).

4.4.2.3   Valves

Prototype valves optimized for the flow conditions encountered in small human veins were fabricated by Schizas et al. (2010) from a zirconium sol-gel–based material. Consecutive layers were cured by TPP to create a 360 μm long by 120 μm wide microvalve. The stair step effect created by layer-by-layer fabrication was eliminated by a combination of feature design to include mainly vertical features and also the optimization of laser fluence to give the best possible material resolution. This study demonstrates how an optimized computer designed model can be converted directly into a prototype by TPP.

4.4.2.4   Bioresorbable Tissue Scaffolds

Despite the success of ORMOCER and similar materials for the fabrication of devices such as microneedle arrays and valves, these materials are limited to the production of permanent devices as they are not biodegradable. Claeyssens et al. (2009) reported the successful fabrication of scaffold-like structures fabricated from a bioresorbable polymer based on poly(ε-caprolactone-co-trimethylenecarbonate)-b-poly(ethylene glycol)-b-poly(ε-caprolactone-co-trimethylenecarbonate). This polymer was reported to degrade at a rate close to that of tissue formation (Mizutani and Matsuda 2002a,b) and was demonstrated to have no negative effect on cell viability.

4.4.2.5   Cell Delivery Vehicles

The injection of stem cells into damaged brain tissue has been suggested as a method for treating brain injury, for example, following a stroke. The use of degradable microparticles that can be injected with the cells and provide a support structure has been shown to facilitate tissue regrowth (Bible et al. 2009). TPP of photocurable PLA has been used to create degradable cell delivery vehicles, which can be loaded with cells prior to injection with the possibility of optimizing the shape and size of the cell delivery vehicles to give maximum protection to the cells during injection, and giving maximum cell loading of the microparticles (Melissinaki et al. 2011).

4.4.3   Experimental Tissue Scaffolds

As well as implantable devices, the resolution and reproducibility of TPP scaffold fabrication allow the production of experimental tissue scaffolds, which may be used to investigate the effect of factors such as three-dimensional culture, pore size, and geometry on the behavior of cells within.

In a work by Tayalia et al. (2008), TPP-fabricated constructs were used to study the effect of scaffold architecture on cell adhesion and migration. Woodpile-based structures with different pore sizes were seeded with cells, and the movement of the cells within was monitored over time to investigate cell migration. It was found that cells reached higher speeds of migration in the three-dimensional constructs compared to a flat substrate. Furthermore, it was found that smaller pore sizes inhibited the movement of the cells, leading to a less uniform distribution, whereas cells moved faster and distributed more evenly in the scaffolds with the largest pores. The ratio of the different oligomers within the curing mixture was varied, allowing control over the mechanical properties of the scaffold, making it possible to investigate the effect of scaffold elasticity on cell behavior in structurally identical scaffolds.

Cell migration studies involving natural matrix materials such as matrigel and ECM proteins derived from tissue have been extensively performed (Friedl and Wolf 2003; Grinnell et al. 2006), as have studies involving traditional porous scaffold fabrication techniques such as gas foaming, particulate leaching, and phase inversion (Hutmacher 2001). Drawbacks with these studies include the variability of pore size within fabricated scaffolds and the highly cell-adhesive nature of natural matrix materials, limiting cell migration. TPP constructs allow greater control over the material properties of the scaffold and the ability to define pore size and shape.

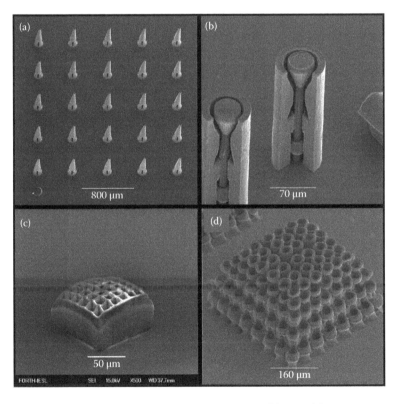

FIGURE 4.5 Example applications of TPP. (a) Microneedle array fabricated from ORMOCER (Adapted from Ovsianikov, A. et al., *Applied Surface Science* 253: 6599–6602, 2007. With permission.), (b) valve fabricated using two-photon methods (Adapted from Schizas, C. et al., *International Journal of Advanced Manufacturing Technology* 48: 435–441, 2010. With permission.), (c) open pore geometry scaffold fabricated from PEGDA (Adapted from Ovsianikov, A. et al., *Acta Biomaterialia* 7: 967–974, 2011. With permission.), and (d) bioresorbable cell scaffold fabricated in PEG-PCL-TMC copolymer. (Reprinted with permission from Claeyssens et al. (2009), 3219–3223. Copyright 2009, American Chemical Society.)

Chapter 4

Hsieh et al. (2010) used TPP to produce three-dimensional scaffolds from a commercially available nondegradable polymer for hepatocyte culture and demonstrated that liver-specific cell functions (urea synthesis and urea secretion) were retained longer in cells cultured in the scaffold compared to cells cultured on spin-coated substrates of the same polymer. This demonstrates the effect of culturing cells in three-dimensional constructs instead of a flat monolayer fashion.

Although nondegradable, PEG has been used extensively in internal medical applications such as constipation aids (Tack 2011). PEGDA was investigated as a suitable material for the creation of TPP structured scaffolds (Ovsianikov et al. 2011b). A resolution of 200 nm was achieved, and the authors suggest that the technique may be used to explore the effect of surface topography on cell–scaffold interaction. Furthermore, a thorough investigation of scaffold material/photoinitiator toxicity was carried out, and it was determined that freshly prepared scaffolds release photoinitiator and monomer material, which is toxic to fibroblasts; however, aging of the samples in distilled water for 6 days led to much improved biocompatibility and negligible cytotoxicity. The swelling of the scaffold materials in water also suggests that although the devices may deviate slightly from their original shape, the ensuing increase in porosity will allow greater nutrient transport within the scaffold, and the scaffolds could be loaded with biological agents such as growth factors, as shown in previous studies highlighted by the authors (Gittard et al. 2010).

4.5 Conclusions

TPP is becoming an established technology for the creation of designer tissue scaffolds. Scaffolds with appropriate dimensions for the engineering of tissues and organs that can be fabricated on a reasonable timescale have been produced (Hsieh et al. 2010). Smaller-scale devices for niche applications have been fabricated, demonstrating the potential of this technology for direct clinical applications (Ovsianikov et al. 2007a).

More efficient photoinitiators (Lemercier et al. 2006) and new materials for microstereolithography (Melchels et al. 2009) combined with new slicing techniques for 3D scan path optimization (Liao et al. 2007), voxel elongation (Li et al. 2009), and parallel processing (Bhuian et al. 2007) are being developed to reduce the reliance of the technique on high-end laser systems while reducing processing times and limitations on the size of scaffold that can be achieved. Demonstrated advantages of custom scaffold features (Melchels et al. 2011) over first-generation scaffolds made by simple bulk processing methods make microstereolithography, particularly TPP, an attractive technique for the creation of a new generation of functional tissue scaffolds.

References

Abbott, A. 2003. Cell culture: Biology's new dimension. *Nature* 424: 870–872.

Albu, S. G., Babighian, G. and Trabalzini, F. 1998. Prognostic factors in tympanoplasty. *American Journal of Otology* 19: 136–140.

Atala, A., Bauer, S. B., Soker, S., Yoo, J. J. and Retik, A. B. 2006. Tissue-engineered autologous bladders for patients needing cystoplasty. *The Lancet* 367: 1241–1246.

Awwad, H. K., El Naggar, M., Mocktar, N. and Barsoum, M. 1986. Intercapillary distance measurement as an indicator of hypoxia in carcinoma of the cervix uteri. *International Journal of Radiation Oncology Biology Physics* 12: 1329–1333.

Baldeck, P. L., Stephan, O. and Andraud, C. 2010. Tri-dimensional micro-structuration of materials by two-photon induced photochemistry. In *Basics and Applications of Photopolymerization Reactions*, eds. J. P. Fouassier and X. Allonas, 199–221: Research Signpost, India.

Bat, E., van Kooten, T. G., Jan, F. J. and Grijpma, D. W. 2011. Resorbable elastomeric networks prepared by photocrosslinking of high-molecular-weight poly(trimethylene carbonate) with photoinitiators and poly(trimethylene carbonate) macromers as crosslinking aids. *Acta Biomaterialia* 7: 1939–1948.

Belfield, K. D., Ren, X. B., Van Stryland, E. W., Hagan, D. J., Dubikovsky, V. and Miesak. E. J. 2000. Near-IR two-photon photoinitiated polymerization using a fluorone/amine initiating system. *Journal of the American Chemical Society* 122: 1217–1218.

Bellucci, D., Cannillo, V., Cattini, A. and Sola, A. 2011. A new generation of scaffolds for bone tissue engineering. *Industrial Ceramics* 31: 59–62.

Bettahalli, N. M. S., Steg, H., Wessling, M. and Stamatialis, D. 2011. Development of poly(L-lactic acid) hollow fiber membranes for artificial vasculature in tissue engineering scaffolds. *Journal of Membrane Science* 371: 117–126.

Bhuian, B., Winfield, R. J., O'Brien, S. and Crean, G. M. 2007. Pattern generation using axicon lens beam shaping in two-photon polymerisation. *Applied Surface Science* 254: 841–844.

Bible, E., Chau, D. Y. S., Alexander, M. R., Price, J., Shakesheff, K. M. and Modo, M. 2009. Attachment of stem cells to scaffold particles for intra-cerebral transplantation. *Nature Protocols* 4: 1440–1453.

Chen, C. S., Mrksich, M., Huang, S., Whitesides, G. M. and Ingber, D. E. 1997. Geometric control of cell life and death. *Science* 276: 1425–1428.

Choi, J. W., Wicker, R. B., Lee, S. H., Choi, K. H., Ha, C. S. and Chung, I. 2009a. Fabrication of 3D biocompatible/biodegradable micro-scaffolds using dynamic mask projection microstereolithography. *Journal of Materials Processing Technology* 209: 5494–5503.

Choi, J. W., Wicker, R. B., Cho, S. H., Ha, C. S. and Lee, S. H. 2009b. Cure depth control for complex 3D microstructure fabrication in dynamic mask projection microstereolithography. *Rapid Prototyping Journal* 15: 59–70.

Chua, C. K., Leong, K. F., Cheah, C. M. and Chua, S. W. 2003a. Development of a tissue engineering scaffold structure library for rapid prototyping. Part 1: Investigation and classification. *International Journal of Advanced Manufacturing Technology* 21: 291–301.

Chua, C. K., Leong, K. F., Cheah, C. M. and Chua, S. W. 2003b. Development of a tissue engineering scaffold structure library for rapid prototyping. Part 2: Parametric library and assembly program. *International Journal of Advanced Manufacturing Technology* 21: 302–312.

Claeyssens, F., Hasan, E. A., Gaidukeviciute, A., Achilleos, D. S., Ranella, A., Reinhardt, C., Ovsianikov, A., Xiao, S., Fotakis, C., Vamvakaki, M., Chichkov, B. N. and Farsari, M. 2009. Three-dimensional biodegradable structures fabricated by two-photon polymerization. *Langmuir* 25: 3219–3223.

Colletti, V. and Fiorino, F. G. 1999. Malleus-to-footplate prosthetic interposition: Experience with 265 patients. *Otolaryngology—Head and Neck Surgery* 120: 437–444.

Colton, C. K. 1995. Implantable biohybrid artificial organs. *Cell Transplantation* 4: 415–436.

Coutu, D. L., Yousefi, A. M. and Galipeau, J. 2009. Three-dimensional porous scaffolds at the crossroads of tissue engineering and cell-based gene therapy. *Journal of Cellular Biochemistry* 108: 537–546.

Dike, L. E., Chen, C. S., Mrksich, M., Tien, J., Whitesides, G. M. and Ingber, D. E. 1999. Geometric control of switching between growth, apoptosis, and differentiation during angiogenesis using micropatterned substrates. *In Vitro Cellular & Developmental Biology-Animal* 35: 441–448.

Doraiswamy, A., Patz, T., Narayan, R., Chichkov, B., Ovsianikov, A., Houbertz, R., Modi, R., Auyeung, R. and Chrisey, D. B. 2005. Biocompatibility of CAD/CAM ORMOCER polymer scaffold structures. In *Nanoscale Materials Science in Biology and Medicine*, eds. C. T. Laurencin and E. A. Botchwey, 51–56: Cambridge Univ. Press, UK.

Doraiswamy, A., Jin, C., Narayan, R. J., Mageswaran, P., Mente, P., Modi, R., Auyeung, R., Chrisey, D. B., Ovsianikov, A. and Chichkov, B. 2006. Two photon induced polymerization of organic-inorganic hybrid biomaterials for microstructured medical devices. *Acta Biomaterialia* 2: 267–275.

Farsari, M., Filippidis, G., Sambani, K., Drakakis, T. S. and Fotakis, C. 2006. Two-photon polymerization of an Eosin Y-sensitized acrylate composite. *Journal of Photochemistry and Photobiology a-Chemistry* 181: 132–135.

Chapter 4

Farsari, M., Vamvakaki, M. and Chichkov, B. N. 2010. Multiphoton polymerization of hybrid materials. *Journal of Optics* 12: 124001–124016.

Friedl, P. and Wolf, K. 2003. Tumour-cell invasion and migration: Diversity and escape mechanisms. *Nature Reviews Cancer* 3: 362–374.

Gan, X. P., Zhou, H. P., Shi, P. F., Wang, P., Wu, J. Y., Tian, Y. P., Yang, J. X., Xu, G. B., Zhou, Y. F. and Jiang, M. H. 2009. A new series of two-photon polymerization initiators: Synthesis and nonlinear optical properties. *Science in China Series B-Chemistry* 52: 2180–2185.

Gandhi, P. S. and Deshmukh, S. 2010. A 2D optomechanical focused laser spot scanner: analysis and experimental results for microstereolithography. *Journal of Micromechanics and Microengineering* 20: 015035–015047.

Gill, A. A. and Claeyssens, F. 2011. 3D structuring of biocompatible and biodegradable polymers via stereolithography. In *3D Cell Culture*, ed. J. W. Haycock, 309–321: Springer Protocols, Clifton, N.J.

Gittard, S. D., Ovsianikov, A., Akar, H., Chichkov, B., Monteiro-Riviere, N. A., Stafslien, S., Chisholm, B., Shin, C. C., Shih, C. M., Lin, S. J., Su, Y. Y. and Narayan, R. J. 2010. Two photon polymerization-micromolding of polyethylene glycol-gentamicin sulfate microneedles. *Advanced Engineering Materials* 12: B77–B82.

Gittard, S. D., Miller, P. R., Boehm, R. D., Ovsianikov, A., Chichkov, B. N., Heiser, J., Gordon, J., Monteiro-Riviere, N. A. and Narayan, R. J. 2011. Multiphoton microscopy of transdermal quantum dot delivery using two photon polymerization-fabricated polymer microneedles. *Faraday Discussions* 149: 171–185.

Glasscock, M. E. III, Jackson, C. G. and Knox, G. W. 1988. Can acquired immunodeficiency syndrome and Creutzfeldt–Jakob disease be transmitted via otologic homografts? *Archives of Otolaryngology—Head and Neck Surgery* 114: 1252–1255.

Goppert-Mayer, M. 1931. Elementary file with two quantum fissures. *Annalen Der Physik* 9: 273–294.

Goulet, R. W., Goldstein S. A., Ciarelli, M. J., Kuhn, J. L., Brown, M. B. and Feldkamp, L. A. 1994. The relationship between the structural and orthogonal compressive properties of trabecular bone. *Journal of Biomechanics* 27: 375–389.

Grinnell, F., Rocha, L. B., Iucu, C., Rhee, S. and Jiang, H. M. 2006. Nested collagen matrices: A new model to study migration of human fibroblast populations in three dimensions. *Experimental Cell Research* 312: 86–94.

Ha, Y., Choi, J. and Lee, S. 2008. Mass production of 3-D microstructures using projection microstereolithography. *Journal of Mechanical Science and Technology* 22: 514–521.

Halstenberg, S., Panitch, A., Rizzi, S., Hall, H. and. Hubbell, J. A. 2002. Biologically engineered protein-graft-poly(ethylene glycol) hydrogels: A cell adhesive and plasmin-degradable biosynthetic material for tissue repair. *Biomacromolecules* 3: 710–723.

Han, L. H., Mapili, G., Chen, S. and Roy, K. 2008. Projection microfabrication of three-dimensional scaffolds for tissue engineering. *Journal of Manufacturing Science and Engineering* 130: 021005.

Haske, W., Chen, V. W., Hales, J. M., Dong, W. T., Barlow, S., Marder, S. R. and Perry, J. W. 2007. 65 nm feature sizes using visible wavelength 3-D multiphoton lithography. *Optics Express* 15: 3426–3436.

Hollister, S. J. 2005. Porous scaffold design for tissue engineering. *Nature Materials* 4: 518–524.

Hsieh, T. M., Wei, C., Ng, B., Narayanan, K., Wan, A. C. A. and Ying, J. Y. 2010. Three-dimensional microstructured tissue scaffolds fabricated by two-photon laser scanning photolithography. *Biomaterials* 31: 7648–7652.

Hsu, S. H. and Ni, H. C. 2009. Fabrication of the microgrooved/microporous polylactide substrates as peripheral nerve conduits and *in vivo* evaluation. *Tissue Engineering Part A* 15: 1381–1390.

Hutmacher, D. W. 2001. Scaffold design and fabrication technologies for engineering tissues—State of the art and future perspectives. *Journal of Biomaterials Science-Polymer Edition* 12: 107–124.

Hutmacher, D. W., Schantz, T., Zein, I., Ng, K. W., Teoh, S. H. and Tan, K. C. 2001. Mechanical properties and cell cultural response of polycaprolactone scaffolds designed and fabricated via fused deposition modeling. *Journal of Biomedical Materials Research* 55: 203–216.

Hutmacher, D. W., Sittinger, M. and Risbud, M. V. 2004. Scaffold-based tissue engineering: rationale for computer-aided design and solid free-form fabrication systems. *Trends in Biotechnology* 22: 354–362.

Jhaveri, S. J., McMullen, J. D., Sijbesma, R., Tan, L. S., Zipfel, W. and Ober, C. K. 2009. Direct three-dimensional microfabrication of hydrogels via two-photon lithography in aqueous solution. *Chemistry of Materials* 21 :2003–2006.

Kaiser, W. and Garrett, C. G. B. 1961. 2-Photon excitation in CAF2-EU2+. *Physical Review Letters* 7: 229–231.

Kato, J., Takeyasu, N., Adachi, Y., Sun, H. B. and Kawata, S. 2005. Multiple-spot parallel processing for laser micronanofabrication. *Applied Physics Letters* 86: 044102.

Kaully, T., Kaufman-Francis, K., Lesman, A. and Levenberg, S. 2009. Vascularization—The conduit to viable engineered tissues. *Tissue Engineering Part B-Reviews* 15: 159–169.

Kuebler, S. M., Rumi, M., Watanabe, T., Braun, K., Cumpston, B. H., Heikal, A. A., Erskine, L. L., Thayumanavan, S., Barlow, S., Marder, S. R. and Perry, J. W. 2001. Optimizing two-photon initiators and exposure conditions for three-dimensional lithographic microfabrication. *Journal of Photopolymer Science and Technology* 14: 657–668.

Kwon, I. K. and Matsuda, T. 2005. Photo-polymerized microarchitectural constructs prepared by microstereolithography (mu SL) using liquid acrylate-end-capped trimethylene carbonate-based prepolymers. *Biomaterials* 26: 1675–1684.

Langer, R. and Vacanti, J. P. 1993. Tissue engineering. *Science* 260: 920–926.

Lee, K. S., Yang, D. Y., Park, S. H. and Kim, R. H. 2006. Recent developments in the use of two-photon polymerization in precise 2D and 3D microfabrications. *Polymers for Advanced Technologies* 17: 72–82.

Lee, K. Y. and Mooney, D. J. 2001. Hydrogels for tissue engineering. *Chemical Reviews* 101: 1869–1879.

Lee, S. H., Moon, J. J. and West, J. L. 2008. Three-dimensional micropatterning of bioactive hydrogels via two-photon laser scanning photolithography for guided 3D cell migration. *Biomaterials* 29: 2962–2968.

Lemercier, G., Martineau, C., Mulatier, J. C., Wang, I., Stephan, O., Baldeck, P. L. and Andraud, C. 2006. Analogs of Michler's ketone for two-photon absorption initiation of polymerization in the near infrared: synthesis and photophysical properties. *New Journal of Chemistry* 30: 1606–1613.

Li, X. F., Winfield, R. J., O'Brien, S. and Crean, G. M. 2009. Application of Bessel beams to 2D microfabrication. *Applied Surface Science* 255: 5146–5149.

Liao, C. Y., Bouriauand, M., Baldeck, P. L., Leon, J. C., Masclet, C. and Chung, T. T. 2007. Two-dimensional slicing method to speed up the fabrication of micro-objects based on two-photon polymerization. *Applied Physics Letters* 91: 033108.

Luo, Y. and Shoichet, M. S. 2004. A photolabile hydrogel for guided three-dimensional cell growth and migration. *Nature Materials* 3: 249–253.

Ma, P. X. and Choi, J. W. 2001. Biodegradable polymer scaffolds with well-defined interconnected spherical pore network. *Tissue Engineering* 7: 23–33.

Matsuda, T., Kwon, I. K. and Kidoaki, S. 2004. Photocurable biodegradable liquid copolymers: Synthesis of acrylate-end-capped trimethylene carbonate-based prepolymers, photocuring, and hydrolysis. *Biomacromolecules* 5: 295–305.

Melchels, F. P. W., Feijen, J. and Grijpma, D. W. 2009. A poly(D,L-lactide) resin for the preparation of tissue engineering scaffolds by stereolithography. *Biomaterials* 30: 3801–3809.

Melchels, F. P. W., Barradas, A. M. C., van Blitterswijk, C. A., de Boer, J., Feijen, J. and Grijpma, D. W. 2010. Effects of the architecture of tissue engineering scaffolds on cell seeding and culturing. *Acta Biomaterialia* 6: 4208–4217.

Melchels, F. P. W., Tonnarelli, B., Olivares, A. L., Martin, I., Lacroix, D., Feijen, J., Wendt, D. J. and Grijpma, D. W. 2011. The influence of the scaffold design on the distribution of adhering cells after perfusion cell seeding. *Biomaterials* 32: 2878–2884.

Melissinaki, V., Gill, A. A., Ortega, I., Vamvakaki, M., Ranella, A., Haycock, J. W., Fotakis, C., Farsari, M. and Claeyssens, F. 2011. Direct laser writing of 3D scaffolds for neural tissue engineering applications. *Biofabrication* 3: 045005.

Mironov, V., Boland, T., Trusk, T., Forgacs, G. and Markwald, R. R. 2003. Organ printing: Computer-aided jet-based 3D tissue engineering. *Trends in Biotechnology* 21: 157–161.

Mironov, V., Kasyanov, V., Drake, C. and Markwald, R. R. 2008. Organ printing: Promises and challenges. *Regenerative Medicine* 3: 93–103.

Mizutani, M. and Matsuda, T. 2002a. Liquid photocurable biodegradable copolymers: *In vivo* degradation of photocured poly(epsilon-caprolactone-co-trimethylene carbonate). *Journal of Biomedical Materials Research* 61: 53–60.

Mizutani, M. and Matsuda, T. 2002b. Photocurable liquid biodegradable copolymers: *In vitro* hydrolytic degradation behaviors of photocured films of coumarin-endcapped poly(epsilon-caprolactone-co-trimethylene carbonate). *Biomacromolecules* 3: 249–255.

Ostendorf, A. and Chichkov, B. N. 2006. Two-photon polymerization: A new approach to micromachining. *Photonics Spectra* 40: 72.

Ovsianikov, A., Chichkov, B., Adunka, O., Pillsbury, H., Doraiswamy, A. and Narayan, R. J. 2007a. Rapid prototyping of ossicular replacement prostheses. *Applied Surface Science* 253: 6603–6607.

Chapter 4

Ovsianikov, A., Chichkov, B., Mente, P., Monteiro-Riviere, N. A., Doraiswamy, A. and Narayan, R. J. 2007b. Two photon polymerization of polymer–ceramic hybrid materials for transdermal drug delivery. *International Journal of Applied Ceramic Technology* 4: 22–29.

Ovsianikov, A., Ostendorf, A. and Chichkov, B. N. 2007c. Three-dimensional photofabrication with femtosecond lasers for applications in photonics and biomedicine. *Applied Surface Science* 253: 6599–6602.

Ovsianikov, A., Schlie, S., Ngezahayo, A., Haverich, A. and Chichkov, B. N. 2007d. Two-photon polymerization technique for microfabrication of CAD-designed 3D scaffolds from commercially available photosensitive materials. *Journal of Tissue Engineering and Regenerative Medicine* 1: 443–449.

Ovsianikov, A., Deiwick, A., Van Vlierberghe, S., Dubruel, P., Moeller, L., Draeger, G. and Chichkov, B. 2011a. Laser fabrication of three-dimensional CAD scaffolds from photosensitive gelatin for applications in tissue engineering. *Biomacromolecules* 12: 851–858.

Ovsianikov, A., Malinauskas, M., Schlie, S., Chichkov, B., Gittard, S., Narayan, R., Lobler, M., Sternberg, K., Schmitz, K. P. and Haverich, A. 2011b. Three-dimensional laser micro- and nano-structuring of acrylated poly(ethylene glycol) materials and evaluation of their cytoxicity for tissue engineering applications. *Acta Biomaterialia* 7: 967–974.

Park, S. H., Lee, S. H., Yang, D. Y., Kong, H. J. and Lee, K. S. 2005. Subregional slicing method to increase three-dimensional nanofabrication efficiency in two-photon polymerization. *Applied Physics Letters* 87: 154108.

Pego, A. P., Vleggeert-Lankamp, C., Deenen, M., Lakke, E., Grijpma, D. W., Poot, A. A., Marani, E. and Feijen, J. 2003. Adhesion and growth of human Schwann cells on trimethylene carbonate (co)polymers. *Journal of Biomedical Materials Research Part A* 67A: 876–885.

Schizas, C., Melissinaki, V., Gaidukeviciute, A., Reinhardt, C., Ohrt, C., Dedoussis, V., Chichkov, B. N., Fotakis, C., Farsari, M. and Karalekas, D. 2010. On the design and fabrication by two-photon polymerization of a readily assembled micro-valve. *International Journal of Advanced Manufacturing Technology* 48: 435–441.

Schlie, S., Ngezahayo, A., Ovsianikov, A., Fabian, T., Kolb, H. A., Haferkamp, H. and Chichkov, B. N. 2007. Three-dimensional cell growth on structures fabricated from ORMOCER (R) by two-photon polymerization technique. *Journal of Biomaterials Applications* 22: 275–287.

Scott, T. F., Kowalski, B. A., Sullivan, A. C., Bowman, C. N. and McLeod, R. R. 2009. Two-color single-photon photoinitiation and photoinhibition for subdiffraction photolithography. *Science* 324: 913–917.

Seck, T. M., Melchels, F. P. W., Feijen, J. and Grijpma, D. W. 2010. Designed biodegradable hydrogel structures prepared by stereolithography using poly(ethylene glycol)/poly(D,L-lactide)-based resins. *Journal of Controlled Release* 148: 34–41.

Seol, Y. J., Kang, T. Y. and Cho, D. W. 2012. Solid freeform fabrication technology applied to tissue engineering with various biomaterials. *Soft Matter* 8: 1730–1735.

Serbin, J. and Chichkov, B. N. 2003. Three-dimensional nanostructuring by two-photon polymerization of hybrid materials. In *Nanotechnology*, eds. R. Vajtai, X. Aymerich, L. B. Kish and A. Rubio, 571–576. Bellingham: Spie-Int Soc Optical Engineering.

Serbin, J., Egbert, A., Ostendorf, A., Chichkov, B. N., Houbertz, R., Domann, G., Schulz, J., Cronauer, C., Fröhlich, L. and Popall, M. 2003. Femtosecond laser-induced two-photon polymerization of inorganic organic hybrid materials for applications in photonics. *Optics Letters* 28: 301–303.

Shin, H., Olsen, B. D. and Khademhosseini, A. 2012. The mechanical properties and cytotoxicity of cell-laden double-network hydrogels based on photocrosslinkable gelatin and gellan gum biomacromolecules. *Biomaterials* 33: 3143–3152.

Stampfl, J., Baudis, S., Heller, C., Liska, R., Neumeister, A., Kling, R., Ostendork, A. and Spitzbart, M. 2008. Photopolymers with tunable mechanical properties processed by laser-based high-resolution stereolithography. *Journal of Micromechanics and Microengineering* 18: 125014.

Stute, U., Serbin, J., Kulik, C. and Chichkov, B. N. 2004. Three-dimensional micro- and nanostructuring with two-photon polymerisation. *International Journal of Materials and Product Technology* 21: 273–284.

Sun, H. B. and Kawata, S. 2003. Two-photon laser precision microfabrication and its applications to micro-nano devices and systems. *Journal of Lightwave Technology* 21: 624–633.

Tack, J. 2011. Current and future therapies for chronic constipation. *Best Practice & Research Clinical Gastroenterology* 25: 151–158.

Takada, K., Sun, H. B. and Kawata, S. 2005. Improved spatial resolution and surface roughness in photopolymerization-based laser nanowriting. *Applied Physics Letters* 86: 071122.

Tan, W. and Desai, T. A. 2003. Microfluidic patterning of cellular biopolymer matrices for biomimetic 3-D structures. *Biomedical Microdevices* 5: 235–244.

Tayalia, P., Mendonca, C. R., Baldacchini, T., Mooney, D. J. and Mazur, E. 2008. 3D cell-migration studies using two-photon engineered polymer scaffolds. *Advanced Materials* 20: 4494–4498.

Van Dorp, W. F. and Hagen, C. W. 2008. A critical literature review of focused electron beam induced deposition. *Journal of Applied Physics* 104: 081301.

Wang, I., Bouriau, M., Baldeck, P. L., Martineau, C. and Andraud, C. 2002. Three-dimensional microfabrication by two-photon-initiated polymerization with a low-cost microlaser. *Optics Letters* 27: 1348–1350.

Wang, Z. Y., He, Y. Z., Yu, X. D., Fu, W., Wang, W. and Huang, H. M. 2012. Rapid vascularization of tissue-engineered vascular grafts *in vivo* by endothelial cells in co-culture with smooth muscle cells. *Journal of Materials Science-Materials in Medicine* 23: 1109–1117.

Whang, K., Healy, K. E., Elenz, D. R., Nam, E. K., Tsai, D. C., Thomas, C. H., Nuber, G. W., Glorieux, F. H., Travers, R. and Sprague, S. M. 1999. Engineering bone regeneration with bioabsorbable scaffolds with novel microarchitecture. *Tissue Engineering* 5: 35–51.

Winfield, R. J. and O'Brien, S. 2011. Two-photon polymerization of an epoxy-acrylate resin material system. *Applied Surface Science* 257: 5389–5392.

Winfield, R. J., Bhuian, B., O'Brien, S. and Crean, G. M. 2007a. Fabrication of grating structures by simultaneous multi-spot fs laser writing. *Applied Surface Science* 253: 8086–8090.

Winfield, R. J., Bhuian, B., O'Brien, S. and Crean, G. M. 2007b. Refractive femtosecond laser beam shaping for two-photon polymerization. *Applied Physics Letters* 90: 111115.

Wu, E. S., Strickler, J. H., Harrell, W. R. and Webb, W. W. 1992. 2-Photon Lithography for Microelectronic Application. In *Optical/Laser Microlithography V, Pts 1 and 2.* ed. J. D. Cuthbert. 776–782. SPIE, USA.

Wylie, R. G., Ahsan, S., Aizawa, Y., Maxwell, K. L., Morshead, C. M. and Shoichet, M. S. 2011. Spatially controlled simultaneous patterning of multiple growth factors in three-dimensional hydrogels. *Nature Materials* 10: 799–806.

Wylie, R. G. and Shoichet, M. S. 2011. Three-dimensional spatial patterning of proteins in hydrogels. *Biomacromolecules* 12: 3789–3796.

Yu, T. Y. and Ober, C. K. 2003. Methods for the topographical patterning and patterned surface modification of hydrogels based on hydroxyethyl methacrylate. *Biomacromolecules* 4: 1126–1131.

Chapter 4

5. *In Situ* Microscopy

Andreas Prediger, Patrick Lindner, Arne Bluma,
Kenneth F. Reardon, and Thomas Scheper

Chapter 5

Optical Techniques in Regenerative Medicine. Edited by Stephen P. Morgan, Felicity R.A.J. Rose, and
Stephen J. Matcher © 2014 CRC Press/Taylor & Francis Group, LLC. ISBN: 978-1-4398-5495-2

5.1   Introduction

Stem cell biology and regenerative science have recently become major fields of interest in modern medicine and biotechnology. A key goal in these areas is the ability to control the proliferation and differentiation of stem cells into desired somatic cells and tissue. Reliable and noninvasive sensor technologies for monitoring of these processes have yet to be developed.

Optical imaging techniques have already been successfully applied for monitoring of various mammalian cell cultivations as well as cultivation processes of unicellular organisms. Since imaging probes can be directly integrated into bioreactors, measurements are possible under sterile conditions, while the measurement process itself remains noninvasive. Similar techniques might be used as monitoring tools in regenerative science.

In this chapter, several optical imaging techniques for direct monitoring of cell culture processes are presented. These systems are capable of *in situ* measurements of various cultivation variables like cell number concentration, cell size distribution, and morphological properties. Image and video acquisition directly from the cultivation process is possible without any risk of contamination.

To demonstrate the potential of these monitoring techniques, several examples of the application of *in situ* microscopy are presented. They include monitoring of different cell cultures, examination of mechanical properties of enzyme carriers, and characterization of crystallization processes. In addition, image processing algorithms used to process the acquired image data are described.

5.2   *In Situ* Imaging Possibilities in Regenerative Science and Medicine

Embryonic and induced pluripotent stem cells play key roles in the development of regenerative therapies as they have the capacity of unlimited proliferation and the ability to differentiate into cell types of all three germ layers. Moreover, they are interesting in the pharmaceutical industry for drug screening and drug target identification (Martin 1981; Thomson et al. 1998).

For the development and application of new therapies, large amounts of stem cells are needed. For example, it is estimated that 1 to 2 billion stem cell–derived cardiomyocytes would be required for a patient after myocardial infarct (Zweigerdt 2009). Thus, culture systems for the generation of these cells under controlled and reproducible conditions are necessary. As stem cells react very sensitively to changes in their environment, monitoring of these systems is crucial. Differentiation processes into desired cell types also require sophisticated monitoring.

In situ imaging techniques, which have proven to be effective for monitoring various biotechnological cultivation processes, could be used for monitoring tasks in regenerative medicine. Their online and noninvasive measuring type makes sampling superfluous and gives insight into processes in real time.

Several *in situ* imaging sensors and analysis software are described in the next section. Finally, application examples of *in situ* microscopy for monitoring of biotechnological processes are presented. Sections 5.3.5.2 and 5.4.3 are of special interest as they describe the monitoring of cell cultures grown on microcarriers. This is a system that is already used for the cultivation of embryonic stem (ES) cells (Fernandes et al. 2007; Abranches et al. 2007).

5.3 *In Situ* Imaging Sensors for Bioprocess Monitoring

Several optical imaging sensors have been developed for different biotechnological applications. In this section, the measuring principles as well as their application ranges and physical properties are described.

5.3.1 Particle Vision and Measurement

Mettler Toledo developed the particle vision and measurement (PVM) probe as a mobile device for in-line characterization of particles as they occur within pharmaceutical and other industrial processes. The illumination system consists of six to eight lasers (class 1 laser diodes, $\lambda = 905$ nm) that are annularly arranged in angles of 60° or 45° around the objective tubus. Six corresponding lenses focus the light through a sapphire window on an adjustable focal plane in the process medium. The remaining two lasers may be reflected off of an optional diffusing backplate, providing lighting for transmission-based imaging. A charge coupled device (CCD) camera is used for the acquisition of images. The system setup is shown in Figure 5.1.

Currently available models include the PVM V819, which has a probe length of 400 mm and a probe diameter of 19 mm, and the PVM V825 Ex with a wetted probe length of 300 mm and a probe diameter of 25 mm. The PVM V825 Ex is designed for installation in pipelines or reactor vessels in industrial process environments. The recordable field of view has an area of 1075 × 825 μm, and particles from approximately 2 μm to 1 mm can be detected. Images are acquired at a rate of 10 frames per second (fps). The PVM has already been applied to in-line monitoring of crystallization processes, characterization of multiphase systems (Barrett and Glennon 2002), and recrystallization processes (O'Sullivan et al. 2003).

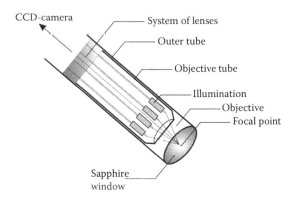

FIGURE 5.1   Schematic of the PVM system. (From Bluma, A. et al., *Analytical and Bioanalytical Chemistry* 398:2429–2438, 2010.)

Chapter 5

5.3.2 Particle Image Analyzer

The particle image analyzer (PIA) detector allows the analysis of the size and morphology of dispersed crystals and particles. It is commercially available from Sequip S+E GmbH and consists of an in-line video microscope and modular image analysis software. These optional modules allow the adaption to various processes. Images are taken with an integrated CCD camera. Background illumination is achieved via fiber optics. A schematic of the system is shown in Figure 5.2.

Images can be acquired at a rate of 25 fps and particles sized 10–500 µm can be measured. Integrated polarization filters allow the analysis of translucent crystals. The illumination system can be equipped with optional color filters. The PIA has been used successfully for the observation of crystals and the determination of their shape. Moreover, it has been used to monitor the influence of additives on crystallization processes (Qu et al. 2006). There are several systems in modular configurations, and versions for reactors from 5 mL up to process scale available.

5.3.3 Envirocam

The Envirocam system, developed by Enviroptics, is a modular *in situ* measurement device that is capable of real-time observation of various processes. The B series models feature back lighting illumination of the measuring zone and are designed for observation of particles in clear liquids or gases. A schematic of the system setup is shown in Figure 5.3.

The F series models have front lighting illumination for the analysis of opaque liquids. All models are designed to operate under harsh process conditions such as high temperature and high pressure. The height of the measuring zone is not adjustable, but different prototypes with measuring zone heights between 6.35 and 25.4 mm have been developed. The control software is compatible with the LabVIEW software platform.

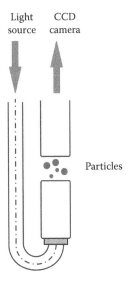

FIGURE 5.2 Schematic of the PIA system.

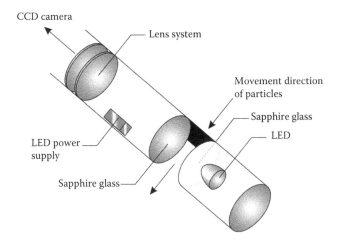

FIGURE 5.3   Measurement principle of the Envirocam probe. The particles are detected while passing through the measuring zone.

The Envirocam system can be used for applications such as droplet analysis and corrosion observation. It has also been used for the characterization of bioreactor systems. Bubble sizes and distributions were measured to analyze oxygen transfer rates under different conditions. Bubbles and particles sizing from 30 to 3.125 mm have been investigated (Junker et al. 2007).

5.3.4   *In Situ* Microscope

5.3.4.1   Instrument Description

The concept of the integration of microscopes into bioreactors, known as *in situ* microscopy, was first described by Suhr et al. in 1991. Since then, various systems have been described (Bittner et al. 1998; Camisard et al. 2002; Joeris et al. 2002; Suhr et al. 1995), and the concept for the most advanced instrument was first described by Frerichs et al. in 2002. This *in situ* microscope (ISM) is a transmitted-light bright-field microscope with finite optics. Depending upon the application, objectives with optical magnifications from 4- to 20-fold can be utilized. Objects with sizes between 0.5 and 400 µm can be monitored with this system. This ISM has been applied to various biotechnological applications (see Section 5.4) in our research group. The patent on the device is currently owned by Sartorius Stedim Biotech GmbH, Goettingen, Germany.

The ISM consists of two segments. The lower segment is autoclavable and consists of the microscope body as well as the flow-through measuring zone. This segment can be integrated into bioreactors via a 25 mm port. The measuring zone is located at the end of the microscope and is immersed into the process medium. Two sapphire glass (artificial Al_2O_3 crystal) slides that are positioned orthogonally in the optical path limit the measuring zone. The size of the measuring zone is variable up to 5 mm at a resolution of 1.25 µm. The illumination system is located under the measuring zone and consists of a light-emitting diode (LED) and a condenser lens.

The upper segment of the ISM consists of two linear translation stages that are used to change the height of the measuring zone as well as to change the position of the objective tube in order to focus the system. Both linear stages are driven by stepping

Chapter 5

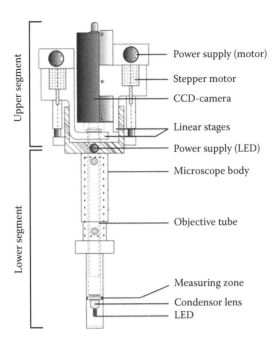

FIGURE 5.4 Schematic of the ISM. (From Bluma, A. et al., *Analytical and Bioanalytical Chemistry* 398:2429–2438, 2010.)

motors. Different CCD cameras can be connected to this segment for image acquisition. A scheme of the entire system is shown in Figure 5.4.

Specially developed control software allows the control of the LED, the stepping motors, and the CCD camera over a single user interface. Different image processing algorithms for the analysis of acquired image data have been developed (see Section 5.3.5).

The measurement principle of this ISM is that the monitored objects are continuously transported through the measuring zone by convection. The assumption for all of the image processing algorithms is that the composition of the process medium in the measuring zone is the same as in the entire reactor. Thus, process parameters such as object size or object density can be calculated directly from the gathered images.

5.3.4.2 ISM Modification for Flow Analytical Applications

A requirement to use the ISM described in the previous section is the presence of a 25 mm standard port through which the microscope can be integrated into the process reactor. These ports often do not exist in small bioreactors, which makes the technique unsuitable for small-scale cultivations and screening applications.

To overcome this limitation, an at-line flow-microscope system for the monitoring of parallel cultivations was developed (Rehbock et al. 2010). The microscope system is based on the ISM, but flow-through measurements are enabled by housing the lower segment of the microscope in a flow cell. This allows the integration of the microscope into flow-injection analysis (FIA) systems, in which small sample volumes are forwarded to the detectors by pumps. A schematic of this flow-through setup is shown in Figure 5.5.

The combination of the modified ISM and FIA systems enables new applications for microscopy. In comparison to the standard ISM instruments, higher cell densities can

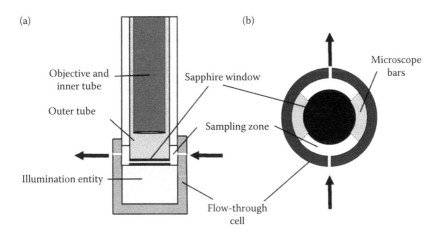

FIGURE 5.5 Schematic of the flow-through cell: (a) lateral view and (b) top view. (From Rehbock, C. et al., *Journal of Biotechnology* 150:87–93, 2010.)

be analyzed as dilution steps can be easily integrated into the FIA system. Moreover, dyes for viability studies (e.g., trypan blue, neutral red, methylene blue) can be injected with the culture sample so that the viability of cell populations can be automatically detected with image processing algorithms.

This at-line flow microscope has also been combined with other sensors to form a multianalyte sensing device (Akin et al. 2011). A simplified setup of the device is shown in Figure 5.6.

For the monitoring of pO_2, a fiber optic sensor (Presens GmbH, Regensburg, Germany) embedded in a flow cell is used. Ethanol and glucose concentrations are

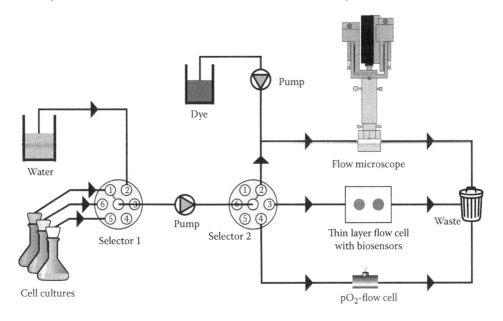

FIGURE 5.6 Simplified schematic of the multianalyte sensing device. Samples from parallel cultivations are forwarded to detection systems with pumps. Dyes can be added to the carrier stream for microscopic analytics.

detected via amperometric measurements with a dual biosensor in a thin layer flow cell. The sensing element of the biosensor consists of separately immobilized alcohol oxidase and pyranose oxidase. The cell density was calculated directly from the data generated with the flow microscope. Results from monitoring a *Saccharomyces cerevisiae* cultivation with the system are shown in Figure 5.7. Ethanol production, glucose consumption, growth of yeast cells, and the decrease of oxygen concentration can be observed.

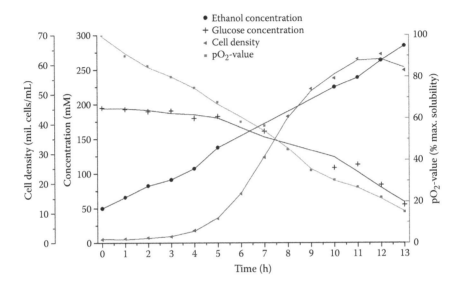

FIGURE 5.7   Monitoring of an *S. cerevisiae* cultivation with the multianalyte sensing system. Cell density was measured with a flow microscope and corresponding image analysis software, pO_2 values were measured via a fiber optic sensor and ethanol, and glucose concentrations were measured using a dual biosensor flow cell. (From Akin, M. et al., *Biosensors and Bioelectronics* 26:4532–4537, 2011.)

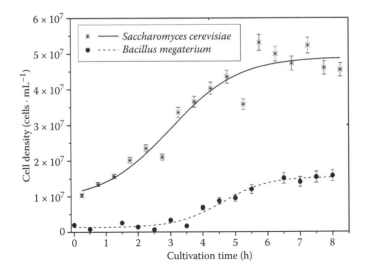

FIGURE 5.8   Results from parallel monitoring of cell densities in cultivations of *S. cerevisiae* and *B. megaterium*. (From Rehbock, C. et al., *Journal of Biotechnology* 150:87–93, 2010.)

Another application is the monitoring of parallel cultivation processes with different microorganisms. In Figure 5.8, the results of a parallel monitoring experiment with *S. cerevisiae* and *Bacillus megaterium* are presented.

Both cultivations showed the typical growth behavior of microorganisms. The *S. cerevisiae* culture grew faster than the *B. megaterium* cells, which experienced a lag phase of about 3.5 h. The *S. cerevisiae* culture reached a final cell density of 5×10^7 cells/ mL, while the *B. megaterium* culture reached 1.5×10^7 cells/mL.

5.3.5   Image Analysis and Processing of *In Situ* Microscopy Images

5.3.5.1   Algorithm for Evaluation of Yeast Cell Cultivations

The image analysis algorithm for yeast cell cultivations was developed with the Delphi programming language (Version 7) from Borland. Cell images must be acquired in a defocused fashion to be suitable for analysis with this algorithm. In this case, yeast cells act as lenses and focus light in the middle of the cell. In the ISM images, each yeast cell has a white kernel and a dark edge, compared to the background, as can be seen in Figure 5.9, which shows cells from three different time steps during cultivation. The term "kernel" is used here to denote the white area inside a yeast cell and not the cell kernel or nucleus in a biological sense.

The algorithm generates the following result values from an input image: number of objects (if two or more cells form a cluster, this cluster is treated as one object), number of pixels in objects that contain cells, number of cells, number of single cells, number of double cells, average pixels per cell, and cells per milliliter. It also classifies the cells into the following categories: small, medium, or large single cell; double cell; cell cluster; and no classification possible. Linder et al. (2007) describe the algorithm in detail.

First the mode value—the most common gray value in the image—is calculated. Then each pixel of the original image that differs from the mode value by at least the gray value threshold (the first parameter of the algorithm) is marked with a flag. In this way, all cells are detected.

Of course, noncell objects that differ significantly from the background and background noise are also marked during this procedure. In the next step, objects are sorted

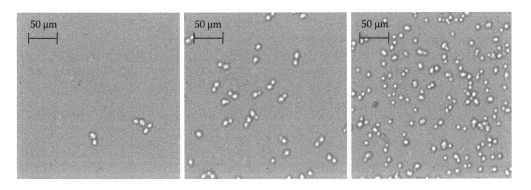

FIGURE 5.9   Images from a yeast cell cultivation after 2, 8.5, and 17.5 h. All images have been collected in a defocused manner so that yeast cells have a white kernel and a dark edge. This simplifies digital image processing.

Chapter 5

into the categories cell object and noncell object according to their properties. To accomplish this, the image is examined with a border detection algorithm. The image is scanned from left to right and row by row until a marked pixel is found, at which point the border detection algorithm is started. The rules of this algorithm are as follows (starting with the direction "right"):

- On a marked pixel, turn left and advance one step.
- On an unmarked pixel, turn right and advance one step.
- After three turns in the same direction, ignore the other rules, turn to the other direction, and advance one step.

This principle is illustrated in Figure 5.10. Within each step, the pixel is marked as a border pixel. When the start location is reached again, all pixels inside this border are marked with an object flag. Then the scanning process starts again until another marked pixel is found and so on. In this way, all pixels that belong to an object are marked (different flags are used each time so that the objects become distinguishable). By counting the number of object flags for an object, its size can be determined. Objects that are smaller than a size threshold value (second parameter of the algorithm) are eliminated.

After these steps have been accomplished, the number of cells in each object is calculated. This is done by scanning the inside area of an object for pixels that are lighter than the mode value by at least the white kernel threshold value (the third parameter of the algorithm). When a kernel pixel has been found, the scanning process is interrupted, and the kernel border and the inside area are detected by the same algorithm that was used to find the cells in the image (see above).

In the final step, the information contained in the flags is transformed into the result values stated above:

- The number of objects is given by the number of different object flags in the image.
- The number of cells is given by the number of white kernels found in the objects.
- The average cell area is given by the total number of object flags divided by the number of cells.
- The cell concentration (cells/mL) is calculated from the cell number and the sampling zone volume of the ISM.
- The classification of the cells into small, medium, and large cells is done by using size threshold values, which are parameters of the algorithm.

To verify the results of the algorithm, 700 images of a yeast cell cultivation were evaluated manually. The number of cells per image was counted several times to minimize counting errors. A comparison of the manual count with the algorithm result is shown in Figure 5.11. The image acquisition during this cultivation was done in 35 measurement cycles with 20 images per cycle. Each dot represents the mean value of the cell number of all 20 images of a cycle. As seen in Figure 5.11, the match between the manual and algorithm cell count was very good. With the algorithm described in this section, 100 images (512×512 pixel, 8-bit gray scale) can be analyzed in 11 s (Intel Pentium 4 CPU with 3.2 GHz, 1 MB DDR-SDRAM and Windows XP). The precision and speed of this yeast cell counting algorithm make it a very useful tool in the evaluation of yeast cultivations.

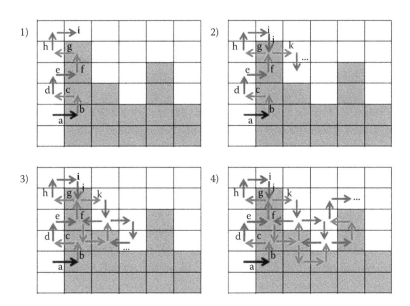

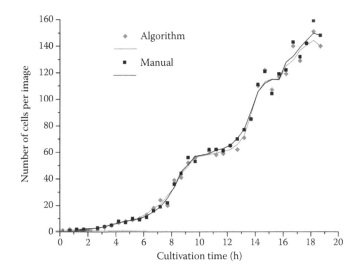

FIGURE 5.10   Illustration of the border detection algorithm. The gray squares are background and the white squares are an example object. Starting at location (1) (black arrow), the algorithm surrounds an object of arbitrary shape as demonstrated with the gray arrows.

FIGURE 5.11   Comparison of the manual count with the ISM algorithm result (number of cells per image versus cultivation time) for a sample from a yeast cultivation.

5.3.5.2   Analysis of Microcarrier-Based Cultivations by Image Analysis and Neural Networks

To analyze microcarrier-based cultivations, an algorithm was developed that consists of standard analysis methods for image segmentation and a neural network to interpret and postprocess the results. The first step of the analysis procedure is the detection of a microcarrier on the image. With the previously described procedure,

Chapter 5

all gray values in the raw image that belong to a microcarrier object can be separated from the background and from other objects. For this separation, a reasonable starting point is the borderline of a carrier because there is an easily recognizable gray value difference that is not dependent on the level of colonization (LOC). The method that was chosen for object segmentation was the convolution of the raw image with the Sobel operator. In Figure 5.12, an example picture shows the results of calculating gray value gradients with this method. There are high gradient values not only on the microcarrier border and in the inside area but also from other small objects and from background noise.

To simplify the following steps in the algorithm and to reduce the influence of noise and nonmicrocarrier objects, all small gradient values are eliminated by thresholding. All values in the Sobel-processed image lower than the threshold value were set to zero. In the next step, a border-tracking algorithm is applied to the Sobel-processed image generating a list of objects and the coordinates of all image pixels that belong to each object. Objects not suitable for further analysis, like the partially visible microcarrier on the upper border of the image in Figure 5.12 and the two small single cells, are excluded in the final step of image segmentation.

The next step after the microcarrier separation procedure is the detection of the area on the microcarrier surface where cells are located; this information is then transformed into the LOC. Since the images are two-dimensional, only a top view of a carrier is available for evaluation. Cells that grow on the sides of a carrier or on the back side cannot be evaluated properly. To account for this partial imaging, it is assumed that the colonization is homogeneous over the entire microcarrier surface such that if 50% of the circular area of a carrier that is visible is covered by cells, then the LOC of the entire microcarrier is assumed to be 50%. While this argument may not hold for each individual microcarrier, on average, over many particles, the LOC that is visible in the top view provides a good estimate for the total LOC.

One starting point to transform the gray value information into the LOC value is to apply the separation method on the microcarrier inside the area as described in the previous section. However, no convincing results could be achieved with this procedure. The cells are often very close to each other and have a wide variety of shapes (e.g., circular,

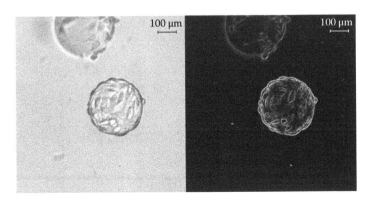

FIGURE 5.12   Left: original image with two microcarriers. Right: image after application of the Sobel operator. (From Rudolph, G. et al., *Biotechnology and Bioengineering* 99:136–145, 2008.)

elliptic). Moreover, the gray values of the cell borders and bodies are different depending on the illumination, reflection, and shadows. All these facts make cell detection very complicated. Many special cases would have to be dealt with by the algorithm, increasing its complexity and the computing time per image.

A simpler way to extract the LOC from the data is by evaluation of the gray value histogram because a direct relationship between histogram shape and LOC, as shown in Figure 5.13, could be found. Microcarriers without cells or with a low LOC have narrow histograms with a maximum at around a gray value of 200. The more a microcarrier is colonized, the more the histogram becomes flat and broad, and the maximum shifts to higher gray values. So in the next step after object detection, the histogram of each image object is calculated. Then all histograms are normalized by dividing each value by the total amount of pixels. After this operation, the sum of all histogram values equals one. Finally, five key values are extracted from the histogram: the value of the maximum and the gray value where it occurs, the value of white (255), and the first two moments of the histogram distribution (mean and variance). These five values are used as input variables into a neural network that was trained previously to predict the LOC from these five histogram values.

The programming and training of the neural network are described in detail by Rudolph et al. (2007). A comparison between the results of the algorithm and a manual determination of the LOC shows that over 90% of the analyzed images are evaluated correctly by the algorithm. This algorithm has also been applied to images of microcarriers (see Section 5.4.3).

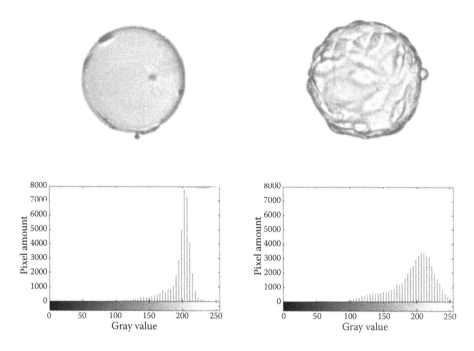

FIGURE 5.13   Left: microcarrier with a low LOC. Right: microcarrier with a LOC of 100%. The corresponding gray value histograms are shown below. The differences in the shapes of the histograms between low and high LOC are evident. (From Rudolph, G. et al., *Biotechnology and Bioengineering* 99:136–145, 2008.)

Chapter 5

5.3.5.3 Image Analysis for Characterization of Aggregates Formed during Dynamic Suspension Cultivation of Embryonic Stem Cells

An image processing algorithm for the characterization of aggregates formed during dynamic suspension cultivation systems for ES cells was developed in C# using Microsoft's Visual Studio 2008 and the .NET Framework 3.5. In the following, the steps of the algorithms are pointed out. At first the images acquired from the microscope are converted to 8-bit gray scale images, and then an edge detection with the method proposed by Canny (1986) is carried out yielding a gradient image. The gradient image is transformed to a binary image using the threshold selection method described by Otsu (1979).

The binary image generated in this fashion is the initial point for a multistep procedure that successively extracts cell aggregates and their shape and size information from the image objects in it. The image objects can be classified into the following groups: cell aggregates with a closed contour, cell aggregates with a broken or opened contour, image objects that are multiple aggregates that overlap or touch each other, and objects much smaller than the aggregates (medium particles, single cells, etc.). All those image objects must either be processed, so that the shape and size of the cell aggregate can be computed, or be eliminated from further processing because it is a nonaggregate object. The aims of the following steps are given as follows:

1. Elimination of objects that are too small to be a cell aggregate and objects that cannot be evaluated because they are located on the image border and thus are only partially visible
2. Repair the broken contours
3. Split objects into single aggregates

Aim 1 is achieved by applying a simple position and size constraint on all image objects. All objects that do not satisfy that constraint are sorted out. To accomplish the second aim, the morphological closing operation is applied on the binary image. Aim 3 is achieved by applying multiple erosion steps on one image object until it breaks apart into two or several pieces and reconstructing the shape and size of the original cell aggregates by applying the same number of dilation steps on the fragments.

All these steps are carried out multiple times on the image. Whenever a closed contour can be generated from an image object by one or more of the aforementioned steps, such that the resulting image object satisfies predefined size and shape criteria, this object is removed from all further processing steps and stored in an output list.

Using the described combination of position and shape constraints and erosion and dilatation steps, over 95% of the cell aggregates on the available images can be detected properly. In most cases, an aggregate could not be detected well when it either was out of focus (no correct contour processing possible) or it overlapped too much with other aggregates and formed a cluster that could not be split with the erosion–dilation sequence. The output of the algorithm is a list containing all detected objects together with their shape information (e.g., compactness—a measure of roundness), their size, and their equivalence diameter in pixel. Given the pixel-to-micron ratio, the computation of the size and diameter in the unit micrometers is also possible.

This algorithm has been applied to images of ES cells, and results are presented in Section 5.4.4.

5.4 Example Applications of *In Situ* Microscopy for Bioprocess Monitoring

Within this chapter, different examples from biotechnology are provided to illustrate the potential of the ISM for mammalian and even stem cell cultivations. In particular, Sections 5.4.2, 5.4.3, and 5.4.4 show that the ISM is already used for monitoring of suspended cell cultivation or microcarrier cultivation of mammalian cells. The other examples, while not directly dedicated to tissue engineering, help to show the broad applicability of this method.

5.4.1 Monitoring of Yeast Cultivations

S. cerevisiae is an important eukaryotic model organism for various applications in the biotechnological industry as well as in molecular biology, and *in situ* microscopy can be used for process monitoring of this organism. Figure 5.14 shows an ISM image of suspended *S. cerevisiae* in a batch cultivation process.

Several single and budding yeast cells can be seen in the different focus areas. Appropriate processing algorithms have been developed for the analysis of these images. The number of single cells, double cells, cell clusters, and information about the cell size and cell volume can be derived with these algorithms (see Section 5.5.1). In Figure 5.15, the results from monitoring a typical *S. cerevisiae* batch culture over 24 h are presented.

In the beginning of the cultivation, about 65% of the observed cells were single cells. Upon entering the exponential growth phase, the portion of single cells dropped to approximately 35% while the percentage of double cells and cell clusters increased. When the stationary phase was reached, the amount of single cells rose to about 55% while the percentage of double cells dropped back to the initial value of about 20%. The proportion of cell clusters decreased as well, but at 20% was about twice as large as the initial value.

Several publications have shown the potential of the online analysis of yeast cultivation processes with the ISM (Bittner et al. 1998; Rehbock et al. 2010; Akin et al. 2011). Cell densities up to 2×10^8 cells/mL can be monitored.

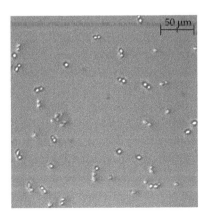

FIGURE 5.14 Image of *S. cerevisiae* cells taken with an ISM during a batch cultivation (cell density approximately 6×10^7 cells/mL).

Chapter 5

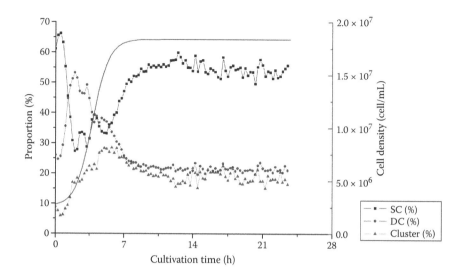

FIGURE 5.15   ISM results from monitoring of a *S. cerevisiae* culture. Cell density development (line) as well as proportions of single cells (SC), double cells (DC), and cell clusters over time are shown.

5.4.2   Monitoring Mammalian Cell Cultivations

In biotechnology, the production of complex proteins is accomplished using mammalian cells. These higher eukaryotes can produce biologically active proteins with the necessary posttranslational modifications. Suitable mammalian cell production systems are Chinese hamster ovary (CHO) and baby hamster kidney (BHK) cells. These cells can be cultivated in suspension, and high cell densities can be achieved. In 2007, about 70% of all recombinant proteins designed for human therapy were produced in CHO cells (Jayapal et al. 2007).

In comparison to prokaryotic or simple eukaryotic cells, such as yeast, the proliferation rates of mammalian cells are low and risks of contamination are high. Accurate monitoring of the culture is necessary to understand and control these complex processes such that the process is stable and efficient. Manual sampling and subsequent off-line analysis is time consuming and involves the risk of contamination. Thus, noninvasive and automatic analytical methods are highly desirable for mammalian cell cultivations. There is considerable motivation for online analysis of different process variables, especially of cell density and cell viability. The ISM offers the possibility to determine cell number, size, and morphology, as well as problems with cell agglomeration. In addition, microbial contamination can be detected at an early stage. Recently, Wiedemann et al. (2011) have shown that online viability studies with ISM systems are possible by assessing the homogeneity of the microscopy images. In Figure 5.16, a typical ISM image of CHO cells acquired during cultivation is shown.

Six single cells with diameters from 15 to 30 μm and different morphologies can be seen in the image. The image processing software detects all of these cells and generates a result image in which detected objects are marked.

To prove that accurate monitoring of the cell concentration is possible, results of other techniques were compared with the results obtained with an ISM. The cell concentration

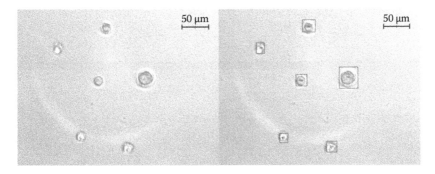

FIGURE 5.16   Image of CHO cells in a bioreactor, 3 h after inoculation. Cells have diameters from 15 to 30 μm. All six cells were detected by the image processing software (right side).

in samples from a CHO cell cultivation was assayed using a Cedex Cell Analyzer (Roche, Indianapolis), via manual cell counting in a Neubauer chamber, and with the ISM. The results of all three methods are compared in Figure 5.17.

The cell concentration measured with the ISM was often slightly higher than those of the other methods, while the Cedex Cell Analyzer measured slightly lower cell densities. Overall similar results for the cell density were obtained with all three methods, demonstrating that *in situ* microscopy is suitable for cell density measurements of mammalian suspension cultures.

The detection of cells with the ISM becomes more difficult if the monitored cells form three-dimensional cell clusters. An ISM image of clustering BHK cells is shown in Figure 5.18. To enable an estimation of the cell density from such images, special image processing algorithms have been developed (Martinez et al. 2005, 2008).

These algorithms allow good estimation of the cell density in the culture process. This is achieved by three-dimensional cell counting since the algorithm allows the detection of cells in three different layers of the cell clusters.

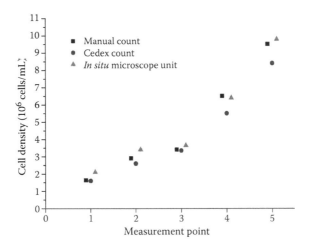

FIGURE 5.17   Comparison of cell density values of CHO cell samples measured with different counting methods.

Chapter 5

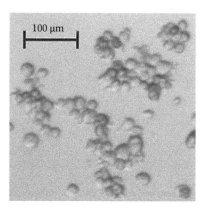

FIGURE 5.18   ISM image of BHK cells and cell clusters.

5.4.3   Monitoring of Microcarrier-Based Fibroblast Cultivations

Most industrial bioprocesses are based on suspension cultivations. However, the use of anchorage-dependent cells is occasionally necessary. The use of microcarriers, providing an increased surface area for cell attachment and growth, is an established method to achieve high cell densities and thus high productivity (van der Velden de Groot 1995). The standard procedure to analyze the cell density during microcarrier cultivation is to detach intact cells or to lyse the cells to free the nucleus and to count the cells (or nuclei) with a hemocytometer. This manual evaluation is very time consuming and only produces the results after a significant delay (Hirtenstein et al. 1980; Mered et al. 1980; Sanford et al. 1951).

Other parameters of interest in microcarrier cultivations are the plating efficiency (PE) and the LOC. The PE is the ratio between colonized and noncolonized microcarriers. The LOC is the percentage of the area of a microcarrier that is covered by cells. Usually these parameters are measured manually, with samples withdrawn from the cultivation.

In situ microscopy allows the in-line observation of microcarrier cultures from inoculation to harvest without any off-line sampling. However, transparent microcarriers such as Cytodex 1 microcarriers (Amersham, Uppsala, Sweden) have to be used in order to monitor the process with an ISM.

Figure 5.19 shows some sample images of NIH-3T3 mouse fibroblasts on Cytodex-1 microcarriers (cropped to show the microcarrier object only) from different phases of a batch cultivation. The images were acquired using an ISM and a monochrome CCD camera. Using a fourfold magnifying objective, each image pixel covers an object area of 0.672 μm^2.

All acquired images were analyzed using a newly developed image processing algorithm (details are presented in Section 5.3.5.2). With this approach, the gray value is used as an indicator for cells growing on the transparent microcarriers. The higher the gray value over the microbead surface, the more cells grow on it. The data are shown in Figure 5.20.

The PE increased only slightly during the course of the cultivation to approximately 90%. This means that most carriers are colonized, but it also shows that carriers that are

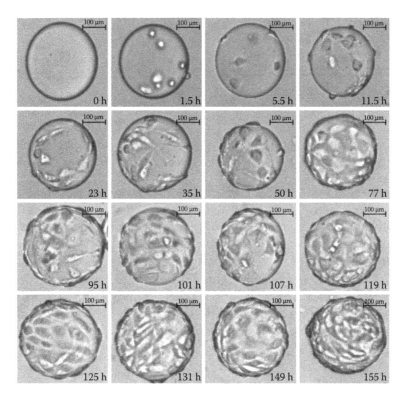

FIGURE 5.19   NIH-3T3 mouse fibroblasts on Cytodex-1 microcarriers. (From Bluma, A. et al., *Analytical and Bioanalytical Chemistry* 398:2429–2438, 2010.)

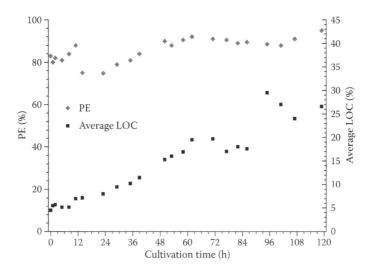

FIGURE 5.20   Results of the analysis of the experimental data of a mouse fibroblast culture on microcarriers. The average LOC and the PE of each measurement cycle are plotted versus the cultivation time.

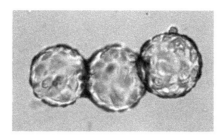

FIGURE 5.21 *In situ* microscopy image of a microcarrier agglomerate. (From Rudolph, G. et al., *Biotechnology and Bioengineering* 99:136–145, 2008.)

uncolonized in the beginning of the cultivation are unlikely to become colonized. After inoculation, the LOC was approximately 5%. In the first 72 h of the cultivation, the LOC value increased to approximately 20%. In the late phase (72–120 h) of the cultivation, the LOC measurements began to vary more. This was due to formation of agglomerates of some of the microcarriers as shown in Figure 5.21.

Within the first 72 h, the cultivation could be monitored satisfactorily by *in situ* microscopy. A microcarrier concentration of 2 g/L was used for the experiments shown in Figures 5.14 through 5.16, but concentrations as high as 4 g/L could also be monitored.

5.4.4 Characterization of Aggregates Formed during Dynamic Suspension Cultivation of Embryonic Stem Cells

ES cells are self-renewing and pluripotent cells that can differentiate into a variety of cell lineages. Optical assays using an off-line microscope system, which is similar to the ISM device, have been carried out to compare five cultivation systems with respect to the size and morphology of generated ES cell aggregates. Therein, the murine ES cell line Brachyury (E14.1, 129/Ola) was used. The generated information is of interest as cells in small aggregates have been found to have higher proliferative capacities than cells in large aggregates (Tomala 2010).

Two spinner flasks with different stirrer geometry were compared with two orbitally shaken culture vessels. A commercially available spinner flask from Integra Biosciences, Fernwald, Germany (referred as to Spinner Flask 1), was equipped with a bulb-shaped pendulum, while an in-house made spinner flask TCI (designated Spinner Flask 2) contained a vertical impeller. An orbitally shaken Erlenmeyer flask (VWR, Darmstadt, Germany) and a cylindrical CultiFlask 50 tube (Sartorius-Stedim Biotech, Goettingen, Germany) were included to characterize the cell's expansion in culture vessels without internal stirrer. A static Petri dish cultivation was performed as reference.

The murine ES cell line Brachyury was cultured and expanded feeder cell free as floating cell aggregates in Petri dishes. Cell expansion in dynamic suspension was analyzed by seeding single cells with a density of 2×10^4 cells/mL each into the different cultivation systems. Cell samples (2% v/v of cultivation working volume) from each experiment were taken daily, transferred into multiwell plates, and examined under a microscope (Olympus IX 50, Olympus Corporation, Puchheim, Germany). Using a camera (Olympus Camedia C-4040 Zoom, Olympus Corporation, Puchheim, Germany), images were taken and saved as .bmp files for image processing. In Figure 5.22, an original microscope image showing several ES cell aggregates is presented.

Morphology and size of generated ES cell aggregates were analyzed in each type of cultivation system using an image-processing algorithm. This algorithm can be implemented onto the ISM system for the monitoring and evaluation of ES cell cultivation and aggregate formation and is described in detail in Section 5.2.5.3.

With regard to proliferation, orbital shaken culture systems and the Spinner Flask 1 were most effective in supporting ES cell expansion. Since the cultures did not show significant differences in viability and metabolic activity, it is assumed that the differences in the proliferation rates were not due to necrotic cell death or physical cell damage. With an extensive analysis using the image processing software, the volume and number of cells per cell aggregate in each cultivation system was determined. In Figure 5.23, the mean aggregate size is shown in the time course of the cultivation.

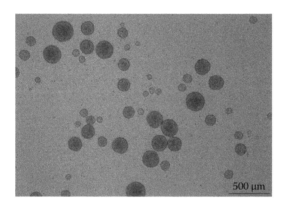

FIGURE 5.22 Microscope image showing ES cell aggregates.

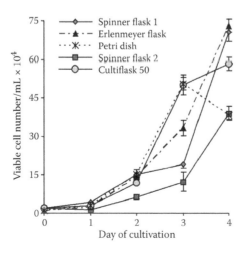

FIGURE 5.23 Mean sizes of aggregates generated in the different cultivation systems at each day of cultivation.

Chapter 5

The number of cells forming an aggregate is indicated as the aggregate size. From the third day of cultivation, slight differences in the aggregate size could already be observed. At day 4 of the cultivation, these differences became more pronounced. Aggregates in the Spinner Flask 2 cultivation contained an average of 501 cells, and aggregates in the Petri dish cultivation comprised an average of 398 cells. Erlenmeyer flask, Spinner Flask 1, and CultiFlask 50 derived aggregates contained significantly fewer cells (206, 268, and 272 cells/aggregate, respectively).

5.4.5 Monitoring of Microalgae Cultivations

Biotechnological applications of microalgae are of interest as their photosynthetic activity enables them to produce lipids (for fuels and other products) directly from CO_2 and light (Pruvost et al. 2009). The economical production of hydrogen from microalgae does not seem possible at the moment due to low cell productivity. In contrast, lipid contents in microalgae have been found to be as high as 0.4 g/g biomass (Li et al. 2008).

The application of an ISM in an algal photobioreactor allows the online observation of changes in cell morphology and cell density. Moreover, the use of *in situ* microscopy enables the user to detect contamination of the microalgae cultivations by predators such as rotifers, other algae, or bacteria, allowing corrective measures to be taken quickly. In the future, online fluorescence imaging might give insights into the lipid productivity of microalgae during a cultivation process, and chlorophyll contents might be determined as well. In this way, the effect of different conditions on the process effectiveness could be measured in real time.

The use of *in situ* microscopy has been demonstrated for several microalgae, including *Chlamydomonas reinhardtii*, *Phaeodactylum tricornutum*, and *Neochloris oleoabundans*. An example image from a cultivation of *P. tricornutum* is shown in Figure 5.24. Larger algae with more complex morphologies can be monitored with the ISM as well. In Figure 5.25, an ISM image of a *Spirulina platensis* cell with its spiraled morphology is displayed.

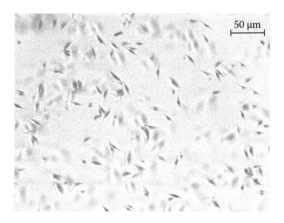

FIGURE 5.24 Image of a *P. tricornutum* cultivation in a photobioreactor, 1 day after inoculation.

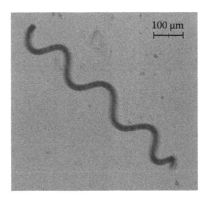

FIGURE 5.25   *In situ* microscopy image of a *S. platensis* cell.

5.4.6   Characterization of the Mechanical Stability of Enzyme Carriers

The immobilization of enzymes is used for industrial processes as the biocatalysts can be reused once the process is finished and the recycling of immobilized enzymes is much easier. The mechanical stability of the carrier material is important because destruction of the carrier eventually results in decreased enzymatic activity. Thus, methods to increase the stability of enzyme carriers, such as coating with silicone, have been investigated (Wiemann et al. 2009). The influence of stabilization procedures and different reactor systems on the mechanical integrity of the carrier material can be monitored during the actual process with an ISM.

To demonstrate the applicability of *in situ* microscopy, Lewatit VP OC 1600 particles were dissolved in water and stirred with a disc plate stirrer in a small vessel. Lewatit VP OC 1600 with Lipase B from *Candida antarctica* adsorbed on its surface is commercially available as Novozym 435. The application of the ISM allowed the noninvasive monitoring of the fragmentation of the particles. An image-processing algorithm was used to detect objects and to classify them according to their eccentricity, which describes their roundness (Prediger et al. 2011). Objects with an eccentricity over a defined threshold are counted as fragments while all others are counted as intact particles. ISM images of intact particles and fragments are shown in Figure 5.26 together with the resulting image from the software.

As seen in Figure 5.26, intact particles are very large relative to the captured field of view. Therefore, they are rarely recorded in their entirety. Since the image-processing algorithm discards all objects that are not whole, an alternative approach to monitor the destruction of the particles more accurately was to capture several hundred images sequentially at one measuring point, forming capture cycles. All detected intact particles and detected fragments in one capture cycle were summed, and the result of this calculation is shown in Figure 5.27.

The number of detected particles decreased until after 96 h almost no intact particles could be detected. The number of detected fragments increased linearly over time. Even when almost no more intact particles were detected toward the end of the experiment, the number of detected fragments continued to rise. This indicates that the fragments themselves were further degraded into smaller fragments over time, resulting in continuously rising fragment numbers.

Chapter 5

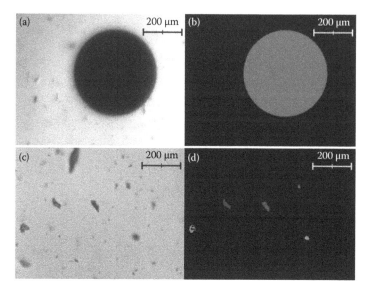

FIGURE 5.26 Two ISM images and the corresponding result image of the processing software. One intact particle is captured (a) and detected by the software (b). Several fragments are captured (c). The five largest are over the size threshold and thus appear in the result image (d). The big particle in the top part is not detected as it is not completely in the image.

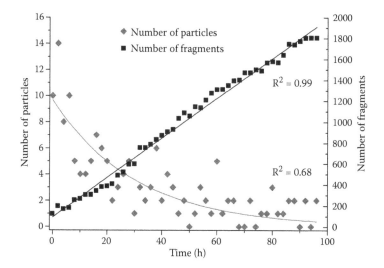

FIGURE 5.27 Two-gram Lewatit VP OC 1600 in 450 mL diH$_2$O was stirred for 96 h at a rate of 160 rpm with a disc plate stirrer. The result of the image processing is shown. The number of detected particles per capture cycle decreases while the number of detected fragments per capture cycle increases. (From Prediger, A. et al., *Chemical Engineering and Technology* 34:837–840, 2011.)

5.4.7   Characterization of Crystallization Processes

Crystallization processes are used for downstream processes in the biotechnological and pharmaceutical industries. The monitoring of crystallization processes with non-invasive sensor technologies would help to improve the efficiency and stability of these processes.

To demonstrate the application potential of *in situ* microscopy on crystallization processes, the ISM was adapted to monitor the batch crystallization process of hen egg white lysozyme (HEWL) continuously (Bluma et al. 2009).

A solution of HEWL was stirred constantly while the temperature was decreased linearly. A custom-made image-processing algorithm for crystal images was used to analyze the acquired image data. For comparison, the concentration of dissolved HEWL was measured at 280 nm with a spectrometer. The results of 35 h of process monitoring are shown in Figure 5.28.

As expected, the relative absorbance of dissolved lysozyme decreased during the crystallization process. The process started with a lysozyme concentration of 5 g/L. At the end of the crystallization, a dissolved lysozyme concentration of about 1.5 g/L remained. From this, it follows that 3.5 g/L of lysozyme was crystallized. There is a good correlation between the crystallized lysozyme mass and the calculated crystal area. The increase in the crystal area during the crystallization process is caused either by an increase in the amount of crystals or by their growth.

Other crystallization processes have been studied using *in situ* microscopy. Figure 5.29 shows a compilation of investigated crystals.

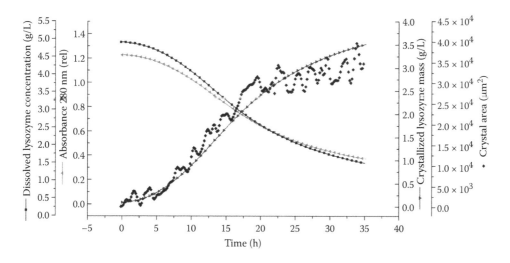

FIGURE 5.28   Monitoring of a protein crystallization process. The curves show the decrease of the dissolved HEWL concentration during the crystallization process, the measured absorption of HEWL, the calculated crystallized HEWL mass, and the calculated crystal area (μm^2) by the image-processing algorithm. (From Bluma, A. et al., *Journal of Crystal Growth* 311:4193–4198, 2009.)

Chapter 5

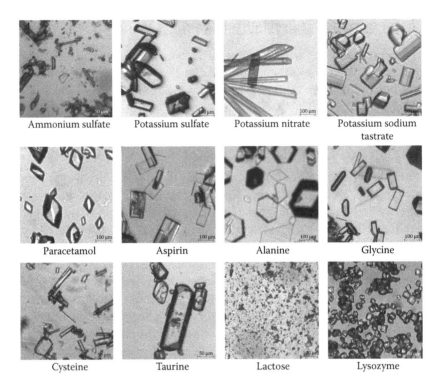

Ammonium sulfate	Potassium sulfate	Potassium nitrate	Potassium sodium tastrate
Paracetamol	Aspirin	Alanine	Glycine
Cysteine	Taurine	Lactose	Lysozyme

FIGURE 5.29　ISM images of typical product crystals of different chemical compounds.

5.5　Conclusions

Optical *in situ* imaging techniques can effectively be used to monitor biotechnological processes. Application areas for these techniques include monitoring different cell cultivation systems, characterization of crystallization processes, monitoring of cells growing on microcarriers, and the investigation of enzyme carrier properties.

Commercially available systems differ mostly in their illumination system (lasers, LED, and optical fibers) and the definition of the sampling volume (via mechanical or optical methods). Images of particles or cells with sizes ranging from 1 µm up to 3 mm can be acquired and analyzed.

Such techniques have potential for use in regenerative medicine. For example, it is important to monitor and control the proliferation and differentiation of stem cells into desired somatic cells and tissue. However, reliable and noninvasive methods for monitoring of these processes have yet to be developed.

The biggest advantage of the optical *in situ* methods presented here is their ability to allow the noninvasive measurement of process variables such as cell density, particle size distributions, and cell morphology. No sampling from the reactor systems is necessary and the process information is available in real time. Thus, optical *in situ* measurements are suitable not only for process monitoring but also for process control.

Despite these tremendous advantages, optical *in situ* imaging techniques are still rarely used in the industry. This will change in the near future as detailed insights into production processes become more important in the pharmaceutical sector due to the process analytical technology (PAT) initiative.

References

Abranches, E., E. Bekman, D. Henrique, and J. M. S. Cabral. 2007. Expansion of mouse embryonic stem cells on microcarriers. *Biotechnology and Bioengineering* 96:1211–1221.

Akin, M., A. Prediger, M. Yuksel et al. 2011. A new set up for multi-analyte sensing: At-line bio-process monitoring. *Biosensors and Bioelectronics* 26:4532–4537.

Barrett, P., and B. Glennon. 2002. Characterizing the metastable zone width and solubility curve using lasentec FBRM and PVM. *Chemical Engineering Research and Design* 80:799–805.

Bittner, C., G. Wehnert, and T. Scheper. 1998. *In situ* microscopy for on-line determination of biomass. *Biotechnology and Bioengineering* 60:24–35.

Bluma, A., T. Hoepfner, G. Rudolph et al. 2009. Adaptation of *in-situ* microscopy for crystallization processes. *Journal of Crystal Growth* 311:4193–4198.

Bluma, A., T. Hoepfner, P. Lindner et al. 2010. *In-situ* imaging sensors for bioprocess monitoring: state of the art. *Analytical and Bioanalytical Chemistry* 398:2429–2438.

Camisard, V., J. P. Brienne, H. Baussart, J. Hammann, and H. Suhr. 2002. Inline characterization of cell concentration and cell volume in agitated bioreactors using *in situ* microscopy: Application to volume variation induced by osmotic stress. *Biotechnology and Bioengineering* 78:73–80.

Canny, J. 1986. A computational approach to edge detection. *IEEE Transactions on Pattern Analysis and Machine Intelligence* 8:679–698.

Fernandes, A. M., T. G. Fernandes, M. M. Diogo, C. L. da Silva, D. Henrique, and J. M. S. Cabral. 2007. Mouse embryonic stem cell expansion in a microcarrier-based stirred culture system. *Journal of Biotechnology* 132:227–236.

Frerichs, J. G., K. Joeris, K. Konstantinov, and T. Scheper. 2002. Use of an *in-situ* microscope for on-line observation of animal cell cultures. *Chemie Ingenieur Technik* 74:1629–1633.

Hirtenstein, M., J. Clark, G. Lindgren, and P. Vretblad. 1980. Microcarriers for animal cell culture: A brief review of theory and practice. *Developments in Biological Standardization* 46:109–116.

Jayapal, K. R., K. F. Wlaschin, W. S. Hu, and M. G. S. Yap. 2007. Recombinant protein therapeutics from CHO cells—20 years and counting. *Chemical Engineering Progress* 103:40–47.

Joeris, K., J. G. Frerichs, K. Konstantinov, and T. Scheper. 2002. *In-situ* microscopy: Online process monitoring of mammalian cell cultures. *Cytotechnology* 38:129–134.

Junker, B., W. Maciejak, B. Darnell, M. Lester, and M. Pollack. 2007. Feasibility of an *in situ* measurement device for bubble size and distribution. *Bioprocess and Biosystems Engineering* 30:313–326.

Li, Y. Q., M. Horsman, B. Wang, N. Wu, and C. Q. Lan. 2008. Effects of nitrogen sources on cell growth and lipid accumulation of green alga *Neochloris oleoabundans*. *Applied Microbiology and Biotechnology* 81:629–636.

Linder, P., C. Krabichler, G. Rudolph, T. Scheper, and B. Hitzmann. 2007. Application of *in-situ*-microscopy and digital image processing in yeast cultivation. *Preprints of the 10th International IFAC Symposium on Computer Applications in Biotechnology* 1:245–250.

Martin, G. R. 1981. Isolation of a pluripotent cell-line from early mouse embryos cultured in medium conditioned by teratocarcinoma stem-cells. *Proceedings of the National Academy of Sciences of the United States of America—Biological Sciences* 78:7634–7638.

Martinez, G., J. G. Frerichs, K. Joeris, K. Konstantinov, and T. Scheper. 2005. Cell density estimation from a still image for *in-situ* microscopy. *2005 IEEE International Conference on Acoustics, Speech, and Signal Processing* (IEEE Cat. No.05CH37625):ii/497–500 Vol. 492|495 vol. (cxviii+5600).

Martinez, G., J. G. Frerichs, G. Rudolph, and T. Scheper. 2008. Three-Dimensional Cell Counting for *In-Situ* Microscopy. *19th International Conference on Pattern Recognition*, Vols 1–6:1454–1457.

Mered, B., P. Albrecht, and H. E. Hopps. 1980. Cell-growth optimization in microcarrier culture. *In Vitro-Journal of the Tissue Culture Association* 16:859–865.

O'Sullivan, B., P. Barrett, G. Hsiao, A. Carr, and B. Glennon. 2003. *In situ* monitoring of polymorphic transitions. *Organic Process Research and Development* 7:977–982.

Otsu, N. 1979. A threshold selection method from gray-level histograms. *IEEE Transactions on Systems Man and Cybernetics* 9:62–66.

Prediger, A., A. Bluma, T. Hoepfner et al. 2011. *In situ* microscopy for online monitoring of enzyme carriers and two-phase processes. *Chemical Engineering and Technology* 34:837–840.

Pruvost, J., G. Van Vooren, G. Cogne, and J. Legrand. 2009. Investigation of biomass and lipids production with *Neochloris oleoabundans* in photobioreactor. *Bioresource Technology* 100:5988–5995.

Chapter 5

Qu, H. Y., M. Louhi-Kultanen, and J. Kallas. 2006. In-line image analysis on the effects of additives in batch cooling crystallization. *Journal of Crystal Growth* 289:286–294.

Rehbock, C., D. Riechers, T. Hoepfner et al. 2010. Development of a flow-through microscopic multitesting system for parallel monitoring of cell samples in biotechnological cultivation processes. *Journal of Biotechnology* 150:87–93.

Rudolph, G., T. Bruckerhoff, A. Bluma, G. Korb, and T. Scheper. 2007. Optical inline measurement procedure for cell count and cell size determination in bioprocess technology. *Chemie Ingenieur Technik* 79:42–51.

Rudolph, G., P. Lindner, A. Gierse et al. 2008. Online monitoring of microcarrier based fibroblast cultivations with *in situ* microscopy. *Biotechnology and Bioengineering* 99:136–145.

Sanford, K. K., W. R. Earle, V. J. Evans, H. K. Waltz, and J. E. Shannon. 1951. The measurement of proliferation in tissue cultures by enumeration of cell nuclei. *Journal of the National Cancer Institute* 11:773–795.

Suhr, H., G. Wehnert, W. Storhas, and P. Speil. 1991. *In situ* microscope probe and measuring system – comprises optical microscope coupled to an image-analysing computer. Patent DE4032002.

Suhr, H., G. Wehnert, K. Schneider et al. 1995. *In-situ* microscopy for online characterization of cell-populations in bioreactors, including cell-concentration measurements by depth from focus. *Biotechnology and Bioengineering* 47:106–116.

Thomson, J. A., J. Itskovitz-Eldor, S. S. Shapiro et al. 1998. Embryonic stem cell lines derived from human blastocysts. *Science* 282:1145–1147.

Tomala, M. 2010. Entwicklung und Anwendung von Strategien zur Expansion pluripotenter Stammzellen. Doctoral Thesis at Institute of Technical Chemistry. Gottfried Wilhelm Leibniz University Hannover, Hannover.

van der Velden de Groot, C. A. M. 1995. Microcarrier technology, present status and perspective. *Cytotechnology* 18:51–56.

Wiedemann, P., J. S. Guez, H. B. Wiegemann et al. 2011. *In situ* microscopic cytometry enables noninvasive viability assessment of animal cells by measuring entropy states. *Biotechnology and Bioengineering* 108:2884–2893.

Wiemann, L. O., P. Weisshaupt, R. Nieguth, O. Thum, and M. B. Ansorge-Schumacher. 2009. Enzyme Stabilization by Deposition of Silicone Coatings. *Organic Process Research and Development* 13:617–620.

Zweigerdt, R. 2009. Large scale production of stem cells and their derivatives. *Engineering of Stem Cells* 114:201–235.

Optical Spectroscopy in Regenerative Medicine

6. Light Scattering as Polarization Spectroscopy

Irene Georgakoudi and Martin Hunter

Optical Techniques in Regenerative Medicine. Edited by Stephen P. Morgan, Felicity R.A.J. Rose, and
Stephen J. Matcher © 2014 CRC Press/Taylor & Francis Group, LLC. ISBN: 978-1-4398-5495-2

Chapter 6

6.1 Introduction

When light falls onto a material sample, the most likely interaction is scattering, resulting from differences in the optical density between the sample and the surrounding medium in which the light is traveling. A number of spectroscopic implementations have evolved that rely on different aspects of this interaction and can be used to assess various characteristics of the morphology and/or organization of the sample, with sensitivity to features from the nanoscale to the macroscale (i.e., from a few nanometers to millimeters). Most of these approaches focus on characterization of the variations of the scattered light intensity as a function of wavelength or scattering angle (Mourant et al. 1995, 2002; Backman et al. 1999; Bigio et al. 2000; Gurjar et al. 2001; Mujat et al. 2008), but some highly sensitive techniques also exploit information in the phase of the scattered light (Wax et al. 2002). In addition, polarization can also be used as a means to select light that has undergone either very few or a large number of scattering events (Backman et al. 1999; Mourant et al. 2002). Finally, some light scattering–based approaches, such as dynamic light scattering (DLS), rely on detection and analysis of the variations in light scattering intensity as a function of time over small time scales (measurement time is typically seconds) (Berne and Pecora 2000). In principle, instruments that perform static light scattering measurements are fairly inexpensive and portable and can provide results in real time. In addition, light scattering–based approaches are typically realized in a noninvasive or minimally invasive manner that does not affect the sample, enabling dynamic assessments over hours, days, or months. This is a significant advantage for applications that would benefit from understanding the evolution of a sample over such time scales, as is typically the case for tissue engineering and regenerative medicine.

Initial biomedical studies of light scattering spectroscopy (LSS) focused on cancer diagnostic applications (Mourant et al. 1995, 1996; Sokolov et al. 1999; Backman et al. 2000; Bigio et al. 2000; Georgakoudi et al. 2001, 2002; Gurjar et al. 2001). Since lipids, and therefore, the membranes that make up cellular organelles, have a higher refractive index (the parameter that is typically used to characterize the optical density of a material) than the surrounding cytoplasm, LSS is sensitive to changes in their sizes, shapes, and organization. For example, it has been shown that LSS can be used to characterize differences both in the overall level of organization and the length scales over which subcellular components organize in normal and dysplastic esophageal and cervical epithelial cells, as well as normal and cancerous white blood cells (Hunter et al. 2006; Mujat et al. 2008; Hsiao et al. 2011). The potential sensitivity of LSS to such changes is extremely high: in colonic epithelia, they have been detected even before molecular stains are able to reliably identify preneoplastic change (Kim et al. 2003; Roy et al. 2004). A number of LSS studies have also focused on assessing changes in the size, shape, and refractive index of precancerous and cancerous cell nuclei in the esophagus, colon, oral cavity, bladder, cervix, and blood (Sokolov et al. 1999; Backman et al. 2000; Georgakoudi et al. 2001, 2002; Müller et al. 2003; Hsiao et al. 2011). Sensitivity to subcellular features of superficial cells is typically present in methods that detect light that has been scattered one or very few times. LSS implementations that rely on multiply scattered light are sensitive to changes in bulk tissue properties and have been used to detect differences in the collagen density and organization between normal and cancerous tissues (Zonios et al. 1999; Georgakoudi et al. 2001, 2002; Chang et al. 2005). Polarization sensitive measurements are also often times highly informative, especially in the context of collagen organization (Pierce et al. 2004; Kerschnitzki et al. 2011).

In this chapter, we will focus on the use and potential applications of static light scattering polarization and spectroscopy (LSPS) in the field of regenerative medicine. We note that approaches such as DLS will not be discussed in this chapter, even though it is widely used to characterize the assembly of numerous biomaterials that are used as tissue engineering scaffolds. We will initially discuss specific tissue features that LSPS can be used to assess in the context of tissue engineering. Then, we will describe the basic design of relevant instruments and provide the theoretical background, describing scattering-based light–matter interactions, and the models that have been used to characterize them so that detailed information about sample morphology and organization can be extracted. We will summarize studies that have started to explore the potential of this technique to assess collagen organization and biomaterial mineralization, and we close with a discussion of potential impact and future directions for LSPS studies in this field.

6.2 Clinical and Biological Measurement Challenges Suited to LSPS: Morphological Characterization of Cells and Tissues

6.2.1 Cellular-Light Scattering

A number of key cellular components can scatter light. Membrane rich organelles, such as the endoplasmic reticulum and mitochondria, lysosomes, and the nucleus are some examples of a cell's potential scattering centers. Since protein complexes and organelles spanning a wide range of sizes are tightly packed within a cell, light scattering–based approaches have been developed to extract information about the packing and organization of cell and matrix components for scales that vary from 20 nm to tens of microns (Backman et al. 2001; Kim et al. 2003; Fang et al. 2007; Mujat et al. 2008). In the context of tissue engineering, light scattering–based approaches could be very well suited to assess morphological and organizational changes that are expected during stem cell differentiation. For example, both the wavelength- and angle-dependent light scattering signals are expected to be sensitive to changes in nuclear size and the organization or packing of subcellular organelles (Backman et al. 2001). In the context of specific tissues, additional, potentially highly useful information may be detectable. For example, LSS techniques could be well suited to characterize changes in size of lipid droplets formed during adipogenic differentiation.

6.2.2 Extracellular Matrix Light Scattering

There are different types of collagen in human tissues, classified as fibrillar or nonfibrillar based on their structure. Fibrillar collagens typically serve as a good source of light scattering contrast, both because of their higher refractive index and their birefringence (i.e., their ability to slow down light by different amounts depending on the

Chapter 6

direction light is traveling). As a result, LSS-based approaches can be used to provide useful information regarding the size, density, organization, and orientation of collagen fibers (Marenzana et al. 2002; Brown et al. 2003; Kostyuk et al. 2003; Kostyuk and Brown 2004; Williams et al. 2005; Morgan et al. 2006; Kerschnitzki et al. 2011). Such information is of great interest during the collagen deposition and remodeling phases that occur throughout the development of numerous tissues, such as bone, heart, tendon, and cartilage.

6.2.3   Biomaterial Scaffold Light Scattering

Many biomaterial scaffolds consist of proteins and/or polymers that have a refractive index that is higher than that of cells, and in the case of polymers, often higher than that of extracellular matrix proteins such as collagen and elastin. Therefore, LSS is well suited to assess a number of important morphological parameters, such as density, porosity, organization, and remodeling that may occur as tissues develop. In this chapter, we will specifically discuss the use of light scattering to assess deposition and organization of mineral deposition on silk films.

6.3   Technical Implementation of LSPS

6.3.1   Experimental Considerations

A number of systems have been developed for the acquisition of LSPS measurements from tissues. They all include (1) a light source that can deliver light at a variety of wavelengths in the visible range (i.e., 400–700 nm), (2) a means to isolate light into distinct wavelength regimes (either a spectrometer or a series of optical bandpass filters), (3) optical elements that deliver and collect light from a specimen (lenses/mirrors or optical fibers), and (4) a detector. In this section, we describe instruments that are designed either on a lab bench or an optical fiber platform.

6.3.1.1   Lab Bench Systems

The design of two types of lab bench systems is presented in Figure 6.1. The simplified schematic of the system presented in Figure 6.1a delivers collimated light onto the specimen. Light from a Xe light source is collimated through a lens, polarized using a linear polarizer, and delivered onto the specimen, typically at an angle to avoid collection of specular reflections (most actual systems include a set of two additional lenses and an iris to collimate well the highly divergent light from the Xe lamp). The backscattered light is directed by a beam splitter to a lens that is positioned a focal length away from the sample, so that all light scattered from the sample at a given angle with respect to the incident beam focuses at a distinct vertical position along the slit of a spectrograph. The light is dispersed as a function of wavelength by the spectrograph, and a charge-coupled device (CCD) detector is used so that intensities detected along different columns represent different wavelengths, while those detected along different rows represent different scattering angles. A representative LSS image is shown in Figure 6.1. An analyzer (i.e., a second linear polarizer) is placed in the detection path, so that data can be collected with it placed along the parallel and perpendicular orientations relative to the polarizer in the

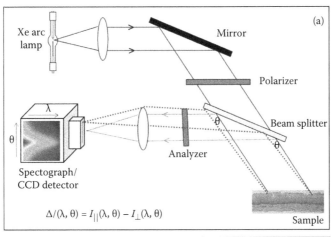

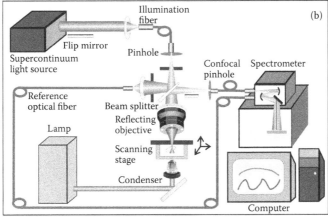

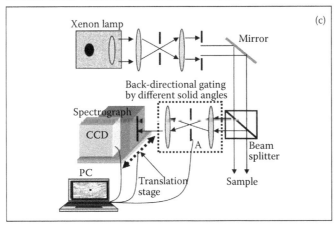

FIGURE 6.1 Schematics of set-ups designed to acquire LSS information (a, c) without, and (b) with depth-resolution capabilities. Images with each pixel containing full LSS spectral information are acquired using set-ups shown in b and c. Schematic in (b) reproduced from Itzkan et al. 2007. Schematic in (c) reproduced from Xu et al. 2010a.

illumination path. As mentioned above, the difference maps between the signal detected along the parallel and perpendicular polarizations include light that has maintained the original polarization and has been scattered only one or a few times. An example of such maps collected from undifferentiated human mesenchymal stem cells (hMSCs) that have undergone adipogenic and osteogenic differentiation is shown in Figure 6.2, along with the corresponding wavelength-dependent spectra collected at 1° (0.5–1.5°) about the exact backscattering direction. The high frequency variations detected only from the adipogenically differentiating cells likely originate from the lipid droplets that the cells have accumulated over 21 days of exposure to a medium that contains factors known to induce adipogenic differentiation (Rice et al. 2010). The osteogenically differentiating cells also exhibit distinct wavelength dependence from the nondifferentiating and adipogenically differentiating cells.

While this type of setup provides detailed wavelength and scattering-angle–dependent LSS information, it does not allow us to select the depth from within the sample where such signals originate, beyond the knowledge that the differential light scattering spectra represent light that has been scattered only once or very few times, and is therefore likely to emanate from the most superficial layers of the specimen. To achieve this type of depth selectivity, a number of schemes have been utilized.

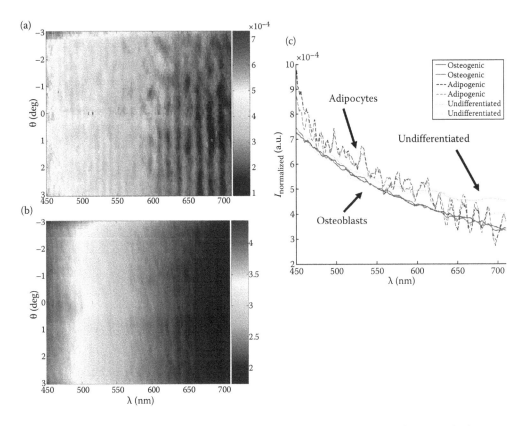

FIGURE 6.2　LSS maps from human mesenchymal stem cell (HMSC) cultures undergoing (a) adipogenic or (b) osteogenic differentiation. (c) LSS spectra acquired at a polar angle of 1 degree from HMSCs in propagation (undifferentiated), adipogenic or osteogenic media culture for 3 weeks.

One scheme has been proposed by Backman's group, who has shown that one can use low coherence illumination (with coherence lengths on the order of 140 µm) to detect an enhancement in the coherently backscattered light over a very small range of angles (less than ±1° about the exact backscattering direction) (Kim et al. 2004). Within this enhanced signal regime, light scattered at slightly larger angles (e.g., 0.25°) emanates from depths that are larger than light scattered at the exact backward direction (i.e., $\theta = 0°$). Analysis of rat colon tissues with this approach has shown that the wavelength dependence of light scattered at $\theta = 0$ is expected to travel through the uppermost 70 µm of the colonic mucosa and is more sensitive to very early cancer changes than light scattered at $\theta = 0.25°$ that traveled through the upper 40 µm of tissue (Roy et al. 2004, 2005; Kim et al. 2005). This technique could indeed be very sensitive to subtle morphological changes that could occur within cells and/or biomaterial scaffolds as tissues develop, but it confers sensitivity to depths that are smaller than the illumination coherence length, that is, usually less than 150 µm.

A different approach has been implemented by Huang et al. (2009), who focused light onto the sample with a microscope objective lens and placed a pinhole in the detection path confocal to the focal plane of the objective in order to eliminate light scattered above and below focus. This setup yielded an axial resolution of 30 µm, but the depth of penetration could be extended to hundreds of microns. Very high levels of axial resolution (on the order of 3 µm) were achieved with the setup shown schematically in Figure 6.1b (Itzkan et al. 2007). A supercontinuum laser source was used for illumination in this case to provide intense illumination, which was focused on a point on the sample. Light scattered from that point was imaged onto a confocal pinhole to eliminate out-of-focus light and was detected using a spectrograph and CCD camera. Spectral angular information is lost in this scheme, but sensitivity to particles as small as 10 nm is achievable by analyzing the acquired wavelength-dependent spectra.

LSS imaging setups have also been demonstrated (Gurjar et al. 2001; Xu et al. 2010a). As shown in Figure 6.1c, a 4f imaging detection scheme is typically employed to create an image of the light scattered from the sample onto a detector. A pinhole or iris positioned at the focal point between the two lenses is used to select the range of scattering angles that is collected by the detector. Wavelength-dependent information is collected by placing a series of bandpass filters in the illumination path and collecting distinct images at each spectral band. Alternatively, the sample can be illuminated by white light, and a spectrograph can be placed at the imaging plane to disperse the wavelength-dependent information onto a CCD camera. The spectrograph and CCD can be scanned along the direction perpendicular to the spectrograph slit to acquire LSS information for a range of x–y locations within the specimen. The angular dependence of backscattered light detected using low coherence interferometric imaging can also be used to provide sensitivity to the size and refractive index of scatterers over depths that extend to the range of a millimeter (Wax et al. 2002; Zhu et al. 2011). This approach has been developed by the Wax group and implemented using both lab bench and optical fiber probe designs.

6.3.1.2   Optical Fiber Probe Systems

Indeed, while often lab bench systems are developed initially to assess the potential of a given approach, one of the advantages of light scattering–based measurements is that it is possible to translate such measurements to platforms that rely on optical fiber probes for light delivery and detection to and from the specimen. As shown in Figure 6.3, such

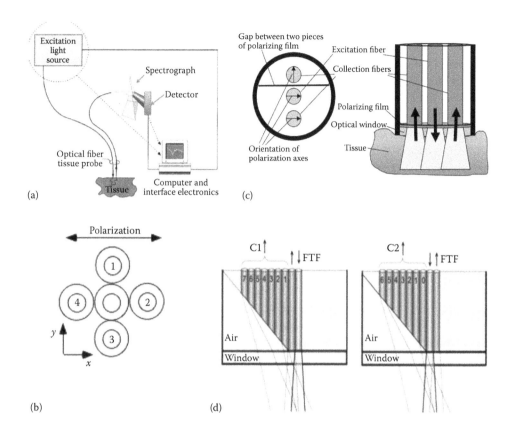

FIGURE 6.3 Schematics of fiber-based LSS set-ups. (a) Single fiber for delivery and single fiber for collection of light. Polarizers are not included in this probe. (From Marenzana, M. et al., *Tissue Eng* 8(3): 409–418, 2002.) (b) Linear polarizer placed on top of probe. Central fiber is used for delivery. Fibers 1–4 are used for collection. (From Johnson, T. M. and J. R. Mourant, *Opt Express* 4: 200–216, 1999.) (c) Central fiber is used for delivery, bottom fiber for collection of light polarized parallel to incident light, and top fiber used for collection of light polarized perpendicular to incident light. (From Myakov, A. et al., *J Biomed Opt* 7: 388–397, 2002.) (d) Beveled fiber for collection of depth-sensitive LSS information. Fiber highlighted in yellow used for delivery. (From Nieman, L. T. et al., *Opt Express* 17(4): 2780–2796, 2009.)

setups are very simple, cheap, and highly portable. The complexity of such probes varies tremendously. In the simplest case, an optical fiber is used to deliver light to the specimen, and a single optical fiber is used to detect the scattered light (Figure 6.3a) (Marenzana et al. 2002). Often, two to six fibers are placed around the delivery fiber to maximize the levels of the detected signal (Müller et al. 2002).

Polarization-sensitive measurements are certainly possible using optical fiber probes. An original design included a linear polarizer attached to the surface of the optical fiber probe that polarizes the incident light along a given direction and allowed light backscattered along the same orientation to be detected by the collection fibers (Johnson and Mourant 1999; Mourant et al. 2001). However, most recent designs include two pieces of linear polarizing films oriented perpendicular to each other that allow detection of light scattered in either parallel or perpendicular orientation relative to the incident light by fibers that are typically symmetrically positioned about the delivery fiber (Myakov et al. 2002; Nieman et al. 2004). This allows extraction of the differential intensity of light backscattered along the parallel and perpendicular orientations relative to the incident polarization, which, as discussed below, allows straightforward extraction of information with regards to the size and refractive index of superficial scatterers.

Depth selectivity is also possible to achieve in fiber-optic LSPS measurements, and a number of probes have been developed (Nieman et al. 2004, 2009). As shown in Figure 6.3, depth sensitivity is achieved either by placing the collection fibers at an angle from the light delivery fiber or by beveling the surface of the probe. The latter allows for more compact probe designs that are often desirable for endoscopic or laparoscopic procedures. In principle, light emanating 1–2 mm below the probe surface could be detected using such probes with depth sensitivity that is on the order of hundred(s) microns. The depth sensitivity of angle-resolved low-coherence interferometry (aLCI) is significantly higher, albeit at the cost of a somewhat more complex and more expensive system design (Zhu et al. 2011).

6.3.2 Theoretical Background on LSPS

LSPS is concerned with the elastic scattering of electromagnetic waves from an arbitrarily sized and shaped particle or medium. The scattering event is characterized by the change in directional momentum of the incident electromagnetic wave, q, as defined in Figure 6.4, where $q = k_s - k_i$; k is the plane electromagnetic wave angular wave number, with magnitude $k = 2\pi/\lambda$; λ is the wavelength of the electromagnetic wave; i and s refer to incident and scattered electromagnetic waves, respectively; d is the size of the scatterer; and θ and φ are the polar and azimuthal scattering angles, respectively. By simple geometry, it follows that the magnitude of scattering momentum change is given by $q = 2k\sin\dfrac{\theta}{2}$. In biomedical optics experiments, the scattered electromagnetic wave is typically visible or near-infrared light.

The essence of LSPS is to derive morphological properties of a scattering object (e.g., size and refractive index) based on the spectral, angular, and polarization dependence of the scattered light intensity, $I(q)$. The simplest approach to solving this problem is to assume the scatterers consist of independent, homogeneous spheres. An exact, numerical solution to the scattering problem can then be evaluated via Mie theory. Alternatively, various analytical approximations to the spherical scattering system can be implemented (e.g., Rayleigh, Rayleigh–Gans, or van de Hulst), depending on the magnitude of the particle diameter relative to the wavelength of the incident light and on the value of the particle relative refractive index.

A more general approach for analyzing LSPS data in biophysical applications is to consider an arbitrarily shaped, inhomogeneous scattering object. One such method is the finite-difference time-domain (FDTD) technique, in which Maxwell's equations are

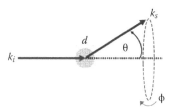

FIGURE 6.4 Schematic diagram of elastic scattering event.

Table 6.1   Range of Validity and Relative Merits of Various LSS Analysis Methods (See Text for Details)

LSS Analysis Method	Limitations	Advantages
Rayleigh	$d \ll \lambda$ (point-like particles)	Analytical methods provide physical insight and ease of calculation
Rayleigh–Gans	$\lvert m - 1 \rvert \ll 1$ (weak scatterer) $kd\lvert m - 1 \rvert \ll 1$ (small phase lag) Spherical, homogeneous scatterer	
Van de Hulst	$1 < m < 2$; allows $d \gg \lambda$ Spherical, homogeneous scatterer Provides only σ_{tot} or $\left(\dfrac{d\sigma}{d\Omega} \right)_{\theta=0^0}$	
Born	Allows irregular shape and density inhomogeneities Weak scatterer (small phase lag and relative density fluctuations)	
Mie	Spherical, homogeneous scatterer Dielectric or metallic Accounts for scattering and absorption	Exact numerical solution Unlimited range of d and m
FDTD	Computationally intensive	Exact numerical solution Allows irregular shape and density inhomogeneities

solved exactly for a well-defined distribution of spatial density fluctuations. Another common approach is to assume a fractal organization of the refractive index spatial inhomogeneities in a scattering object and to apply the first Born approximation for solving the scattering problem. These various analysis methods will be discussed below. Their limitations and relative merits are summarized in Table 6.1.

6.3.2.1   LSPS Analysis Methods

6.3.2.1.1   Analytical Approximations for Spherical Scatterers

6.3.2.1.1.1 Rayleigh Approximation ($d \ll \lambda$)   The Rayleigh approximation applies to scattering objects whose length scales, d, are significantly shorter than the wavelength of the scattered light. The intensity of Rayleigh scattered light has a well-known λ^{-4} wavelength dependence, as well as a polarization-sensitive $\cos^2 \theta$ polar angle dependence associated with electric dipole radiation. For a spherical scatterer of diameter $d \ll \lambda$, illuminated by a linearly polarized plane electromagnetic wave of intensity I_0, the scattered light intensity along a plane parallel to the incident polarization is given by

$$I(\lambda,\theta) = I_0 \frac{8\pi^4 d^6}{r^2 \lambda^4} \left(\frac{m^2 - 1}{m^2 + 2} \right)^2 \cos^2 \theta \tag{6.1}$$

where r is the distance from the scatterer to the observer, and m is its refractive index relative to the surrounding medium (Kerker 1969).

It is rare to find Rayleigh scattering to be the dominant mechanism in LSPS experiments in biological cells and tissues. However, in certain tissues rich in collagen fibrils, such as human skin, a Rayleigh scattering component has been detected at short visible wavelengths (Saidi et al. 1995). Increased Rayleigh scattering has also been correlated to the severity of human oral cancers and associated with an increase in density of small intracellular particles due to epithelial proliferation (McGee et al. 2009).

6.3.2.1.1.2 Rayleigh–Gans Approximation The Rayleigh–Gans approximation is an extension of Rayleigh scattering, which allows for interference effects, in the far-field, between scattered electromagnetic fields generated at different locations within a homogeneous sphere. It assumes that each infinitesimal volume element within the scattering objects acts as an electric dipole oscillator, and that the accumulated phase shift of the incident wave front traversing it is negligible (i.e., weak scatterer approximation). The Rayleigh–Gans model is a particular case (spherical scatterer) of the first Born approximation (Born and Wolf 1999). For incident light polarized parallel to the scattering plane, the Rayleigh–Gans model predicts scattered light intensities, which are a combination of a Rayleigh scattering term and an oscillatory form factor, $F(\theta)$:

$$I(\lambda,\theta) = I_0 \frac{8\pi^4 d^6}{r^2\lambda^4}\left(\frac{m^2-1}{m^2+2}\right)^2 \cos^2\theta . F(\theta) \tag{6.2}$$

$$F(\theta) = \left(\frac{9\pi}{2u^3}\right) J_{\frac{3}{2}}^2(u) = \left[\left(\frac{3}{u^3}\right)(\sin u - u\cos u)\right]^2 \tag{6.3}$$

where $J_{\frac{3}{2}}(u)$ is a Bessel function of order 3/2 and $u = \frac{1}{2}qd$ (Kerker 1969).

In order for the Rayleigh–Gans approximation to apply, the following inequalities must hold:

$$|m - 1| \ll 1 \text{ (weak scatterer)} \tag{6.4}$$

$$kd|m - 1| \ll 1 \text{ (negligible phase lag)} \tag{6.5}$$

The Rayleigh–Gans approximation is thus valid for the scattering of visible light from submicron biological features or organelles, whose typical relative refractive index is around $m = 1.05$.

6.3.2.1.1.3 van de Hulst Approximation ($d \gg \lambda$) For homogeneous spherical particles whose diameters are significantly larger than the wavelength of scattered light, the van de Hulst formula provides a useful approximation that encompasses scattering over a broad range of relative refractive indices ($1 < m < 2$) (van de Hulst 1982). The

Chapter 6

model is based on evaluating the far-field forward differential scattering cross section, $\left(\dfrac{d\sigma}{d\Omega}\right)_{\theta=0^0}$, as the interference between a diffractive term (treating the sphere as an absorptive disc of diameter d) and a refractive term (considering only the phase lag of central rays, which pass virtually undeflected through the center of the sphere). Application of the optical theorem then yields the following expression for the total scattered power, $P(\rho)$:

$$P(\rho) = I_0\sigma = I_0\left(\frac{\pi d^2}{4}\right)\left[2 - \frac{4}{\rho}\sin\rho + \frac{4}{\rho^2}(1-\cos\rho)\right] \tag{6.6}$$

where σ is the total scattering cross section and $\rho = kd|m - 1|$ is the phase lag acquired by a light ray passing through the sphere along a full diameter (van de Hulst 1982). Salient features of this equation include highly oscillatory wavelength dependence at large values of ρ, with oscillation frequency proportional to the particle size. The total scattering cross section, σ, exhibits a limiting value equal to twice the geometric cross section of the particle, which accounts for the term "anomalous diffraction" used to describe this scattering regime. The van de Hulst approximation has been used to correlate spectral oscillations in LSPS spectra to variations in cell nucleus diameter of normal and cancerous colonic cells (Perelman et al. 1998).

6.3.2.1.2 First Born Approximation   The analytical approximations given above provide physical insight into how the size and refractive index of an object can affect its angular and spectral light scattering properties. However, aside from a limited subset of biological specimens or subcellular organelles (e.g., cell suspensions, cell nuclei, lysosomes), the assumption of a homogeneous spherical object is seldom a valid representation of biological cell or tissue scatterers. A more realistic approach is to define a biological scattering object by its continuous refractive index in space, $n(r)$, and to consider how fluctuations in $n(r)$ affect the far-field intensity of light scattered from that object. In the case of weak scattering, for which the incident light intensity may be considered uniform throughout the scattering object's extent and the maximum accumulated phase lag is negligible, the first Born approximation (also known as Rayleigh–Debye theory) may be applied. According to this model, if the fluctuations in $n(r)$ are small relative to the mean value, $\bar{n}(r)$, the scattered light intensity is proportional to the power spectral density (PSD) of these fluctuations (Born and Wolf 1999):

$$I(q) = \frac{k^4}{16\pi^2}\left|\int [n^2(r')-1]e^{-iq\cdot r'}d^3r'\right|^2 \tag{6.7}$$

This equation can also be recast, via the Wiener–Khinchin theorem, as a function of the autocorrelation function $C(r) = \int [n^2(r')-1][n^2(r'+r)-1]d^3r'$:

$$I(q) = \frac{k^4}{16\pi^2} \int C(r') e^{-iq\cdot r'} \mathrm{d}^3 r' \tag{6.8}$$

Thus, LSPS measurements of $I(q)$ may be used to quantify spatial correlations in the refractive index fluctuations of a scattering object by means of a simple inverse Fourier transformation. An illustrative example would be the case of a monochromatic plane wave scattering from an infinite sinusoidal grating, for which Equation 6.8 would predict a pair of angular intensity delta functions in the far field (corresponding to the angular positions of the first-order diffraction lines).

In biophysical LSPS experiments, Equation 6.8 is often invoked to associate a submicron fractal organization of cells and tissues to a measured inverse power law in the wavelength dependence of LSPS spectra, $I(\lambda) \propto \lambda^{-\gamma}$, where γ is a constant typically in the range $1 < \gamma < 2$ (Moscoso et al. 2001; Xu and Alfano 2005; Hunter et al. 2006). The rationale behind these models is that fractal objects exhibit scale invariance and are thus characterized by inverse power law autocorrelation functions, $C(r) \propto r^{-\delta}$ (whose Fourier transforms are also inverse power laws, with power exponent $\gamma = \delta + 1$) (Raine 1981; Hunter et al. 2006). We note, however, that the inversion of Equation 6.8 is an ill-posed problem without a unique solution for $n(r)$, that is, there is more than one possible refractive index distribution, $n(r)$, which could account for a particular inverse power law exponent, δ, in the autocorrelation function. Various types of fractal models have, therefore, been proposed by different researchers to account for inverse power law LSPS spectra, and these are summarized below.

6.3.2.1.3 Numerical Methods

6.3.2.1.3.1 Mie Theory
Mie theory involves the numerical solution of Maxwell's equations for a plane electromagnetic wave incident on a homogeneous sphere of arbitrary size and refractive index (Mie 1908). This scattering problem can be characterized in terms of the Jones matrix, which relates the incident (i) and far-field scattered (s) electric field amplitudes parallel and perpendicular to the scattering plane, $E_\parallel$ and $E_\perp$, via (Bohren and Hoffman 1998)

$$\begin{pmatrix} E_{\parallel,s} \\ E_{\perp,s} \end{pmatrix} = \begin{pmatrix} S_1 & S_3 \\ S_4 & S_2 \end{pmatrix} \begin{pmatrix} E_{\parallel,i} \\ E_{\perp,i} \end{pmatrix} \tag{6.9}$$

The scattering plane is defined by the incident wave and the scattered wave directions. For an object with spherical symmetry, the amplitude functions $S_3 = S_4 = 0$, while S_1 and S_2 are given by an expansion in associated Legendre polynomials, $P_n^m(\cos\theta)$ (Kerker 1969):

$$S_1 = \sum_{n=1}^{\infty} \frac{2n+1}{n(n+1)} [a_n \pi_n(\cos\theta) + b_n \tau_n(\cos\theta)](-1)^{n+1} \tag{6.10}$$

$$S_2 = \sum_{n=1}^{\infty} \frac{2n+1}{n(n+1)} [a_n \tau_n(\cos\theta) + b_n \pi_n(\cos\theta)](-1)^{n+1} \tag{6.11}$$

where the angular functions are defined via

$$\pi_n(\cos\theta) = \frac{P_n^1(\cos\theta)}{\sin\theta} \tag{6.12}$$

$$\tau_n(\cos\theta) = \frac{d}{d\theta} P_n^1(\cos\theta) \tag{6.13}$$

Equations 6.10 through 6.13 may be solved numerically on a standard personal computer; several software packages are publically available to perform this task (Grainger et al. 2004; Bohren and Hoffman 1998; Laven 2011). The far-field scattered light intensities, parallel and perpendicular to the scattering plane, may then be evaluated as follows:

$$I_\parallel(\lambda) = \frac{\lambda^2}{4\pi^2 r^2} |S_1|^2 \sin^2\varphi \tag{6.14}$$

$$I_\perp(\lambda) = \frac{\lambda^2}{4\pi^2 r^2} |S_2|^2 \cos^2\varphi \tag{6.15}$$

In biomedical optics, Mie theory is often applied to model the light scattering properties of cells in suspension as well as *in vivo* cell nuclei. These are typically described by a Gaussian distribution of spherical diameters and can accurately account for the small (less than a few percent) oscillatory LSPS spectral component often observed in the backscattered light from cell monolayers and animal tissues (Backman et al. 1999; Mujat et al. 2008; Hsiao et al. 2011). In addition, Mie theory has also been invoked to account for an inverse power law wavelength dependence of LSPS spectra, which often dominates the backscattered light signal in biological specimens (see discussion below).

6.3.2.1.3.2 FDTD Methods FDTD simulations can examine the time-dependent evolution of electromagnetic fields inside a scattering object via Maxwell's equations. Although much more computationally intensive than Mie theory, FDTD methods can be applied to a scatterer of arbitrary size and spatial refractive index distribution, $n(r)$, and thus provide realistic scattering amplitudes that reflect the internal inhomogeneities of biological tissues and subcellular organelles at the submicron scale. It has been applied, for example, in examining how changes in the subnuclear chromatin density of precancerous epithelial cells, in the human ovary and cervix, can affect their angular scattering properties relative to those of normal cells (Drezek et al. 1999, 2003).

6.3.2.2 LSPS of Animal Cells and Tissues

6.3.2.2.1 Polarization Gating The anisotropy parameter, g, is a measure of the average polar scattering angle due to a single scattering event from an arbitrary object and is defined as the expectation value of $\cos\theta$:

$$g = 2\pi \int_0^\pi p(\theta)\cos\theta\sin\theta\,d\theta = \int_{-1}^1 p(\cos\theta)\cos\theta\,d(\cos\theta) \equiv <\cos\theta> \qquad (6.16)$$

where $p(\theta)$ is the scattering phase function. In biological cells and tissues, elastic scattering is predominantly clustered around the forward scattering direction, giving rise to typical anisotropy parameter values in the range $0.8 < g < 0.97$ (Flock et al. 1987). This implies that most of the light that is illuminated onto the surface of a biological specimen will retain its original directionality, and only after multiple scattering events (typically >10) will such scattered photons resurface from the specimen. This high number of collisions randomizes the direction and polarization of the initially launched photons; thus, the majority of reflected light from biological tissues is diffusely scattered light and can be successfully analyzed by photon diffusion models (Patterson et al. 1989; Fantini et al. 1994). Such studies are particularly well suited for monitoring changes in blood oxygenation in subepithelial or deep tissues (depth of between 1 mm and several centimeters from the tissue surface) or for the detection of cancerous tumors (Yu et al. 2010; Sassaroli et al. 2011).

In many biomedical optics applications, however, it is of interest to examine only the singly scattered photons from the topmost layer of a biological specimen. This applies to the early detection of epithelial neoplasia, which is primarily localized to the proliferating cell layer immediately above the basal membrane (typically at depths of 0.1–1 mm from the tissue surface). Another such area of interest would be the morphological characterization of biomineralized deposits, such as bone and teeth surfaces. LSPS spectroscopy allows these studies to be performed by discriminating against the (dominant) diffusely scattered light component. This is achieved by polarization gating, that is, by imparting linear polarization onto the incident electromagnetic waves, and then selectively detecting the backscattered light intensities at polarizations parallel and perpendicular to the incident beam, $I_\parallel(q)$ and $I_\perp(q)$. Single-scattering events, especially for particles of a size smaller than the wavelength of light, tend to strongly preserve the initial polarization state. Thus, by monitoring the residually polarized scattered light intensity, $\Delta I(q) = I_\parallel(q) - I_\perp(q)$, the diffusely scattered component vanishes and the singly scattered light from the topmost layer becomes the dominant contributor to the detected light signal.

Although large particles, such as cell nuclei, scatter predominantly in the forward direction, a small but detectable fraction of their scattering will cluster closely around the exact backward direction (the so-called "glory scattering"). In these cases, the residually polarized backscattered component, if averaged over all azimuthal collection angles, tends to zero. However, for spherical particles, $I_\parallel(q)$ and $I_\perp(q)$ have complementary azimuthal angular dependences, that is, a maximum in intensity for one polarized component corresponds to a minimum in the other. This is in contrast with the

Chapter 6

backscattered light from particles of a small size compared to the wavelength of the light, for which the azimuthal scattering function is isotropic. Thus, by performing both polarization gating and azimuthal angle selection in backscattered light studies, it is possible to not only discriminate against diffusely scattered light from the deeper regions of a biological specimen but also minimize or eliminate altogether the contribution from submicron particles close to the specimen surface. This variant of the LSPS technique has been shown to discriminate between normal and cancerous *ex vivo* colon tissues by characterizing the increase in nuclear diameter associated with cancerous progression in cells (Yu et al. 2006).

6.3.2.2.2 LSPS Characterization of Submicron Cell and Tissue Morphology

As predicted by Mie theory, scattering of visible light from dilute cell suspensions or cell monolayers will be dominated, in both forward and backward geometries, by spectral oscillations associated with whole-cell or nuclear scattering. The residually polarized backscattered light from most animal tissue surfaces, however, exhibits a characteristic inverse power law spectral dependence, $\Delta I(\lambda) \propto \lambda^{-\gamma}$, with power exponent γ typically in the range $1 < \gamma < 2$ (Backman et al. 2001; Hunter et al. 2006). The lack of oscillatory structures in such spectra is commonly understood to indicate that small (submicron) particles are the dominant contributors to $\Delta I(\lambda)$ spectra in biological specimens. The precise morphology of such small tissue scatterers, though, remains under dispute.

It is possible to account for an inverse power law $\Delta I(\lambda)$ wavelength dependence using Mie theory. An ensemble of independent, homogenous spheres of equal refractive index and an inverse power law size distribution, $N(d) \propto d^{-\beta}$, will exhibit inverse power law backscattering spectra with exponent $\gamma = \beta - 3$, provided the range of diameters is sufficiently large. Several researchers have modeled LSPS spectra from cell monolayers and animal tissues in this manner, invoking an inverse power law size distribution of submicron, subcellular spheres, typically in the range 10 nm $< d < 2$ μm, as the dominant scatterers in these systems (Backman et al. 2001; Mujat et al. 2008). Except for exceptional cases, however, such as adipocytic cells rich in subcellular lipid droplets, the assumption of a broad distribution of spherical droplets of equal refractive index vastly oversimplifies the morphological complexity of typical biological cells.

A more realistic approach considers the continuous spatial fluctuations in subcellular or tissue refractive index and how these interact with an incident electromagnetic wave. The typically low relative refractive index of biological specimens allows implementation of the first Born approximation to these scattering systems. Under these conditions, an inverse power law LSPS spectrum can be associated with an inverse power law autocorrelation function, $C(r) \propto r^{-\delta}$ (see Equation 6.8). The power exponent, δ, is related to the fractal dimension of the scattering system, which quantifies its scale-invariant morphological organization. Inversion of Equation 6.8, however, is an ill-specified problem, and various distinct fractal models may be invoked to account for the same inverse power law $\Delta I(\lambda)$ spectrum.

One of the earliest fractal models used in interpreting these results is based on a mass fractal (self-similar) organization of subcellular density fluctuations (Wax et al. 2002)— where self-similarity refers to the quality of an object to remain invariant in form under varying scale or magnification. More precisely, a self-similar object is one whose (fractal) dimension, D, is given by $D = \log N/\log r$, where N is the length of the object measured with a ruler of length r (Voss 1986). The inverse power law wavelength dependence is a well-established phenomenon in the light scattering properties of diffusion-limited aggregates, which exhibit a mass fractal spatial organization of its constituent building units, that is, the total mass of aggregate enclosed within a sphere of radius R scales as $M(R) \propto R^D$, with the fractal dimension given by $D = 3 - \delta$ (Teixeira 1986). However, the validity of this model for biological cells and tissues remains unclear, as there are no independent, direct determinations of any type of mass fractal organization in these systems.

Several studies, however, have shown that submicron density inhomogeneities in biological specimens can exhibit statistically self-affine fractal correlations, that is, the ensemble-averaged variance, $S(x)$, of the specimen refractive index, $n(x)$, is given by $S(x) = |n(x + a) - n(x)|^2 \propto |a|^{2H}$, where the Hurst parameter, H, is limited to the range $0 < H < 1$. Schmitt and Kumar (1996), for example, have observed self-affine fluctuations of various mouse and human tissues, at the submicron scale, via phase contrast microscopy. Einstein et al. (1998a,b) have demonstrated self-affine organization of chromatin in benign and cancerous human breast epithelial cells via bright-field microscopy, and reduced nicotineamide adenine dinucleotide (NADH) autofluorescence microscopy imaging studies have shown a self-affine spatial organization of mitochondrial networks in normal and precancerous engineered tissues (Levitt et al. 2007). Trabecular bone has also been found to exhibit a self-affine morphology (Dougherty and Henebry 2001).

In light of these findings, several researchers have opted for a forward modeling approach for LSPS, in which cell and tissue inhomogeneities are assumed to have a self-affine organization. Moscoso et al. (2001) first implemented a limited self-affine scattering model to account for depolarization properties of biological tissue; the model assumed an exponential form of the tissue autocorrelation function, which is a specific case of the more general von Karman autocorrelations that characterize self-affine systems. This restriction in the model was later removed by Xu and Alfano (2005), who considered a range of possible exponential correlation functions weighted by an inverse power law distribution of correlation lengths. The weighting function is chosen arbitrarily, however, and the physical interpretation of some of their derived fractal dimensions in human tissues is unclear, given that $D > 4$ [i.e., greater than the Euclidean dimension (D_E) + 1, which is the accepted upper range for a self-affine object (Voss 1986)].

Hunter et al. (2006) have proposed a simpler generalization of the Moscoso approach, in which the first Born approximation is solved using the generalized von Karman autocorrelation function to model the tissue inhomogeneities. Under these conditions, the spectral dependence of residually polarized light can be shown to follow

$$\Delta I(\lambda) \propto \lambda^{-4} \frac{1}{[1+(4\pi L/\lambda)^2]^\alpha} \tag{6.17}$$

where the exponent α is related to the Hurst parameter, H, via $H = \alpha - D_E/2$; H is limited to the range $0 < H < 1$; L is the fractal upper scale of the scattering system (the upper bound of fractal correlation lengths); and the fractal dimension may be evaluated as $D = D_E + 1 - H$ (Voss 1986). If the fractal upper scale is significantly larger than the wavelength of the scattered light, Equation 6.17 reduces to $\Delta I(\lambda) \propto \lambda^{2\alpha-4}$ and may thus account for an inverse power law LSPS spectrum observed from a biological specimen. A Hurst parameter of $H = 0.5$ would be associated with Brownian noise statistics—in particular, the autocorrelation would be given by the running integral of a purely uncorrelated, Gaussian noise function. Values of $H < 0.5$ would imply spatially anticorrelated fractional Brownian organization of the tissue refractive index, whereas values of $H > 0.5$ would indicate persistent, longer-range correlations to be present. Previous determinations of Hurst parameters in various normal human cells and tissues have shown values $H < 0.5$ (Schmitt and Kumar 1996; Levitt et al. 2007).

6.4   Representative Applications of LSPS to Assess Important Morphological and Organizational Parameters of Engineered Constructs

6.4.1   Use of LSPS to Assess Collagen Content and Organization

Collagen plays a key role in the development and repair of multiple tissues. The concentration, diameter, and organization of collagen fibers often dictate the mechanical, and in tissues like the cornea also the optical, properties of tissues. A series of LSS measurements have been performed that suggest that important collagen characteristics can be assessed using ultimately noninvasive or minimally invasive measurements (Figure 6.3). Therefore, in principle, this information can be obtained *in vivo* or *in situ*. The instrument used in several of these studies was a simple fiber-optic setup, described in the previous section and shown in Figure 6.3a (Marenzana et al. 2002). The light delivery and collection fibers had core diameters of 400 and 200 µm, respectively, and were placed 50 µm apart. Measurements were performed using type I rat tail collagen gels (2.03 mg/ml) embedded with 10^6 human fibroblasts. Gels were prepared under different tension conditions to simulate specimens with varying degrees of collagen alignment. The fiber optic probe was placed in contact with the samples and was used to make measurements dynamically over multiple days. The intensity scattered at 500 nm normalized to the intensity and slope of the spectrum in the 615–810 nm region was highly anticorrelated with the area of gels allowed to contract over 60 h while floating, that is, not being subjected to any tension. When gels were allowed to contract under tension, there was an initial increase in the normalized scattering intensity at 500 nm, which reached a plateau following 400 min. When the gel was released from the substrate, there was an abrupt increase in the measured scattering as the gel contracted, reaching equilibrium within 15 min. This increase in light scattering intensity was shown to correlate highly with an increase in tensile forces generated by the construct (Marenzana et al. 2002). This study highlighted the use of this approach to make dynamic measurements of changes in collagen gel contraction and the corresponding tensile forces without altering the sample in any way.

Follow-up studies were performed with gels that were tethered on two opposing ends that defined the direction of principal strain of the gel and could be subjected to tensile loading (Kostyuk and Brown 2004). LSS measurements were acquired with a very similar setup and with a probe that had similar size fibers but significantly higher source-detector separation (2.75 mm instead of 350 µm). This increase in the distance between the fibers effectively increased the tissue volume that was probed by the measurements. In addition, this probe was rotated about the axis of principal strain at 45° intervals. As the gels contracted over 72 h, there was a significant increase in the LSS signal collected perpendicular to the principal strain axis compared to that collected along the parallel direction in areas of the gel within which the fibers were aligned (Figure 6.5). The LSS signal was isotropic in noncontracted gels or areas of the contracted gels where the fibers were randomly organized. An anisotropy factor (AF), defined as the backscattered signal collected at 90° over that collected at 0° relative to the principal strain axis, was defined as a simple metric of this effect. This factor was wavelength-dependent and exhibited the highest values at longer wavelengths. There was a strong correlation between the AF calculated at 400 nm and the applied strain as well as the time of contraction. Comparisons between acellular gels that were contracted by applied tension and gels contracted by fibroblasts suggested that collagen fiber alignment (and not cell alignment) was the main contributor to the observed anisotropy (Kostyuk and Brown 2004).

Interestingly, similar measurements performed on excised equine superficial digital flexor tendon samples yielded a significant backscatter anisotropy along the axis of fiber alignment, which was enhanced by applying strain parallel to this axis (Kostyuk et al. 2003). On the other hand, LSS measurements performed with a source-detector fiber separation that was 300 μm exhibited anisotropy along the perpendicular direction. This

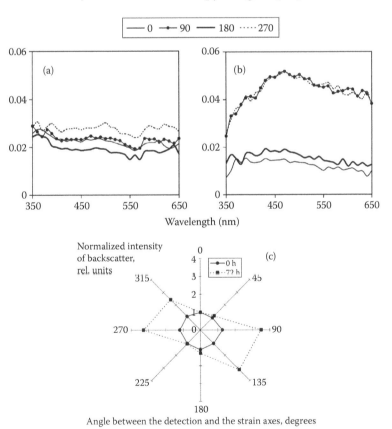

FIGURE 6.5 Backscattered light spectra acquired from the central zone of: (a) a non-contracted gel, and; (b) a 72 h contracted gel. (c) Radial diagram of the intensity of backscattered light at 700 nm from the noncontracted (circles) and contracted (squares) gels. (From Kostyuk, O. and R.A. Brown, *Biophys J* 87(1): 648–655, 2004.)

discrepancy was attributed to the fact that the LSS measurements performed with the large source-detector separation fibers were sensitive to collagen fibers of the interior tendon residing up to 3 mm from the surface, while the small source-detector separation probe was sensitive to the superficial tendon layer, where the fibers align at least in part along the circumference of the tendon. Enhancements in the anisotropy factor at 600 nm along the direction perpendicular to the tendon axis upon loading were also observed in measurements of cadaveric rabbit tendons with a probe that had the light delivery-collection fibers placed approximately 350 µm apart (Morgan et al. 2006). This enhancement was attributed to changes in the collagen fibril diameter and fibril straightening.

A strong correlation between backscattered intensity and collagen concentration as assessed by trichrome staining has also been reported in a diabetic rat wound healing model (Weingarten et al. 2008). The probe in that study consisted of 600 randomly mixed light delivery and collection fibers with 50 µm core diameters covering an area of approximately 3 mm in diameter. Light scattering data in the 630 to 700 nm were collected. It was assumed that this probe is sensitive to the upper 100 to 300 µm of tissue.

A somewhat more sophisticated LSS-based imaging setup, which employs spatial filtering to suppress diffusely scattered light and enhance the detection of light scattered from a specimen in the backward direction after undergoing one or very few scattering events, has been used to detect nanoscale changes in the structure of mineralized collagen fibers in excised bovine cortical bone specimens (Xu et al. 2010b). The specimens were approximately 200 µm thick with an area of 15 × 15 mm. LSS data were recorded in the 500 to 700 nm range, and both wavelength- and intensity-dependent changes were noted for the bone area that experienced maximum deformation upon loading. These changes were attributed most likely to changes in refractive index induced by the formation of nanoscale voids. Thinning of the mineralized collagen fibers could also contribute to the observed LSS changes.

Overall, these studies demonstrate that the intensity and wavelength dependence of backscattered light can serve as a sensitive indicator of changes in the concentration and organization of collagen fibers. Changes in the size, organization, and orientation of the fibers all contribute to the observed LSS signals in a manner that is at least partially dependent on the design of the probe that is used. Further studies combining more detailed theoretical and experimental studies are needed to elucidate further the contributions of each component. Nevertheless, the use of a simple probe and instrument to acquire rapidly and dynamically over time information on collagen content and organization certainly promises to serve as a useful monitoring tool in several tissue engineering applications that involve collagen deposition and remodeling. The ability to extend these measurements into a minimally invasive platform for *in vivo* data acquisition is also an exciting and realistic prospect.

6.4.2   Optical Characterization of Hydroxyapatite Mineral Growth on Silk Films

Deposition of hydroxyapatite on aligned collagen fibers is the basis for biomineralization in teeth and bone. The development of engineered tissue for bone repair and replacement, therefore, requires a detailed understanding and control of the mineralization process in tissue scaffolds. Current approaches for assessing the morphology of biomineral deposits, such as scanning electron microscopy (SEM), are highly invasive and preclude a dynamic investigation of the deposition process. LSPS can thus provide a valuable, noninvasive alternative for real-time assessment of the extent of mineralization, as well as for characterizing the submicron morphology of natural and engineered bone.

Gupta et al. (2008, 2009) have recently explored the use of engineered silk scaffolds as templates for hydroxyapatite deposition and characterized the film morphology via LSPS and SEM imaging. In that work, mineralization was induced inorganically by treatment of the films with $CaCl_2(aq)$ and $Na_2HPO_4(aq)$ solutions. Three types of silk films were used in order to explore the possible effect of protein crystallinity (β-sheet content) and film morphology on the mineralization process: water-annealed, methanol-treated, and polyaspartic acid (PAA)–mixed silk films. The films were imaged by LSPS repeatedly over seven cycles of mineralization, and a subset of each type of film was set aside for SEM analysis, for comparison with LSPS results (Figure 6.6a).

Fourier analysis of the SEM images showed an inverse power law dependence of their radial PSD, $\Phi(k_r) \propto k_r^{-\beta}$, at high spatial frequencies in the range $0.8~\mu m^{-1} < k_r < 1.4~\mu m^{-1}$ (Figure 6.6b). The power exponents were typically in the range $1 < \beta < 2$, with water-annealed films showing markedly higher values of β than methanol-treated and PAA-mixed films. Through the Wiener–Khinchin theorem, these results imply an inverse power law autocorrelation of the mineral density fluctuations, $C(r) \propto r^{-\delta}$, at submicron

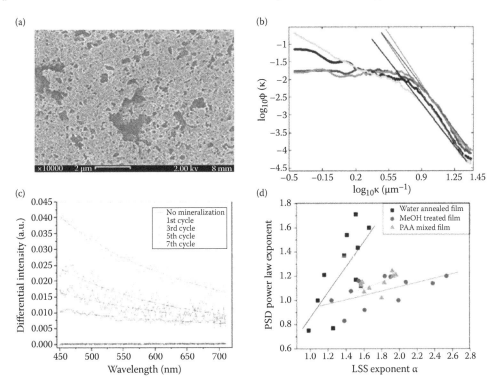

FIGURE 6.6 LSPS and SEM analysis of inorganic hydroxyapatite deposition on water-annealed silk films (From Gupta, S. et al., *Biomaterials* 29(15): 2359–2369, 2008.) (a) SEM image of mineral film after 5th deposition cycle; (b) PSD spectra of SEM images after 1st, 3rd, 5th and 7th mineralization cycles (purple, green black and cyan curves, respectively). Note high frequency power law behavior; (c) LSPS spectra of films after 1st, 3rd, 5th and 7th mineralization cycles. Smooth lines are fits using a self-affine fractal scattering model (see text for details); (d) correlation between LSPS and SEM power law exponents.

correlation lengths. As described in Section 6.3.2.1.2, such scale invariance is evidence of a regime of fractal organization. Although the specific type of fractal morphology (e.g., mass fractal vs. self-affine) cannot be unambiguously assigned from the PSD spectra alone, it is reasonable to assume that it is associated with a statistically self-affine mineral microstructure. Self-affinity is observed in a broad variety of depositional processes due to the inherent asymmetry in lateral versus vertical deposition dynamics (i.e., in the plane of film deposition vs. perpendicular to it) (Biscarini et al. 1996; Stoliar et al. 2010). Furthermore, trabecular human bone has been shown to exhibit a self-affine fractal architecture (Dougherty and Henebry 2001).

LSPS analysis of the films provided noninvasive characterization of the relative amounts of mineralization in each film type, as well as differences in their submicron morphology. LSPS spectra from PAA films showed the greatest backscattered light intensities, while the water-annealed films showed the least. Thermogravimetric analysis (TGA) indicated this result correlated well with the level of mineralization on each film. In addition, a portion of the films was also analyzed via temperature modulated differential scanning calorimetry (TMDSC) in order to determine their relative β-sheet content. A higher fraction of β-sheet in the silk film protein provides a higher degree of crystallinity and thermal stability, which can be measured via TMDSC. The results indicated that PAA-mixed films had the highest β-sheet content, followed by methanol-treated and water-annealed films.

The wavelength dependence of the LSPS spectra consistently showed a monotonic decrease in intensity at longer wavelengths. Figure 6.6c shows a set of LSPS spectra collected from a water-annealed silk film after one, three, five, and seven cycles of mineralization. The spectral dependence was well modeled by a modified inverse power law associated with scattering from a self-affine object (Equation 6.17, smooth lines in Figure 6.6c). Fractal upper scales extracted from the LSPS data in all films were limited to the range $150 < L < 350$ nm, corroborating the SEM finding of self-affine organization at submicron length scales. In the case of water-annealed films, these values of L are of the same order of the size of individual mineral flakes, which constitute the film surface layer (Figure 6.6a). The self-affine LSPS power exponents, α, were well correlated with the self-affine PSD power law exponents, β, indicating that the LSPS technique has significant potential as a noninvasive tool for characterizing surface layer morphologies at submicron scales (Figure 6.6d).

Finally, quantitative differences in the LSPS power exponents can be used to ascertain subtle changes in the type of self-affine organization of the various films. The Hurst parameter, H, is a common parameter used to quantify self-affine fractals and is related to the LSPS power exponent via $H = \alpha - D_E/2$ (Section 6.3.2.2.2). SEM images of water-annealed films show a surface morphology dominated by mineral flakes; thus, one may consider their dominant scatterer Euclidean dimension to be $D_E = 2$. This would imply that the measured LSPS power exponents in water-annealed films, which cover the range $1.75 < \alpha < 2.0$, can be associated with submicron mineral platelets having Hurst parameters in the range $0.75 < H < 1$ (corresponding to persistently, or positively, correlated density fluctuations of the platelets). By contrast, SEM images of the methanol-treated and PAA-mixed films show a more densely packed surface morphology more likely associated with a bulk scatterer ($D_E = 3$). The Hurst parameters derived for these films would thus fall in the range $0.25 < H < 0.5$, corresponding to antipersistent, or negatively, correlated three-dimensional fluctuations in film density. These results are in line with the sharp difference in SEM-LSPS correlations observed between water-annealed films on one hand and methanol-treated and PAA-mixed films on the other (Figure 6.6d).

In conclusion, Gupta et al. showed the LSPS technique to compare favorably with SEM as a tool for quantitative morphological assessment of the topmost layers of mineral deposits with the advantage of being a noninvasive optical method. Its high sensitivity allowed for early detection of hydroxyapatite mineralization on various engineered silk films, and the magnitude of the backscattered light signal was a good measure of the amount of mineral deposited on the films. Fourier analysis of the SEM images indicated a fractal morphology of the films at submicron scales. The spectral dependence of LSPS corroborated this finding, being well modeled by a modified inverse power law, which is associated with scattering from a self-affine object. Fractal parameters derived via SEM and LSPS were well correlated but were significantly different for water-annealed films than for methanol-treated and PAA-mixed films. LSPS may therefore serve as a valuable tool for the study of biomineralization processes in engineered tissues and ultimately for guiding the formation of artificial bone with the desired mechanical properties.

6.5 Conclusions

In summary, we have described here some of the theoretical and experimental foundations for performing light scattering measurements using polarized or unpolarized light to assess content and organization of collagen and mineral deposition, which are key components of several tissue engineering and regenerative medicine applications. The approaches described can be implemented using fairly simple and inexpensive instruments that can acquire data in a noninvasive or minimally invasive manner, repeatedly within the same sample. As these measurements are compatible with a fiber probe light delivery/collection geometry, they could be implemented using a highly portable platform that can be modified to access or enhance sensitivity to specific tissue depths. There are of course depth limitations, typically to within the upper 1–3 mm of tissue. The probe measurements we have described typically do not provide detailed imaging information but rather bulk estimates from the volume of the tissue that is sampled. However, LSS and LSPS measurements can be and have been performed in imaging mode, if more detailed spatial information is required. Certainly, the apparent exquisite sensitivity of these approaches, especially to tissue nanoscale organizational aspects that are not easy to examine, the simplicity of the devices, and their noninvasive nature make them promising tools for monitoring engineered tissues both *in vitro* and *in vivo*.

References

Backman, V., R. Gurjar et al. (1999). "Polarized light scattering spectroscopy for quantitative measurement of epithelial cellular structures." *IEEE J Sel Top Quantum Electron* **5**: 1019–1026.

Backman, V., M. B. Wallace et al. (2000). "Detection of preinvasive cancer cells." *Nature* **406**(6791): 35–36.

Backman, V., V. Gopal et al. (2001). "Measuring cellular structure at submicrometer scale with light scattering spectroscopy." *IEEE J Sel Top Quantum Electron* **7**(6): 887–893.

Berne, B. J. and R. Pecora (2000). *Dynamic Light Scattering: With Applications to Chemistry, Biology, and Physics*. Mineola, New York.

Chapter 6

Bigio, I., S. Bown et al. (2000). "Diagnosis of breast cancer using elastic-scattering spectroscopy: preliminary clinical results." *J Biomed Opt* **5**: 221–228.

Biscarini, F., P. Samori et al. (1996). "Scaling behavior of anisotropic organic thin films grown in high vacuum." *Phys Rev Lett* **78**(12): 2389–2392.

Bohren, C. and D. Hoffman (1998). *Absorption and Scattering of Light by Small Particles.* John Wiley & Sons, Inc, New York.

Born, M. and E. Wolf (1999). *Principles of Optics.* Cambridge University Press, Cambridge.

Brown, E., T. McKee et al. (2003). "Dynamic imaging of collagen and its modulation in tumors *in vivo* using second-harmonic generation." *Nat Med* **9**(6): 796–800.

Chang, S. K., Y. N. Mirabal et al. (2005). "Combined reflectance and fluorescence spectroscopy for *in vivo* detection of cervical pre-cancer." *J Biomed Opt* **10**(2): 024031.

Dougherty, G. and G. M. Henebry (2001). "Fractal signature and lacunarity in the measurement of the texture of trabecular bone in clinical CT images." *Med Eng Phys* **23**(6): 369–380.

Drezek, R., A. Dunn et al. (1999). "Light scattering from cells: finite-difference time-domain simulations and goniometric measurements." *Appl Optics* **38**: 3651–3661.

Drezek, R., M. Guillaud et al. (2003). "Light scattering from cervical cells throughout neoplastic progression: influence of nuclear morphology, DNA content, and chromatin texture." *J Biomed Opt* **8**(1): 7–16.

Einstein, A., H.-S. Wu et al. (1998a). "Self-affinity and lacunarity of chromatin texture in benign and malignant breast epithelial cell nuclei." *Phys Rev Lett* **80**(2): 397–400.

Einstein, A. J., H. S. Wu et al. (1998b). "Fractal characterization of chromatin appearance for diagnosis in breast cytology." *J Pathol* **185**(4): 366–381.

Fang, H., L. Qiu et al. (2007). "Confocal light absorption and scattering spectroscopic microscopy." *Appl Opt* **46**(10): 1760–1769.

Fantini, S., M. Franceschini et al. (1994). "Semi-infinite-geometry boundary problem for light migration in highly scattering media: a frequency-domain study in the diffusion approximation." *J Opt Soc Am B* **11**(10): 2128–2138.

Flock, S. T., B. C. Wilson et al. (1987). "Total attenuation coefficients and scattering phase functions of tissues and phantom materials at 633 nm." *Med Phys* **14**(5): 835–841.

Georgakoudi, I., B. Jacobson et al. (2001). "Fluorescence, reflectance and light scattering spectroscopy for evaluating dysplasia in patients with Barrett's esophagus." *Gastroenterology* **120**: 1620–1629.

Georgakoudi, I., E. E. Sheets et al. (2002). "Trimodal spectroscopy for the detection and characterization of cervical precancers *in vivo*." *Am J Obstet Gynecol* **186**(3): 374–382.

Grainger, R. G., J. Lucas, G. E. Thomas, and G. Ewan. (2004). "The calculation of Mie derivatives." *Appl Opt* **43**(28): 5386–5393.

Gupta, S., M. Hunter et al. (2008). "Non-invasive optical characterization of biomaterial mineralization." *Biomaterials* **29**(15): 2359–2369.

Gupta, S., M. Hunter et al. (2009). "Optical characterization of the nanoscale organization of mineral deposits on silk films." *Appl Opt* **48**(10): D45–D51.

Gurjar, R. S., V. Backman et al. (2001). "Imaging human epithelial properties with polarized light-scattering spectroscopy." *Nat Me* **7**(11): 1245–1248.

Hsiao, A., M. Hunter et al. (2011). "Noninvasive identification of subcellular organization and nuclear morphology features associated with leukemic cells using light-scattering spectroscopy." *J Biomed Opt* **16**(3): 037007.

Huang, P., M. Hunter et al. (2009). "Confocal light scattering spectroscopic imaging system for *in situ* tissue characterization." *Appl Opt* **48**(13): 2595–2599.

Hunter, M., V. Backman et al. (2006). "Tissue self-affinity and polarized light scattering in the born approximation: a new model for precancer detection." *Phys Rev Lett* **97**(13): 138102.

Itzkan, I., L. Qiu et al. (2007). "Confocal light absorption and scattering spectroscopic microscopy monitors organelles in live cells with no exogenous labels." *Proc Natl Acad Sci U S A* **104**(44): 17255–17260.

Johnson, T. M. and J. R. Mourant (1999). "Polarized wavelength-dependent measurements of turbid media." *Opt Express* **4**: 200–216.

Kerker, M. (1969). *The Scattering of Light and Other Electromagnetic Radiation.* Academic Press, Inc, New York.

Kerschnitzki, M., W. Wagermaier et al. (2011). "The organization of the osteocyte network mirrors the extracellular matrix orientation in bone." *J Struct Biol* **173**(2): 303–311.

Kim, Y., Y. Liu et al. (2003). "Simultaneous measurement of angular and spectral properties of light scattering for characterization of tissue microarchitecture and its alteration in early precancer." *IEEE J Sel Top Quantum Electron* **9**: 243–256.

Kim, Y. L., Y. Liu et al. (2004). "Coherent backscattering spectroscopy." *Opt Lett* **29**(16): 1906–1908.

Kim, Y. L., Y. Liu et al. (2005). "Depth-resolved low-coherence enhanced backscattering." *Opt Lett* **30**(7): 741–743.

Kostyuk, O., H. Birch et al. (2003). "Structural changes in loaded equine tendons can be monitored by novel spectroscopic technique." *J Physiol* **554**(3): 791–801.

Kostyuk, O. and R. A. Brown (2004). "Novel spectroscopic technique for *in situ* monitoring of collagen fibril alignment in gels." *Biophys J* **87**(1): 648–655.

Laven, P. Retrieved November 3, 2011, from http://philiplaven.com/mieplot.htm. Page updated on August 9, 2011.

Levitt, J., M. Hunter et al. (2007). "Diagnostic cellular organization features extracted from autofluorescence images." *Opt Lett* **32**: 3305–3307.

Marenzana, M., D. Pickard et al. (2002). "Optical measurement of three-dimensional collagen gel constructs by elastic scattering spectroscopy." *Tissue Eng* **8**(3): 409–418.

McGee, S., V. Mardirossian et al. (2009). "Anatomy-based algorithms for detecting oral cancer using reflectance and fluorescence spectroscopy." *Ann Otol Rhinol Laryngol* **118**(11): 817–826.

Mie, G. (1908). "Beiträge zur Optik trüber Medien, speziell kolloidaler Matallösungen." *Ann Physik* **25**: 377–445.

Morgan, M., O. Kostyuk et al. (2006). "*In situ* monitoring of tendon structural changes by elastic scattering spectroscopy: Correlation with changes in collagen fibril diameter and crimp." *Tissue Eng* **12**(7): 1821–1831.

Moscoso, M., J. Keller et al. (2001). "Depolarization and blurring of optical images by biological tissue." *J Opt Soc Am A* **18**(4): 948–960.

Mourant, J. R., I. Bigio et al. (1995). "Spectroscopic diagnosis of bladder cancer with elastic scattering spectroscopy." *Lasers Surg Med* **17**: 350–357.

Mourant, J. R., I. Bigio et al. (1996). "Elastic scattering spectroscopy as a diagnostic for differentiating pathologies in the gastrointestinal tract: preliminary testing." *J Biomed Opt* **1**: 1–8.

Mourant, J. R., T. M. Johnson et al. (2001). "Characterizing mammalian cells and cell phantoms by polarized backscattering fiber-optic measurements." *Appl Opt* **40**: 5114–5123.

Mourant, J. R., T. M. Johnson et al. (2002). "Polarized angular dependent spectroscopy of epithelial cells and epithelial cell nuclei to determine the size scale of scattering structures." *J Biomed Opt* **7**: 378–387.

Mujat, C., C. Greiner et al. (2008). "Endogenous optical biomarkers of normal and human papillomavirus immortalized epithelial cells." *Int J Cancer* **122**: 363–371.

Müller, M. G., A. Wax et al. (2002). "A reflectance spectrofluorimeter for real-time spectral diagnosis of disease." *Rev Sci Instrum* **73**: 3933–3937.

Müller, M. G., T. Valdez et al. (2003). "Spectroscopic detection and evaluation of morphologic and biochemical changes in early human oral carcinoma." *Cancer* **97**: 1681–1692.

Myakov, A., L. Nieman et al. (2002). "Fiber optic probe for polarized reflectance spectroscopy *in vivo*: Design and performance." *J Biomed Opt* **7**: 388–397.

Nieman, L., A. Myakov et al. (2004). "Optical sectioning using a fiber probe with an angled illumination-collection geometry: Evaluation in engineered tissue phantoms." *Appl Opt* **43**(6): 1308–1319.

Nieman, L. T., M. Jakovljevic et al. (2009). "Compact beveled fiber optic probe design for enhanced depth discrimination in epithelial tissues." *Opt Express* **17**(4): 2780–2796.

Patterson, M. S., B. Chance et al. (1989). "Time resolved reflectance and transmittance for the non-invasive measurement of tissue optical properties." *Appl Opt* **28**(12): 2331–2336.

Perelman, L., V. Backman et al. (1998). "Observation of periodic fine structure in reflectance from biological tissue: A new technique for measuring nuclear size distribution." *Phys Rev Lett* **80**: 627–630.

Pierce, M. C., R. L. Sheridan et al. (2004). "Collagen denaturation can be quantified in burned human skin using polarization-sensitive optical coherence tomography." *Burns* **30**(6): 511–517.

Raine, D. (1981). *The Isotropic Universe: An Introduction to Cosmology*. Adam Hilger, Ltd, Bristol.

Rice, W. L., D. L. Kaplan et al. (2010). "Two-photon microscopy for non-invasive, quantitative monitoring of stem cell differentiation." *PLoS One* **5**(4): e10075.

Roy, H. K., Y. Liu et al. (2004). "Four-dimensional elastic light-scattering fingerprints as preneoplastic markers in the rat model of colon carcinogenesis." *Gastroenterology* **126**(4): 1071–1081; discussion 1948.

Roy, H. K., Y. L. Kim et al. (2005). "Spectral markers in preneoplastic intestinal mucosa: an accurate predictor of tumor risk in the MIN mouse." *Cancer Epidemiol Biomarkers Prev* **14**(7): 1639–1645.

Saidi, I., S. Jacques et al. (1995). "Mie and Rayleigh modeling of visible-light scattering in neonatal skin." *Appl Opt* **34**: 7410–7418.

Chapter 6

Sassaroli, A., F. Zheng et al. (2011). "Phase difference between low-frequency oscillations of cerebral deoxy- and oxy-hemoglobin concentrations during a mental task." *J Innov Opt Health Sci* **4**(2): 151–158.

Schmitt, J. and G. Kumar (1996). "Turbulent nature of refractive-index variations in biological tissue." *Opt Lett* **21**: 1210–1312.

Sokolov, K., R. Drezek et al. (1999). "Reflectance spectroscopy with polarized light: is it sensitive to cellular and nuclear morphology?" *Opt Express* **5**: 302–317.

Stoliar, P., A. Calo et al. (2010). "Fabrication of fractal surfaces by electron beam lithography." *IEEE Trans Nanotech* **9**(2): 229–236.

Teixeira, J. (1986). Experimental methods for studying fractal aggregates. *On Growth and Form, Fractal and Non-Fractal Patterns in Physics.* H. S. a. N. Ostrowsky. Nijhoff, Boston, MA.

van de Hulst, H. (1982). *Light Scattering by Small Particles.* Dover Publications, Inc, New York.

Voss, R. (1986). "Characterization and measurements of random fractals." *Phys Scripta* **T13**: 27–30.

Wax, A., C. Yang et al. (2002). "Cellular organization and substructure measured using angle-resolved low-coherence interferometry." *Biophys J* **82**(4): 2256–2264.

Weingarten, M., E. Papazoglou et al. (2008). "Correlation of near infrared absorption and diffuse reflectance spectroscopy scattering with tissue neovascularization and collagen concentration in a diabetic rat wound healing model." *Wound Rep Reg* **16**: 234–242.

Williams, R. M., W. R. Zipfel et al. (2005). "Interpreting second-harmonic generation images of collagen I fibrils." *Biophys J* **88**(2): 1377–1386.

Xu, M. and R. R. Alfano (2005). "Fractal mechanisms of light scattering in biological tissue and cells." *Opt Lett* **30**(22): 3051–3053.

Xu, Z., J. Liu et al. (2010a). "Back-directional gated spectroscopic imaging for diffuse light suppression in high anisotropic media and its preclinical applications for microvascular imaging." *IEEE J Sel Top Quantum Electron* **16**(4): 815–823.

Xu, Z., X. Sun et al. (2010b). "Spectroscopic visualization of nanoscale deformation in bone: interaction of light with partially disordered nanostructure." *J Biomed Opt* **15**(6): 060503.

Yu, C. C., C. Lau et al. (2006). "Assessing epithelial cell nuclear morphology by using azimuthal light scattering spectroscopy." *Opt Lett* **31**(21): 3119–3121.

Yu, Y., A. Sassaroli et al. (2010). "Near infrared, broad-band spectral imaging of the human breast for quantitative oximetry: Applications to healthy and cancerous breasts." *J Innov Opt Health Sci* **3**: 267–277.

Zhu, Y., N. G. Terry et al. (2011). "Design and validation of an angle-resolved low-coherence interferometry fiber probe for *in vivo* clinical measurements of depth-resolved nuclear morphology." *J Biomed Opt* **16**(1): 011003.

Zonios, G., L. T. Perelman et al. (1999). "Diffuse reflectance spectroscopy of human adenomatous colon polyps *in vivo*." *Appl Opt* **38**: 6628–6637.

7. Fluorescence Spectroscopy

William R. Lloyd, Leng-Chun Chen, and Mary-Ann Mycek

Chapter 7

Optical Techniques in Regenerative Medicine. Edited by Stephen P. Morgan, Felicity R.A.J. Rose, and Stephen J. Matcher © 2014 CRC Press/Taylor & Francis Group, LLC. ISBN: 978-1-4398-5495-2

7.1   Introduction

Fluorescence spectroscopy is a widely employed technique in chemical and biomedical applications for noninvasive specimen interrogation and monitoring that can be performed on both living and stained/sectioned samples (Lakowicz 2006). A variety of spectroscopic techniques have been developed for fluorescence applications in quantitative clinical tissue diagnostics (Pitts and Mycek 2001; Chandra et al. 2009; Pfefer et al. 2003b), including several involving cancer diagnostics (Chandra et al. 2010; Mycek et al. 1998; Volynskaya et al. 2008; Uehlinger et al. 2009). The methods developed for clinical fluorescence tissue sensing serve as an excellent starting point for applications in regenerative medicine, since similar excitation sources, optical delivery systems, and photon detectors can be employed for sensing in tissue-engineered constructs with minimal modifications (Mycek and Pogue 2003). Furthermore, these techniques can be employed to optically measure human tissues, providing the experimental groundwork to study tissue-engineered constructs both during development and *in vivo*, after human implantation.

7.2   Clinical and Biological Measurement Challenges That the Method Is Suited to Address

Biomedical applications of fluorescence spectroscopy include bulk tissue diagnostics (e.g., distinguishing healthy from nonviable tissue), determination of the relative quality of a sample (e.g., measuring cellular growth rate), and assessment of local biochemical changes (e.g., characterizing the binding state of nicotinamide adenine dinucleotide [NADH]) (Lakowicz 2006). Advantages of fluorescence sensing include the ability to assess and monitor tissue-engineered constructs in a noninvasive, nondestructive, and locally selective manner.

Further, fluorescence measurements cause minimal to no damage; therefore, measurements can be taken multiple times, at multiple sites, and performed during culturing of a sample, prior to release of a tissue-engineered construct for clinical use, and after implantation. In particular, fluorescence spectroscopy is able to meet the specific needs for the monitoring and analysis of tissue-engineered cell-scaffold constructs, including analyzing a complex three-dimensional structure with local heterogeneities, being adaptable to measuring a variety of constructs—including living samples for label-free optical assessment of tissue-engineered constructs—and translatable to *in vivo* measurements post-implantation (Lee et al. 2010). Additionally, fluorescent dyes have been investigated as a noninvasive method to track stem cells in small animal imaging (Frangioni and Hajjar 2004; Sutton et al. 2008). Applications involving small animals and human *in vivo* measurements are beyond the scope of this chapter.

7.3   Technical Implementation of the Method

7.3.1   Fluorescence

Several physical processes can occur when light is absorbed by a sample (Lakowicz 2006). Figure 7.1 shows a Jablonski diagram detailing the stages of one such process, fluorescence, which occurs when an excitation photon is absorbed by a sample fluorophore.

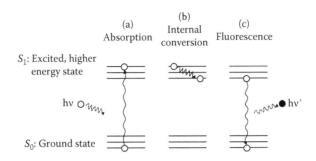

S_1: Excited, higher energy state

(a) Absorption

(b) Internal conversion

(c) Fluorescence

hv

hv'

S_0: Ground state

FIGURE 7.1 Jablonski diagram sketch demonstrating the simplified three-step fluorescence process. (a) Excitation photon is absorbed by a fluorophore, raising the energy of an electron to an excited state. (b) Electron undergoes internal energy conversion to the base level of the excited state. (c) Excited electron decays to the ground state, emitting a fluorescence photon with energy equal to the energy difference between the states.

The fluorescence photon is emitted isotropically (emitted in a random direction) with energy equal to the difference between the ground and excited energy levels of the electron. Fluorescent molecules contain different characteristic energy levels and electrons that can absorb excitation energy; therefore, each fluorescent molecule has unique and characteristic optical absorption and fluorescence emission spectra.

7.3.1.1 Endogenous versus Exogenous Fluorophores

Fluorophores in biological systems can be divided conveniently into two groups: endogenous and exogenous. Endogenous fluorophores are naturally occurring biological molecules that are native to cells and tissues, including the biological materials employed to develop tissue-engineered constructs and intracellular metabolites. Typically, endogenous fluorophores are low-light level emitters (Wagnieres et al. 1998; Vishwanath and Mycek 2004a) compared to exogenous fluorophores, molecules that are not naturally occurring in a sample and are added to provide contrast for fluorescence measurements. Exogenous fluorophores are widely used to image cells and thin tissue sections, but introducing exogenous molecules into thick, tissue-engineered constructs involves a more invasive process, which could alter metabolic viability and compromise construct sterility. Therefore, when analyzing tissue-engineered constructs via fluorescence spectroscopy, it is common to measure only endogenous fluorescence. Careful application of exogenous fluorophores is required to provide useful fluorescence contrast and characterization in thick tissue constructs.

7.3.1.1.1 **Endogenous Fluorophores** Endogenous fluorophores typically report on two common components of a tissue-engineered construct: the extracellular matrix and cellular metabolism. Figure 7.2 shows the emission spectra for several endogenous fluorophores, including metabolic coenzymes, extracellular matrix components, amino acids, vitamins, pigments, and some proteins (Wagnieres et al. 1998; Zellweger 2000). Endogenous fluorophores that report on the extracellular matrix can be monitored to study the growth and development of a tissue-engineered construct. These molecules include collagen, elastin, and keratin.

Endogenous fluorophores that report on cellular metabolism include NADH and flavin adenine dinucleotide (FAD), cellular biochemicals that play a role in redox

Chapter 7

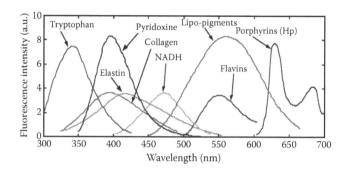

FIGURE 7.2 Emission spectra of common endogenous tissue fluorophores, including extracellular matrix components (collagen, elastin), cellular metabolites (NADH, flavins), amino acids (tryptophan), pigments (lipo-pigments), vitamins (pyridoxine), and other aromatic organic compounds (porphyrins). Emission spectra of these fluorophores range from the ultraviolet through the visible portion of the electromagnetic spectrum. Additionally, the fluorescence emission for each fluorophore is distinct, although emission spectra do overlap with one another. (From Wagnieres, G. et al., *Photochemistry and Photobiology*, 68:603–632, 1998; Zellweger, M., Fluorescence Spectroscopy of Exogenous, Exogenously-Induced and Endogenous Fluorophores for the Photodetection and Photodynamic Therapy of Cancer. PhD dissertation, Ecole Polytechnique Federale de Lausanne. With permission.)

metabolism and are naturally fluorescent in their reduced (NADH) and oxidized (FAD) states. NADH is produced during cellular glycolysis, an alternative pathway to oxidative phosphorylation for adenosine triphosphate (ATP) generation that occurs in cellular mitochondria, and is transported into mitochondria (Heikal 2010). Then, the high energy electron carried by NADH is utilized during oxidative phosphorylation and the electron transport chain to create a proton gradient, converting NADH to NAD^+ and $FADH_2$ to FAD, while generating cellular energy, ATP. In addition, acetyl coenzyme A (acetyl CoA) produced during glycolysis and transported into the mitochondria enters the citric acid cycle and converts NAD^+ to NADH, converts FAD into $FADH_2$, and produces ATP. Having fluorescent molecules at local concentrations dependent upon cellular metabolic processes allows for monitoring of energy metabolism within tissue-engineered constructs. Reported applications are discussed in Section 7.4.3.

The emission spectrum of each fluorophore is characteristic, with a distinct spectral line shape, but many fluorophores emit in the wavelength range from ~325 to 600 nm. Often, several fluorophores are excited with the same excitation light source, and each contributes to the measured fluorescence spectra.

7.3.1.1.2 Exogenous Fluorophores

Exogenous fluorophores, or fluorescent probes, are fluorescent molecules that can be employed to report on spatial localization and environmental conditions of a sample (Starly and Choubey 2008). Exogenous fluorescent drugs have been developed for use in photodynamic therapy (Celli et al. 2010), monitoring pH (Kuwana and Sevick-Muraca 2003), and cancer diagnostics (Brown et al. 2009), where the introduced fluorophore accumulates in the tumor region. A combined approach can also be employed with both endogenous and exogenous fluorophores for enhanced contrast (Andersson-Engels et al. 1990).

Introducing exogenous fluorophores into tissue-engineered constructs can help to characterize the molecules present and their microenvironment, but often compromises the sterility and integrity of the sample, thereby introducing safety and toxicity concerns (Ramanujam 2000). Therefore, discussed applications will typically focus on endogenous fluorophores that maintain tissue-engineered construct viability.

7.3.1.2   Experimental Design

The design of the experimental setup determines the sample volume that will be studied, the fluorophores that will be excited, and how the emitted photons will be detected. Therefore, careful consideration needs to be given to the experimental setup employed, the excitation source chosen, the detector employed, as well as the sample optical properties to measure the most useful data.

7.3.1.2.1 Excitation Source   All fluorescence measurements require an excitation source, most commonly a laser, light-emitting diode, or lamp. Excitation sources have a variety of characteristic properties that determine the fluorescent data measured. First, sources can be continuous, for collection of wavelength-resolved data, or pulsed, for collection of wavelength- and time-resolved data. Pulsed sources can be on the order of nanoseconds for single photon excitation or on the order of femtoseconds for multiphoton excitation or time-correlated single photon counting (TCSPC). Second, each source has a characteristic wavelength corresponding to its peak power output. Lastly, pulsed sources have a pulse repetition rate (i.e., number of pulses per second) and corresponding duty cycle (i.e., the pulse duration relative to the time between successive pulses).

Laser wavelength must be selected to preferentially excite the fluorophores of interest in a construct (Wagnieres et al. 1998), with consideration given to sources of background signal or artifacts that may be present. Common excitation source wavelengths used are in the UV-visible range to excite collagen, elastin, FAD, and NADH.

To collect time-resolved fluorescence, the laser pulse repetition rate and duty cycle need to be considered. A common practice in fluorescence measurements is to average a set number of measurements to increase the output signal-to-noise ratio. Therefore, to acquire data in a rapid fashion, higher pulse repetition rate sources allow for more measurements to be taken in a shorter time, permitting averaging of more fluorescence data collections and thus higher signal-to-noise ratios (Lloyd et al. 2010). However, the detector must be optimized to work with such an excitation source. The duty cycle must also be considered, including excitation pulse intensity and duration, so that the excitation source generates sufficient fluorescence without harming or damaging the sample (Niemz 2007).

7.3.1.2.2 Emission Detector   Emitted fluorescence light can be collected with wavelength resolution or time resolution (or both). Each modality provides information about the concentration and microenvironment of molecular fluorophores present in a sample and each employs different detection schemes.

For wavelength-resolved data, common detector configurations are spectrographs coupled to a charge-coupled device (including intensifiers [ICCDs] or electron multiplying charge-coupled device [EMCCD] [Michalet et al. 2007]) for detection. For time-resolved data, common detectors include photomultiplier tubes (Lloyd et al. 2010), avalanche photodiodes (Chandra et al. 2006), and streak cameras (Buhler and Graf 1998). Two common collection techniques are TCSPC (Urayama et al. 2003) and direct fluorescence decay recording via a digitizing oscilloscope (Marcu et al. 2009; Mycek et al. 1998; Pitts and Mycek 2001).

Additionally, to collect both wavelength- and time-resolved fluorescence simultaneously, two common methods exist that employ either one or a combination of the previously mentioned detectors. First, the fluorescence emission light can be beam-split and directed at two separate detectors, one each for wavelength and time detection (Chandra et al. 2007a). Second, an emission monochromator can selectively step through the desired wavelength regime with a measured time-resolved decay at each selected wavelength (Lloyd et al. 2010).

7.3.1.2.3 Optical Sample Properties

All tissue-engineered constructs and tissues have optical properties that will affect the fluorescence signal measured from the sample. In particular, we will discuss five optical properties: the scattering coefficient (μ_s), absorption coefficient (μ_a), fluorophore absorption coefficient (μ_{afx}), anisotropy (g), and absorption quantum yield (Φ) (Vishwanath and Mycek 2005; Vishwanath et al. 2002). These properties all influence photon propagation within the sample.

Figure 7.3 shows a photon path with a center-to-center source-to-detector spacing, ρ. This source-to-detector spacing preferentially detects scattered photons with longer average photon paths (photons that have traveled deeper within a tissue) than would smaller spacing. The scattering coefficient is the reciprocal of the average path length a photon travels between scattering events. The absorption coefficient is the reciprocal of the average path length a photon travels before being absorbed by a nonfluorophore. The fluorophore absorption coefficient is the reciprocal of the average path length an excitation photon travels before being absorbed by a fluorophore. Multiple fluorophore absorption coefficients are needed for a sample with multiple fluorophores.

The tissue anisotropy is the mean cosine of the scattering angle of a photon within a sample. Anisotropy is determined by the phase function used to model a tissue, and anisotropy values range from 0 to 1, with higher values representing more forward scattering media (Richards-Kortum and Sevick-Muraca 1996). The quantum yield of a fluorophore defines the efficiency with which a fluorophore absorbing an excitation photon

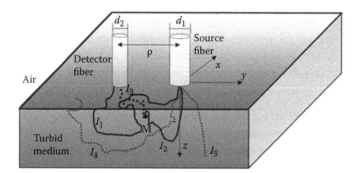

FIGURE 7.3 Model depicting photon propagation in turbid media. The sample is modeled as one-layer tissue-engineered construct with different paths a photon could travel illustrated. Path I_1 shows an excitation photon that is scattered, eventually reaching the detector. Path I_2 outlines a photon that is absorbed and becomes a fluorescence photon at M, following dotted path I_3 upon being detected. The dashed path I_4 is not detected because the photon reaches the surface at a point where no detector is present. The dashed path I_5 is not detected because the photon is absorbed by the tissue and is not reemitted. The source-detector center-to-center spacing, ρ, plays a role in determining the average depth a photon travels in the tissue before it is detected at the surface. (From Vishwanath, K. et al., *Physics in Medicine and Biology* 47:3387–3405, 2002. With permission.)

Table 7.1 Approximate Ranges for Optical Properties of Biological Tissue

	NADH	FAD	Keratin	Collagen	Tissue
Φ	0.02–0.1	0.03	0.25	0.1–0.4	–
τ (ns)	0.4–2.5	3.0	4.0	4.5–6	–
μ_{afx} (cm^{-1})	~0.2–6.1	~0.9	~0.5	~0.3–1.5	–
μ_a (cm^{-1})	–	–	–	–	0.1–30
μ_s (cm^{-1})	–	–	–	–	40–500
g					0.85–0.98

Sources: Vishwanath, K. and Mycek, M. A., *Optics Letters* 29:1512–1514, 2004; Chandra, M. et al., *Proceedings of the International Society for Optics and Photonics* 6628:66280B, 2007; Cheong, W. F. et al., *IEEE Journal of Quantum Electronics* 26:2166–2185, 1990.

Note: Φ, fluorescent quantum yield; τ, fluorescent lifetime; μ_a, absorption coefficient; μ_s, scattering coefficient; μ_{afx}, fluorophore absorption coefficient; g, anisotropy.

undergoes the fluorescence process. Each fluorophore has a different quantum yield, which also depends on the excitation wavelength employed. In addition, each fluorophore has a fluorescence lifetime, τ, which relates to the radiative decay of a fluorophore, discussed further in Section 7.3.2.

Table 7.1 introduces common values of the optical properties that can be employed to mathematically model a tissue construct that contains any combination of four common endogenous fluorophores discussed in the applications of this chapter: NADH, FAD, keratin, and collagen. These values can be measured experimentally or obtained from literature. These optical parameters can be employed in mathematical models to predict photon propagation and optimize experimental setups. The values for each optical property are dependent on the components of the sample and the wavelength of light employed to interrogate the sample, with different values used for modeling excitation and fluorescence light.

7.3.1.2.4 Additional Experimental Considerations

To avoid altering or damaging specimens, careful consideration must be given to the experimental conditions employed for fluorescence excitation and detection. Photobleaching is a phenomenon that can cause a fluorophore to be permanently damaged and unable to undergo fluorescence events (Bernas et al. 2005). In fluorescence spectroscopy measurements, photobleaching is seen as a reduction in detected fluorescence over repeated measurements of a sample (Pogue et al. 2001). Methods to reduce photobleaching include reducing the excitation laser energy or limiting the excitation time.

Another consideration is the pressure with which the optical probe is brought into physical contact with the tissue specimen. Probe pressure has been shown to impact optical spectroscopic measurements including measured fluorescence (Rivoire et al. 2004; Lim et al. 2010) and reflectance (Reif et al. 2008). Measurements are impacted from increased probe pressure resulting from tissue compression, possibly influencing local construct scattering and absorption properties. Therefore, a setup with minimal, controlled probe pressure is best suited to minimize possible effects for fluorescence measurements.

Chapter 7

7.3.1.3 Environmental Conditions

Controlled environmental parameters, like the background signal, temperature, pH, and oxygenation levels, are required to create repeatable measurements that accurately reflect the sample conditions. All of these experimental variables can change over time; therefore, to accurately compare measurements, measurement time should be kept to a minimum.

7.3.1.3.1 Background Artifacts—Absorption and Florescence Background artifacts can be inherent to a sample or originate due to the experimental setup, including absorbing chromophores, room lights, and culture media. Chromophores are molecules that do not fluoresce, but instead absorb excitation or emission photons, thus distorting measured fluorescence spectra. To limit background artifacts, spectra should be background corrected to account for room lights prior to each measurement (Chandra et al. 2007a) and acquisition time should be minimized.

Culture media present when measuring cells or constructs can produce background fluorescence or cause excitation attenuation due to absorption from any of the variety of nutrients, cofactors, and other molecules present. There are two common techniques to limit the influence of culture media: employ phenol-red free media (Pitts et al. 2001; Zhong et al. 2007) or, prior to measurement, wash the sample construct in phosphate buffered saline (PBS) (Palmer et al. 2003), although this alters the sample's physiological conditions and could impact measured fluorescence compared to native culture conditions.

Additional background fluorescence can occur when trying to measure cells atop a highly fluorescent collagen scaffold or once a construct is implanted into a patient, when blood absorption becomes a determining factor in the amount of signal attenuation. Analyzing the effects of both hemoglobin concentration and oxygenation quenching has been developed on tissue samples with spectral filtering modulation (Liu and Vo-Dinh 2009). Similar background signal artifacts can be present when studying exogenous fluorophores, where endogenous fluorescence is a background artifact that should be accounted for during analysis (Alberti et al. 1987).

7.3.1.3.2 Temperature, pH, and Oxygenation Common protocols developed for tissue-engineered constructs include culturing constructs in an incubator at 37°C with controlled pH and oxygenation to promote cell growth (Izumi et al. 2004). Often, measurements are made under sufficiently different or unmonitored environmental conditions. These environmental factors can significantly impact the subsequent fluorescence measurements, including temperature variations (Chance et al. 1979), pH (Kirkpatrick et al. 2006; Chang et al. 2009), and oxygenation in cell studies (Skala et al. 2007) and tissue studies (Vishwasrao et al. 2005). Oxygenation has also been shown to be an effective quencher of fluorescence (Lakowicz and Weber 1973).

7.3.2 Spectroscopic Fluorescence Data

Often, qualitative and quantitative differences are observed in both wavelength- and time-resolved fluorescence data. Traditionally, most researchers collect wavelength-resolved data, because wavelength-resolved instrumentation is readily commercially available and spectral data are often sufficient for accurate sample analysis. However, time-resolved data are very useful when measured fluorophores have overlapping emission spectra (i.e., free/bound NADH and FAD) or when isolating time-resolved fluorescence decays from a wavelength-regime where the fluorophore of interest has a higher signal contribution than the background signal. Both modalities can be employed to extract useful biological information from the studied sample, which is often being employed in parallel to extract maximum fluorescence information from a sample.

7.3.2.1   Wavelength-Resolved Data

Wavelength-resolved spectra can highlight local changes in fluorophore concentration and environmental changes (Lakowicz 2006). Each fluorophore has a distinct absorption spectrum and a corresponding emission spectrum. In samples with multiple fluorophores, the measurement is a linear combination of each fluorescence emission spectrum. In tissue-engineered constructs, many fluorophores present have well-characterized wavelength-resolved fluorescence spectra (see Figure 7.2). Therefore, these molecules can be preferentially excited with prior knowledge of their absorption spectra and analyzed postmeasurement with published literature spectra or measurements on purified endogenous fluorophores (Wilson et al. 2010) when photon propagation effects are also considered (Chandra et al. 2006).

7.3.2.2   Time-Resolved Data

Each fluorophore has characteristic time-resolved kinetics, represented by the fluorescence lifetime, τ. The fluorescence lifetime is the average time spent by a fluorophore in the excited state and has been well characterized (Chorvat and Chorvatova 2009) for common biological components. Fluorescence is a radiative decay process that occurs typically on the order of nanoseconds (Lakowicz 2006). Measured fluorescence decay data are the convolution of the intrinsic fluorescence decay and the instrument response function (a measurement characterizing the pulse profile of a detected excitation laser pulse). The most common fluorescence decay model is a linear combination of exponential decays, although alternative fitting algorithms can be employed, including stretched exponential (Elson et al. 2004), Laguerre deconvolution technique (Jo et al. 2006), and phasor analysis (Digman et al. 2008).

Extracting lifetime parameters can be rather straightforward, but interpretation can be difficult. Each fluorophore has a natural variability in its measured lifetime due to the complex sample environmental conditions that can affect decay behavior. Additionally, the number of parameters used to fit data can lead to high-quality fits, which may or may not be unique. In a sample containing multiple fluorophores, the resulting lifetime measurement is a linear combination of each individual lifetime and the fluorescence quantum yield in the wavelength range studied (Muretta et al. 2010).

7.3.2.3   Wavelength- and Time-Resolved Data

Wavelength- and time-resolved fluorescence data can be collected simultaneously in order to provide additional sample information without adding to measurement time (Chandra et al. 2006). Depending on the application, both methodologies can be employed initially to determine if one data domain is sufficient, although both methodologies can be advantageous to improve sample analysis and characterization (Chandra et al. 2007a; Marcu et al. 2009).

Figure 7.4 shows simultaneously collected wavelength- and time-resolved fluorescence data (Lloyd et al. 2010), containing a resulting wavelength-resolved spectra and

Chapter 7

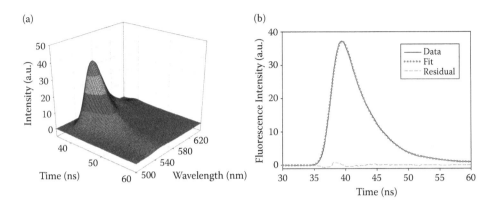

FIGURE 7.4 Wavelength- and time-resolved fluorescence data from a standard calibration fluorophore (Rhodamine 6G) is shown. (a) Three-dimensional matrix is composed of fluorescence time-resolved decays at user-selected wavelengths and a resulting wavelength-resolved spectrum. Time-resolved decays can be extracted and analyzed at each user-input wavelength, including the peak fluorescence intensity wavelength. In a complex sample with multiple fluorophores, decays can be extracted and analyzed at several wavelengths, yielding different decay dynamics vs. wavelength that reveal the relative proportion of each fluorophore in the sample. (b) Fluorescence decay at the wavelength of peak fluorescence intensity was extracted and analyzed via a least-squares iterative fit to a single exponential decay. Minimal residual shows a high-quality fit. (From Lloyd, W. et al., *Biomedical Optics Express* 1:574–586, 2010. With permission.)

numerous time-resolved decays. Figure 7.4a is a three-dimensional matrix composed of both wavelength- and time-resolved fluorescence intensity data. Time-resolved decays can be extracted at each measured wavelength. In Figure 7.4b, the fluorescence decay was extracted at the wavelength of peak fluorescence intensity and fit to a single-exponential decay with a least-squares iterative fit. Extracted lifetime parameters compared very well to expected literature lifetimes (Lloyd et al. 2010). The breadth of collected fluorescence data and high quality, accurate fit illustrates the promise for using time-resolved fluorescence instruments to monitor tissue-engineered constructs with multiple fluorophores present and varying environmental conditions.

7.3.2.4 Second Harmonic Generation: Complementary Collagen Sensing

Second harmonic generation (SHG) is detailed in Chapter 3. Here, we discuss SHG as a complementary technique for comparison with fluorescence spectroscopy measurements. SHG is not a fluorescence process; it is a nonlinear optical process that occurs when two photons interact with a noncentrosymmetric material, combining the energy from the two photons and emitting a single photon with twice the energy. Under the right experimental conditions, strong SHG signals arise from collagen fibrils that are highly organized (Bayan et al. 2009). Uses for SHG signals in tissue-engineered constructs include confirming the presence, location, type, and approximate concentration of organized collagen in a specimen. Since collagen (organized or not) is a bright, endogenous fluorophore, SHG measurements can provide a complementary means of sensing organized collagen.

7.3.2.5 Fluorescence Data Analysis

A variety of analysis techniques can be employed to correct fluorescence data and extract information potentially useful for sample classification and viability determination, including the tissue fluorophores present, their relative concentrations, their microenvironment, and the scattering and absorption coefficients of the sample. Many classification techniques have been applied to tissue fluorescence spectroscopy such as principal component analysis (Zhu et al. 2009), linear discriminate analysis (Pfefer et al. 2003b), partial least squares discriminate analysis (Sahar et al. 2009), semiempirical algorithms (Wilson et al. 2010), and fluorescence ratio algorithms (Chandra et al. 2007a). Techniques range in complexity from relatively simple to technically sophisticated in nature, and which to employ is application dependent. Often, several techniques are evaluated before identifying the algorithm optimal for a specific application.

7.3.2.5.1 Correcting Fluorescence Data   Many factors can impact the quality of fit and extraction of sample parameters from measured data. Of these, particularly notable are fluorescence attenuation artifacts that can affect the spectral line shape or influence the time-resolved fluorescence decay from a sample. Therefore, many techniques have been developed to correct measured tissue fluorescence data for common attenuation artifacts (Bradley and Thorniley 2006), including blood absorption and oxygen quenching with fluorescence measurements (Liu and Vo-Dinh 2009) and in combination with reflectance measurements (Kim et al. 2010; Wilson et al. 2010; Wilson and Mycek 2011). Carefully applying these techniques can enable the reconstruction of intrinsic fluorescence data, thereby improving the quality and capabilities of data analysis.

7.3.2.5.2 Quantitative Models for Probe Design   Mathematical models have been developed to quantify fluorescence measurements from tissue samples (Wilson and Mycek 2011) and tissue-engineered samples (Wilson et al. 2011), including models based on the diffusion approximation (Kienle and Patterson 1997), semiempirical models (Yudovsky and Pilon 2010; Wilson et al. 2010), and Monte Carlo (MC) simulations (Pfefer et al. 2002; Vishwanath and Mycek 2004b, 2005; Vishwanath et al. 2002). Accurate models provide the means of predicting measured fluorescence prior to experimentation, affording the opportunity to optimize probe design. These models are only as accurate as the input parameters (Bhowmick et al. 2009) and only useful if the optimal probe design can be manufactured. For label-free sensing in tissue engineering applications, endogenous fluorophores have been well characterized and literature values can be employed as model inputs (Wagnieres et al. 1998; Chorvat and Chorvatova 2009).

MC codes have been especially important in the development of fluorescence spectroscopy applications, instrumentation setups, and appropriate fiber-optic probes (Wang et al. 1995; Vishwanath et al. 2002). An accurate MC code provides a flexible framework that can be employed to easily and accurately model a new optical setup or to predict the expected fluorescence data obtained from measuring a new sample with input excitation and collection parameters. The MC approach can follow photon propagation for small path lengths (as small as 10–100 μm) (Wang et al. 1995), making it possible to model configurations in which the same fiber is employed for both excitation and detection. MC models can be successfully applied to photon propagation in small tissue volumes, a regime where other models based on transport equations lose accuracy (Wilson and Mycek 2011). MC codes for fluorescence have been successfully developed to model human tissue (Vishwanath and Mycek 2004b) as well as tissue-engineered constructs (Wilson et al. 2008). Several of the studies in Section 7.3.3 aimed to develop specialized fibers with MC models alone. Before spending the time and resources for

Chapter 7

probe development, having the ability to predict the optimal probe design can be an invaluable tool.

7.3.3   Fiber–Optic Probe Spectroscopy

Fiber-optic probes are employed for fluorescence spectroscopy applications that are incompatible with free space light delivery and detection. Under these conditions, remote sensing with fiber-optic probes is convenient because it is noninvasive or minimally invasive, nondestructive, portable, and adaptable to measure a variety of samples. Many fiber-optic probes have been developed for use in biological tissue diagnostics, offering capabilities for tissue-construct monitoring (Utzinger and Richards-Kortum 2003). Additionally, probes allow repeated measurements with selective sample interrogation, can be scanned over the sample volume to characterize a large tissue surface relatively quickly, and require little user training prior to use.

7.3.3.1   Probe Design

Fiber-optic probe geometries have been developed and employed for a wide range of biomedical optical spectroscopy sensing applications (Mignani and Baldini 1996; Utzinger and Richards-Kortum 2003; Fang et al. 2011b). Probe design is dependent upon the optical technique being employed; however, many probe geometry concepts can be adapted for different modalities. For example, ball lenses have been used for both fluorescence (Schwarz et al. 2008) and Raman spectroscopy (Mo et al. 2010). Two modalities with similar probe designs are fluorescence and reflectance spectroscopies (described in Chapter 6), principally because both techniques probe broad spectrum changes in the UV-visible wavelength range.

Several important choices are made when designing a probe, including the number of excitation and emission fibers and their diameters, the excitation and collection geometry, and the method for coupling the fiber-optic probes to the sample. Also, careful consideration must be given to the fluorophores under investigation, where they are located in the sample, the sample volume to study, background signal present, environmental factors (temperature, pH, oxygenation, probe pressure), and any external size constraints on the fiber-probe geometry.

Figure 7.5 provides an overview of several fiber-probe geometries that can be employed for the detection of fluorescence from tissue-engineered constructs. In particular, arrangements are included with fiber probes normal to the tissue surface with varying configurations for excitation and collection (Figure 7.5a through d), with angled fibers (Figure 7.5e), with a ball lens (Figure 7.5f), and with beveled fibers (Figure 7.5g and h). Probes best used for measuring fluorescence from bulk tissues are located in Figure 7.5a–d whereas those best used for measuring fluorescence from controlled depths within a tissue are located in Figure 7.5e–h.

Figure 7.5a and b illustrate the two simplest fiber-probe geometries to employ for experiments in scattering media. The excitation photons enter the tissue according to the optical fiber's numerical aperture, but due to light scattering in the medium, this creates an optical glow ball dependent on the scattering coefficient of the sample, but on the same order of size as the fiber diameter. Detected fluorescence photons are most likely to originate within this glow ball volume (Pfefer et al. 2001a; Vishwanath and Mycek 2005).

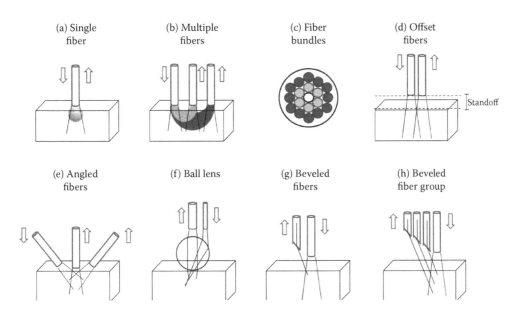

FIGURE 7.5 Fiber-optic probes for tissue spectroscopy have been developed using a variety of geometries, as illustrated here (Utzinger and Richards-Kortum 2003). (a) Probe that uses a single fiber for both excitation and detection. The excitation light propagates into the turbid media and creates a glow ball of excitation light; detected fluorescence photons also originate within the glow ball. (b) Probe with a single source fiber and multiple detector fibers positioned at various distances from the excitation fiber. This probe can detect photons that traveled deeper into the tissue than in (a), with larger depth penetration for increased center-to-center spacing. Average classical photon paths are shown between each fiber by the parabolic volumes. (c) Ring fiber bundle is employed to detect the same fluorescence information as in (b), but more fluorescence is detected with multiple source or detection fibers. Additionally, several center-to-center fiber spacings can be employed to detect different sample depths. Each concentric ring of fibers (gray and black) represents collection fibers with a fixed center-to-center spacing relative to the central white fiber. Employing concentric rings of fibers for excitation or emission detection can increase the measured fluorescence intensities. Fiber rings can be employed for either excitation or detection, with modifications interrogating different photon paths. (d) Vertical offset is introduced between the fiber probes and the sample surface to predominately excite and detect from a more superficial depth than in either (a) or (b). (e) Probe in which the fibers are angled relative to the sample. Angled excitation and detection promote the detection of photons to propagate to shallower paths and thus detected fluorescence originates from shallower tissue depths. (f) Ball lens is employed to excite fluorescence in the upper layer of a sample. The detected fluorescence has been shown to be from a more superficial layer than that in (a) and (b) (Schwarz et al. 2005). (g, h) Beveled fibers are employed that function similar to the angled fiber in (e) without needing to change probe orientation relative to tissue surface. (h) Beveled fibers can be designed to detect fluorescence from different sample layers or depths (Nieman et al. 2009).

When employing the optical setup in Figure 7.5b, the source-to-detector fiber separation has a significant impact on the path detected photons travel. Here, the detected fluorescence photons are most likely to travel along a parabolic shape, with the depth traveled related to the source-to-detector fiber separation, and also dependent on the scattering coefficient.

7.3.3.2 Fiber-Optic Probe Geometries

An important concern when selecting an experimental probe geometry is the sample depth interrogated. The exact sample volume interrogated can be determined by employing MC or diffusion photon propagation models. However, rough approximations can be provided for the upper and lower rows of Figure 7.5. For the upper row, sample depth analyzed is on the order of the center-to-center fiber spacing employed (Fang et al. 2011a; Utzinger and Richards-Kortum 2003) (i.e., ~1 mm^3 sample volume analyzed with 600 μm diameter probes at small center-to-center spacing). For the bottom row, experimental setups are designed to mimic confocal microscopes with thin sectioning capabilities (Fang et al. 2011a; Nieman et al. 2009) (i.e., preferentially isolating a measurement top-layer thickness on the order of <500 μm).

7.3.3.2.1 Fibers Perpendicular to Tissue   One of the simplest probe designs employs a single optical fiber (or multiple optical fibers) oriented perpendicular to the tissue surface (Pfefer et al. 2002, 2003a; Mycek and Pogue 2003). When such a probe is placed in contact with tissue, excitation photons create a glow ball (Figure 7.5a and b), with photon excited volume diameter related to the scattering coefficient of the sample that has a diameter on the order of magnitude of the probe diameter, dependent on the scattering coefficient. Common applications involve bulk samples in which fluorescence is detected from a large sample volume.

For a single fiber, a shallower sample layer is studied and more signal is generally detected, as the light photons travel a shorter path before detection (Pfefer et al. 2002). For multiple probe geometries, the excitation photons generating detected fluorescence photons travel a longer and relatively deeper path prior to detection, encountering more scatterers, absorbers, and fluorophores in optical paths that are parabolic in shape, based on the sample optical scattering properties (Pfefer et al. 2001b). Detection depth depends on the diameter of the fibers employed and the center-to-center probe spacing (Figure 7.5b), with smaller fiber diameters and smaller center-to-center spacing detecting shallower depths.

An MC study to analyze the effects of numerical aperture, fiber diameter, source-collection fiber separation distance, and fiber-tissue spacer thickness in multifiber probe geometries designed for fluorescence spectroscopy (Pfefer et al. 2002) predicted several experimental effects. These include that increases in numerical aperture could increase the detected signal without changing the origin of fluorescence. For example, increasing the probe to tissue surface standoff distance promotes probing a more superficial depth, shrinking both excitation and collection fibers promotes detecting fluorescence from a superficial depth, and increased center-to-center source detector fiber-optic probe spacing results in homogenous sample fluorescence measurements because detected photons have a wide range of optical paths before being collected. Each finding should be considered when designing or implementing a fluorescence spectroscopic measurement with a multifiber setup.

7.3.3.2.2 Angled Fibers   As with the probes described in Section 7.3.3.2.1, angled probes can be designed using MC codes and validated on tissue-simulating phantoms before use on tissues. Angled excitation or collection fibers can be employed to achieve depth selectivity within a sample (Wilson et al. 2008; Liu and Ramanujam 2004) and to increase depth selectivity in multilayered epithelial tissues (Skala et al. 2004; Liu and Ramanujam 2004). Such probe geometries employed are easily translatable to tissue-engineered constructs, which mimic similar tissue geometries.

An MC code was employed to simulate a spectroscopy setup with 200 μm excitation and collection fibers in a two-layered model with a 450 μm intestinal epithelium and semi-infinite stromal layer (Skala et al. 2004; Liu and Ramanujam 2004). The excitation fibers were simulated at tilt angles relative to the tissue normal of 0°, 15°, 30°, and 45°, while the collection fiber was oriented at the tissue normal. The center-to-center spacing between excitation and collection fibers was varied between 200 and 1000 µm in 200 µm steps. The computational results showed that at a tilt angle of 45° and 200 µm center-to-center spacing, the detected fluorescence was primarily from the upper half of the epithelium as compared to the 0° collection fiber. Additionally, for all simulated collection fiber tilt angles, the percent of total fluorescence detected originating in the epithelial layer decreased for increased center-to-center spacing. These results show that employing angled collection fibers with minimal center-to-center probe spacing can be used to selectively detect fluorescence from superficial layers, advantageous when employed to limit detected background fluorescence if the sample area to be studied is isolated at a superficial depth.

Another approach to angle the excitation light is to employ beveled fibers, similar to the sketch in Figure 7.5h. This model was employed for elastic light scattering (Nieman et al. 2009), but the probe design could be employed for fluorescence spectroscopy with minimal alterations. The model tested a developed probe with 110 µm diameter beveled fibers (35°, 40°, 45°). Predictions showed that depth resolution between ~350 and ~1200 µm could be achieved and confirmed with a three-layer scattering phantom and that a 40° bevel angle was optimal for depth resolving measurements on tissue-simulating phantoms mimicking oral precancer and *in vivo* on normal human oral mucosa. Beveled fibers could be manufactured to mimic the angled fiber results while not increasing the outer diameter of a stand-alone probe due to the bending radius of the tilted fiber, a setup that can be employed when sectioning is desired but the probe outer diameter is a limiting design factor.

7.3.3.2.3 Lens-Coupled Fibers   Fiber-optic probes coupled with lenses (such as an objective lens or ball lens) can be employed to achieve fluorescence depth selectivity. These setups often attempt to mimic the depth selection of a microscope without requiring the additional expense and bulk of microscopy equipment. Advantages to employing such setups include a smaller fiber bundle setup that is desirable for clinical measurements. Similar probes have been employed for Raman spectroscopy (Mo et al. 2010).

Figure 7.6 shows a probe developed for selective excitation and fluorescence detection from the epithelial layer of a biological tissue (Schwarz et al. 2005, 2008). Figure 7.6a illustrates the design, with two illumination fibers located off-axis in symmetrical positions and the collection fiber on axis, with a 0.5 mm separation between the distal fiber ends and the top of the ball lens. Coupling illumination fibers through the ball lens created an excitation photon path angled to the tissue normal at an oblique angle, traversing to a shallow sample depth. A two-layered tissue-simulating phantom was studied with each layer 400 µm thick composed of different fluorescent beads embedded in a collagen matrix.

Figure 7.6b depicts the depth sectioning capabilities of this probe design with two fluorescence spectra, one each for excitation of 370 and 450 nm. Measured spectra were significantly different, resulting from the characteristic differences between the absorption spectra of the fluorophores and the nonhomogenous distribution of fluorophores in the sample layers. When compared to measurements with fibers perpendicular to

Chapter 7

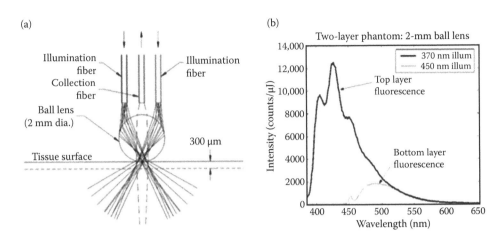

FIGURE 7.6 (a) Fiber-optic probe coupled with a ball lens (Schwarz et al. 2005) was employed for improved depth selectivity. The ball lens couples the illumination fiber to the surface upon which excitation photons propagate into the sample at a shallower angle relative to the surface normal. This setup accomplishes depth selectivity up to 300 μm. (b) Two fluorescence spectra were measured from a two-layer tissue-simulating phantom and compared to a fiber perpendicular to the sample surface configuration (not shown). Results showed that more fluorescence was detected from the top layer when employing the ball lens setup. In addition, the fluorescence intensity was 3.8 times greater from the top layer when compared to a perpendicular fiber source. Intensities were corrected to highlight the significant relative increase in signal detected from the top layer as compared to the bottom layer. (From Schwarz, R. A. et al., *Optics Letters* 30:1159–1161, 2005. With permission.)

the sample, employing a ball lens was found to extract 3.8 times and 0.74 times as much fluorescence from the top and bottom layers, respectively. This experiment highlights two distinct advantages of fluorescence spectroscopy: (1) excitation source choice can influence the detected fluorescence, and (2) employing a ball lens can lead to greater fluorescence detection from a superficial depth.

A second probe was developed that simultaneously sensed three collection depths, resolving fluorescence from depths of ~100, ~325, and ~1000 μm in tissues to detect each of the epithelial and stromal layers, as demonstrated with measurements in the oral cavity (Schwarz et al. 2008). Preliminary *in vivo* data measured from oral mucosa samples showed that the probe could differentiate dysplastic and cancerous tissue from normal tissues and that the short path length of detected photons in the shallow channel appeared to mitigate the absorption effects of hemoglobin. Similar probes could be developed for applications in regenerative medicine, including simultaneously monitoring the growth of multiple layers in tissue-engineered constructs.

Fiber-optic probes incorporating both beveled fibers and ball lenses have been designed and simulated for fluorescence spectroscopy measurements with improved depth-resolution (Jaillon et al. 2008a). In the design, a nonbeveled excitation fiber was offset from the center of the probe, and the beveled detection fiber was centrally located. Both fibers were offset from the ball lens by a fixed distance. MC results predicted that a larger beveled angle probes deeper tissue depths and that a larger source-detector separation predominantly sampled deeper tissue layers. In a later study (Jaillon et al. 2008b), the beveled collection fiber employed for fluorescence spectroscopy was coupled in the MC model to a half ball lens. Here, results predicted that an increased bevel angle selectively probes a more superficial sample layer. Therefore, two possible probe designs are presented, although work employing half ball lenses is very limited.

Depth-selective probes have also been employed to analyze samples at a constant, superficial depth for reflectance spectroscopy (Fang et al. 2011b). With minimal to no changes in design, this probe could be employed for fluorescence. In the study, 200 μm diameter excitation fibers were coupled with a ball lens and achieved a constant depth penetration of 200 μm. This focused penetration depth was modeled with an MC code to be a decrease from both a standard fiber-optic probe (photon incidence normal to tissue surface) and an angled fiber-optic probe. The depth-selectivity-with-fiber-optic probe setups are impacted more by the maximal overlap between the illumination and collection spots rather than by changing the fiber-optic probe's diameter. However, in an absorbing sample, smaller probes minimized effects from the sample optical properties, which are both useful effects to consider when designing a fluorescence spectroscopy probe.

7.3.4 Confocal and Multiphoton–Excitation Fluorescence Spectroscopy

Inhomogeneous tissue-engineered samples, such as layered constructs, require spectroscopic techniques that preferentially excite fluorophores in a single layer of the construct, possibly at some preferred depth. These specifications are difficult to achieve using fiber-optic probes. Conventional applications to achieve sample sectioning include histology (destructive in nature) or other methods of sample staining. However, living tissue-engineered constructs can be analyzed with confocal and multiphoton-excitation fluorescence spectroscopy in a nondestructive and nonharming manner.

Therefore, confocal and multiphoton spectroscopies have been employed to optically section thick biological samples down to thicknesses comparable to a cellular layer (a few micrometers) (Chen et al. 2011), thereby reducing the presence of background fluorescence from additional layers (Gareau et al. 2004; Pena et al. 2005). Confocal and multiphoton-excitation fluorescence spectroscopies are newly applied in the field of regenerative medicine, with a few reported studies. Both techniques are of importance because many tissue-engineered constructs aim to mimic human tissues, including engineered constructs involving layered samples with at least two layers (Izumi et al. 2004). Therefore, the studies discussed here include monitoring human epithelial tissues, because the optical sectioning techniques developed for layered samples with comparable fluorophores are translatable to applications in regenerative medicine.

7.3.4.1 Confocal Fluorescence Spectroscopy

Figure 7.7 depicts the optical excitation and fluorescence detection paths for both fiber-optic probe geometries (Figure 7.5) and for a confocal fluorescence spectroscopy setup. Confocal fluorescence spectroscopy systems are capable of obtaining high-quality, depth-resolved fluorescence measurements with improved axial spatial resolution as compared to fiber-optic probe spectroscopy. The confocal spectroscopy system illuminates a focal point within a sample and collects fluorescence from that focal point while eliminating out-of-focus signals with a spatial pinhole placed in a conjugate focal plane to the specimen in front of the detector. Several groups have reported studies on biological samples employing similar confocal spectroscopy setups (Gareau et al. 2004;

Chapter 7

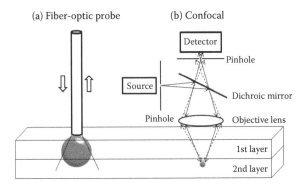

FIGURE 7.7   (a) Fiber-optic probe is employed for fluorescence excitation and collection, yielding detected emission photons from each of the first and second layers. (b) In comparison, a confocal system employs a focusing lens and pinhole geometry, which detect emission photons from the focused plane, blocking those photons that arise from out-of-focus planes. (From Gareau, D. S. et al., *Journal of Biomedical Optics* 9:254–258, 2004.)

Pezzotti et al. 2006). Additional systems have been developed employing a fiber-optic probe as a pinhole to monitor epithelial tissues (Bigelow et al. 2003; Jean et al. 2007).

7.3.4.2   Multiphoton-Excitation Fluorescence Spectroscopy

Multiphoton-excitation fluorescence spectroscopy has significant sectioning capabilities as compared to confocal spectroscopy due to the nonlinear, two (or three)-photon absorption that is locally confined in space without the use of a spatial pinhole (Teuchner et al. 1999; Wu and Qu 2005). In two-photon absorption, two photons at λ_{MP} are absorbed simultaneously to mimic the energy of one absorbed photon at $\lambda_{MP}/2$. Advantages are that two-photon absorption occurs in a confined sample area, providing high spatial resolution at the focal point and little out-of-focus photobleaching. Additionally, long-wavelength excitation photons at λ_{MP} cause deeper penetration depth, and they have less energy than single photon excitation, causing less damage to biological samples (Huang et al. 2002). Moreover, all fluorescent photons originating from the sample are detected, including those scattered by turbid biological samples.

7.4   Example Applications

7.4.1   Extracellular Matrix (Single Photon Fluorescence)

Fluorescence spectroscopy has been successfully employed to sense the alterations in extracellular matrix collagen development resulting from polymerization in tissue-engineered constructs that contain matrix metalloproteases (MMPs) (Kirkpatrick et al. 2006). Type I rat tail collagen was studied at a pH of 7.4 in a controlled 37°C measurement environment with three excitation wavelengths: 270, 320, and 360 nm. Excitation at 270 nm detected changes in amino acid fluorescence (specifically, tyrosine) and at both 320 and 360 nm monitored collagen fluorescence. Emission spectra were collected from 300 to 650 nm. Figure 7.8 shows the changes induced by polymerization as characterized by collagen fluorescence over culture days (Figure 7.8a), as well as for varying excitation wavelengths with two pH values (Figure 7.8b). Two observations made

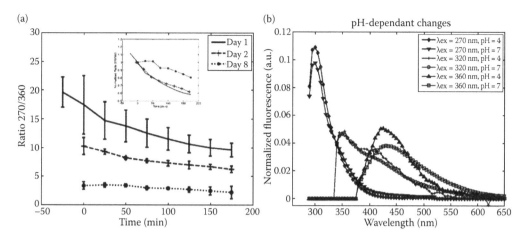

FIGURE 7.8 (a) Time-dependent fluorescence of cultured collagen gels with induced polymerization was analyzed. A fluorescence intensity ratio was calculated by integrating spectra across all wavelengths and dividing the resulting total intensity values from 270 nm excitation with those from 360 nm excitation. The observed trend showed that the measured intensity ratio decreases between culture days and with consecutive measurements within 175 min, detecting the decrease in collagen polymerization. Additionally, the fluorescence ratio was significantly lower with increasing culture days, noting higher fluorescence intensity at 360 nm as compared to 270 nm. (b) Effect of pH was analyzed by measuring a collagen solution at two different pH values with three excitation wavelengths. At each wavelength, the measured fluorescence spectra differed resulting from the pH change. (From Kirkpatrick, N. D. et al., *Journal of Biomedical Optics* 11:054021, 2006. With permission.)

showed that as the gel polymerized over time, the peak fluorescence intensity increased with 360 nm excitation wavelength. Conversely, the peak fluorescence intensity with 270 nm excitation showed a decreasing trend. To quantify this change, the integrated fluorescence intensity for 270 nm excitation was divided by the integrated fluorescence intensity for 360 nm excitation, as seen in Figure 7.8a. In addition, the developmental processes of collagen were studied with respect to three experimental variations: polymerization, pH, and temperature. Figure 7.8b shows the effect of excitation wavelength and pH on measurements, in which a pH change from 4 to 7 impacts measured fluorescence spectra.

Significant changes in fluorescence spectra were observed when comparing unpolymerized and polymerized collagen. Polymerized collagen constructs measured on day 1 had high variation in measured fluorescence, attributed to alterations in collagen cross-linking and a decrease in fluorescence characteristic of amino acids. The effects of polymerization became more consistent between constructs as the incubation time increased to day 8. Noticeable fluorescence differences were also seen after polymerization (spectral red shift) and after increasing temperature (decrease in fluorescence intensity). Results showed that the integrated fluorescence intensity with 270 nm excitation decreased relative to the integrated fluorescence intensity with 360 nm excitation over time on the same measurement day and over measurement days, which was hypothesized to occur due to changes in the nonenzymatic association of the collagen fibrils. The same trend was also observed in a study measuring the fluorescence of bovine muscle (Sahar et al. 2009). This study highlights several advantages of employing fluorescence

Chapter 7

spectroscopy, including monitoring changes in collagen polymerization over time (and other environmental influences that could be studied) and the ability to follow these changes noninvasively over time.

In addition to polymerization effects, thermal stressing of collagen constructs has been analyzed by fluorescence spectroscopy (Theodossiou et al. 2002). When collagen is thermally stressed, conformational changes occur that can irreversibly or reversibly alter the fluorescence properties of the sample. To study these conformational changes, type I collagen from Achilles bovine tendon was studied with a 337 nm excitation source at three emission wavelengths near the type I collagen emission peak (395 nm): 405, 410, and 415 nm. Samples were thermally stressed in a 40°C to 80°C water bath before being returned to 4°C prior to measurement. Results showed that the normalized fluorescence relative to the control samples (heated to 30°C) had similar normalized intensity values among the three wavelengths for each thermal stressing preparations and linearly decreased with increasing temperature, consistent with SHG signals measured on specimens under the same thermal preparations up to 60°C. At high temperatures (≥ 65°C), where SHG shows no measured signal due to complete denaturing, fluorescence sensing measured the random collagen cross-links present even after being denatured.

Similarly, stem cell development of osteogenic extracellular matrix was studied with wavelength- and time-resolved fluorescence spectroscopy by monitoring changes in collagen types I, III, IV, and V (Ashjian et al. 2004). The sample was excited with a 1.5 mm spot size from a 337.1 nm pulsed nitrogen laser (10 Hz, 3 ns pulse width Full-Width Half Maximum [FWHM]) delivered via a 600 µm diameter fiber. Eighteen 200 µm collection fibers were integrated to maximize detected fluorescence, similar to Figure 7.5c. Detected fluorescence was directed into a scanning monochromator coupled to a photomultiplier tube, amplified, and detected at 5 GS/s with a digitizing oscilloscope. Fluorescence decays were collected at 5 nm intervals from 360 to 510 nm; fluorescence spectra were calculated as the time-integrated fluorescence at each wavelength. For all fluorescence decays, the data were best fit with a least-squares minimization to a biexponential decay, where two tau (see Section 7.3.2.2) values and their corresponding amplitudes are extracted. The fast decay component is denoted as the smaller tau value, whereas the slow decay component is denoted as the larger tau value. Fractional contributions for each tau value are calculated as the percentage contribution. Four samples with differing levels of cell differentiation were studied: nondifferentiated cells and cells differentiated for 3, 5, and 7 weeks.

Figure 7.9 shows results from the study analyzing wavelength- and time-resolved fluorescence data. For wavelength-resolved fluorescence data (Figure 7.9a), a small red shift was measured for differentiated samples at 3, 5, and 7 weeks relative to the nondifferentiated sample with no other significant results. Only the time-resolved fluorescence was successful in monitoring the collagen changes over time (Figure 7.9b and c). Figure 7.9b shows extracted fluorescence lifetime parameters at multiple wavelengths. Similarly, Figure 7.9c extracted the fluorescence lifetime parameters at three wavelengths: 380, 430, and 470 nm. For all three extracted wavelengths in Figure 7.9c, the fast decay lifetime increased after cell differentiation (comparing to non-differentiated control), while the amplitude contribution of the fast tau (A1 in Figure 7.9) decreased after differentiation was induced (weeks 3, 5, and 7 compared to control). The opposite was true for the slow decay constant. After 5 weeks of osteoinduction, the slow decay constant decreased relative to the nondifferentiated control. In addition, the fast decay constant was the

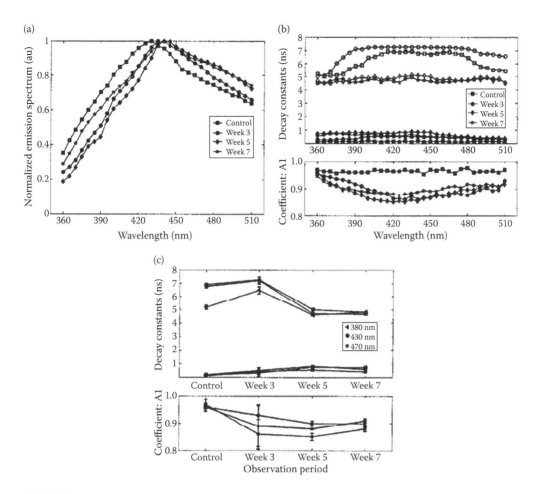

FIGURE 7.9 Fluorescence was measured on control and osteoinduced PLA cells at 3, 5, and 7 weeks into the culture. (a) Mean fluorescence spectra are shown for three trials. (b) Top plot shows the extracted fluorescence decay constants for the slow (hollow symbols) and fast (closed symbols) decay constants. Bottom plot shows the fast decay constant (A1) fractional contribution of amplitude fitting parameters. (c) Lifetime constants were extracted at three wavelengths to show variation of fluorescence decays over time. Error bars are the mean ± standard deviation. (From Ashjian, P. et al., *Tissue Engineering* 10:411–420, 2004. With permission.)

primary contributor to the overall decay, calculated to be ~95% of the nondifferentiated control and ~80% of the differentiated samples at 3, 5, and 7 weeks. Results also indicated that there was a CNI (calf skin collagen type I) overexpression after 3 weeks, confirmed through observing no significant changes in collagen type IV or V expression during differentiation. For analyzing tissue-engineered constructs primarily developed on collagen or involving osteoinduction (or a similar process), fluorescence sensing in the time-resolved domain may be a useful technique to study minor sample differences when wavelength-resolved data show no significant differences.

Endogenous fluorescence spectroscopy has also been employed to characterize tissue alterations based on the introduction of a photothermodynamic balloon (Shimazaki et al. 2009), the effect of UV radiation on elastin amidst collagen (Sionkowska et al.

Chapter 7

2007), chemical and rheological properties (bovine muscle) (Sahar et al. 2009), and tissue welding (porcine aorta samples) (Liu et al. 2010).

7.4.2   Cartilage (Single Photon Fluorescence)

The viscoelasticity of tissue-engineered cartilage samples (Kutsuna et al. 2010; Ishihara et al. 2009) was investigated by monitoring fluorescence from type II collagen, a major component of the viscoelastic properties of cartilage. Cartilage samples were developed with cells harvested from Japanese white rabbits that had similar properties to native cartilage including cell distribution, composition, and concentration of both glycosaminoglycans and collagen. In such samples, type II collagen is the dominant endogenous fluorophore. A ~5 mm sample spot size was excited via angled fibers (45°, 400 μm diameter) for excitation (20 mm standoff, lens coupled) and emission. Measurements were taken over a 5 week culture period. Results showed that over time, the peak wavelength of the fluorescence emission was blue shifted (with statistical significance). This effect was first detected after 3 weeks of culture and was magnified after five culture weeks. This blue shift in the fluorescence emission peak was found to be consistent with chondrogenesis, while the opposite would indicate osteogenesis, resulting from the increasing deposition of type II collagen of the samples studied. These results show that development of collagen composition in constructs, including sample viscoelasticity, can be monitored through changes in measured fluorescence.

Articular cartilage tissue samples have been studied with fluorescence spectroscopy in the wavelength range 360–700 nm, using a 355 nm laser source to preferentially excite extracellular collagen and intracellular NADH (Chandra et al. 2006). Figure 7.10 shows

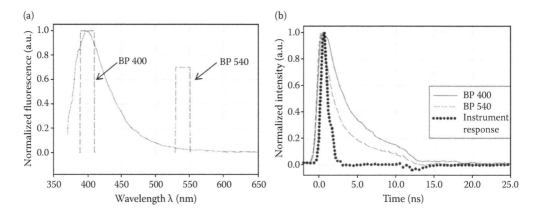

FIGURE 7.10   Endogenous fluorescence from articular cartilage samples was studied with spectral and temporal resolution. (a) Representative measured fluorescence spectrum is shown. The peak near 400 nm is representative of collagen. The fluorescence decay parameters over all emission wavelengths are characteristic of collagen. However, band-pass (BP) filtered emission (b) has noticeably different fluorescence decay depending on the emission wavelength selected. In particular, fluorescence decays measured at BP 540 nm are significantly faster, resulting from an increased contribution from the fast decay of NADH, as compared to the slow decay of collagen. (From Chandra, M. et al., *Optics Express* 14:6157–6171, 2006. With permission.)

the importance of selecting wavelength bands to change the dominant fluorophore contributing to the time-resolved fluorescence decay. The wavelength ranges selected were 400 ± 10 nm to maximize the collagen contribution and 540 ± 10 nm to maximize the NADH contribution. Figure 7.10 shows the measured data from articular cartilage samples. In Figure 7.10a, the wavelength-resolved spectrum is shown with the wavelength bands used to measure time-resolved data shown in Figure 7.10b. The fluorescence decays in Figure 7.10b highlight the distinct differences in lifetime between collagen (slow decay component) and NADH (fast decay component).

Selecting the desired wavelength regime can impact the dominant fluorophore measured and the information that can be extracted from a sample. When no band-pass filters were employed, the measured fluorescence decay was representative of collagen. To study the impact of NADH fluorescence, time-resolved fluorescence decays were measured using band-pass filters to select emission wavelength ranges. As expected, the time-resolved decay centered on 540 nm had a significant NADH (shorter-lived fluorophore) contribution, as seen by the faster decay measured relative to the decay at 400 nm (from the longer-lived fluorophore, collagen). This study on articular cartilage samples illustrates the importance of selecting fluorescence emission wavelengths and shows the potential for monitoring and characterizing similar tissue constructs with wavelength- and time-resolved fluorescence spectroscopy.

7.4.3 Extracellular Matrix (Multiphoton Fluorescence)

Confocal fluorescence spectroscopy was employed for detection of subcutaneous cartilage, located in the dermis below the stratum corneum and epidermis, expressing exogenous green fluorescent protein (GFP) underlying the skin of a mouse both *ex vivo* and *in vivo* (Gareau et al. 2004). Measurements employed a topical fiber-optic probe and a confocal fluorescence spectroscopy setup. Measured spectra were fitted with the spectrum of enhanced GFP and collagen, which exists extensively in human tissues. Results showed that fiber-optic probe measurements without confocal detection suffered greatly from unwanted collagen background signal, whereas the corresponding confocal measurements had little collagen signal. Therefore, the subdermal fluorophore of interest was predominantly excited and isolated from the background construct with a confocal setup. A similar approach can be used for measuring additional tissue-engineered constructs in which minimizing background fluorescence is desired.

In another report, confocal fluorescence spectroscopy was employed to study residual stress fields stored in artificial hip prostheses during their implantation *in vivo* (Pezzotti et al. 2006). Residual stress fields play a major role in tissue wear and thus studying residual stresses helps one to understand the causes behind the failure of artificial joints. Fluorescence signals from the native Cr^{3+} impurity in polycrystalline alumina (Al_2O_3) as a biomaterial and its piezospectroscopic behavior enabled quantitative characterization of residual stress fields of artificial hip joints. Confocal fluorescence microscopy was utilized to perform axial scanning. Results indicated that fluorescence spectra can be used to evaluate the residual stress fields in Al_2O_3, either on the surface or in the subsurface, which will enable optimization for biomaterial processing in order to develop reliable artificial hip joints. Potential translational use of the method in other tissue-engineered constructs and artificial joints is also possible.

Chapter 7

7.4.4 Epithelial Tissues (Multiphoton Fluorescence)

Confocal and multiphoton fluorescence spectroscopies have been employed to study layered epithelial tissues, which can contain multiple endogenous fluorophores such as keratin, melanin, NADH, flavoproteins, collagen, and elastin. The fluorescence properties of these fluorophores have been previously reported (Figure 7.2) (Wagnieres et al. 1998) and widely studied as a means to noninvasively monitor a sample via local endogenous variability (Mycek and Pogue 2003).

Fluorescence properties of purified keratin under two-photon excitation as compared to one-photon excitation were characterized (Pena et al. 2005). Multiphoton excitation was performed using a femtosecond Ti:sapphire laser operating in the 700–1000 nm range. The one-photon absorption spectrum of keratin has a maximum excitation wavelength of ~300 nm. Figure 7.11a shows the keratin fluorescence spectrum under different two-photon excitation conditions. Results show that fluorescence is measured with excitation wavelengths up to 900 nm, and fluorescence emission overlaps with the fluorescence spectra from collagen, NADH, and FAD. Figure 7.11b shows the action cross section measured from purified keratin, which was found to be comparable to other endogenous fluorophores. Measured fluorescence spectra from histology cuts of human skin show that the primary fluorophore with 860 nm excitation is characteristic of keratin. These results on keratin are of practical importance when studying human skin, as well as tissue-engineered constructs derived from keratinocytes.

Fluorescence of melanin under multiphoton excitation was characterized (Teuchner et al. 1999). Melanin samples were studied with 800 nm two-photon excitation and were compared to 400 nm one-photon excitation. Similar to the previous measurements on keratin, measurements on melanin showed that the two-photon fluorescence spectra

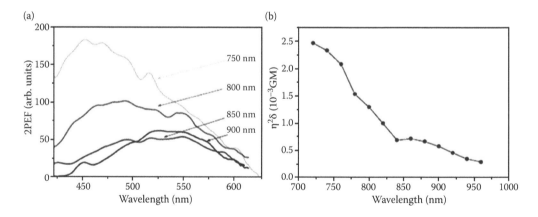

FIGURE 7.11 (a) Two-photon excited fluorescence (2PEF) spectra from purified keratin excited with varying excitation wavelength, normalized to the same excitation intensity. Results show that the spectrum shape and intensity change as the excitation wavelength increases. Additionally, two-photon excited keratin samples exhibit fluorescence at longer wavelengths compared to one-photon excitation. Keratin fluorescence excited at greater than 800 nm (two-photon) shows absorption of photons beyond the ~300 nm maximum absorption wavelength for one-photon excitation. (b) 2PEF excitation spectrum calibrated with comparable fluorescein measurements from Xu and Webb (1996). Keratin cross section was shown to increase with decreasing excitation wavelength and peaks at excitation <700 nm (within ±42% error). (From Pena, A. et al., *Optics Express* 13:6268–6274, 2005. With permission.)

exhibited greater fluorescence at longer wavelengths. Measured synthetic melanin in dimethyl sulfoxide (DMSO) and excised healthy human skin tissue excited at the NIR tail of melanin absorption exhibited melanin fluorescence. Results showed that melanin fluorescence can be selectively excited with femtosecond pulses out of other fluorophores due to its resonant two-photon absorption, a property that may allow for melanin to be used as a means for identifying malignant changes in tissue or applied to tissue-engineered constructs to study similar destructive processes.

A multiphoton-excitation fluorescence spectroscopy system to characterize autofluorescence and SHG of multilayered epithelial tissues was reported (Wu and Qu 2005, 2006). A Ti:sapphire femtosecond laser was employed to sample a 150 × 150 μm area with a 200 μm fiber probe coupled to a spectrometer. Three types of fresh biological samples were harvested and assessed from rabbits: oral, esophageal, and colonic tissue specimens. Results showed that the layered structure of epithelial tissues could be differentiated as the focal plane was moved deeper into the stroma. Measuring the stromal layer, fluorescence spectra were broad and correlated with strong SHG signals, attributed to the collagen fluorescence. As the focal plane was moved up to the epithelium, the fluorescence spectra had similar properties as those from NADH and flavoproteins, but no SHG signal was measured. As the focal plane was directed to the keratin surface, the fluorescence curves were again broad, caused by keratin fluorescence, as no localized SHG signals were detected.

Figure 7.12 shows the optical measurements of esophageal tissue excited at 750 nm. Characteristic differences were observed from measurements of the epithelium (0–80 μm depth) and the underlying tissue (100–200 μm depth). Results showed that the endogenous fluorescence signal could differentiate the layers of the tissue with complementary SHG sensing to determine if collagen was present (strong SHG signals were only measured from the underlying tissue layer). The layer differentiation observed matched the Masson-stained section results (a destructive technique performed subsequent to imaging), indicating that multiphoton-excitation fluorescence spectroscopy has the ability to differentiate and investigate different layers within the epithelial tissues

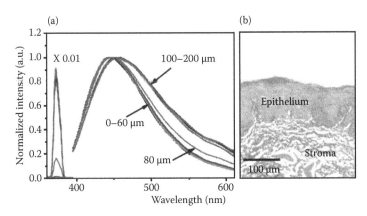

FIGURE 7.12 (a) 2PEF and SHG signals of keratinized esophageal tissue at different tissue depths (b) and its corresponding Masson-stained section. 2PEF spectra were normalized to the peak intensity. Results show that the autofluorescence from the epithelial layer (0–80 μm depth) is significantly different than that of the underlying stromal layer (100–200 μm depth), characteristic of increased collagen signal, verified with SHG measurements. (From Wu, Y. C. and Qu, J. N. Y., *Optics Letters* 30:3045–3047, 2005. With permission.)

Chapter 7

nondestructively. Future applications could employ a similar procedure to characterize the autofluorescence from layered constructs.

A second study by the same group monitored epithelial tissues with a one-photon confocal fluorescence spectroscopy system to detect endogenous fluorescence time-resolved decays from freshly excised esophageal and oral tissue samples from rabbits (Wu and Qu 2006). The employed system was similar to that in another study by Wu and Qu (2005), except a 405 nm source was employed to excite epithelial tissue samples. Fluorescence was collected via a 100 μm optical confocal fiber-optic pinhole. Two wavelength bands were studied: a blue band (440–480 nm) and a yellow band (530–570 nm). Fluorescence from each band at the same site had similar temporal characteristics. Results show that the time-resolved decays from nonkeratinized epithelial samples were significantly different than those from stromal and keratinized epithelial tissues when fit to a two-component exponential decay. Measured fluorescence decays were significantly different when comparing the keratinized, epithelial, and stromal layers. Results showed that even without the SHG signals generated by the multiphoton excitation, a confocal fluorescence spectroscopy system is capable of differentiating different tissue layers via fluorescence lifetime detection.

One-photon and two-photon confocal fluorescence spectroscopy was employed to analyze rabbit esophageal epithelial tissue with a three-layered structure (Zheng et al. 2008). A Ti:sapphire laser (excitation of 355–415 nm for one-photon spectroscopy and 710–830 nm for two-photon spectroscopy) was coupled to a 40× objective lens to study a $100 \times 100 \ \mu m^2$ sample area. Results indicated that measured spectra from one- and two-photon excitation have characteristic differences, with two-photon excitation providing a more accurate representation of tissue morphology, with double peaks that are broader than one-photon measurements in the keratinized epithelium. Similar differences were seen in each of the remaining two layers. Ratios of fluorescence intensities in two wavelength bands were shown to be a possible means to identify epithelial structure. This instrumentation and technique show promise for future applications of fluorescence spectroscopy with optical sectioning in tissue-engineered construct monitoring.

7.4.5   Metabolic Sensing (Multiphoton Fluorescence)

An important application of fluorescence spectroscopy is to monitor cellular or tissue metabolism. It has been reported that NADH and FAD are natural biomarkers for cellular metabolism. In addition, a ratiometric method to evaluate cellular metabolism was developed with NADH and FAD fluorescence (Heikal 2010). Therefore, characterizing the fluorescence properties of these endogenous fluorophores is the first step toward the accurate quantification of cellular viability. Both confocal detection and multiphoton excitation techniques can optically resolve layered biological samples, allowing a thin cellular layer or even a monolayer of cells at a fixed depth within the sample to be investigated to monitor sample viability and metabolic function, both on living tissue-engineered constructs and live tissue.

Two-photon excited fluorescence spectroscopy was employed to characterize the fluorescence properties of NAD(P)H and flavoproteins, including FAD and lipoamide dehydrogenase (LipDH) in solution (Huang et al. 2002). A Ti:sapphire laser source (720–1000 nm) was employed to excite fluorophore solutions in quartz cuvettes.

Figure 7.13 shows the two-photon excitation cross-section spectra of NAD(P)H, FAD, and LipDH. Reported results showed that below 720 nm excitation, all the molecules have maximal absorptions in the experimental range. Above 800 nm excitation, there is little NAD(P)H absorption while the FAD and LipDH had peak absorption around 900 nm. Also, the measured LipDH concentration is much lower than NAD(P)H concentration; however, LipDH has a larger excitation cross section and its emission can be detected.

Living cardiomyocytes were imaged with two-photon excitation at 750, 800, and 900 nm to monitor NADH (410–490 nm) and FAD fluorescence emission (510–650 nm), with each fluorophore predominating in its associated wavelength band, indicated by fluorescence changes induced by mitochondrial inhibitor and uncoupler. In addition, the result of emission spectra of living cardiomyocytes further confirmed that at 750 nm excitation, the cellular spectrum resembled the NADH *in vitro* spectrum except for an enhancement at wavelengths > 490 nm, attributed to the flavoprotein fluorescence; at 900 nm excitation, the cellular spectrum peaked at about the same wavelength as the *in vitro* LipDH spectrum, but it was broader, which was attributed to the complexity of cellular flavoprotein components; at 800 nm excitation, the cellular spectrum resembled that at 900 nm excitation, which was attributed to NADH fluorescence leakage. The potential to monitor both NADH and FAD, with highly characterized cellular spectra, shows promise as a future technique in tissue-engineered applications to monitor and quantify cellular metabolism.

A ratiometric method was developed to quantify the proportion of NADH that is enzyme-bound or free, each of which has different characteristic fluorescence lifetimes. A confocal fluorescence spectroscopy system (365 nm source) equipped with a time-resolved detector was employed to measure endogenous fluorescence in living SiHa and

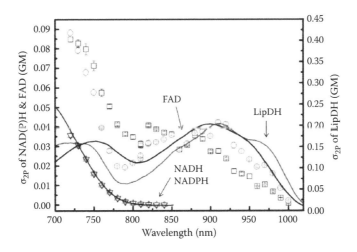

FIGURE 7.13 Two-photon excitation action cross section spectra (symbols) of NADH (triangles), NADPH (inverted triangles), FAD (squares), and LipDH (circles) are compared to the single-photon excitation action cross section spectra of NADH, FAD, and LipDH (solid lines, plotted with excitation wavelength doubled for comparison with two-photon excitation). In practice, fluorescence from NADH and NADPH are often considered together as NAD(P)H. Characterizing fluorescence from NAD(P)H, FAD, and LipDH is the first step towards creating a ratiometric algorithm to quantify cellular energy metabolism. Error bars represent standard deviation (Huang 2002). (Used with permission.)

Ect1/E6E7 cells (Wu et al. 2006). All fluorescence decay data were modeled as a two-component exponential decay, extracting both short and long lifetime components, corresponding to free and enzyme-bound NADH, respectively. Results demonstrated that cancer cells (SiHa) had a higher free to enzyme-bound NADH ratio and could be differentiated from normal cells using fluorescence lifetime sensing. Additionally, employing 385 or 405 nm excitation decreased the sensitivity of characterizing normal and cancerous samples. Ratiometric techniques that characterize the metabolic activity of cells or the enzyme-bound proportion of NADH or FAD could be employed to characterize tissue engineered constructs that are developing normally from nonnormal samples.

7.5   Conclusions

Fluorescence spectroscopy has been widely investigated as a diagnostic technique to assess tissue status, but only recently has been applied to studies in regenerative medicine. In tissue engineering, a primary obstacle for the clinical translation of "cell-scaffold products" is known to be "the development of appropriate *in vitro* and *in vivo* testing and characterization methods" (Lee et al. 2010). In particular, it is of great importance to develop targeted, selective, quantitative, and noninvasive techniques to assess tissue-engineered constructs during their development and culturing, as well as post-implantation, for a variety of regenerative medicine applications.

In this chapter, we have shown that fluorescence spectroscopy of endogenous or exogenous tissue fluorophores provides a useful tool for interrogating tissue-engineered constructs under a variety of experimental conditions. Applications include quantitative analysis of extracellular matrix materials (including cartilage), depth-sectioned investigation of layered tissues, as well as metabolic sensing in living tissue systems. The fluorescence spectroscopy techniques described in this chapter can be translated to applications beyond the scope of this chapter, including preclinical (small animal) and clinical (*in vivo* human) research. With some modifications, the instrumentation and data analysis methods described above may be employed to monitor construct/tissue viability, cellular metabolism, and extracellular matrix development *in vivo*.

The studies highlighted in this review chapter include successful applications of wavelength- and time-resolved fluorescence spectroscopies to model the extracellular matrix in bulk constructs, as well as to selectively excite fluorescence from endogenous molecular biomarkers to extract biochemical information in layered tissue constructs. Therefore, fluorescence spectroscopy has the potential to serve as a useful diagnostic method for tissue engineers and clinical tool for physicians working to develop successful tissue constructs for applications in regenerative medicine.

Acknowledgments

The authors acknowledge support in part from the National Institutes of Health (R01-DE019431, to M.-A.M) and the US Department of Education (GAANN Fellowship, to W.R.L.). The authors thank Robert Wilson and Seung Yup Lee for helpful discussions.

References

Alberti, S., Parks, D. R. and Herzenberg, L. A. 1987. A Single laser method for subtraction of cell autofluorescence in flow-cytometry. *Cytometry* 8:114–119.

Andersson-Engels, S., Johansson, J., Stenram, U., Svanberg, K. and Svanberg, S. 1990. Time-resolved laser-induced fluorescence spectroscopy for enhanced demarcation of human atherosclerotic plaques. *Journal of Photochemistry and Photobiology B* 4:363–369.

Ashjian, P., Elbarbary, A., Zuk, P. et al. 2004. Noninvasive in situ evaluation of osteogenic differentiation by time-resolved laser-induced fluorescence spectroscopy. *Tissue Engineering* 10:411–420.

Bayan, C., Levitt, J. M., Miller, E., Kaplan, D. and Georgakoudi, I. 2009. Fully automated, quantitative, non-invasive assessment of collagen fiber content and organization in thick collagen gels. *Journal of Applied Physics* 105:102042–102042-11.

Bernas, T., Robinson, J. P., Asem, E. K. and Rajwa, B. 2005. Loss of image quality in photobleaching during microscopic imaging of fluorescent probes bound to chromatin. *Journal of Biomedical Optics* 10:064015.

Bhowmick, G. K., Gautam, N. and Gantayet, L. M. 2009. Design optimization of fiber optic probes for remote fluorescence spectroscopy. *Optics Communications* 282:2676–2684.

Bigelow, C. E., Conover, D. L. and Foster, T. H. 2003. Confocal fluorescence spectroscopy and anisotropy imaging system. *Optics Letters* 28:695–697.

Bradley, R. S. and Thorniley, M. S. 2006. A review of attenuation correction techniques for tissue fluorescence. *Journal of the Royal Society Interface* 3:1–13.

Brown, J. Q., Vishwanath, K., Palmer, G. M. and Ramanujam, N. 2009. Advances in quantitative UV-visible spectroscopy for clinical and pre-clinical application in cancer. *Current Opinion in Biotechnology* 20:119–131.

Buhler, C. A. and Graf, U. 1998. Multidimensional fluorescence spectroscopy using a streak camera based pulse florometer. *Review of Scientific Instruments* 63:1512–1518.

Celli, J. P., Spring, B. Q., Rizvi, I. et al. 2010. Imaging and photodynamic therapy: mechanisms, monitoring, and optimization. *Chemical Reviews* 110:2795–2838.

Chance, B., Schoener, B., Oshino, R., Itshak, F. and Nakase, Y. 1979. Oxidation-reduction ratio studies of mitochondria in freeze-trapped samples—NADH and flavoprotein fluorescence signals. *Journal of Biological Chemistry* 254:4764–4771.

Chandra, M., Vishwanath, K., Fichter, G. D. et al. 2006. Quantitative molecular sensing in biological tissues: an approach to non-invasive optical characterization. *Optics Express* 14:6157–6171.

Chandra, M., Scheiman, J., Heidt, D. et al. 2007a. Probing pancreatic disease using tissue optical spectroscopy. *Journal of Biomedical Optics* 12:060501.

Chandra, M., Wilson, R. H., Lo, W. L. et al. 2007b. Sensing metabolic activity in tissue engineered constructs. *Proceedings of the International Society for Optics and Photonics* 6628:66280B.

Chandra, M., Wilson, R. H., Scheiman, J. et al. 2009. Optical spectroscopy for clinical detection of pancreatic cancer. *Proceedings of the International Society for Optics and Photonics* 7368:73681G.

Chandra, M., Scheiman, J., Simeone, D. et al. 2010. Spectral areas and ratios classifier algorithm for pancreatic tissue classification using optical spectroscopy. *Journal of Biomedical Optics* 15:010514.

Chang, C.-W., Wu, M., Merajver, S. D. and Mycek, M. A. 2009. Physiological fluorescence lifetime imaging microscopy improves Fo¨rster resonance energy transfer detection in living cells. *Journal of Biomedical Optics* 14:060502.

Chen, L. C., Lloyd, W. R., Kuo, S. et al. 2011. Nonlinear optical molecular imaging enables metabolic redox sensing in tissue-engineered constructs. *Molecular Imaging III* 8089.

Cheong, W. F., Prahl, S. A. and Welch, A. J. 1990. A review of the optical-properties of biological tissues. *IEEE Journal of Quantum Electronics* 26:2166–2185.

Chorvat, D. and Chorvatova, A. 2009. Multi-wavelength fluorescence lifetime spectroscopy: a new approach to the study of endogenous fluorescence in living cells and tissues. *Laser Physics Letters* 6:175–193.

Digman, M. A., Caiolfa, V. R., Zamai, M. and Gratton, E. 2008. The phasor approach to fluorescence lifetime imaging analysis. *Biophysical Journal* 94:L14–L16.

Elson, D., Requejo-Isidro, J., Munro, I. et al. 2004. Time-domain fluorescence lifetime imaging applied to biological tissue. *Photochemical & Photobiological Sciences* 3:795–801.

Fang, C., Brokl, D., Brand, R. E. and Liu, Y. 2011a. Depth-selective fiber-optic probe for characterization of superficial tissue at a constant physical depth. *Biomedical Optics Express* 2:838–849.

Chapter 7

Fang, C., Brokl, D., Brand, R. E. and Liu, Y. 2011b. Depth-selective fiber-optic probe for characterization of superficial tissue at a constant physical depth. *Biomedical Optics Express* 2:838–849.

Frangioni, J. V. and Hajjar, R. J. 2004. *In vivo* tracking of stem cells for clinical trials in cardiovascular disease. *Circulation* 110:3378–3383.

Gareau, D. S., Bargo, P. R., Horton, W. A. and Jacques, S. L. 2004. Confocal fluorescence spectroscopy of sub-cutaneous cartilage expressing green fluorescent protein versus cutaneous collagen autofluorescence. *Journal of Biomedical Optics* 9:254–258.

Heikal, A. A. 2010. Intracellular coenzymes as natural biomarkers for metabolic activities and mitochondrial anomalies. *Biomarkers in Medicine* 4:241–263.

Huang, S. H., Heikal, A. A. and Webb, W. W. 2002. Two-photon fluorescence spectroscopy and microscopy of NAD(P)H and flavoprotein. *Biophysical Journal* 82:2811–2825.

Ishihara, M., Bansaku, I., Sato, M., Mochida, J. and Kikuchi, M. 2009. Multifunctional characterization of engineered cartilage using nano-pulsed laser. *World Congress on Medical Physics and Biomedical Engineering,* 25(Pt 10):25:69–70.

Izumi, K., Song, J. and Feinberg, S. E. 2004. Development of a tissue-engineered human oral mucosa: from the bench to the bed side. *Cells Tissues Organs* 176:134–152.

Jaillon, F., Zheng, W. and Huang, Z. 2008a. Beveled fiber-optic probe couples a ball lens for improving depth-resolved fluorescence measurements of layered tissue: Monte Carlo simulations. *Physics in Medicine and Biology* 53:937.

Jaillon, F., Zheng, W. and Huang, Z. 2008b. Half-ball lens couples a beveled fiber probe for depth-resolved spectroscopy: Monte Carlo simulations. *Applied Optics* 47:3152–3157.

Jean, F., Bourg-Heckly, G. and Viellerobe, B. 2007. Fibered confocal spectroscopy and multicolor imaging system for *in vivo* fluorescence analysis. *Optics Express* 15:4008–4017.

Jo, J. A., Fang, Q., Papaioannou, T. et al. 2006. Laguerre-based method for analysis of time-resolved fluorescence data: application to in-vivo characterization and diagnosis of atherosclerotic lesions. *Journal of Biomedical Optics* 11:021004.

Kienle, A. and Patterson, M. S. 1997. Improved solutions of the steady-state and the time-resolved diffusion equations for reflectance from a semi-infinite turbid medium. *Journal of the Optical Society of America A* 14:246–254.

Kim, A., Khurana, M., Moriyama, Y. and Wilson, B. C. 2010. Quantification of *in vivo* fluorescence decoupled from the effects of tissue optical properties using fiber-optic spectroscopy measurements. *Journal of Biomedical Optics* 15:067006.

Kirkpatrick, N. D., Hoying, J. B., Botting, S. K., Weiss, J. A. and Utzinger, U. 2006. *In vitro* model for endogenous optical signatures of collagen. *Journal of Biomedical Optics* 11:054021.

Kutsuna, T., Sato, M., Ishihara, M. et al. 2010. Noninvasive evaluation of tissue-engineered cartilage with time-resolved laser-induced fluorescence spectroscopy. *Tissue Engineering Part C-Methods* 16:365–373.

Kuwana, E. and Sevick-Muraca, E. M. 2003. Fluorescence lifetime spectroscopy for pH sensing in scattering media. *Analytical Chemistry* 75:4325–4329.

Lakowicz, J. R. 2006. *Principles of Fluorescence Spectroscopy*. New York: Springer.

Lakowicz, J. R. and Weber, G. 1973. Quenching of fluorescence by oxygen. A probe for structural fluctuations in macromolecules. *Biochemistry* 12:4161–4170.

Lee, M. H., Arcidiacono, J. A., Bilek, A. M. et al. 2010. Considerations for tissue-engineered and regenerative medicine product development prior to clinical trials in the United States. *Tissue Engineering Part B* 16:41–54.

Lim, L., Rajaram, N., Nichols, B., and Tunnell, J. 2010. Time resolved study of probe pressure effects on skin fluorescence and reflectance spectroscopy measurements, in Biomedical optics and 3-D imaging, OSA technical digest (CD) (Optical Society of America), paper BTuD102.

Liu, C. H., Wang, W. B., Kartazaev, V., Savag, H. and Alfano, R. R. 2010. Changes of collagen, elastin and tryptophan contents in laser welded porcine aorta tissues studied using fluorescence spectroscopy. *Optical Biopsy VII* 7561:181.

Liu, Q. and Ramanujam, N. 2004. Experimental proof of the feasibility of using an angled fiber-optic probe for depth-sensitive fluorescence spectroscopy of turbid media. *Optics Letters* 29:2034–2036.

Liu, Q. and Vo-Dinh, T. 2009. Spectral filtering modulation method for estimation of hemoglobin concentration and oxygenation based on a single fluorescence emission spectrum in tissue phantoms. *Medical Physics* 36:4819–4829.

Lloyd, W., Wilson, R. H., Chang, C.-W., Gillispie, G. D. and Mycek, M.-A. 2010. Instrumentation to rapidly acquire fluorescence wavelength-time matrices of biological tissues. *Biomedical Optics Express* 1:574–586.

Marcu, L., Jo, J. A., Fang, Q. Y. et al. 2009. Detection of rupture-prone atherosclerotic plaques by time-resolved laser-induced fluorescence spectroscopy. *Atherosclerosis* 204:156–164.

Michalet, X., Siegmund, O. H. W., Vallerga, J. V. et al. 2007. Detectors for single-molecule fluorescence imaging and spectroscopy. *Journal of Modern Optics* 54:239–281.

Mignani, A. G. and Baldini, F. 1996. Biomedical sensors using optical fibres. *Reports on Progress in Physics* 59:1–28.

Mo, J., Zheng, W. and Huang, Z. 2010. Fiber-optic Raman probe couples ball lens for depth-selected Raman measurements of epithelial tissue. *Biomedical Optics Express* 1:17–30.

Muretta, J. M., Kyrychenko, A., Ladokhin, A. S. et al. 2010. High-performance time-resolved fluorescence by direct waveform recording. *Review of Scientific Instruments* 81:103101–103101-8.

Mycek, M.A. and Pogue, B. W., eds. 2003. *Handbook of Biomedical Fluorescence,* New York: Marcel Dekker, Inc.

Mycek, M.A., Schomacker, K. and Nishioka, N. 1998. Colonic polyp differentiation using time resolved autofluorescence spectroscopy. *Gastrointestinal Endoscopy* 48:390–394.

Nieman, L. T., Jakovljevic, M. and Sokolov, K. 2009. Compact beveled fiber optic probe design for enhanced depth discrimination in epithelial tissues. *Optics Express* 17:2780–2796.

Niemz, M. H. 2007. *Laser–Tissue Interactions: Fundamentals and Applications.* Germany: Springer.

Palmer, G. M., Keely, P. J., Breslin, T. M. and Ramanujam, N. 2003. Autofluorescence spectroscopy of normal and malignant human breast cell lines. *Photochemistry and Photobiology* 78:462–469.

Pena, A., Strupler, M., Boulesteix, T. and Schanne-Klein, M. 2005. Spectroscopic analysis of keratin endogenous signal for skin multiphoton microscopy. *Optics Express* 13:6268–6274.

Pezzotti, G., Tateiwa, T., Zhu, W. et al. 2006. Fluorescence spectroscopic analysis of surface and subsurface residual stress fields in alumina hip joints. *Journal of Biomedical Optics* 11:024009.

Pfefer, T. J., Schomacker, K. T., Ediger, M. N. and Nishioka, N. S. 2001a. Light propagation in tissue during fluorescence spectroscopy with single-fiber probes. *IEEE Journal of Selected Topics in Quantum Electronics* 7:1004–1012.

Pfefer, T. J., Schomacker, K. T. and Nishioka, N. S. 2001b. Effect of fiber optic probe design on fluorescent light propagation in tissue. *Laser-Tissue Interaction XII: Photochemical, Photothermal, and Photomechanical* 2:410–416.

Pfefer, T. J., Schomacker, K. T., Ediger, M. N. and Nishioka, N. S. 2002. Multiple-fiber probe design for fluorescence spectroscopy in tissue. *Applied Optics* 41:4712–4721.

Pfefer, T. J., Matchette, L. S., Ross, A. M. and Ediger, M. N. 2003a. Selective detection of fluorophore layers in turbid media: the role of fiber-optic probe design. *Optics Letters* 28:120–122.

Pfefer, T. J., Paithankar, D. Y., Poneros, J. M., Schomacker, K. T. and Nishioka, N. S. 2003b. Temporally and spectrally resolved fluorescence spectroscopy for the detection of high grade dysplasia in Barrett's esophagus. *Lasers in Surgery and Medicine* 32:10–16.

Pitts, J., Sloboda, R., Dragnev, K., Dmitrovsky, E. and Mycek, M. A. 2001. Autofluorescence characteristics of immortalized and carcinogen-transformed human bronchial epithelial cells. *Journal of Biomedical Optics* 6:31–40.

Pitts, J. D. and Mycek, M. A. 2001. Design and development of a rapid acquisition laser-based fluorometer with simultaneous spectral and temporal resolution. *Review of Scientific Instruments* 72:3061–3072.

Pogue, B. W., Pitts, J. D., Mycek, M. A. et al. 2001. In vivo NADH fluorescence monitoring as an assay for cellular damage in photodynamic therapy. *Photochemistry and Photobiology* 74:817–824.

Ramanujam, N. 2000. Fluorescence spectroscopy of neoplastic and non-neoplastic tissues. *Neoplasia (New York, N Y)* 2:89–117.

Reif, R., Amorosino, M. S., Calabro, K. W. et al. 2008. Analysis of changes in reflectance measurements on biological tissues subjected to different probe pressures. *Journal of Biomedical Optics* 13:010502.

Richards-Kortum, R. and Sevick-Muraca, E. 1996. Quantitative optical spectroscopy for tissue diagnosis. *Annual Review of Physical Chemistry* 47:555–606.

Rivoire, K., Nath, A., Cox, D. et al. 2004. The effects of repeated spectroscopic pressure measurements on fluorescence intensity in the cervix. *American Journal of Obstetrics and Gynecology* 191:1606–1617.

Sahar, A., Boubellouta, T., Lepetit, J. and Dufour, E. 2009. Front-face fluorescence spectroscopy as a tool to classify seven bovine muscles according to their chemical and rheological characteristics. *Meat Science* 83:672–677.

Schwarz, R. A., Arifler, D., Chang, S. K. et al. 2005. Ball lens coupled fiber-optic probe for depth-resolved spectroscopy of epithelial tissue. *Optics Letters* 30:1159–1161.

Schwarz, R. A., Gao, W., Daye, D. et al. 2008. Autofluorescence and diffuse reflectance spectroscopy of oral epithelial tissue using a depth-sensitive fiber-optic probe. *Applied Optics* 47:825–834.

Chapter 7

Shimazaki, N., Tokunaga, H., Arai, T. and Sakurada, M. 2009. Development of novel short-term heating angioplasty: assessment of artery collagen/elastin ratio and its contribution to artery dilatation. In Dössel, O. and Schlegel, W. C., eds., *World Congress on Medical Physics and Biomedical Engineering, September 7–12, 2009, Munich, Germany.* Berlin: Springer, 263–266.

Sionkowska, A., Skopinska, J., Wisniewski, M. and Leznicki, A. 2007. Spectroscopic studies into the influence of UV radiation on elastin in the presence of collagen. *Journal of Photochemistry and Photobiology B-Biology* 86:186–191.

Skala, M. C., Palmer, G. M., Zhu, C. et al. 2004. Investigation of fiber-optic probe designs for optical spectroscopic diagnosis of epithelial pre-cancers. *Lasers in Surgery and Medicine* 34:25–38.

Skala, M. C., Riching, K. M., Bird, D. K. et al. 2007. *In vivo* multiphoton fluorescence lifetime imaging of protein-bound and free nicotinamide adenine dinucleotide in normal and precancerous epithelia. *Journal of Biomedical Optics* 12:024014.

Starly, B. and Choubey, A. 2008. Enabling sensor technologies for the quantitative evaluation of engineered tissue. *Annals of Biomedical Engineering* 36:30–40.

Sutton, E. J., Henning, T. D., Pichler, B. J., Bremer, C. and Daldrup-Link, H. E. 2008. Cell tracking with optical imaging. *European Radiology* 18:2021–2032.

Teuchner, K., Freyer, W., Leupold, D. et al. 1999. Femtosecond two-photon excited fluorescence of melanin*. *Photochemistry and Photobiology* 70:146–151.

Theodossiou, T., Rapti, G. S., Hovhannisyan, V. et al. 2002. Thermally induced irreversible conformational changes in collagen probed by optical second harmonic generation and laser-induced fluorescence. *Lasers in Surgery and Medicine* 17:34–41.

Uehlinger, P., Gabrecht, T., Glanzmann, T. et al. 2009. In vivo time-resolved spectroscopy of the human bronchial early cancer autofluorescence. *Journal of Biomedical Optics* 14:024011.

Urayama, P., Zhong, W., Beamish, J. A. et al. 2003. A UV-visible-NIR fluorescence lifetime imaging microscope for laser-based biological sensing with picosecond resolution. *Applied Physics B-Lasers and Optics* 76:483–496.

Utzinger, U. and Richards-Kortum, R. R. 2003. Fiber optic probes for biomedical optical spectroscopy. *Journal of Biomedical Optics* 8:121–147.

Vishwanath, K. and Mycek, M. A. 2004a. Do fluorescence decays remitted from tissues accurately reflect intrinsic fluorophore lifetimes? *Optics Letters* 29:1512–1514.

Vishwanath, K. and Mycek, M. A. 2004b. *Simulations of Time-Resolved Autofluorescence Decays in Epithelial Tissue for Differing Probe Geometries,* translated by The Optical Society of America, ThF14.

Vishwanath, K. and Mycek, M. A. 2005. Time-resolved photon migration in bi-layered tissue models. *Optics Express* 13:7466–7482.

Vishwanath, K., Pogue, B. W. and Mycek, M.-A. 2002. Quantitative fluorescence lifetime spectroscopy in turbid media: comparison of theoretical, experimental and computational methods. *Physics in Medicine and Biology* 47:3387–3405.

Vishwasrao, H. D., Heikal, A. A., Kasischke, K. A. and Webb, W. W. 2005. Conformational dependence of intracellular NADH on metabolic state revealed by associated fluorescence anisotropy. *Journal of Biological Chemistry* 280:25119–25126.

Volynskaya, Z., Haka, A. S., Bechtel, K. L. et al. 2008. Diagnosing breast cancer using diffuse reflectance spectroscopy and intrinsic fluorescence spectroscopy. *Journal of Biomedical Optics* 13:024012.

Wagnieres, G., Star, W. and Wilson, B. 1998. *In vivo* fluorescence spectroscopy and imaging for oncological applications. *Photochemistry and Photobiology* 68:603–632.

Wang, L., Jacques, S. L. and Zheng, L. 1995. MCML—Monte Carlo modeling of photon transport in multi-layered tissues. *Computer Methods and Programs in Biomedicine* 47:131–146.

Wilson, R. H. and Mycek, M. A. 2011. Models of light propagation in human tissue applied to cancer diagnostics. *Technology in Cancer Research and Treatment* 10:121–134.

Wilson, R. H., Chandra, M., Lo, W. L. et al. 2008. Simulated fiber-optic interrogation of autofluorescence from superficial layer of tissue-engineered construct. *Frontiers in Optics 2008 Technical Digest (Optical Society of America)*:FTuK6.

Wilson, R. H., Chandra, M., Chen, L. C. et al. 2010. Photon-tissue interaction model enables quantitative optical analysis of human pancreatic tissues. *Optics Express* 18:21612–21621.

Wilson, R. H., Chen, L.C., Shiuhyangkuo, Lloyd, W., Marcelo, Cynthia, Feinberg, Stephen E., and Mycek, M.-A. et al. 2011. Mesh-based Monte Carlo code for fluorescence modeling in complex tissues with irregular boundaries. *Proc.SPIE* 8090, Novel Biophotonic Techniques and Applications, 80900E. doi:10.1117/12.889718.

Wu, Y., Zheng, W. and Qu, J. Y. 2006. Sensing cell metabolism by time-resolved autofluorescence. *Optics Letters* 31:3122–3124.

Wu, Y. C. and Qu, J. N. Y. 2005. Two-photon autofluorescence spectroscopy and second-harmonic generation of epithelial tissue. *Optics Letters* 30:3045–3047.

Wu, Y. C. and Qu, J. Y. 2006. Combined time- and depth-resolved autofluorescence spectroscopy for tissue diagnosis—art. no. 608018. *Advanced Biomedical and Clinical Diagnostic Systems IV* 6080:8018.

Xu, C. and Webb, W. W. 1996. Measurement of two-photon excitation cross sections of molecular fluorophores with data from 690 to 1050 nm. *Journal of the Optical Society of America B-Optical Physics* 13:481–491.

Yudovsky, D. and Pilon, L. 2010. Modeling the local excitation fluence rate and fluorescence emission in absorbing and strongly scattering multilayered media. *Applied Optics* 49:6072–6084.

Zellweger, M. 2000. Fluorescence Spectroscopy of Exogenous, Exogenously-Induced and Endogenous Fluorophores for the Photodetection and Photodynamic Therapy of Cancer. PhD dissertation, Ecole Polytechnique Federale de Lausanne.

Zheng, W., Li, D., Wu, Y. and Qu, J. Y. 2008. *Single- and Two-Photon Excited Autofluorescence of Epithelial Tissue,* translated by Optical Society of America, BTuF38.

Zhong, W., Wu, M., Chang, C. W. et al. 2007. Picosecond-resolution fluorescence lifetime imaging microscopy: a useful tool for sensing molecular interactions *in vivo* via FRET. *Optics Express* 15:18220–18235.

Zhu, C. F., Burnside, E. S., Sisney, G. A. et al. 2009. Fluorescence spectroscopy: an adjunct diagnostic tool to image-guided core needle biopsy of the breast. *IEEE Transactions on Biomedical Engineering* 56:2518–2528.

Chapter 7

8. Raman Spectroscopy

Flavius C. Pascut, Andy Downes, and Ioan Notingher

8.1 Introduction

Since its discovery by Raman and Krishnan in 1926, the Raman effect has enjoyed an explosion of new applications. The past 20 years in the Raman spectroscopy history are perhaps the most exciting for regenerative medicine with the demonstration of the first Raman application on live cells (Puppels et al. 1990). Since then, a plethora of new applications have emerged that demonstrate the true potential of Raman technology in the field of regenerative medicine, as will be discussed in this chapter.

Chapter 8

8.1.1 The Raman Effect

The Raman effect belongs to the class of light–matter interaction mechanisms. While infrared spectroscopy is based on the absorption of light, the Raman effect is attributed to the inelastic scattering of light. If the photons that make up the light interact with molecules inside the specimen in such a way that distorts its electron cloud (i.e., inducing a polarizability change), then the Raman effect can be observed. In contrast to ultraviolet light, photons in the visible and near-infrared part of the spectrum do not have enough energy to activate the electronic transition of the molecules but will create an unstable and short-lived state called "virtual state." As the molecules reach the thermal equilibrium, following photon interaction, three distinctive wavelengths will be observed. Quantum mechanics dictates that a photon is born with a specific energy that cannot be modified during its life. Hence, during scattering processes, it is assumed that the incident photon is absorbed and then reemitted with either (1) almost the same energy Rayleigh scattering, (2) less energy (Raman–Stokes generating a spectrum on the longer wavelengths compared to incident light), or (3) higher energy (Raman anti-Stokes generating spectrum on the shorter wavelengths compared to incident light). An illustration of the three processes is shown in Figure 8.1, together with the process for coherent anti-Stokes Raman scattering (CARS).

During the Raman effect, the inelastic energy transfer that takes place is used by the molecule to change the position of its nuclei. As the amount of energy required for such motion is specific to individual molecules, the Raman emission spectrum is directly related to the molecular vibrations of the individual molecules.

The mathematical treatment of the Raman effect is beyond the scope of this chapter. An extensive treatment of both classical and quantum theory of Raman can be found in Long (2002). If the sample is in thermal equilibrium at room temperature, the number of molecules in the ground state is significantly higher than those in the excited state, with the probability of being in a particular state being given by the Maxwell–Boltzmann distribution. As a result, the Stokes lines are much stronger than the anti-Stokes lines. Hence, in practice, the Raman–Stokes lines are routinely preferred.

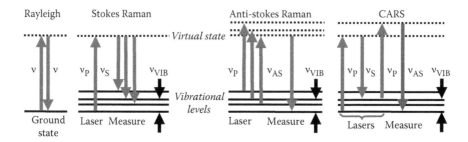

FIGURE 8.1 Energy level diagrams of various types of light scattering by molecules. Rayleigh scattering is elastic, with scattered photons having the same energy as incident photons. In Stokes scattering, the molecule gains energy and a photon with a lower frequency $\nu_S = \nu_P - \nu_{VIB}$, where ν_{VIB} is the vibrational frequency of the molecular bond. In anti-Stokes scattering, the molecule loses energy and a photon with a higher frequency $\nu_{AS} = \nu_P + \nu_{VIB}$. In CARS, two pump photons (ν_P) are absorbed, one Stokes photon undergoes stimulated emission (ν_S), and a higher energy photon is emitted (ν_{AS}).

The Raman effect is quite inefficient for typical biological molecules as only one in a million photons of the incident light is Raman scattered, and it is related to the fact that the Raman scattering cross section of individual molecules is very small. However, with the advent of lasers and high sensitivity detectors like charge coupled devices (CCDs), Raman spectroscopy has become a powerful tool for use in regenerative medicine.

CARS (Maker and Terhune 1965; Zumbusch, Holtom, and Xie 1999; Cheng and Xie 2012) is a variant of Raman spectroscopy. Instead of exciting a vibrational bond with a single laser frequency (ν), two different frequencies (ν_P and ν_S) are used. This process is orders of magnitude more efficient than standard Raman spectroscopy by exciting at the lasers' "beat" or difference frequency, so offers the potential for increased sensitivity or improved imaging speed. Figure. 8.1 shows that three photons are absorbed: two "pump" photons at ν_P and one "Stokes" photon at ν_S. Typically, one laser frequency is fixed at ν_S and the other is tuned to ν_P to excite the required vibrational mode at a frequency $\nu_P - \nu_S = \nu_{VIB}$. The process is near-instantaneous, occurring within the vibrational lifetime of the molecule—typically between 1 and 10 ps. This restricts efficient illumination to lasers with picosecond or femtosecond pulse durations. Picosecond lasers used in CARS have transform-limited spectral widths of a few wavenumbers, so can only excite one peak in the vibrational (Raman) spectrum. Femtosecond lasers have far larger spectral widths: a 100 fs laser has a spectral width of ~10 nm and is able to excite several vibrational modes at once (typically a range of ~200 cm^{-1}). However, this is not enough to excite the full Raman spectrum: for this, an ultrabroadband source of spectral width ~200 nm is required. Supercontinuum lasers (Kano and Hamaguchi 2006; Birks et al. 2002), which employ a femtosecond laser to pump an optical fiber, have recently produced such broad sources, but their intensity profile across the spectrum is far from flat. So we have three variants of CARS microscopy: picosecond CARS (Zumbusch, Holtom, and Xie 1999; Cheng et al. 2001) for imaging of one spectral peak; femtosecond "multiplex" CARS (Cheng et al. 2002; Müller and Schins 2002) for spectral imaging of part of the vibrational spectrum; and "ultrabroadband" multiplex CARS (Kano and Hamaguchi 2005; Kee and Cicerone 2004) for full spectral imaging.

8.1.2 Advantages and Key Features of Raman Microspectroscopy (RMS)

Raman spectroscopy has several particularly useful features, including being noninvasive, being nondestructive, not requiring labels, having high spatial resolution, having minimum sample preparation, and having the ability to investigate biological samples in their physiological conditions.

Perhaps the most attractive feature of Raman spectroscopy lies in its ability to detect spectroscopic information at a molecular level without the use of chemical labels or cell fixation. Such high specificity enables the identification and characterization of individual biochemical molecules inside complex biological samples such as cells (Puppels et al. 1990). Recent studies have demonstrated its potential for measuring the biochemical properties of live cells (Notingher and Hench 2006). RMS can measure intrinsic chemical differences between different cells types, without the use of cellular markers and molecular probes or other invasive procedures that in themselves might alter the very process under investigation. Biochemical differences between cell types are known to exist, as cells are highly specialized entities that are required to perform specific

Chapter 8

functions and consequently will produce their own specific biomolecules. For example, cardiomyocytes (CMs) contain a large number of myofibrils, osteoblasts secrete a collagen-1 matrix that is subsequently mineralized, β-cells secrete insulin, neurons produce neurotransmitters, and so on. Detecting those biochemical changes related to various cellular processes has enabled RMS to be applicable to the study of numerous cellular processes such as proliferation (Short et al. 2005), mitosis (Matthaus et al. 2006), stem cell differentiation (Swain et al. 2008; Schulze et al. 2010), and drugs' and toxins' interaction with cells by monitoring the DNA fragmentation in live lung cells exposed to cancer drugs (Owen et al. 2006; Notingher, Green et al. 2004; Pyrgiotakis et al. 2008). The high chemical specificity of RMS has enabled discrimination between healthy and tumor-derived bone cells (Notingher, Jell et al. 2004), discrimination between undifferentiated hESC and hESC-derived CMs (Chan et al. 2009), the study of biochemical changes related to cell death (Zoladek et al. 2011), the monitoring of recombinant proteins in transgenic microbial cells (Xie et al. 2007), probing of dynamic changes during stimulation and cell remodeling in CM (Inya-Agha et al. 2007), cell cycle dynamics in single living cells (Swain, Jell, and Stevens 2008), and rapid phenotype identification of nosocomial pathogen distinctive species (Maquelin et al. 2002).

The same laser used for the spectroscopic interrogation can be employed for optical trapping and manipulation of cells, thus creating laser tweezers Raman spectroscopy (LTRS). This method has been used to monitor real-time changes of structure and bioactivity of a single mitochondrion (Tang et al. 2007), discriminate between normal and transformed human hematopoietic cells (Chan et al. 2006), distinguish red and white blood cells (Bankapur et al. 2010), and study nonadherent cells (Jess et al. 2007), and is combined with an optofluidic platform for Raman-activated cell sorting (Lau, Lee, and Chan 2008).

The combination of Raman spectroscopy and optical microscopy assures diffraction limited performance of RMS for a detailed analysis at the micrometric scale. This has enabled the analysis of organelles in cells (Matthaus et al. 2007) and in bacteria (Huang et al. 2005), imaging of DNA and protein distribution in apoptotic cells (Uzunbajakava et al. 2003), lipid vesicles (Krafft et al. 2005), and deposits (Wood et al. 2008) in individual fixed cells. Applications to live cells maintained in buffer solutions include analysis of subcellular components, such as nucleus (Puppels et al. 1990), heme moieties in erythrocytes (Wood and McNaughton 2002), and cytotoxic granules in human killer T cells (Takai, Masuko, and Takeuchi 1997). More recently, time-course RMS imaging has been demonstrated, both for short time periods on cells in saline buffers (20 min) (Hamada et al. 2008) and on cells maintained under physiological conditions for extended periods of times (Zoladek et al. 2011).

Figure 8.2a shows a typical application in which RMS has been used to image biological interactions that take place at the junction between live dendritic cells (DC) and T cells (Zoladek et al. 2010). Actin polarization plays a key role in the formation process of immunological synapse (IS). The process of actin accumulation on the edges of the DCs is enhanced when the cells are exposed to an extracellular matrix (ECM) protein (e.g., laminin). The Raman spectral images corresponding to proteins indicate a homogeneous distribution of actin in the control group. However, when the DCs are pretreated with laminin, an increase in protein content takes place when the IS is formed. The accumulation of actin is confirmed using retrospective immunostaining of the same

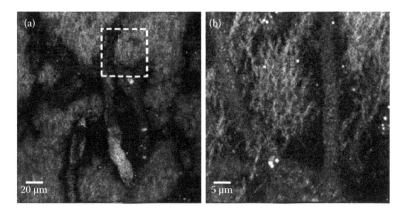

FIGURE 1.4 Overlay CARS (yellow) and SHG (blue) microscopy images of smooth muscle cells grown on a cellulose scaffold. Images were obtained using intrinsic contrast, with the lipids within the cells providing high contrast for CARS and the fibers of the cellulose scaffold providing SHG contrast. Cells could be tracked in real time in 3D, revealing that while cells migrate into the scaffold, they extrude filopodia on the surface. The close-up image in (b) shows the extending filopodia interspersed in the cellulose fiber network. (From Brackmann, C. et al., *Journal of Biomedical Optics*, 16: 021115, 2011.)

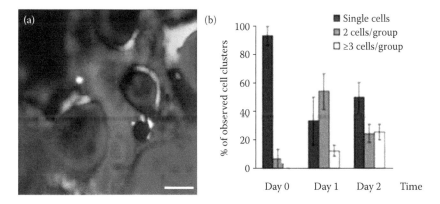

FIGURE 1.5 (a) Image of DiD-labeled labeled HSCs (white) in the calvarium at 1 day after injection into a recipient (Col2.3 GFP mouse) showing a single cell on the left and a cluster of four cells on the right that have undergone cell division. Bone (blue), osteoblasts (green), vasculature (red), bar = 50 μm. (b) At day 0 (on the day of transplantation), most HSCs were observed as single cells, with more clusters observed in subsequent days. (From Lo Celso, C. et al., *Nature*, 457:72–76, 2009a.)

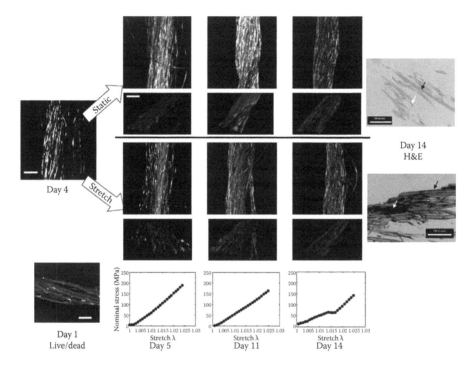

FIGURE 2.7 GFP-transfected dermal fibroblasts (green) grown to near confluence over 5 days on silk fibers (red). Daily mechanical loading was then applied until day 14 and confocal scans taken daily. Images are shown for day 4, before the effect of mechanical loading and for days 5, 11, and 14 after stretching was applied. Live/dead staining is shown for day 1 after seeding. Scale bar is 300 μm. H+E staining is shown for the end point following stretched (top) or static (bottom) conditions. Scale bar is 100 μm. Black arrows indicate silk fibers; white arrows indicate cells. Mechanical test data at each time point are also shown. (With kind permission from Springer Science+Business Media: *Ann. Biomed. Eng.*, Bioreactor system using noninvasive imaging and mechanical stretch for biomaterial screening, 39, 2011, 1390–1402, Kluge, J. A., Leisk, G. G., Cardwell, R. D. et al.)

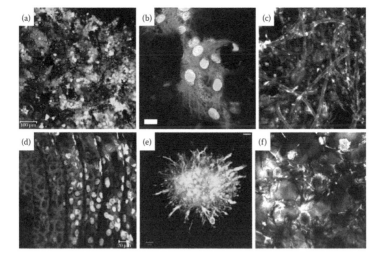

FIGURE 2.8 Fluorescent images of fixed and stained tissue constructs. (a) Live (green) and dead (red) stained chondrocyte-seeded cryogels after 15 days culture. (Bolgen, N., Yang, Y., Korkusuz, P. et al. 3D ingrowth of bovine articular chondrocytes in biodegradable cryogel scaffolds for cartilage tissue engineering. *J. Tissue Eng Regen. Med.* 2011. 5. 770–779. Copyright Wiley-VCH Verlag GmbH & Co. KGaA. Reproduced with permission.) (b) Osteoblasts on EBPADMA-300 scaffolds after 3 days culture, showing nuclei (green), actin fibers (red), and photoscaffold (white). (From Landis, F. A. et al., *Biomacromolecules.* 7: 1751–1757, 2006.) (c) Colocalization of endothelial cells (green) and laminin matrix (red) on a hyaluronan-based scaffold. (Reprinted from *Biomaterials*, 25, Turner, N. J., Kielty, C. M., Walker, M. G., and Canfield, A. E., A novel hyaluronan-based biomaterial (Hyaff-11) as a scaffold for endothelial cells in tissue-engineered vascular grafts, 5955–5964, Copyright 2004, with permission from Elsevier.) (d) Corneal cells on an acellular fish-scale derived scaffold after 7 days culture, showing nuclei (green) and actin fibers (red). (From Lin, C. C. et al., *Eur. Cell Mater.* 19: 50–57, 2010.) (e) Projection of z-stack (1 mm thick) of live (green) or dead (red) staining of neural cells in degradable hydrogels after 14 days culture. (Reprinted from *Biomaterials*, 27, Mahoney, M. J., and Anseth, K. S., Three-dimensional growth and function of neural tissue in degradable polyethylene glycol hydrogels, 2265–2274, Copyright 2006, with permission from Elsevier) (f) Chondrogenic cells cultured on BMP-6 loaded chitosan scaffolds cultured in dynamic conditions after 4 weeks, showing actin fibers (green) and DNA (red) and background view of scaffold morphology. (Tigli, R. S., and Gumusderelioglu, M., Chondrogenesis on BMP-6 loaded chitosan scaffolds in stationary and dynamic cultures. *Biotechnol. Bioeng.* 2009. 104. 601–610. Copyright Wiley-VCH Verlag GmbH & Co. KGaA. Reproduced with permission.)

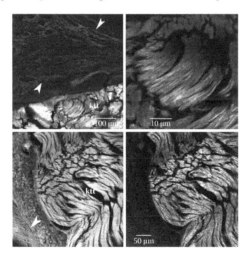

FIGURE 3.5 SHG and two-photon fluorescence give complementary information as shown by these SHG (cyan) and fluorescence (red) micrographs from picrosirius-stained sections of kangaroo tail tendon (ktt) embedded in rat paravertebral muscle (arrowed). The tendon is predominantly type I collagen, which generates intense SHG. The paravertebral tissue is dominated by type III collagen, which shows much weaker SHG activity, but which binds strongly to the picrosirius red. Note also the much greater contrast between these collagen types when imaging in the forward direction (bottom-right) compared with the backward direction (bottom-left). This is because forward directed emission from type I collagen is dominated by intense SHG. (Reproduced from Cox, G. et al., *J Struc Biol* 141: 53–62, 2003. With permission.)

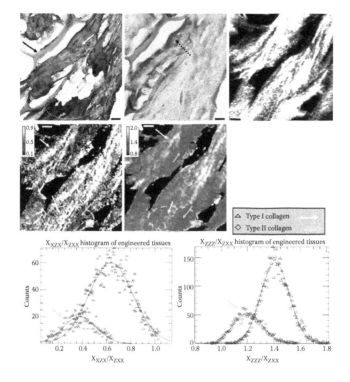

FIGURE 3.7 Polarimetry can potentially be used to discriminate between different collagen types, as shown by these results. By fitting the polarization response on a pixel-by-pixel basis and extracting estimates of rho by fitting (3.1 to the data, histograms of rho values can be assembled for collagens of known type. It is found that type II collagen produces significantly different values to type I. Color-coded images may then be produced, which map these collagen subtypes without the need for immunohistochemical labeling. (Reproduced from Su, P. et al., *Biomaterials* 31: 9415–9421, 2010. With permission.)

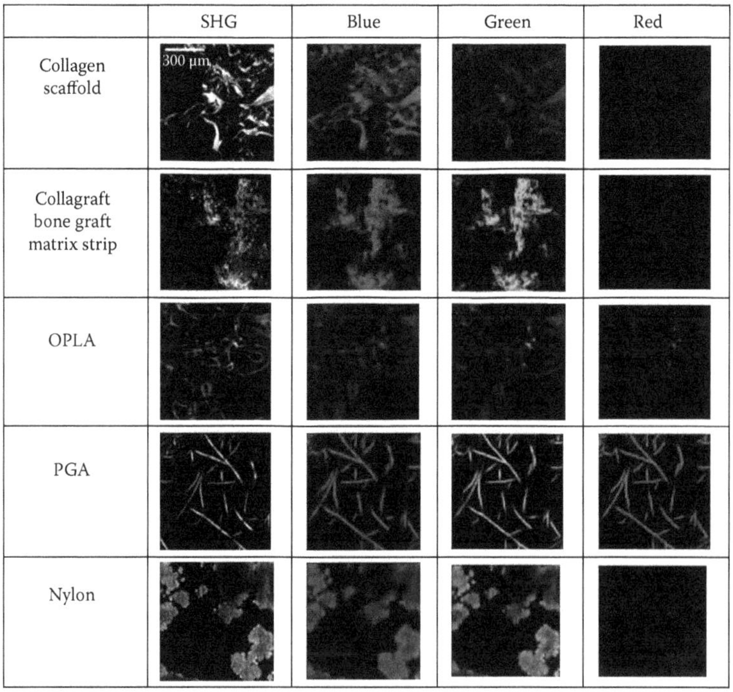

	SHG	Blue	Green	Red
Collagen scaffold	300 μm			
Collagraft bone graft matrix strip				
OPLA				
PGA				
Nylon				

FIGURE 3.8 Scaffold materials can also generate significant levels of SHG, as shown by the images in the left column. These signals must be carefully characterized as they will potentially interfere with weak signals generated by low levels of collagen production. Note also that the SHG and autofluorescence from composite Collagraft material are not colocalized and hence are complementary. (Reproduced from Sun, Y. et al., *Microsc Res Tech* 71: 140–145, 2008. With permission.)

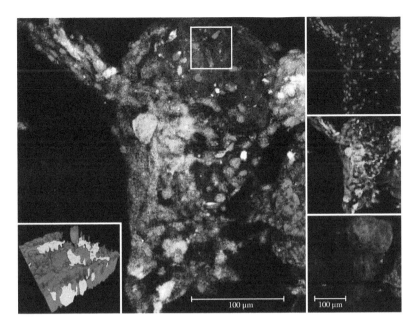

FIGURE 3.9 Simultaneous 3D mapping of procollagen and collagen deposition in a porous polycaprolactone scaffold can be achieved using a combination of immunolabeled TPEF and SHG imaging. The three images on the right show, from the top down respectively, cell nuclei labeled using propidium iodide, procollagen labeled using antibody-conjugated Cy3, and mature collagen generating epidetected SHG. (Reproduced from Filová, E. et al., *J Biomed Opt* 15: 066011, 2010. With permission.)

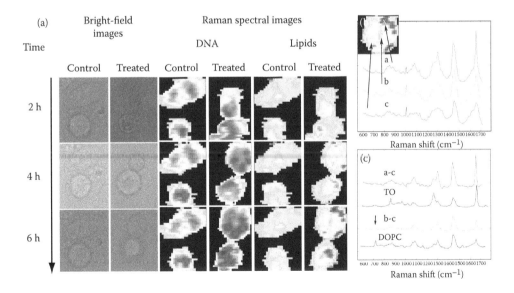

FIGURE 8.5 Time course imaging of apoptotic cells. (a) MDA-MB-231 cells are followed over a 6 h period. Raman spectral images of DNA and lipids show significant changes in distribution when the cells are exposed to etoposide. (b) Raman spectra at different location showing different types of lipids that are present in the treated cells. (c) Identification of lipids by comparison of spectral differences with pure trilinolenin (TO) and phosphatidyl choline (DOPC). (From Zoladek, A. et al., *Journal of Raman Spectroscopy* 42(3):251–258, 2011. Reproduced by permission of the *Journal of Raman Spectroscopy*.)

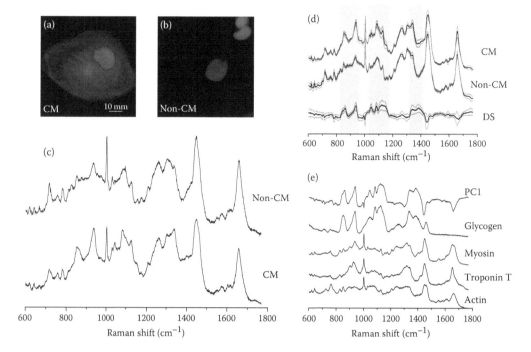

FIGURE 8.6 Cardiac phenotype identification. (a) Typical immunostaining image of a typical CM cell nuclei DAPI (blue) and cardiac α-actinin (red). (b) Immunostaining image of a typical noncardiomyocyte. (c) Raman spectra of the cells shown in (a) and (b). (d) Average Raman spectra of CM and non-CM populations and their computed difference spectrum (DS). (e) Comparison between the discriminant factor (PC1) and selected chemicals that are likely to be more abundant in CMs compared with other phenotypes. (From Pascut, F. C. et al., *Biophysical Journal* 100 (1):251–259, 2011. Reproduced by permission of the *Biophysical Journal*.)

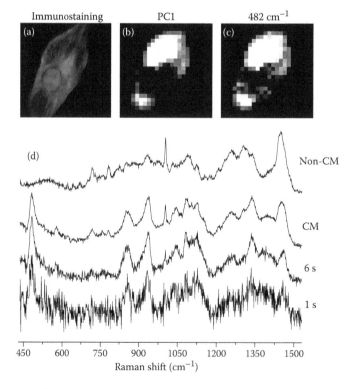

FIGURE 8.7 (a) Immunofluorescence for α-actinin. (b) Raman map of the first PC. (c) Raman map of the 482 cm^{-1} band attributed to glycogen for a typical CM derived from hESC. (d) Average Raman spectrum for the same CM compared to a typical average spectrum for a non-CM. The line scans over the CM for acquisition times from 1 to 6 s are recorded by scanning the cell through the laser spot in the horizontal direction at the position indicated by the arrow in (a). (From Pascut, F. C. et al., *Journal of Biomedical Optics* 16 (4), 2011.)

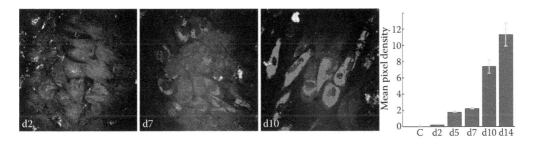

FIGURE 8.10 Multimodal images of ADSCs induced toward adipocytes at different stages postinduction: (2, 7, and 10 days). Red: CARS images of the C–H$_2$ stretch frequency, which is dominated by lipids at 2845 cm^{-1}. Green: TPEF of flavoproteins acquired using a BP filter at 609 nm. Image size 212 × 212 μm, laser powers are 12 mW (ν_p) and 8 mW (ν_s), and the acquisition time is 21 s for all images. Right: Quantitative CARS evaluation of lipid droplet accumulation. CARS pixel density from adipogenic cells was significantly increased at each time point, showing the augmentation of lipid contents of adipo-induced cells in time (compared to the previous day). No collagen deposition was observed in SHG imaging of the adipo-induced cells (*C* is the control sample before induction). These results are published in another form elsewhere. (From Mouras, R. et al., *Journal of Biomedical Optics* 17 (11):116011, 2012.)

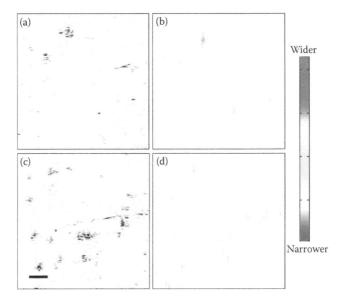

FIGURE 10.5 OCT and SOCT images of two types of cells in engineered tissues. (a) Structural OCT image of fibroblasts in a 3D scaffold. (b) SOCT analysis of (a). (c) Structural OCT image of macrophages in a 3D scaffold. (d) SOCT analysis of (c). The SOCT images are constructed using the HSV color scale. Central color differences are noted for each cell type, representing differing degrees of optical scattering from cell size, shape, nuclei size, and organelle distribution. Scale bar is 50 μm.

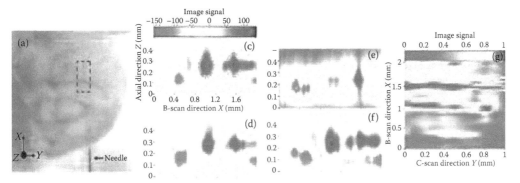

FIGURE 13.10 *Ex vivo* US modulated tomography of *ex vivo* tissue: (a) photograph of blood vasculature in rat ear highlight region of interest; (c–f) B-scans obtained 2 mm apart; (g) image obtained using maximum image projection processing.

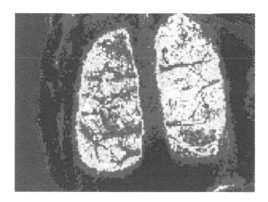

FIGURE 14.6 Middle cerebral artery occlusion imaged with a moorFLPI through a thinned rat skull. The image was obtained as part of a sequence taken at 25 Hz, at a spatial resolution of 152 × 113 pixels to study fast dynamic responses. Higher-resolution images, 760 × 565 pixels, can be obtained at 1 Hz. (Courtesy of Dr. Thaddeus Nowak, University of Tennessee, USA.)

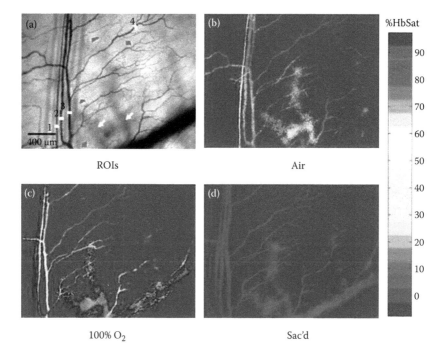

FIGURE 14.7 Example of transmission hyperspectral imaging on a nude mouse dorsal skinfold chamber model. (a) Conventional bright-field image, showing blood vessels as dark shadows due to strong light absorption by hemoglobin. (b) Hemoglobin saturation image extracted from a hyperspectral data set collected in parallel with the bright-field image, while the animal was breathing air. (c) Identical image field while breathing 100% oxygen. Note the increase in hemoglobin saturation, as expected under conditions of unaltered cellular metabolism. (d) Following pentobarbital-induced sacrifice. Continued cellular respiration and O_2 extraction following failure of the circulation is consistent with the observed very low saturation values. (Adapted from Sorg, B. S. et al., *Journal of Biomedical Optics* 10(4):044004, 2005.)

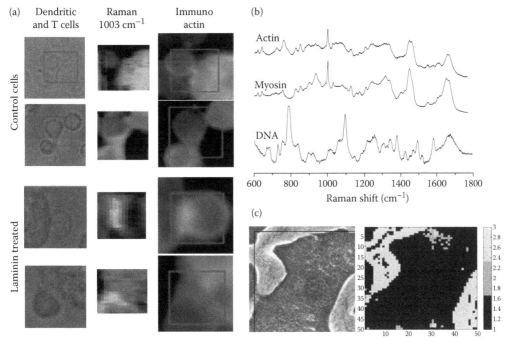

FIGURE 8.2 Examples of Raman spectra for cells and tissue imaging. (a) Images of actin distribution at ISs when dendritic cells were exposed to laminin. (b) Comparison of Raman spectra between actin, myosin, and DNA. (Zoladek, A. B., R. K. Johal, S. Garcia-Nieto, F. Pascut, K. M. Shakesheff, A. M. Ghaemmaghami, and I. Notingher, Label-free molecular imaging of immunological synapses between dendritic and T cells by Raman micro-spectroscopy. *Analyst* 135 (12):3205–3212, 2010. Reproduced by permission of The Royal Society of Chemistry.) (c) Imaging the BCC distribution inside tissue sections using Raman. (From Larraona-Puy, M. et al., *Journal of Biomedical Optics* 14 (5), 2009.)

cells after the Raman measurements were taken. Typical Raman spectra actin, myosin, and DNA are shown in Figure 8.2b for comparison.

Raman spectroscopy has also emerged as a novel diagnostic tool for cancer detection. While diffraction limited spots can be used to study processes at a cellular level, larger spot sizes (>10 μm) are required to cover large blocks of tissues. Indeed, RMS has been extensively used to distinguish tissue pathologies in many cancers, including breast (Frank, McCreery, and Redd 1995; Haka et al. 2005), head and neck cancer (Harris et al. 2010), bladder (Crow et al. 2003), esophageal (Stone et al. 2002), cervical (Mahadevan-Jansen et al. 1998), lung (Huang et al. 2003), and skin cancer (Gniadecka et al. 1997).

Figure 8.2c shows the potential of RMS for automated evaluation of excised skin tissues and detection of basal cell carcinomas (BCC) (Larraona-Puy et al. 2009). In this case, selected Raman bands, which provide the greatest discrimination between BCC and normal skin tissue, are used in building a multivariate linear discriminant analysis (LDA) model (Johnson and Wichern 1998) for automated classification. This model is then applied on the Raman images obtained from new patients, and the results are correlated with a retrospective hematoxylin and eosin stain of the same tissue sections. The RMS images show an excellent correlation with the gold standard of histopathology sections.

Chapter 8

The fact that RMS measurements are noninvasive makes this technique particularly attractive as the level of sample preparation can be kept to a minimum. Regardless of the physical state of the sample, solid, liquid, or vapor, the only requirement is that the interface layer between the biological material and the detection system is Raman inactive and biocompatible. Common suitable substrates used in Raman investigations are quartz, magnesium fluoride, calcium fluoride, and zinc sulfide as they are known to be weak Raman scatterers. However, little is known about the biocompatibility of such substrates. Recently, it was shown that cell physiology could be affected by using quartz substrates (Meade et al. 2007). Uncoated quartz substrates could lead to a decrease in cellular proliferation and protein content in the case of HaCaT keratinocytes, and a gelatin coating of such substrates has been recommended. The substrate biocompatibility is likely to be cell dependent; therefore, the choice of substrates and coating has to be carefully investigated for individual cell lines.

Water is an integral part of living organisms. In contrast with infrared spectroscopy, in which water is a strong absorber, in the visible and NIR region, water is a weak Raman scatterer. Therefore, contribution from water molecules to the Raman measurements is minimal, and measurements of live cells in their physiological and sterile conditions can be performed. In addition, the use of water as a solvent is possible in the study of organic molecules.

8.1.3 Advantages and Key Features of CARS Microscopy

CARS microscopy requires the use of pulsed laser systems, with wavelengths typically in the range 780–1064 nm, so good penetration into tissue is achieved. These wavelengths are also used in other multiphoton microscopies (Chen et al. 2009; Downes, Mouras, and Elfick 2009) with pulsed lasers, namely, two-photon excitation fluorescence (TPEF), second harmonic generation (SHG), and sum frequency generation (SFG). All modes can be excited well enough with the two laser frequencies (ν_P and ν_S), and as these imaging modes produce signals at different wavelengths, they can be recorded on separate spectral channels. All modes can usefully be applied to unlabeled living tissue: TPEF records autofluorescence of elastin and intracellular proteins, SHG is specific to collagen, and SFG is like CARS in that it images chemistry (but only in structural proteins or at interfaces).

Picosecond CARS systems are used to excite just one vibrational peak within the Raman spectrum and simply measure the total intensity of photons within a narrow range (2–10 cm^{-1}), which is similar to the widths of peaks in the Raman spectrum of cells. Multiplex or ultrabroadband multiplex CARS systems require the acquisition of a spectrum at each pixel. This means that for the same laser powers, a spectral CARS image can contain ~100 separate spectral channels of information. If each channel is to have the same signal-to-noise ratio (SNR) as the single picosecond image, spectral imaging with CARS will always be significantly slower than picosecond imaging. At 5–50 ms pixel dwell time, this is significantly faster than confocal Raman but comparable to Raman line illumination systems (where spectra from imaging points along one axis are acquired simultaneously). By comparison, picosecond CARS systems tend to have pixel dwell times in the range 1–100 µs, which enables fast imaging and the acquisition of 3D image stacks.

Such laser systems are relatively easy to launch into laser scanning microscopes, and confocal imaging is inherent due to the low chance of multiple photons being absorbed outside the focal spot. Hence, no pinhole is required in detection, for CARS or any other type of multiphoton microscopy. The CARS signal is forward propagated, and for cell cultures, this transmitted signal is far stronger than the backward-propagated signal (epi-detection). In fact, the epi-detected signal is only a result of reflections of the forward propagated signal at interfaces. However, epi-detected CARS has been performed *in vivo* at video rates (Evans et al. 2005).

Photodamage is a concern for some cells—long exposures of live cells to laser illumination can restrict time-lapse studies with Raman imaging. Photodamage in cells is the result of absorption of visible light, and the chance of photodamage by two-photon absorption of two near-infrared photons is significantly higher with picosecond lasers than for constant power lasers. Femtosecond lasers produce even higher levels of photodamage (König et al. 1999). When taking into account the required exposure times, both standard Raman imaging and multiplex/ultrabroadband CARS produce levels of photodamage that may limit time-lapse studies of some living cell types. Picosecond CARS produces significantly lower amounts of photodamage per image and so is more suitable for time lapse imaging and 3D stacks.

There are two drawbacks with CARS microscopy: firstly, a process called four-wave mixing can occur without any molecules excited—resulting in a "nonresonant background" signal (Cheng et al. 2001) in addition to the required CARS signal. Secondly, because all molecules emit the anti-Stokes photon in phase, their amplitudes add coherently so the CARS signal is dependent on the square of the number of molecules in the focal spot. These two drawbacks limit the minimum observable concentration of molecules within the focal spot to strong peaks in the Raman spectrum.

The nonresonant background can be removed in several ways: by controlling the polarization (Cheng, Book, and Xie 2001); subtracting an image acquired "off resonance"—away from the center of the Raman peak; or with modulation techniques (Chen, Sung, and Lim 2010; Ganikhanov et al. 2006). Femtosecond CARS schemes can also use time-resolved CARS (Volkmer, Book, and Xie 2002) or dual quadrature spectral interferometry (Lim et al. 2006) to remove this unwanted background. Intensity in images and spectra are then background-free but are still quadratically dependent on concentration.

More recently, methods have been devised that measure a signal that is linear in concentration. Heterodyne CARS (h-CARS) (Potma, Evans, and Xie 2006) uses interferometry of the signal with another beam of the same frequency. Stimulated Raman scattering (SRS) (Freudiger et al. 2008) requires amplitude modulation (AM) of one beam and transfers some modulation to the other beam. The amount of transferred modulation in SRS is then measured with a lock-in amplifier, giving vast improvements in the minimum observable concentration to a sensitivity of below 100 µM (Saar et al. 2010).

Picosecond CARS, and these two variants (SRS and h-CARS), are ideally suited to regenerative medicine. The advantages include (1) 3D imaging capabilities, (2) time lapse imaging, (3) its label-free nature, and (4) the separate imaging of cells and scaffold or ECM.

Chapter 8

8.2   Clinical and Biological Measurement Challenges

Raman microscopy is uniquely placed to address several challenges in regenerative medicine. Although the list is extensive, by far the most important features of RMS that are relevant to regenerative medicine are that it is *in situ* nondestructive and allows *in vitro* analysis, label-free time-course imaging of various biomarkers inside cells that allows monitoring of a variety of cellular processes, and detection and localization of specific chemical information present inside cells or tissues.

Perhaps the two most relevant challenges in regenerative medicine are safety and efficacy of the cell and tissue-based therapies. Raman technology with its unique attributes can have a significant impact in overcoming such obstacles and speeding up the translation of laboratory research into clinical applications. For instance, it has been reported that undifferentiated hESC can form teratomas when injected into an immunocompromised mouse (Reubinoff et al. 2000). Therefore, the presence of undifferentiated hESC inside a therapeutic cell population, even in small quantities, could lead to uncontrolled tissue growth. It has been shown that RMS can be used to monitor the differentiation of murine embryonic stem cells (mESCs). By measuring the ratio between RNA and protein level as an indicator of mRNA translation, the degree of differentiation of mESCs can be assessed (Notingher, Bisson et al. 2004). More recently, similar findings have been confirmed using CA1 hESCs (Schulze et al. 2010). Unwanted phenotype in the therapeutic populations not only reduces the efficacy but compromises the suitability of transplants. For instance, transplant of skeletal muscle cells instead of CMs can cause harmful arrhythmias (Menasché et al. 2003). The use of Raman spectroscopy for the phenotypic identification of CMs will be presented in detail in Section 8.4.2.

To fully realize the potential of cell-based therapies, the number of cells available for therapeutic products will have to be increased dramatically. Two-dimensional cell cultures in tissue culture flasks will have to be replaced by three-dimensional cell cultures grown in bioreactors that reproduce the specific physiological conditions associated with cell and tissue cultures. To ensure the consistency, uniformity, quality, and phenotype stability of the final products, methods for automated cell expansion need to be critically controlled. Standard laboratory methods in which a sample is taken and then analyzed are time consuming, increase the risk of contamination, and could perturb the physiology of the entire biomass. Strictly monitoring the conditions and processes inside the bioreactor requires sensors for real-time, noninvasive monitoring of nutrients, oxidants, catabolites, pH, etc. In some bioreactors, fluorescence imaging can be difficult to use and common fluorescent labels are known to interfere with the long-term development of cells and tissues. Therefore, noninvasive interrogation techniques such as Raman spectroscopy are needed that do not alter the normal biological processes inside the cell culture. Raman applicability to bioprocess monitoring will be discussed in more depth in Section 8.4.6.

The use of Raman-based technologies will significantly aid the realization of cell-based therapies for organ and tissue regeneration. At the same time, its use will speed up the transition from current medical practice to a complete restoration of organs and tissue functions in the body.

8.3 Technical Implementation and Considerations of the Method

8.3.1 RMS

One of the most flexible RMS setups for regenerative medicine applications is built around an inverted optical microscope. Such a configuration allows the study of both living cells in their physiological conditions and tissue samples alike. Designs based on an upright configuration are available and routinely used for the characterization of tissue samples; however, dipping microscope objectives into cell cultures over long periods of time is impractical as it exposes the cells to the risk of infection. A typical diagram of a Raman system is shown in Figure 8.3. The laser is directed inside the microscope via a series of mirrors and used for excitation of Raman photons. The microscope objective is used both for sample irradiation as well as for the collection of the Raman photons. The two beam paths, irradiation and emission, are separated by a dichroic beam-splitter after which the Raman photons are focused onto the entrance slit of a dispersive spectrograph. There are a number of variations from the standard design including the use of single optical fibers or fiber bundles, or different geometries for illumination and detection; all those cases can be found elsewhere (McCreery 2000).

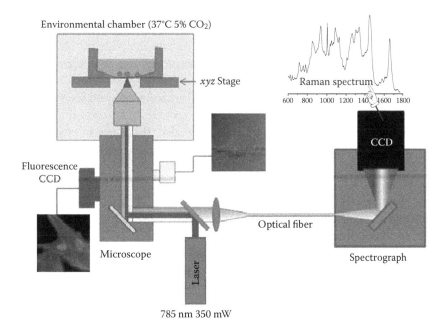

FIGURE 8.3 Typical diagram of Raman-based apparatus for investigating cells in their natural physiological conditions. (From Zoladek, A. et al., *Journal of Raman Spectroscopy* 42(3):251–258, 2011. Reproduced by permission of the *Journal of Raman Spectroscopy*.)

Chapter 8

The absolute intensity of Raman bands, ignoring resonance and surface effects, depends on several parameters as indicated in the following equation:

$$P_R \approx K P_0 \sigma_0 \nu_0 (\nu_0 - \nu_j)^3 C dz$$

There is a linear dependence between the incident photon flux at the sample (P_0 in photons per second) and the number of Raman photons detected (P_R). The laser frequency (ν_0) will influence the Raman process, as the number of emitted photons is proportional to the fourth power of the laser frequency. σ_0 is the frequency-independent cross section, ν_j is the vibrational frequency of the Raman mode, and C is the density of scatterers in molecules per cubic centimeter and is one of the sample specific parameters that often are fixed for a particular sample, especially for the case of biological samples. It can be seen that there are very few experimental parameters available to the design of the experiment: P_0, ν_0, dz (the instrument depth of field), and K (accounts for the instrument response that includes the spectral response of collection and detection optics, spectrograph, and the detector). The choice will depend for the particular application. The damage threshold of each sample will impose an upper limit to the maximum incident photon flux that can be used. Using lasers in the UV range of the spectrum would be more advantageous as it would speed up the measurement process. However, living organisms, such as cells, have low tolerance to this type of radiation limiting the amount of power that can be used. In addition, the Raman resonance effects will change the relative intensity of particular Raman bands, making quantitative analysis particularly difficult. Lasers in the visible region of the spectrum have the advantages of being the most accessible, having high beam quality, and being compact and low cost. Working in this region brings several other benefits, such as the fact that commercially available instruments are optimized for this spectral region as standard, the CCDs used in the construction of Raman instruments have the highest quantum efficiency, and high-quality microscope objectives are routinely available. This allows for shorter imaging times with reported typical acquisition times in the range of 100–500 ms (Miljkovic et al. 2010). However, live cells maintained into their physiological conditions tend to exhibit strong fluorescence background, often caused by cell culture media making Raman measurements a challenging task. If the use of visible lasers is a must, data sampling techniques like shifted subtracted Raman spectroscopy are available to address this issue (Bell, Bourguignon, and Dennis 1998).

By using NIR lasers, fluorescence background can be routinely avoided in most of the cases at the expense of higher acquisition times. Using a 785 nm laser, 1–20 s integration time per pixel is reported (Krafft et al. 2006) providing an image acquisition time of typically 10 to 300 min, depending on the overall sample size and scanning resolution. Such lasers (700–850 nm) are known to induce less cell toxicity compared to visible lasers (Puppels et al. 1991). Several studies investigating the damage induced by the lasers in the 700–1000 nm range have shown little effect of such lasers on cell viability (Neuman et al. 1999; Notingher et al. 2002).

The use of longer NIR laser wavelength (1064 nm) combined with Fourier transform Raman (FT-Raman) is limited to the investigation of living cells as the CCDs lack sensitivity above 1000 nm and the InGaAs detectors available at such long wavelengths are less efficient. In contrast with dispersive Raman, FT-Raman is an interferometer-based technique where an interferogram is recorded by the detector and a fast Fourier transform algorithm is used to restore the sample Raman spectrum (Hendra, Jones, and Warnes 1991). Very few samples will fluoresce at these wavelengths and most samples have a very low absorption coefficient at this wavelength, thus allowing high power lasers to be used. This wavelength is very attractive in bioprocess monitoring, where strong fluorescence from the cell culture broth will interfere with weak Raman signals. Another advantage is the availability of vast databases of pure chemical compounds that were measured using this technique.

8.3.2 CARS

A CARS system is pictured schematically in Figure 8.4. Such a microscope will be designed for use with one pair of laser sources, from the choice of the most common four shown. Repetition rates tend to be around 80 MHz, which permits a short pixel dwell time. The first option for picosecond CARS is to synchronize two independent tunable Ti:sapphire oscillators. The second, most common, is to pump a tunable optical parametric oscillator (OPO) with 532 nm pulses created by frequency doubling a 1064 nm Nd:Vanadate laser. Multiplex CARS requires 1 fs source to excite a broad range of vibrational frequencies, and ultrabroadband multiplex CARS requires a supercontinuum spectrum created within a photonic crystal fiber (PCF). Laser sources are currently far more expensive than standard visible lasers required for fluorescence imaging, but are always improving in terms of cost as well as power and ease of use. As an upgrade to a standard confocal fluorescence microscope, the only other modifications required concern the choice of optical filters to accurately separate the signals.

For SRS imaging, AM of one beam (ν_S) is required. Then the small modulation transferred to the other beam (ν_p) is measured with a silicon (Si) detector and fed into a lock-in amplifier. A TPPL image is recorded in tandem with this SRS image. For h-CARS imaging, PM of an LO is required. This requires using a separate reference sample with a high concentration of molecules with the required vibrational bond to generate a strong, constant CARS amplitude. This is passed through a long pass (LP) filter, then a phase

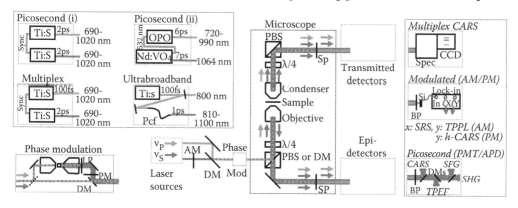

FIGURE 8.4 CARS and multimodal microscopy system. An inverted laser-scanning microscope (center) requires minor modification with a quarter wave plate (λ/4), polarizing beam splitter (PBS) or dichroic mirror (DM), and short pass (SP) filters. The laser source (left) is one of four variants pictured, which may undergo AM for SRS and two-photon photothermal lensing (TPPL) imaging, or by phase modulation (PM) of a local oscillator (LO) for h-CARS. Detectors for transmitted or backscattered signals are photomultiplier tubes (PMTs) or avalanche photodiodes (APDs) for picosecond CARS, silicon detectors for SRS, TPPL, and h-CARS, or a spectrometer and CCD for multiplex and ultrabroadband CARS.

Chapter 8

modulator, and constructively interfered with the signal from the sample. The required h-CARS signal is recorded by the lock-in amplifier.

Normal picosecond CARS images (at $2\,\nu_S - \nu_P$) are recorded alongside SHG (at $2\,\nu_S$ or $2\,\nu_P$), SFG (at $\nu_S + \nu_P$), and TPEF signals by PMTs or APDs. The signals are separated spectrally with DMs and band-pass (BP) filters. Low wavenumber CARS signals are more difficult to detect for many reasons. Firstly, longer wavelength excitation is required, and these have higher losses in microscope systems. Secondly, the detected CARS wavelengths are longer and both PMTs and APDs have reduced detection efficiency at longer wavelengths. Thirdly, the Raman scattering effect depends inversely on the fourth power of the wavelength.

8.4 Example Applications

The Raman spectrum, particularly in the fingerprint region, is a convolution of many molecular vibrations typically associated with live cells. For instance, Raman bands associated to nucleic acids include backbone vibrations O-P-O at 788 cm^{-1} and PO$_2^-$ at

TABLE 8.1 Peak Assignment for Raman Spectra

Peak Position (cm^{-1})	Assignment			
	DNA/RNA	Proteins	Lipids	Carbohydrates
719			C–N$^+$(CH$_3$)$_3$ str.	
760		Ring br. Trp		
782	T,C,U ring br.			
788	O-P-O str. DNA			
842				C–O–C str.
853		C–C str. Pro, ring br. Tyr		
860				CH bending, CH and CH$_2$ def.
936–940		C–C BK str α-helix C–C str. (Lys,Asp), CH$_3$ str. (Leu,Val)		C–O–C glycosidic bond
1004		Sym ring br. Phe		
1045–1125	PO$_2^-$ str. C–N str.		Chain C–C str.	C–O and C–C str. with some C–O–H contribution
1301		Amide III α-helix	CH$_2$ twist	
1340	A,G	CH def.		C–O–H bending
1381				CH bending in plane
1450–1460		CH def.	CH def.	CH$_2$ bending in plane
1658–1662		Amide I α-helix	C = C str.	
1768			C = O str. ester	

Note: str., stretch; def., deformation; br., breathing; see text for references.

1095 cm^{-1} and vibration specific to nucleotides thymine (T), cytosine (C), and uracil (U) at 782 cm^{-1}. These will have higher relative intensities in the nucleus region as compared to cytoplasm. The Raman bands routinely associated with proteins are found in the ranges 1650–1680 cm^{-1} (amide I), 1450 cm^{-1} (C–H bending), 1200–1350 cm^{-1} (amide III), 1005 cm^{-1} phenylalanine (Phe), 853 cm^{-1} tyrosine (Tyr), and 760 cm^{-1} tryptophan (Try). Lipids are characterized by Raman bands at 1449 cm^{-1} (C–H bending vibrations), 1301 cm^{-1} (CH$_2$ twisting and rocking), and 1000–1100 cm^{-1} spectral range (C–C stretching), while Raman bands of unsaturated lipids could be found at 1658 cm^{-1} (C=C stretching) (Tu 1982). Phospholipids exhibit Raman bands in the 700–900 cm^{-1} assigned to different residues at the phosphate ester head group (Takai, Masuko, and Takeuchi 1997). C–O–C vibrations of the glycosidic bonds in carbohydrates can be found in the 800–1100 cm^{-1} range. Table 8.1 contains a summary of molecular vibration assignments used throughout this chapter.

8.4.1 Time Course Imaging of Apoptotic Cells

Apoptosis is a process of programmed cell death during which the cell commits suicide without affecting the neighboring cells. This occurs naturally and is part of the essential processes that are required for natural cell turnover within the body. Any malfunction to this important process can have dramatic consequences for the cells and surrounding tissue.

In tissue engineering, apoptosis plays a key role in a variety of processes. For instance, in bone formation, apoptosis is an integral part of bone turnover, repair, and regeneration during which 50%–70% of osteoblasts undergo apoptosis (Bran et al. 2008). Mounting evidence suggests that apoptosis could play a key role in the development of heart failure, with an apoptotic rate of 2%–12% in the border zone of human myocardial infarcts (Olivetti et al. 1996). Maintaining hemopoietic homeostasis in a host is precisely regulated by the number of hemopoietic stem cells; apoptosis is once again a key ingredient in such processes (Alenzi et al. 2009).

In spite of significant scientific breakthroughs, the observation and understanding of the temporal events that take place during apoptosis still remain a challenge. Noninvasive, molecular specific techniques, such as RMS, can follow the same group of cells undergoing apoptosis and monitor any biochemical changes that take place during the entire process. Molecular-based techniques such as immunostaining require cell fixation and membrane permeabilization to study processes inside cells; therefore, they are invasive by nature and provide only a single snapshot in the lifetime of the process. Transgenic methods (Bouchier-Hayes et al. 2008) that express green fluorescent protein (GFP) from specific promoters are available but require laborious and expensive protocols to ensure molecular specificity. Even at its best, fluorescence imaging is not quantitative as processes associated with emission intensity suffer from a high variability mainly due to photobleaching and intrinsic variability associated with standard immunostaining protocols.

Figure 8.5 shows typical label-free, time-course imaging of apoptotic cells by RMS (Zoladek et al. 2011). Human breast cancer cells (MDA-MB-231) are used as a cellular model of the tumor. The cells are exposed to 300 µM etoposide (an anticancer agent) to induce apoptosis. Raman images of both control and drug exposed cells are acquired at 2 h intervals for the duration of the experiment, 6 h in total. At the same time, bright field images are acquired to monitor any morphological changes that take place during

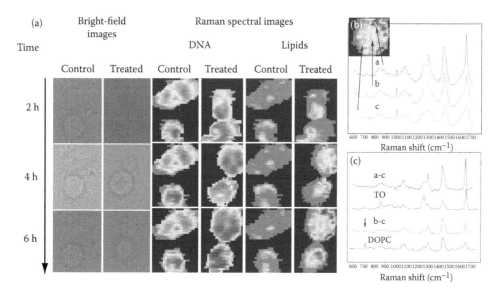

FIGURE 8.5 **(See color insert.)** Time course imaging of apoptotic cells. (a) MDA-MB-231 cells are followed over a 6 h period. Raman spectral images of DNA and lipids show significant changes in distribution when the cells are exposed to etoposide. (b) Raman spectra at different location showing different types of lipids that are present in the treated cells. (c) Identification of lipids by comparison of spectral differences with pure trilinolenin (TO) and phosphatidyl choline (DOPC). (From Zoladek, A. et al., *Journal of Raman Spectroscopy* 42(3):251–258, 2011. Reproduced by permission of the *Journal of Raman Spectroscopy*.)

apoptosis. At the end of the process, 2 h after the completion of the last RMS spectral image, a viability assay is performed using 6-CDFA and Anexin V. The combination of the two stains allows discrimination between viable cells (Anexin V negative, 6-CDFA positive), early apoptotic cells (Anexin V positive, 6-CDFA positive), and necrotic cells (Anexin V positive, 6-CDFA negative).

A common method in obtaining Raman chemical maps, independent of measurement conditions, is to normalize the spectrum using one of the spectral features that is found to be constant for the duration of the experiment. The phenylalanine (symmetric ring stretching at 1005 cm^{-1}) was not affected by the etoposide treatment; therefore, it was used as a reference.

In contrast with control cells, the RMS spectral image, 6 h after etoposide treatment, shows a significant increase in signal strength of the DNA (ratio between 788 and 1005 cm^{-1} band) for MDA-MB-231 cells, indicating that nuclear condensation takes place. This increase concurs with previous reports in apoptotic HeLa cells (Uzunbajakava et al. 2003).

The most dramatic changes in the Raman images between control and etoposide treated cells were found to be caused by a significant increase in the signal strength of lipids (the ratio between 1659 and 1005 cm^{-1} bands). While the control cells show no variation in the lipid content and distribution for the duration of the experiment, the treated cells show a significant lipid buildup in the cytoplasm of apoptotic cells and a change in the lipid spatial distribution. Figure 8.5b shows that regions rich in lipids can be found in healthy cells, and close inspection of individual spectra of those regions indicates the presence of the 719 cm^{-1} Raman band associated with the choline group and a moderate level of unsaturation (1659 cm^{-1} band is lower than 1449 cm^{-1}) common to cell-membrane phospholipids (Takai, Masuko, and Takeuchi 1997). These can be mainly found in the vicinity of the cell nuclei. Given their proximity, these regions were attributed to endoplasmic reticulum and Golgi apparatus typically rich in plasma membranes (Krafft et al. 2006).

In contrast, for the etoposide treated cells, the location of lipid rich features is completely different from the regions found in control cells. A closer inspection of individual spectra taken from these regions reveals some interesting findings, shown in Figure 8.5c. By subtracting the contribution of cytoplasm (spectrum c) from the spectra that corresponds to lipid rich regions, a and b for instance, significant differences in the type of lipids can be identified. For instance, position b contains phospholipids with a moderate level of unsaturation and contains the 719 cm^{-1} band (choline group) similar to the control cells. However, the comparison between the Raman difference spectra (b–c) and the Raman spectra of purified phosphatidylcholine indicates that these regions have a higher concentration of plasma membrane phospholipids than healthy cells. In addition, inspecting the Raman spectra from another lipid rich region, position a, shows the presence of lipids with a higher level of unsaturation (the 1659 cm^{-1} band is higher than the 1449 cm^{-1} band), and this position lacks any Raman bands below 800 cm^{-1}. Comparing this location with the spectrum of pure trilinolenin suggests that this, and similar lipid regions, consists mostly of unsaturated nonmembrane lipids (triglyceride). While accumulation of membrane phospholipids is associated with processes preceding the formation of apoptotic bodies (Zhang et al. 1999), RMS also shows an increased accumulation of nonmembrane lipids. Similar findings were detected using other methods such as ^{1}H NMR (Schmitz et al. 2005) and TEM (Zhang et al. 1999). However, RMS has the advantage of being able to image these regions at a subcellular level and in a noninvasive way. These are important features given the high heterogeneity of cell culture.

8.4.2 Phenotype Identification of Live Cardiomyocytes Derived from Human Embryonic Stem Cells

Stem cells have enormous potential in treating a variety of illnesses, ranging from Alzheimer's and Parkinson's to diabetes and cardiovascular disorders (Klimanskaya, Rosenthal, and Lanza 2008), and are also an important ingredient in tissue engineering (Marot, Knezevic, and Novakovic 2010). Pluripotency is the most sought after ability of human embryonic stem cells (hESCs), as it can provide an unlimited supply of virtually any cell type, making them attractive for many applications, including regenerative medicine, tissue engineering, drug screening, and cancer treatments. In spite of significant scientific advances, high efficiency differentiation and enrichment protocols required for clinical application that produce homogeneous cell populations of the desired phenotype still remain a challenge. This creates numerous concerns regarding the safety of hESC therapy. For instance, the presence and proliferation of undifferentiated cells after transplantation could cause teratoma and uncontrolled tissue growth (Reubinoff et al. 2000). In addition, the presence of unwanted phenotype in the therapeutic cell population, like the transplant of skeletal muscle cells instead of CMs into the left ventricular wall of the heart, can cause harmful arrhythmias (Menasché et al.

2003). Such concerns highlight an immediate need for noninvasive techniques that are capable of identifying specific cellular phenotype. RMS is uniquely placed to address such challenges. Phenotypic identification of CMs is a typical example of the strength and simplicity of this technique.

During differentiation, CMs become specialized to perform specific duties. Given the nature of the activity expected to be undertaken during the life cycle of such cells, a large number of myofibrils are produced. In addition, CMs will require a large amount of glycogen as the energy source that fuels the beating process of these cells. Identifying such molecular markers that are specific to CM derived from hESCs makes RMS extremely attractive for noninvasive, label-free phenotype identification. RMS can discriminate between CMs and other phenotypes with over 95% specificity (the ability of RMS to correctly identify all non-CM) and over 95% sensitivity (the ability to correctly identify CM) (Pascut, Goh, Welch et al. 2011) that is typically required for clinical applications. This is a significant improvement over a previous study in which the phenotypic identification of CMs against undifferentiated hESC was 66% (Chan et al. 2009).

Minute biochemical differences often define the cellular phenotype. In addition, such differences are often hard to discriminate against the large molecular background, which is common to all cells. The identification, detection, and validation of the biochemicals that make up the cellular phenotype of specific cells within cell culture require careful consideration of all the factors involved during the measurement process. Addressing challenges such as heterogeneity of cell populations and sampling techniques in conjunction with multivariate statistical models may allow accurate phenotype identification and discrimination of cells at levels that are clinically relevant.

In cases of high heterogeneity of cell populations, retrospective immunostaining assay for each individual cell is vital as it provides a gold standard for comparison. In this particular example, the percentage of CMs in the cell population that was available for the Raman measurements was below 10%. Although, by inspecting the beating status of the cell before the Raman measurement, CM identification could be achieved, this is highly subjective and the nonbeating or resting CMs are left out. A typical immunostaining image for one of the commonly used cardiac phenotypes, α-actinin, is shown in Figure 8.6a in red together with DAPI, which stains for the cell nucleus, shown in blue. In Figure 8.6b, the α-actinin is not present, indicating the absence of the cardiac phenotype. This is typical to non-CM cells. Figure 8.6c shows typical Raman spectra of the cells shown in Figure 8.6a and b.

Figure 8.6d shows the average Raman spectrum of 50 CMs and 40 non-CMs derived from hESCs following a typical differentiation protocol described by (Burridge et al. 2007). The computed standard deviation is indicated by the gray lines. Variation between the two cell populations is shown in the computed DS. Although minute differences are present across the whole spectrum, significant visual differences are emphasized by the shaded vertical bars.

It is well known that the molecular vibrations of proteins, carbohydrates, and lipids contribute to the 825–950 cm^{-1} region of the spectrum, whereas nucleic acids, lipids, and carbohydrates often dominate the 1035–1170 cm^{-1} region. The narrowest of the three regions, 1320–1420 cm^{-1}, shows slightly less variation between the two classes; nevertheless, in this region, the contribution is dominated by proteins, carbohydrates, and nucleic acids.

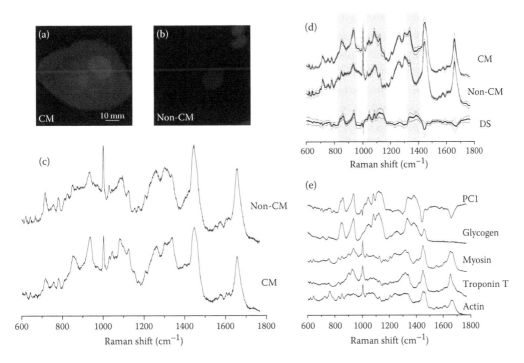

FIGURE 8.6 (See color insert.) Cardiac phenotype identification. (a) Typical immunostaining image of a typical CM cell nuclei DAPI (blue) and cardiac α-actinin (red). (b) Immunostaining image of a typical noncardiomyocyte. (c) Raman spectra of the cells shown in (a) and (b). (d) Average Raman spectra of CM and non-CM populations and their computed difference spectrum (DS). (e) Comparison between the discriminant factor (PC1) and selected chemicals that are likely to be more abundant in CMs compared with other phenotypes. (From Pascut, F. C. et al., *Biophysical Journal* 100 (1):251–259, 2011. Reproduced by permission of the *Biophysical Journal*.)

Minute spectral differences are present across multiple Raman bands. Multivariate statistical methods are extremely valuable for detecting minute biochemical differences, which often are difficult to observe by visual inspection. Principal component analysis (PCA) (Johnson and Wichern 1998) is perhaps one of the simplest and widely used methods. It involves the decomposition of a multivariate dataset into individual uncorrelated principal components (PCs). The first PC will capture the highest amount of variation in the dataset, the second PC captures the highest variation uncaptured by the first, and so on. Using this type of analysis, most of the variation (65%) in the spectral differences between the CM and non-CM was captured by the first PC. This component is shown in Figure 8.6e. Just by using the first PC as a classifier, discrimination between CM and non-CM can be achieved with an accuracy of 96% for sensitivity and 100% for specificity. More than one component can be used; however, in this case, adding additional components did not increase the model performance significantly enough to warrant the increase in the complexity of analysis.

Despite the high accuracy discrimination between CM and non-CM provided by multivariate analysis, such analysis gives little indication of the model's ability to classify cells for which the measurements have yet to be performed. In such situations, cross-validation (CV) (Devijver and Kittler 1982) is another statistical tool that is routinely used to estimate the performance of predictive models. A *k*-fold CV splits the

Chapter 8

dataset into k equally sized parts: k-1 parts are used to produce a model and the remaining part is used to obtain the predictions. The different CV methods can also be labeled with the percentages corresponding to the splitting, for example, the fivefold CV can also be referred to as the 80%/20% CV. The case in which k is equal to the total number of elements in the dataset is called leave-one-out CV (LOOCV). In this method, all spectra except one are used to build a model and then to classify the left-out spectrum. This method is repeated so that each spectrum is predicted once. The use of CV is essential as it can test a realistic scenario in which a highly specific regime is targeted, as required by regenerative medicine. As such, a target of 100% specificity is imposed by adjusting the classification boundary to the PCA model. This emulates the scenario in which no unwanted cells are predicted as CM, at the expense of misclassifying a few true CMs. Using such analysis, 97% specificity and 96% sensitivity can be achieved, but more importantly, it shows the robustness of the model as the results are independent of the CV method used.

Both PCA and CV are extremely powerful tools for extracting and validating the Raman fingerprint (first PC) that provides maximum discrimination between the two (CM and non-CM) classes. However, they give no information about the biomolecular identity of the fingerprint itself. Figure 8.6e shows two of the most common biomolecules that are expected to be present in such cells: glycogen and myosin. There is a striking resemblance between PC1 and glycogen, particularly in the shaded areas. This might suggest that glycogen is the primary contributor to the discriminant fingerprint that allows the classification. Indeed, glycogen Raman bands corresponding to CH bending, CH and CH_2 deformation vibrations at 860 cm^{-1}, C–O–C glycosidic bond vibration at 938 cm^{-1}, C–O and C–C stretching in the region 1084–1123 cm^{-1}, C–O–H bending vibration that contributes to 1340 cm^{-1}, CH in plane bending at 1381 cm^{-1}, and CH_2 in plane bending at 1450 cm^{-1} overlap strongly with the Raman discriminant fingerprint (Galat 1980; Cael, Koenig, and Blackwell 1975). Myosin also has several bands that overlap with the fingerprint bands: 853 cm^{-1} assigned to C–C stretching of proline (Pro) and ring breathing of tyrosine together with 936 cm^{-1} associated to the protein α-helix carbon backbone stretch, CH_3 stretching vibrations of leucine (Leu) and valine (Val), and C–C stretching lysine (Lys) and aspartame (Asp) amino acids (Tu 1982; Carew, Asher, and Stanley 1975).

8.4.3   Cell Sorting Potential of Raman Spectroscopy

The previous section has shown how biomolecules such as glycogen and proteins can be for the phenotypic identification of CM derived from hESC. However, the reported measurement times required for the acquisition of high-quality Raman spectra from live cells are being reported in the range of 2–10 min per cell (Pascut, Goh, Welch et al. 2011). Such long acquisition times make the sorting of a meaningful number of cells impractical. The identification speed has to increase significantly if Raman activated cell sorting (RACS) is to be used in clinical applications. The following application presents the current state of RACS and discusses several ways of how the speed of phenotype identification can be improved (Pascut, Goh, George et al. 2011).

Using the dataset of Raman spectra collected from 50 CMs and 40 non-CMs, the shortest integration time per cell required for the identification of CMs can be established while maintaining the high accuracy of the method (e.g., 95% sensitivity/specificity). Measuring the same dataset with different acquisition times per cell is highly impractical, especially as the data will have to be acquired over several weeks, even months, to account for as much variability as possible (cellular, differentiation protocol, and user variability, to name just a few factors). For this purpose, the same combination of PCA model and CV technique was used with a slight modification. While the model was performed using data with a very high SNR, the model predictions were performed on data with a varying level of SNR that would correspond to much lower acquisition times. In a cell sorting scenario, this is standard practice as the prediction model is built well in advance and has at its basis a significant number of spectra with very high SNR. The predictions are then made on signals that have a much lower SNR. Despite imposing a highly specific regime with a 100% specificity target, the spectral model shows a high resilience to noise. Only after decreasing the acquisition time below 6 s per cell does the average predicted specificity drop below 95% (Pascut, Goh, George et al. 2011). Similar findings are confirmed by taking a single random line scan across a CM, as data show in Figure 8.7. As it appears that glycogen is important in the identification of CMs, measurements in the lower frequency spectral range are shown where glycogen is expected to have a strong band at 482 cm^{-1}. Figure 8.7b and c shows the internal distribution of

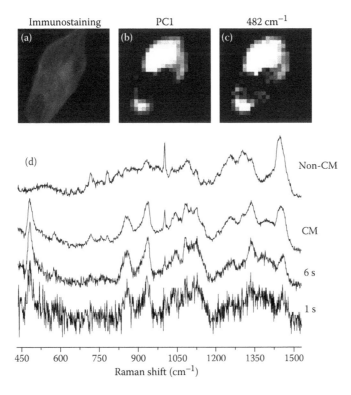

FIGURE 8.7 (**See color insert.**) (a) Immunofluorescence for α-actinin. (b) Raman map of the first PC. (c) Raman map of the 482 cm^{-1} band attributed to glycogen for a typical CM derived from hESC. (d) Average Raman spectrum for the same CM compared to a typical average spectrum for a non-CM. The line scans over the CM for acquisition times from 1 to 6 s are recorded by scanning the cell through the laser spot in the horizontal direction at the position indicated by the arrow in (a). (From Pascut, F. C. et al., *Journal of Biomedical Optics* 16 (4), 2011.)

Chapter 8

glycogen using both the PCA model and the 482 cm^{-1} band area. This band is isolated from other major components typically found inside cells. In addition, this region shows no significant features in the case of non-CM despite long integration time (10 min). As such, it could be exploited in a single wavelength setup similar to fluorescence activated cell sorting (FACS).

Currently, RMS cannot compete with the cell sorting speed usually associated with FACS (>10^5 cells/s) and magnetic activated cell sorting (>10^8 cells/s). In its current form, RMS requires a measurement time of ~5 s per cell to achieve high classification accuracy (>95% specificity and sensitivity required by clinical applications) for hESC-derived CMs. Nevertheless, by using high power lasers and single photon counters that are routinely used in commercial FACS instruments, sorting speeds could be significantly increased. For example, a Ti:Sapphire laser with an output power >5 W at 785 nm would bring a speed improvement by a factor of over 30 times. The current results were obtained with a CCD that generates one event (count) for every 13 photons that reach the detector. Therefore, the use of a photon counter will again increase the speed by another order of magnitude. In the end, the sorting speeds that could be attained by RACS using the currently available technologies compare extremely favorably with FACS during its early beginnings (Mansberg, Saunders, and Groner 1974). In addition, the morphology of the cells inside the flowing stream is significantly different. The cells are more compact, yielding a higher SNR and better discrimination from the background, increasing further the sorting speed. Even more, the discrimination performance using the glycogen band at 488 cm^{-1} has not yet been investigated. However, a simple visual comparison between the 1 s spectrum of CM and the 10 min spectrum of an entire non-CM cell could suggest that this band alone could provide a better discrimination than the 860, 938, 1084, and 1123 cm^{-1} bands combined.

8.4.4   Extracellular Matrix Characterization Using Raman Spectroscopy

Designing a laboratory-grown tissue construct for clinical applications requires a deep understanding not only of the physical properties of engineered tissues but also their biochemical properties. The engineered tissue needs to replicate both the properties and functions of the part/organ it replaces. For instance, synchronization and transmission of contractile motion are vital for cardiac tissue, sound wave transmission and electrical conversion are important in the case of the inner ear mechanism, and high physical strength needs to be replicated if bone or connective tissue is required. With more than 2 million replacement procedures a year (Lewandrowski et al. 2000), bone tissue is in high demand for regenerative medicine. Therefore, the next focus is to describe applications of RMS in assessing the biochemical properties of engineered bone tissue.

In this application, mineralized bone nodules were formed *in vitro* using cells from a variety of sources such as mouse ESC, neonatal calvarial osteoblasts (OBs), and adult bone marrow–derived mesenchymal stem cells (MSCs) (Gentleman et al. 2009). The physical and biochemical properties of such constructs were then compared with each other and, more importantly, compared with native bone to emphasize the importance for selection of an appropriate cell source in such applications. Early differences between the three cell sources have been noticed; ESC formed nodules quickly with dense mineralized structures, while OB and MSC formed nodules more slowly in more isolated domains. However, no differences were found using standard histological stains for detecting calcified tissue, scanning electron microscopy for nodule structure, or x-ray spectroscopy for the measurement of the Ca/P ratio.

Figure 8.8a shows typical Raman spectra of the three cell cultures before mineralization in the fingerprint region. The Raman spectrum is extremely rich in molecular vibration information, and distinctive spectral features can be observed between the three cell cultures. While the ESC spectrum is dominated by proteins, nucleic acids, and phospholipids, the OB and MSC spectra are dominated by proteins like type II collagen, whereas the contribution of nucleic acids and phospholipids is less visible. The pure collagen spectrum is shown for comparison, indicating that type II collagen is a precursor for bone mineralization for OB and MSC but not in ESC, findings confirmed using real-time polymerase chain reaction (RT-PCR) by monitoring the Col2a1 and Sox9 genes.

By the end of the bone forming process, day 28, bone nodules were present in all three cultures. Figure 8.8b shows typical Raman spectra of ESC, OB, and MSC culture together with a bone spectrum for comparison. The striking similarities between the four spectra indicate the completion of the mineralization process. There are a few dominant Raman features: PO_4^{3-} symmetric stretch around 960 cm^{-1} and substituted CO_3^{2-} vibration at 1070 cm^{-1}; both these peaks are associated with hydroxyapatite, a mineral constituent of the bone. RMS allows for the estimation of several biochemical parameters specific to each individual cell culture such as mineral-to-matrix ratio, carbonate-to-phosphate ratio, and phosphate peak position. Mineral-to-matrix ratio, calculated as PO_4^{3-} peak area divided by protein peak area in the 1600–1720 cm^{-1} region, was lower for ESC than native bone or OB and MSC. Carbonate/phosphate ratio is an indication of carbonate substitution into an apatite lattice and was found to be higher in ESC than

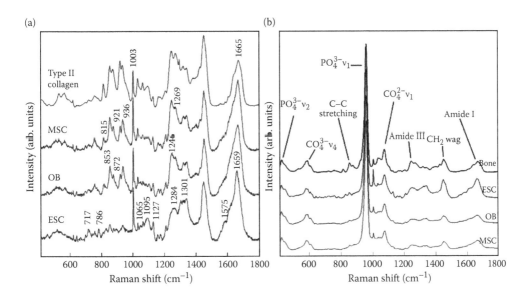

FIGURE 8.8 Raman spectroscopy of bone nodule derived from three different cell types. (a) Raman spectra of premineralization ESC, OB, and MSC nodules plotted with type II collagen spectrum. (b) Raman spectra of mouse bone together with ESC, OB, and MSC after 28 days in culture. (Reprinted by permission from Macmillan Publishers Ltd: *Nature Materials*, Gentleman et al., copyright 2009.)

Chapter 8

other cultures. The PO_4^{3-} band shape and position could be used as an indicator of mineral crystallinity (Tarnowski, Ignelzi, and Morris 2002), which is higher in ESC than OB, MSC, or bone. Increased carbonate substitution and higher mineral crystallinity have both been positively correlated with a deterioration of mechanical properties of bone (Akkus, Adar, and Schaffler 2004). By combining the PCA with factor analysis (FA) of the Raman spectra, the complexity of the mineral and matrix environment can be investigated. For instance, the mineral environment developed by the OB and MSC derived nodules is more similar to the native bone than ESC. While ESC derived nodules contained less mineral species, the Raman FA analysis of OB and MSC nodules revealed the presence of more diverse mineral species.

This application demonstrates that RMS can have a key role in assessing the biocompatibility of cell-based constructs with native bone tissue. All the parameters retrieved from the Raman spectrum can be used as an indicator of the biochemical and mechanical properties of the engineered bone. RMS is a powerful addition to the more familiar bone tissue characterization technologies such as Fourier-transform infrared (FTIR) spectroscopy, x-ray diffraction (XRD), and electron microscopy (EM).

8.4.5   3D Scaffold Characterization Using Raman Spectroscopy

Although cells are an integral part of all tissue types, cells alone cannot form useful tissue specimens for clinical applications. A 3D connective matrix is required to form a cohesive assembly and to provide a template for tissue growth. By combining individual cells and biodegradable 3D scaffolds, engineered tissue constructs can be developed for the restoration of damaged or diseased tissue. The main function of such scaffolds is to offer a temporary replacement for the ECM, serving as a frame that stimulates and aids cell proliferation and tissue growth.

More recently, scaffolds made of polymer based biomaterials poly(lactic-co-glycolic acid) (PLGA) have been used as a delivery tool for plasmid DNA containing SOX trio genes to aid in the transfection of adipose stem cells toward chondrogenic cells (Im, Kim, and Lee 2011). Important processes, such as degradation and absorption rate, are required to be understood before the tissue is implanted in the target host. Hydrolysis of the ester bonds of the PLGA molecule causes it to degrade into naturally occurring metabolites like glycolic acid and lactic acid. For successful integration, the degradation rate of the implant has to match the body's ability to regenerate and interface the engineered construct. The implantation of such biomaterial will most likely result in a foreign body reaction involving macrophages, which engulf and digest small particles of the degrading materials. Raman spectroscopy has been used to study the chemical bonds involved in the degradation of PLGA microspheres (van Apeldoorn et al. 2004). The microspheres added to a macrophage cell line (RAW 264.7) showed signs of degradation after 1 week. The intensity of the Raman band corresponding to the ester group at 1768 cm^{-1} was used to estimate the degradation of PLGA. The decrease in intensity of this band was associated with signs of intracellular degradation where the created void is replaced by both proteins (presence of the bands Phe at 1004 cm^{-1}, amide I at 1662 cm^{-1}) and lipids (presence of band at 1440 cm^{-1}) specific to cell cytoplasm that have traveled through the formed pores inside the PLGA microspheres. To further aid the understanding of the degradation process, Raman spectroscopy has also been used to quantify the volume fraction of biodegradable polymers like tyrosine-derived poly-carbonate-based biodegradable polymer (DTE) and an embedding polymer-polymethylmethacrylate (PMMA) in histological preparations (Nandagawali, Yerramshetty, and Akkus 2007).

Another class of biomaterials that are suitable candidates for regenerative medicine are bioactive glasses. Biocompatibility in clinical application as synthetic bone replacement has already been proven (Cordioli et al. 2001). In addition, these glasses have the ability to influence proliferation, differentiation, and attachments of cells. Particularly attractive for bone replacement is their ability to form hydroxycarbonate apatite (HCA) when in contact with biological fluids. Using Raman spectroscopy, the symmetric stretch vibration of the phosphate group (960 cm^{-1}) can be followed; thus, the degradation of bioactive glass scaffolds with different chemical compositions that are immersed in simulated body fluids can be investigated (Moimas et al. 2006). In addition, Raman spectroscopy has been used in the monitoring of chondrocyte response to bioactive glasses, confirming the fact that cell phenotype is preserved and cartilage cellular matrix is created over the 4 weeks of cell culture (Jones et al. 2005).

8.4.6 Bioprocesses and Bioreactor Monitoring Using Raman Sensors

Raman spectroscopy has been demonstrated as a valuable tool for the study of cell proliferation, differentiation, and apoptosis at single cell levels. However, translation of cell cultures from the laboratory into large-scale manufacturing bioreactors requires intermediate steps to validate the feasibility of such a transition.

The first building blocks in this transition are already in place as Raman spectroscopy has long been used for *in situ* bioprocess monitoring and concentration estimation of glucose, acetate, formate, lactate, phenylalanine, creatine, etc. (Cannizzaro et al. 2003; Lee et al. 2004; Shaw et al. 1999). More recently, an in-line Raman probe, immersed inside a 500 L bioreactor, has been used for real-time monitoring of multiple parameters (glutamine, glutamate, glucose, lactate, ammonium, viable cell density, and total cell density) in mammalian cell cultures (Abu-Absi et al. 2011).

Real-time assessment of the cell culture status (viability and phenotype) inside small-scale bioreactors is needed to justify the transition to industrial-scale bioreactors. The evaluation and validation using a microbioreactor model is vital, as laboratory cell culture conditions do not replicate the conditions inside bioreactors where dynamic processes are continuously taking place (constant supply of nutrients, continuous waste removal, dynamic adjustments of culture conditions). Recently, the combination of Raman microscopy and microbioreactors has been demonstrated in the study of human bone marrow stromal cell (hBMSC) proliferation and differentiation (Pully et al. 2010). The cells were cultured in both basic medium for control cells and osteogenic medium in several microbioreactors. Raman observations were made through a CaF2 window fitted to the bioreactor's chamber. The control cells displayed standard Raman bands of biomolecules normally associated with these cells with no change over the 21-day experiment, confirming phenotype stability. The Raman spectra taken from the cells in the presence of osteogenic media showed the presence of the characteristic phosphate band (960 cm^{-1}) associated with hydroxyapatite, thus confirming the differentiation of hBMSC toward the osteogenic phenotype.

Chapter 8

8.4.7 Applications of CARS Microscopy in Regenerative Medicine

CARS microscopy is able to image myelin sheaths and so is particularly useful in the study of neuorodegenerative diseases. The process of myelination and demyelination has been monitored by CARS imaging (Fu et al. 2007) at the C–H$_2$ stretch frequency (2845 cm^{-1}), which highlights dense structural proteins such as myelin. Figure 8.9 shows an accelerated process of myelin degradation in spinal tissue, monitored quantitatively by picosecond CARS microscopy. The authors also performed *in vivo* CARS imaging of myelin.

CARS and multimodal imaging are also ideally suited to monitoring stem cell differentiation. Human adipose-derived stem cells (ADSCs) are multipotent, and their differentiation into osteoblasts (bone cells) and adipocytes (fat cells) has been monitored with CARS and multimodal microscopy (Downes et al. 2011; Mouras et al. 2011, 2012). The differentiation into adipocytes is shown in Figure 8.10. CARS imaging of the C–H$_2$ stretch frequency (at 2845 cm^{-1}) is highly sensitive to lipid droplets—appearing intensely bright even after 2 days of the 14-day differentiation process. These images of lipid droplets have been used as a quantitative assay of lipid production and hence as a noninvasive, label-free method of monitoring stem cell differentiation. Multimodal imaging means that any autofluorescence in cells can be acquired simultaneously. The distribution of the autofluorescent flavoproteins observed in Figure 8.10 varies dramatically over time and so can be used as another measurement to monitor the quality of differentiation.

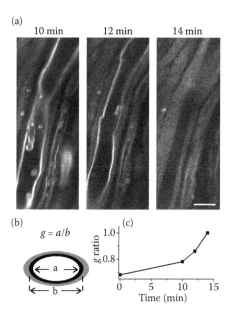

FIGURE 8.9 Real-time CARS imaging of myelin degradation. (a) Time-lapse CARS images of myelin swelling in the spinal tissue incubated with a Krebs' solution containing 10 mg/mL lyso-PtdCho. (b) Diagram of measuring the *g* ratio of a partially swollen myelin fiber based on the remaining compact region. (c) Increase of *g* ratio during the process of myelin swelling. Scale bar = 10 μm. (Reproduced from Fu, Y. et al., *Journal of Neuroscience Research* 85 (13):2870–2881, 2007.)

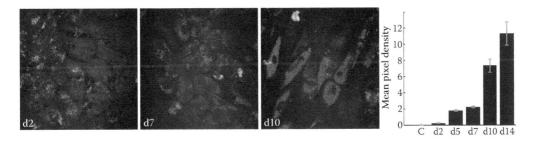

FIGURE 8.10 **(See color insert.)** Multimodal images of ADSCs induced toward adipocytes at different stages postinduction: (2, 7, and 10 days). Red: CARS images of the C–H$_2$ stretch frequency, which is dominated by lipids at 2845 cm^{-1}. Green: TPEF of flavoproteins acquired using a BP filter at 609 nm. Image size 212 × 212 μm, laser powers are 12 mW (ν_P) and 8 mW (ν_S), and the acquisition time is 21 s for all images. Right: Quantitative CARS evaluation of lipid droplet accumulation. CARS pixel density from adipogenic cells was significantly increased at each time point, showing the augmentation of lipid contents of adipo-induced cells in time (compared to the previous day). No collagen deposition was observed in SHG imaging of the adipo-induced cells (*C* is the control sample before induction). These results are published in another form elsewhere. (From Mouras, R. et al., *Journal of Biomedical Optics* 17 (11):116011, 2012.)

The differentiation from ADSCs into osteoblasts is shown in Figure 8.11. CARS imaging can be tuned to the vibrational peaks specific to the phosphate in hydroxyapatite (at 960 cm^{-1}) and to carbonate (at 1070 cm^{-1}). In addition, CARS can be used to highlight cells (at the C–H$_2$ stretch frequency of 2845 cm^{-1})—lower densities of these bonds in the nucleus compared to the cytoplasm means that CARS imaging can be used as a label-free nuclear stain. The combination of imaging of cells (CARS), autofluorescent flavoproteins (TPEF), and collagen (SHG) is presented in Figure 8.11c, demonstrating its superiority compared to the transmitted light image in Figure 8.11d. Also clear in Figure 8.11c is the high axial (sectioning) resolution of multimodal imaging—the collagen is imaged only within a <1 μm slice, whereas all the collagen "layers" are projected onto the transmitted light image.

Scaffolds can also be easily observed with multimodal imaging. For example, the growth of bovine smooth muscle cells on cellulose scaffolds was observed by CARS

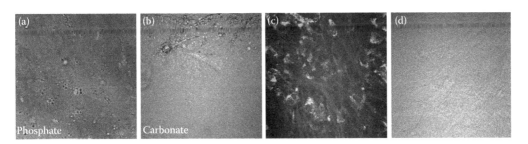

FIGURE 8.11 CARS images of (a) phosphate in hydroxyapatite (at 960 cm^{-1}). (b) Both CARS images were acquired in 120 s and are superimposed on the transmitted light images. (c) Multimodal images of ADSCs induced toward osteoblasts after 10 days, with CARS (C–H$_2$ at 2845 cm^{-1}), TPEF of flavoprotein, and SHG of collagen. Acquisition time was 21 s. (d) Corresponding transmitted light image to (c). Image sizes are all 212 × 212 μm and laser powers are 12 mW (ν_P) and 8 mW (ν_S). (Adapted from Downes, A. et al., *Journal of Raman Spectroscopy* 42(10):1864–1870, 2011; Mouras, R. et al., Non linear optical microscopy of adipose-derived stem cells induced towards osteoblasts and adipocytes. *Proceedings of SPIE* 8086, 2011.)

Chapter 8

imaging to identify cells (C–H$_2$ stretch frequency, 2845 cm^{-1}) and simultaneous SHG imaging to reveal the scaffold (Brackmann et al. 2012). As image rates are fast, 3D imaging was achieved thanks also to the inherent sectioning ability of multiphoton excitation. Such 3D label-free imaging of the growth and differentiation of cells within scaffolds and the production of ECM are powerful tools for regenerative medicine. Although normal picosecond CARS microscopy is suited to high concentrations (such as lipid droplets, mineralization, ECM, and DNA in the nucleus), SRS and h-CARS will enable imaging of small peaks in the Raman spectrum—in the micromolar range. Also, spectral imaging by multiplex CARS will offer improved sensitivity over the imaging of the single peaks imaged in Figures 8.9 through 8.11.

8.5 Conclusion

Developments in Raman spectroscopic instrumentation have led to a large number of applications in regenerative medicine. Despite the low photon conversion efficiency compared to other optical techniques, RMS offers immense benefits in terms of being noninvasive, being nondestructive, offering label-free chemical specificity, offering minimum sample preparation, and having the ability to investigate biological samples in their physiological conditions. Steady progress on improving this technology is being made, such as the use of electron multiplying CCDs, improving the photon collection efficiency by using high efficiency collection optics and the design of high optical throughput spectrometers. In the past 20 years, the acquisition speed and detection limit have improved by over two orders of magnitude.

Although RMS is capable of high spatial resolution, the biomolecular mapping using Raman spectroscopy for large samples remains a challenging task. However, more advanced methods for sampling (Rowlands et al. 2011) and compressed sensing (Candes, Romberg, and Tao 2006) can decrease the acquisition times significantly and will most likely become routine in the future. The few examples shown here of Raman spectroscopy relevant to regenerative medicine are only a selection of a much wider range available in the literature.

CARS (and multimodal) microscopy has already demonstrated itself as a method highly applicable to regenerative medicine. As these label-free multimodal techniques become more sensitive and quantitative, they will be increasingly exploited for exciting new measurements in regenerative medicine.

References

Abu-Absi, N. R., B. M. Kenty, M. Ehly Cuellar, M. C. Borys, S. Sakhamuri, D. J. Strachan, M. C. Hausladen, and Z. J. Li. 2011. Real time monitoring of multiple parameters in mammalian cell culture bioreactors using an in-line Raman spectroscopy probe. *Biotechnology and Bioengineering* 108(5):1215–1221.

Akkus, O., F. Adar, and M. B. Schaffler. 2004. Age-related changes in physicochemical properties of mineral crystals are related to impaired mechanical function of cortical bone. *Bone* 34(3):443–453.

Alenzi, F. Q., B. Q. Alenazi, S. Y. Ahmad, M. L. Salem, A. A. Al-Jabri, and R. K. H. Wyse. 2009. The haemopoietic stem cell: between apoptosis and self renewal. *Yale Journal of Biology and Medicine* 82(1):7–18.

Bankapur, A., E. Zachariah, S. Chidangil, M. Valiathan, and D. Mathur. 2010. Raman tweezers spectroscopy of live, single red and white blood cells. *Plos One* 5(4):e10427.

Bell, S. E. J., E. S. O. Bourguignon, and A. Dennis. 1998. Analysis of luminescent samples using subtracted shifted Raman spectroscopy. *Analyst* 123(8):1729–1734.

Birks, T. A., D. Bahloul, T. P. M. Man, W. J. Wadsworth, and P. S. J. Russell. 2002. Supercontinuum generation in tapered fibres. In: Lasers and Electro-Optics, 2002. CLEO '02. Technical Digest. Summaries of Papers Presented. *Conference on Lasers & Electro-Optics (CLEO)* 486–487.

Bouchier-Hayes, L., C. Munoz-Pinedo, S. Connell, and D. R. Green. 2008. Measuring apoptosis at the single cell level. *Methods* 44(3):222–228.

Brackmann, C., J.-O. Dahlberg, N. E. Vrana, C. Lally, P. Gatenholm, and A. Enejder. 2012. Non-linear microscopy of smooth muscle cells in artificial extracellular matrices made of cellulose. *Journal of Biophotonics* 5(5–6):404–414.

Bran, G. M., J. Stern-Straeter, K. Hormann, F. Riedel, and U. R. Goessler. 2008. Apoptosis in bone for tissue engineering. *Archives of Medical Research* 39(5):467–482.

Burridge, P. W., D. Anderson, H. Priddle, M. D. B. Munoz, S. Chamberlain, C. Allegrucci, L. E. Young, and C. Denning. 2007. Improved human embryonic stem cell embryoid body homogeneity and cardiomyocyte differentiation from a novel V-96 plate aggregation system highlights interline variability. *Stem Cells* 25(4):929–938.

Cael, J. J., J. L. Koenig, and J. Blackwell. 1975. Infrared and Raman-spectroscopy of carbohydrates part VI: normal coordinate analysis of V-amylose. *Biopolymers* 14(9):1885–1903.

Candes, E. J., J. K. Romberg, and T. Tao. 2006. Stable signal recovery from incomplete and inaccurate measurements. *Communications on Pure and Applied Mathematics* 59(8):1207–1223.

Cannizzaro, C., M. Rhiel, I. Marison, and U. von Stockar. 2003. On-line monitoring of Phaffia rhodozyma fed-batch process with *in situ* dispersive Raman spectroscopy. *Biotechnology and Bioengineering* 83(6):668–680.

Carew, E. B., I. M. Asher, and H. E. Stanley. 1975. Laser Raman spectroscopy—new probe of myosin substructure. *Science* 188(4191):933–936.

Chan, J. W., D. S. Taylor, T. Zwerdling, S. M. Lane, K. Ihara, and T. Huser. 2006. Micro-Raman spectroscopy detects individual neoplastic and normal hematopoietic cells. *Biophysical Journal* 90(2):648–656.

Chan, J. W., D. K. Lieu, T. Huser, and R. A. Li. 2009. Label-free separation of human embryonic stem cells and their cardiac derivatives using Raman spectroscopy. *Analytical Chemistry* 81(4):1324–1331.

Chen, B.-C., J. Sung, and S.-H. Lim. 2010. Chemical imaging with frequency modulation coherent anti-Stokes Raman scattering microscopy at the vibrational fingerprint region. *The Journal of Physical Chemistry B* 114(50):16871–16880.

Chen, H., H. Wang, M. N. Slipchenko, Y. Jung, Y. Shi, J. Zhu, K. K. Buhman, and J.-X. Cheng. 2009. A multimodal platform for nonlinear optical microscopy and microspectroscopy. *Optics Express* 17(3):1282–1290.

Cheng, J. X., and X. S. Xie. 2012. *Coherent Raman Scattering Microscopy*. Boca Raton, FL: CRC Press Inc.

Cheng, J.-X., L. D. Book, and X. S. Xie. 2001. Polarization coherent anti-Stokes Raman scattering microscopy. *Optics Letters* 26(17):1341–1343.

Cheng, J.-X., A. Volkmer, L. D. Book, and X. S. Xie. 2001. An epi-detected coherent anti-Stokes Raman scattering (E-CARS) microscope with high spectral resolution and high sensitivity. *The Journal of Physical Chemistry B* 105(7):1277–1280.

Cheng, J.-X., A. Volkmer, L. D. Book, and X. S. Xie. 2002. Multiplex coherent anti-Stokes Raman scattering microspectroscopy and study of lipid vesicles. *The Journal of Physical Chemistry B* 106(34):8493–8498.

Cordioli, G., C. Mazzocco, E. Schepers, E. Brugnolo, and Z. Majzoub. 2001. Maxillary sinus floor augmentation using bioactive glass granules and autogenous bone with simultaneous implant placement. Clinical and histological findings. *Clinical Oral Implants Research* 12(3):270–278.

Crow, P., N. Stone, C. A. Kendall, J. S. Uff, J. A. M. Farmer, H. Barr, and M. P. J. Wright. 2003. The use of Raman spectroscopy to identify and grade prostatic adenocarcinoma *in vitro*. *British Journal of Cancer* 89(1):106–108.

Devijver, P. A., and J. Kittler. 1982. *Pattern Recognition: A Statistical Approach*. London: Prentice-Hall.

Downes, A., R. Mouras, and A. Elfick. 2009. A versatile CARS microscope for biological imaging. *Journal of Raman Spectroscopy* 40(7):757–762.

Downes, A., R. Mouras, P. Bagnaninchi, and A. Elfick. 2011. Raman spectroscopy and CARS microscopy of stem cells and their derivatives. *Journal of Raman Spectroscopy* 42(10):1864–1870.

Evans, C. L., E. O. Potma, M. Puoris'haag, D. Côté, C. P. Lin, and X. S. Xie. 2005. Chemical imaging of tissue *in vivo* with video-rate coherent anti-Stokes Raman scattering microscopy. *Proceedings of the National Academy of Sciences of the United States of America* 102(46):16807–16812.

Frank, C. J., R. L. McCreery, and D. C. B. Redd. 1995. Raman spectroscopy of normal and diseased human breast tissues. *Analytical Chemistry* 67(5):777–783.

Chapter 8

Freudiger, C. W., W. Min, B. G. Saar, S. Lu, G. R. Holtom, C. He, J. C. Tsai, J. X. Kang, and X. S. Xie. 2008. Label-free biomedical imaging with high sensitivity by stimulated Raman scattering microscopy. *Science* 322(5909):1857–1861.

Fu, Y., H. Wang, T. B. Huff, R. Shi, and J.-X. Cheng. 2007. Coherent anti-Stokes Raman scattering imaging of myelin degradation reveals a calcium-dependent pathway in lyso-PtdCho-induced demyelination. *Journal of Neuroscience Research* 85(13):2870–2881.

Galat, A. 1980. Study of the Raman-scattering and infrared-absorption spectra of branched polysaccharides. *Acta Biochimica Polonica* 27(2):135–142.

Ganikhanov, F., C. L. Evans, B. G. Saar, and X. S. Xie. 2006. High-sensitivity vibrational imaging with frequency modulation coherent anti-Stokes Raman scattering (FM CARS) microscopy. *Optics Letters.* 31(12):1872–1874.

Gentleman, E., R. J. Swain, N. D. Evans, S. Boonrungsiman, G. Jell, M. D. Ball, T. A. V. Shean, M. L. Oyen, A. Porter, and M. M. Stevens. 2009. Comparative materials differences revealed in engineered bone as a function of cell-specific differentiation. *Nature Materials* 8(9):763–770.

Gniadecka, M., H. C. Wulf, O. F. Nielsen, D. H. Christensen, and J. Hercogova. 1997. Distinctive molecular abnormalities in benign and malignant skin lesions: studies by Raman spectroscopy. *Photochemistry and Photobiology* 66(4):418–423.

Haka, A. S., K. E. Shafer-Peltier, M. Fitzmaurice, J. Crowe, R. R. Dasari, and M. S. Feld. 2005. Diagnosing breast cancer by using Raman spectroscopy. *Proceedings of the National Academy of Sciences of the United States of America* 102(35):12371–12376.

Hamada, K., K. Fujita, N. I. Smith, M. Kobayashi, Y. Inouye, and S. Kawata. 2008. Raman microscopy for dynamic molecular imaging of living cells. *Journal of Biomedical Optics* 13(4):044027.

Harris, A. T., A. Rennie, H. Waqar-Uddin, S. R. Wheatley, S. K. Ghosh, D. P. Martin-Hirsch, S. E. Fisher, A. S. High, J. Kirkham, and T. Upile. 2010. Raman spectroscopy in head and neck cancer. *Head and Neck Oncology* 2:26.

Hendra, P., C. Jones, and G. Warnes. 1991. *Fourier Transform Raman Spectroscopy: Instrumentation and Chemical Applications*. Ellis Horwood. New York.

Huang, Y. S., T. Karashima, M. Yamamoto, and H. Hamaguchi. 2005. Molecular-level investigation of the structure, transformation, and bioactivity of single living fission yeast cells by time- and space-resolved Raman spectroscopy. *Biochemistry* 44(30):10009–10019.

Huang, Z. W., A. McWilliams, H. Lui, D. I. McLean, S. Lam, and H. S. Zeng. 2003. Near-infrared Raman spectroscopy for optical diagnosis of lung cancer. *International Journal of Cancer* 107(6):1047–1052.

Im, G. I., H. J. Kim, and J. H. Lee. 2011. Chondrogenesis of adipose stem cells in a porous PLGA scaffold impregnated with plasmid DNA containing SOX trio (SOX-5,-6 and-9) genes. *Biomaterials* 32(19):4385–4392.

Inya-Agha, O., N. Klauke, T. Davies, G. Smith, and J. M. Cooper. 2007. Spectroscopic probing of dynamic changes during stimulation and cell remodeling in the single cardiac myocyte. *Analytical Chemistry* 79(12):4581–4587.

Jess, P. R. T., V. Garces-Chavez, A. C. Riches, C. S. Herrington, and K. Dholakia. 2007. Simultaneous Raman micro-spectroscopy of optically trapped and stacked cells. *Journal of Raman Spectroscopy* 38(9):1082–1088.

Johnson, R. A., and D. W. Wichern. 1998. *Applied Multivariate Statistical Analysis*. Fourth ed. New Jersey: Prentice Hall.

Jones, J. R., A. Vats, L. Notingher, J. E. Gough, N. S. Tolley, J. M. Polak, and L. L. Hench. 2005. *In situ* monitoring of chondrocyte response to bioactive scaffolds using Raman spectroscopy. *Bioceramics 17* 284–286:623–626.

Kano, H., and H. Hamaguchi. 2005. Vibrationally resonant imaging of a single living cell by supercontinuum-based multiplex coherent anti-Stokes Raman scattering microspectroscopy. *Optics Express* 13(4):1322–1327.

Kano, H., and H. Hamaguchi. 2006. Dispersion-compensated supercontinuum generation for ultrabroad-band multiplex coherent anti-Stokes Raman scattering spectroscopy. *Journal of Raman Spectroscopy* 37(1–3):411–415.

Kee, T. W., and M. T. Cicerone. 2004. Simple approach to one-laser, broadband coherent anti-Stokes Raman scattering microscopy. *Optics Letters* 29(23):2701–2703.

Klimanskaya, I., N. Rosenthal, and R. Lanza. 2008. Derive and conquer: sourcing and differentiating stem cells for therapeutic applications. *Nature Reviews Drug Discovery* 7(2):131–142.

König, K., T. W. Becker, P. Fischer, I. Riemann, and K. J. Halbhuber. 1999. Pulse-length dependence of cellular response to intense near-infrared laser pulses in multiphoton microscopes. *Optics Letters* 24(2):113–115.

Krafft, C., T. Knetschke, R. H. W. Funk, and R. Salzer. 2005. Identification of organelles and vesicles in single cells by Raman microspectroscopic mapping. *Vibrational Spectroscopy* 38(1–2):85–93.

Krafft, C., T. Knetschke, R. H. W. Funk, and R. Salzer. 2006. Studies on stress-induced changes at the subcellular level by Raman microspectroscopic mapping. *Analytical Chemistry* 78(13):4424–4429.

Larraona-Puy, M., A. Ghita, A. Zoladek, W. Perkins, S. Varma, I. H. Leach, A. A. Koloydenko, H. Williams, and I. Notingher. 2009. Development of Raman microspectroscopy for automated detection and imaging of basal cell carcinoma. *Journal of Biomedical Optics* 14(5):054031.

Lau, A. Y., L. P. Lee, and J. W. Chan. 2008. An integrated optofluidic platform for Raman-activated cell sorting. *Lab Chip* 8(7):1116–1120.

Lee, H. L. T., P. Boccazzi, N. Gorret, R. J. Ram, and A. J. Sinskey. 2004. *In situ* bioprocess monitoring of Escherichia coli bioreactions using Raman spectroscopy. *Vibrational Spectroscopy* 35(1–2):131–137.

Lewandrowski, K. U., J. D. Gresser, D. L. Wise, and D. J. Trantolo. 2000. Bioresorbable bone graft substitutes of different osteoconductivities: a histologic evaluation of osteointegration of poly(propylene glycol-co-fumaric acid)-based cement implants in rats. *Biomaterials* 21(8):757–764.

Lim, S.-H., A. G. Caster, O. Nicolet, and S. R. Leone. 2006. Chemical imaging by single pulse interferometric coherent anti-Stokes Raman scattering microscopy. *The Journal of Physical Chemistry B* 110(11):5196–5204.

Long, D. A. 2002. *The Raman Effect: A Unified Treatment of the Theory of Raman Scattering by Molecules.* John Wiley & Sons, Inc. Chichester, UK.

Mahadevan-Jansen, A., M. F. Mitchell, N. Ramanujam, A. Malpica, S. Thomsen, U. Utzinger, and R. Richards-Kortum. 1998. Near-infrared Raman spectroscopy for *in vitro* detection of cervical precancers. *Photochemistry and Photobiology* 68(1):123–132.

Maker, P. D., and R. W. Terhune. 1965. Study of optical effects due to an induced polarization third order in the electric field strength. *Physical Review* 137(3A):A801–A818.

Mansberg, H. P., A. M. Saunders, and W. Groner. 1974. The hemalog D white cell differential system. *Journal of Histochemistry & Cytochemistry* 22(7):711–724.

Maquelin, K., L. P. Choo-Smith, H. P. Endtz, H. A. Bruining, and G. J. Puppels. 2002. Rapid identification of Candida species by confocal Raman micro spectroscopy. *Journal of Clinical Microbiology* 40(2):594–600.

Marot, D., M. Knezevic, and G. V. Novakovic. 2010. Bone tissue engineering with human stem cells. *Stem Cell Research & Therapy* 1(2):10.

Matthaus, C., S. Boydston-White, M. Miljkovic, M. Romeo, and M. Diem. 2006. Raman and infrared microspectral imaging of mitotic cells. *Applied Spectroscopy* 60(1):1–8.

Matthaus, C., T. Chernenko, J. A. Newmark, C. M. Warner, and M. Diem. 2007. Label-free detection of mitochondrial distribution in cells by nonresonant Raman microspectroscopy. *Biophysical Journal* 93(2):668–673.

McCreery, R. L. 2000. *Raman Spectroscopy for Chemical Analysis.* John Wiley & Sons, Inc. New York.

Meade, A. D., F. M. Lyng, P. Knief, and H. J. Byrne. 2007. Growth substrate induced functional changes elucidated by FTIR and Raman spectroscopy in *in-vitro* cultured human keratinocytes. *Analytical and Bioanalytical Chemistry* 387(5):1717–1728,

Menasché, P., A. A. Hagège, J. T. Vilquin, M. Desnos, E. Abergel, B. Pouzet, A. Bel, S. Sarateanu, M. Scorsin, K. Schwartz, P. Bruneval, M. Benbunan, J. P. Marolleau, and D. Duboc. 2003. Autologous skeletal myoblast transplantation for severe postinfarction left ventricular dysfunction. *Journal of American College of Cardiology* 41(7):1078–1083.

Miljkovic, M., T. Chernenko, M. J. Romeo, B. Bird, C. Matthaus, and M. Diem. 2010. Label-free imaging of human cells: algorithms for image reconstruction of Raman hyperspectral datasets. *Analyst* 135(8):2002–2013.

Moimas, L., G. De Rosa, V. Sergo, and C. Schmid. 2006. Bioactive porous scaffolds for tissue engineering applications: investigation on the degradation process by Raman spectroscopy and scanning electron microscopy. *Journal of Applied Biomaterials and Biomechanics* 4(2):102–109.

Mouras, R., P. Bagnaninchi, A. Downes, M. Muratore, and A. Elfick. 2011. Non linear optical microscopy of adipose-derived stem cells induced towards osteoblasts and adipocytes. *Proceedings of SPIE* 8086.

Mouras, R., P. O. Bagnaninchi, A. R. Downes, and A. P. D. Elfick. 2012. Label-free assessment of adipose-derived stem cell differentiation using coherent anti-Stokes Raman scattering and multiphoton microscopy. *Journal of Biomedical Optics* 17(11):116011.

Chapter 8

Müller, M., and J. M. Schins. 2002. Imaging the thermodynamic state of lipid membranes with multiplex CARS microscopy. *The Journal of Physical Chemistry B* 106(14):3715–3723.

Nandagawali, S. T., J. S. Yerramshetty, and O. Akkus. 2007. Raman imaging for quantification of the volume fraction of biodegradable polymers in histological preparations. *Journal of Biomedical Materials Research Part A* 82A(3):611–617.

Neuman, K. C., E. H. Chadd, G. F. Liou, K. Bergman, and S. M. Block. 1999. Characterization of photodamage to Escherichia coli in optical traps. *Biophysical Journal* 77(5):2856–2863.

Notingher, I., and L. L. Hench. 2006. Raman microspectroscopy: a noninvasive tool for studies of individual living cells *in vitro*. *Expert Review of Medical Devices* 3(2):215–234.

Notingher, I., S. Verrier, H. Romanska, A. E. Bishop, J. M. Polak, and L. L. Hench. 2002. *In situ* characterisation of living cells by Raman spectroscopy. *Spectroscopy-an International Journal* 16(2):43–51.

Notingher, I., I. Bisson, A. E. Bishop, W. L. Randle, J. M. P. Polak, and L. L. Hench. 2004. *In situ* spectral monitoring of mRNA translation in embryonic stem cells during differentiation *in vitro*. *Analytical Chemistry* 76(11):3185–3193.

Notingher, I., C. Green, C. Dyer, E. Perkins, N. Hopkins, C. Lindsay, and L. L. Hench. 2004. Discrimination between ricin and sulphur mustard toxicity *in vitro* using Raman spectroscopy. *Journal of the Royal Society Interface* 1(1):79–90.

Notingher, I., G. Jell, U. Lohbauer, V. Salih, and L. L. Hench. 2004. *In situ* non-invasive spectral discrimination between bone cell phenotypes used in tissue engineering. *Journal of Cellular Biochemistry* 92(6):1180–1192.

Olivetti, G., F. Quaini, R. Sala, C. Lagrasta, D. Corradi, E. Bonacina, S. R. Gambert, E. Cigola, and P. Anversa. 1996. Acute myocardial infarction in humans is associated with activation of programmed myocyte cell death in the surviving portion of the heart. *Journal of Molecular and Cellular Cardiology* 28(9):2005–2016.

Owen, C. A., J. Selvakumaran, I. Notingher, G. Jell, L. L. Hench, and M. M. Stevens. 2006. *In vitro* toxicology evaluation of pharmaceuticals using Raman micro-spectroscopy. *Journal of Cellular Biochemistry* 99(1):178–186.

Pascut, F. C., H. T. Goh, V. George, C. Denning, and I. Notingher. 2011. Toward label-free Raman-activated cell sorting of cardiomyocytes derived from human embryonic stem cells. *Journal of Biomedical Optics* 16(4).

Pascut, F. C., H. T. Goh, N. Welch, L. D. Buttery, C. Denning, and I. Notingher. 2011. Noninvasive detection and imaging of molecular markers in live cardiomyocytes derived from human embryonic stem cells. *Biophysical Journal* 100(1):251–259.

Potma, E. O., C. L. Evans, and X. S. Xie. 2006. Heterodyne coherent anti-Stokes Raman scattering (CARS) imaging. *Optics Letters* 31(2):241–243.

Pully, V. V., A. Lenferink, H. J. van Manen, V. Subramaniam, C. A. van Blitterswijk, and C. Otto. 2010. Microbioreactors for Raman microscopy of stromal cell differentiation. *Analytical Chemistry* 82(5):1844–1850.

Puppels, G. J., F. F. de Mul, C. Otto, J. Greve, M. Robert-Nicoud, D. J. Arndt-Jovin, and T. M. Jovin. 1990. Studying single living cells and chromosomes by confocal Raman microspectroscopy. *Nature* 347(6290):301–303.

Puppels, G. J., J. H. F. Olminkhof, G. M. J. Segersnolten, C. Otto, F. F. M. Demul, and J. Greve. 1991. Laser irradiation and Raman spectroscopy of single living cells and chromosomes: sample degradation occurs with 514.5nm but not with 660nm laser light. *Experimental Cell Research* 195(2):361–367.

Pyrgiotakis, G., T. K. Bhowmick, K. Finton, A. K. Suresh, S. G. Kane, J. R. Bellare, and B. M. Moudgil. 2008. Cell (A549)-particle (Jasada Bhasma) interactions using Raman spectroscopy. *Biopolymers* 89(6):555–564.

Reubinoff, B. E., M. F. Pera, C. Y. Fong, A. Trounson, and A. Bongso. 2000. Embryonic stem cell lines from human blastocysts: somatic differentiation *in vitro*. *Nature Biotechnology* 18(4):399–404.

Rowlands, C. J., S. Varma, W. Perkins, I. Leach, H. Williams, and I. Notingher. 2011. Rapid acquisition of Raman spectral maps through minimal sampling: applications in tissue imaging. *Journal of Biophotonics*.

Saar, B. G., C. W. Freudiger, J. Reichman, C. M. Stanley, G. R. Holtom, and X. S. Xie. 2010. Video-rate molecular imaging *in vivo* with stimulated Raman scattering. *Science* 330(6009):1368–1370.

Schmitz, J. E., M. I. Kettunen, D. E. Hu, and K. M. Brindle. 2005. H-1 MRS-visible lipids accumulate during apoptosis of lymphoma cells *in vitro* and *in vivo*. *Magnetic Resonance in Medicine* 54(1):43–50.

Schulze, H. G., S. O. Konorov, N. J. Caron, J. M. Piret, M. W. Blades, and R. F. B. Turner. 2010. Assessing differentiation status of human embryonic stem cells noninvasively using Raman microspectroscopy. *Analytical Chemistry* 82(12):5020–5027.

Shaw, A. D., N. Kaderbhai, A. Jones, A. M. Woodward, R. Goodacre, J. J. Rowland, and D. B. Kell. 1999. Noninvasive, on-line monitoring of the biotransformation by yeast of glucose to ethanol using dispersive Raman spectroscopy and chemometrics. *Applied Spectroscopy* 53(11):1419–1428.

Short, K. W., S. Carpenter, J. P. Freyer, and J. R. Mourant. 2005. Raman spectroscopy detects biochemical changes due to proliferation in mammalian cell cultures. *Biophysical Journal* 88(6):4274–4288.

Stone, N., C. Kendall, N. Shepherd, P. Crow, and H. Barr. 2002. Near-infrared Raman spectroscopy for the classification of epithelial pre-cancers and cancers. *Journal of Raman Spectroscopy* 33(7):564–573.

Swain, R. J., G. Jell, and M. A. Stevens. 2008. Non-invasive analysis of cell cycle dynamics in single living cells with Raman micro-spectroscopy. *Journal of Cellular Biochemistry* 104(4):1427–1438.

Swain, R. J., S. J. Kemp, P. Goldstraw, T. D. Tetley, and M. M. Stevens. 2008. Spectral monitoring of surfactant clearance during alveolar epithelial type II cell differentiation. *Biophysical Journal* 95(12):5978–5987.

Takai, Y., T. Masuko, and H. Takeuchi. 1997. Lipid structure of cytotoxic granules in living human killer T lymphocytes studied by Raman microspectroscopy. *Biochimica Et Biophysica Acta-General Subjects* 1335(1–2):199–208.

Tang, H., H. Yao, G. Wang, Y. Wang, Y. Q. Li, and M. Feng. 2007. NIR Raman spectroscopic investigation of single mitochondria trapped by optical tweezers. *Optics Express* 15:12708–12716.

Tarnowski, C. P., M. A. Ignelzi, and M. D. Morris. 2002. Mineralization of developing mouse calvaria as revealed by Raman microspectroscopy. *Journal of Bone and Mineral Research* 17(6):1118–1126.

Tu, A. T. 1982. *Raman Spectroscopy in Biology: Principles and Applications.* New York: John Wiley and Sons.

Uzunbajakava, N., A. Lenferink, Y. Kraan, E. Volokhina, G. Vrensen, J. Greve, and C. Otto. 2003. Nonresonant confocal Raman imaging of DNA and protein distribution in apoptotic cells. *Biophysical Journal* 84(6):3968–3981.

van Apeldoorn, A. A., H. J. van Manen, J. M. Bezemer, J. D. de Bruijn, C. A. van Blitterswijk, and C. Otto. 2004. Raman imaging of PLGA microsphere degradation inside macrophages. *Journal of the American Chemical Society* 126(41):13226–13227.

Volkmer, A., L. D. Book, and X. S. Xie. 2002. Time-resolved coherent anti-Stokes Raman scattering microscopy: Imaging based on Raman free induction decay. *Applied Physics Letters* 80(9):1505–1507.

Wood, B. R., and D. McNaughton. 2002. Micro-Raman characterization of high- and low-spin heme moieties within single living erythrocytes. *Biopolymers* 67(4–5):259–262.

Wood, B. R., T. Chernenko, C. Matthaus, M. Diem, C. Chong, U. Bernhard, C. Jene, A. A. Brandli, D. McNaughton, M. J. Tobin, A. Trounson, and O. Lacham-Kaplan. 2008. Shedding new light on the molecular architecture of oocytes using a combination of synchrotron Fourier transform-infrared and Raman spectroscopic mapping. *Analytical Chemistry* 80(23):9065–9072.

Xie, C. G., N. Nguyen, Y. Zhu, and Y. Q. Li. 2007. Detection of the recombinant proteins in single transgenic microbial cell using laser tweezers and Raman Spectroscopy. *Analytical Chemistry* 79(24):9269–9275.

Zhang, J. D., M. C. Reedy, Y. A. Hannun, and L. M. Obeid. 1999. Inhibition of caspases inhibits the release of apoptotic bodies: Bcl-2 inhibits the initiation of formation of apoptotic bodies in chemotherapeutic agent-induced apoptosis. *Journal of Cell Biology* 145(1):99–108.

Zoladek, A. B., R. K. Johal, S. Garcia-Nieto, F. Pascut, K. M. Shakesheff, A. M. Ghaemmaghami, and I. Notingher. 2010. Label-free molecular imaging of immunological synapses between dendritic and T cells by Raman micro-spectroscopy. *Analyst* 135(12):3205–3212.

Zoladek, A., F. C. Pascut, P. Patel, and I. Notingher. 2011. Non-invasive time-course imaging of apoptotic cells by confocal Raman micro-spectroscopy. *Journal of Raman Spectroscopy* 42(3):251–258.

Zumbusch, A., G. R. Holtom, and X. S. Xie. 1999. Three-dimensional vibrational imaging by coherent anti-Stokes Raman scattering. *Physical Review Letters* 82(20):4142–4145.

Chapter 8

Optical Tomography in Regenerative Medicine

9. Optical Transillumination and Projection Tomography for Multidimensional Mesoscopic Imaging

Jim Swoger, James Sharpe, and Mark A. Haidekker

Chapter 9

Optical Techniques in Regenerative Medicine. Edited by Stephen P. Morgan, Felicity R.A.J. Rose, and Stephen J. Matcher © 2014 CRC Press/Taylor & Francis Group, LLC. ISBN: 978-1-4398-5495-2

9.1   Introduction

Computed tomography (CT) is a 3D volumetric imaging technique in which multiple projections through an object are measured, and subsequently, its 3D structure is reconstructed using computational algorithms. First implemented in the early 1970s using x-ray radiation (Beckmann 2006), x-ray CT has become a standard tool for medical imaging. Although in medical applications, millimeter-scale resolutions are typical, micro-CT has improved the resolution to the micron level, and under cryogenic conditions, even nanometer resolutions have been demonstrated (Larabell and Le Gros 2004).

One of the drawbacks to x-ray CT is the lack of specific contrast agents for biological imaging. Contrast arises from the absorption of x-rays as they pass through the sample, and although different tissues can exhibit different levels of absorption (e.g., bone is generally a stronger absorber than soft tissue), it can be difficult to distinguish between various soft tissues or to specifically label a particular biomolecule of interest.

In addition to x-rays, the concepts of CT have been extended to other types of radiation. Single photon emission computed tomography (SPECT) detects gamma rays that are emitted from radioisotopes introduced into the sample (Mariani et al. 2010). The radioisotope can be bound to a biologically relevant molecule to specifically target a tissue or feature of interest, which is one way to overcome the lack of specific contrast agents from which x-ray CT suffers. A related form of CT is positron emission tomography (PET), in which the radioisotope emits a positron (rather than directly emitting a gamma radiation) (Kim et al. 2006). When the positron annihilates with an electron in the tissue, a pair of gamma rays are emitted in opposite directions. The detection of nearly simultaneous gamma-rays ensures that these were generated in the same annihilation event, and this information can be used to determine the line (projection) through the sample along which the event occurred. Electron tomography (ET) is the application of CT techniques to an electron microscopy framework (Downing et al. 2007) and can provide molecular-level resolution of subcellular features, although it is currently limited to very thin samples.

Another option is to perform CT at visible wavelengths of the electromagnetic spectrum. The main motivations for this approach are as follows: (1) it allows the use of the many fluorescent and nonfluorescent contrast agents developed for light microscopy over the last two centuries; (2) it does not involve ionizing radiation and in principle can be used on living tissue; and (3) a CT system at visible wavelengths can be constructed relatively easily. Optical transillumination tomography (OTT) and optical projection tomography (OPT) are practical implementations of CT using visible light and are the subject of this chapter.

The CT variations discussed above can be grouped into two broad categories. In *transmission CT*, the radiation source is external to the sample itself, generally consists of a directed, spatially coherent beam, and results in a contrast primarily due to absorption. X-ray CT, ET, and OTT are examples of transmission CT. SPECT and PET can be termed *emission CT*; here the radiation detected is emitted by the sample itself, is incoherent, and is generally without a preferred direction. As will be discussed below, OPT can be implemented in both transmission and emission modes.

9.2   Clinical and Biological Measurement Challenges OTT and OPT Are Suited to Address

9.2.1   Imaging of Mesoscopic–Scale 3D and 4D Samples

Traditionally, light microscopy has been performed on thin, slab-like samples that can be treated as quasi-2D in nature. Improved optics, as well as other technological developments [such as the confocal principle and multiphoton fluorescence excitation (Pawley 2006; see also Chapter 2), and, more recently, superresolution techniques (Klar et al. 2000; Betzig et al. 2006; Rust et al. 2006)], have helped to push the boundaries toward higher, subcellular resolution. However, in the last few years, there has been increasing recognition of the benefits of going in a different direction, of being able to capture the entirety of larger specimens as well—global, or comprehensive, imaging of whole organs and tissues. These benefits include (1) recording the full 3D geometry of an organ to understand the functional relationships between the component tissues, (2) pinpointing the 3D position or distribution of one cell type with respect to another, and (3) accurate geometric phenotyping. The benefits of whole-organ imaging have only been recognized recently due to the advent of new imaging technologies. Thus, progress in mesoscopic imaging procedures such as OTT and OPT has unlocked the door to a class of samples that was previously inaccessible to optical imaging. These are mesoscopic bioimaging methods aimed to obtain structural 3D information from specimens between 0.5 and 15 mm in diameter, that is, precisely the range of interest for embryology, whole organs in adult small animals, and human biopsies.

The optical microscope has been in existence since the beginning of the 17th century (Hecht 1987) and has proved an invaluable tool in the development of modern biological and medical sciences. Resolutions down to about 0.3 μm can be achieved with relatively simple (as compared to, say, electron microscopy or magnetic resonance imaging) systems, and a variety of contrasts (e.g., bright-field, dark-field, fluorescence, phase, and polarization) are available. However, traditional optical microscopy is usually limited to relatively thin samples: these may be extended in two dimensions on the centimeter scale, but high-resolution imaging is typically limited to a surface layer perhaps a few hundred microns deep. There are several reasons for this.

The primary limiting factor for 3D optical microscopy is optical attenuation, either due to absorption or scattering. Although many biological tissues are transparent on the scale of microns, they become translucent or opaque on millimeter scales. One way to avoid this problem is through the use of chemical clearing agents, although this generally precludes live studies (see Section 9.3.5). A second significant factor is the presence of out-of-focus light. Even in an optically transparent sample, if it is thicker than the depth of field of the microscope (the depth around the focal plane within which the sample can be considered in-focus; see Figure 9.1), any image will contain both the well-resolved features within the depth of field and defocused light from above and below. This can make quantitative analysis of the images difficult, especially for applications in which high depth resolution is critical. In the last few decades, optical sectioning microscopes such as confocal and multiphoton systems have been developed that to a large extent overcome the issue of out-of-focus light, and allowed 3D imaging to depths approaching a millimeter in some cases (Pawley 2006).

Chapter 9

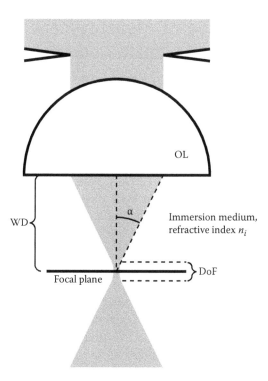

FIGURE 9.1   Schematic of the objective lens and focal region of an optical microscope. The gray regions represent the beam of light that is focused into the sample. OL, objective lens; WD, working distance; DoF, depth of field; α, opening angle (see text).

Even for optical sectioning microscopes working in transparent samples, there is generally still a practical limitation to the depths at which they work well. This is because the resolution, and particularly the axial (depth) resolution, improves with increasing numerical aperture (NA) of the objective lens used, where $NA = n_i \sin(\alpha)$, where n_i is the refractive index of the medium between the objective lens and the sample, and α is the "opening angle" of the focused light (see Figure 9.1). So for high resolution, we need a high NA; unfortunately, practical objective lenses with high NA also have small working distances (the distance between the bottom of the objective lens and the focal plane), typically a few tens to hundreds of microns. So a trade-off exists: we want high NA for high resolution, but this brings with it small working distances that are not suitable for mesoscopic-scale samples. Since the axial resolution varies with NA^2 (the lateral resolution varies linearly with NA), reducing the NA to achieve a long working distance means that the depth resolution will be very poor.

It is clear that this depth limitation in optical microscopy can be overcome by physically slicing the sample into thin sections and treating these as separate 2D specimens for imaging purposes. With care it is possible to make a 3D reconstruction from these 2D images; however, the process is extremely time-intensive, susceptible to registration artifacts, and, of course, destructive to the sample. In contrast, with OTT and OPT, intact mesoscopic samples can be imaged nondestructively in a matter of minutes, which permits reasonably high throughput studies involving tens or hundreds of samples (e.g., Alanentalo et al. 2010).

Another advantage of these techniques over traditional optical microscopies is the fact that their depth resolution is not determined by the NA of the objective lens used. In fact, the resolution of optical CT reconstructions is nearly isotropic and equal to the (high) lateral resolution of the original projections in all three dimensions. Thus, image acquisition can be performed using a relatively low NA, long working distance objective without sacrificing 3D image quality.

9.3   Technical Implementation of OTT and OPT

9.3.1   X-Ray CT and OTT

OTT and x-ray–based CT are closely related. Both methods are based on the Radon transform, that is, the integration of the local absorption of photons along straight lines. Figure 9.2 illustrates the acquisition of projection information. A very thin photon beam (pencil beam) is emitted by the source, and the intensity is recorded by a detector across the object. The object is completely covered by scanning it along parallel lines. At any angle θ, the absorption as a function of the spatial coordinate t is called the *projection* $P(\theta)$. When the attenuation of the incident beam is purely caused by absorption, the projected intensity $I(\theta,t)$ can be described by Lambert–Beer's law:

$$I(\theta,t) = I_0 \cdot \exp\left(-\int_s \mu(X)\,\mathrm{d}X\right) \tag{9.1}$$

where I_0 is the incident beam intensity, s describes the path of the beam, and $\mu(X)$ is the (unknown) absorption at any location X along the path s. In two dimensions, the path can be related to the fixed coordinate system (x,y) through

$$s(\theta,t): \begin{pmatrix} x(\rho) \\ y(\rho) \end{pmatrix} = \begin{pmatrix} \rho\sin\theta + t\cos\theta \\ -\rho\cos\theta + t\sin\theta \end{pmatrix} \tag{9.2}$$

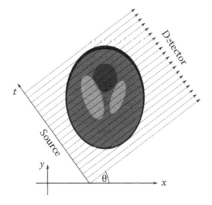

FIGURE 9.2   Illustration of the Radon transform and the acquisition of projection information. The object is scanned along straight lines subtending an angle θ with an arbitrary reference coordinate system. The spatial coordinate t rotates with the scan direction and coincides with the y-axis only when $\theta = 0$. Along the scan lines, photons are absorbed, and the detector records total absorption along each line.

Chapter 9

where ρ is a spatial coordinate parallel to the x-ray beam and perpendicular to the projection coordinate *t*.

Projection data $I(\theta,t)$ is sampled in discrete steps Δt and at discrete angular increments $\Delta\theta$. The spatial map of photon absorption $\mu(x,y)$—this is the desired tomographic slice—can be obtained from the projection data by tomographic reconstruction. One of the earliest reconstruction techniques uses the Fourier slice theorem, that is, the relationship between Radon transform and Fourier transform. The goal is to fill a frequency-domain placeholder $F(u,v)$ with sampled projection data until $\mu(x,y)$ can be obtained with reasonable quality from the inverse Fourier transform of $F(u,v)$. In this case, "reasonable" refers to a compromise between the number of projections, determined by Δt and $\Delta\theta$, and the need to interpolate pixels in $F(u,v)$ that are not "touched" by any path $s(\theta,t)$. The need to interpolate data is exacerbated by the fact that acquisition data exist in polar coordinates, but the reconstruction takes place in Cartesian coordinates. Larger increments of Δt and $\Delta\theta$ imply that more pixels have to be filled by interpolation with the associated degradation of image quality. This fundamental dilemma applies to other reconstruction techniques as well. As an alternative to the Fourier space reconstruction, projection data can be back-projected over a spatial-domain placeholder, whereby the projection data need to be filtered to compensate for the inherent point-spread function of the backprojection. This process is referred to as filtered back projection and is arguably the most widespread reconstruction technique. As a third alternative, the spatial-domain placeholder can be interpreted as a vector of unknown values [i.e., the absorption map $\mu(x,y)$], and each path provides one equation in a linear equation system that can be solved for $\mu(x,y)$. This process is called arithmetic reconstruction technique (ART). Usually, the equation system is not fully determined but rather overdetermined in the presence of noise, which requires special iterative solution methods. The mathematical basis for projection and reconstruction is particularly well described in the comprehensive book by Kak and Slaney (1988).

The principle of projection acquisition and reconstruction is similar for x-ray CT and OTT. In both cases, photons are absorbed or scattered in relation to material properties. Atomic processes, predominantly Compton scattering and the photoelectric effect, affect high-energy x-ray photons along their path. Conversely, visible or near-infrared light that is used for OTT is affected by larger molecules (Rayleigh and Mie scattering) in addition to absorption by chromophores. For x-ray tomography, the absorption is recovered as a bulk absorption coefficient μ (often normalized to the absorption of water), whereas the apparent absorption in visible-light tomography reflects both absorption μ_A and scattering μ_S. In tissue samples, frequently $\mu_S \gg \mu_A$.

Unlike x-rays, visible light is also influenced by the refractive index of the medium. Tomographic reconstruction always assumes that photons propagate along a straight line, but in the presence of refractive index mismatches inside the sample, the straight ray principle no longer holds. The magnitude of this problem can be demonstrated with ray tracing simulations (Yao and Haidekker 2005; Haidekker 2005). Two situations are shown in Figure 9.3. Figure 9.3a is a ray tracing simulation of a water-filled Teflon tube in air, that is, without surrounding index-matching fluid. It becomes obvious that a reconstruction is not possible in this case. In Figure 9.3b, an index-matching fluid has been added, and most rays passing through the sample can be used for reconstruction.

(a)   (b)

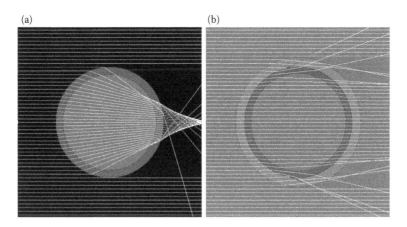

FIGURE 9.3 Ray tracing simulation of a water-filled Teflon tube in air without index-matching (a) and of a thin tissue layer wrapped around a Teflon tube (b). In (b), all air has been replaced by water, which reduces the refractive index mismatch along the sample. The gray shades correspond to the refractive indices (black represents $n = 1.0$ for air, and the highest refractive index is approximately $n = 1.37$ for tissue). In (b), most of the rays (white lines) passing through the sample can be used for reconstruction, and those rays that suffer from significant refraction can be eliminated by collimation or spatial filtering.

Rays that are minimally affected by refraction carry the full information about absorption in the sample. Refracted rays that strongly deviate from the incident angle θ can be removed by collimation, but the recorded intensity would indicate an artifactually high absorption. Because of refraction, the incident laser beam needs to be directed onto the sample in index-matching media. Furthermore, if light source and detector are outside of the media, refraction needs to be minimized by using a parallel-plate container with the laser beam introduced normal to the surface. Refraction is the single most important challenge for OTT.

On the other hand, a number of challenges that affect x-ray tomography do not apply for laser light. The emission of x-ray sources is polychromatic and poorly collimated. Collimators absorb a considerable amount of the emitted photons, which can be compensated for by using a stronger x-ray source. The polychromatic character of the beam leads to beam hardening artifacts, because low-energy components of the x-ray beam are predominantly absorbed, and the x-ray beam does not strictly obey Equation 9.1. Conversely, laser light is monochromatic and highly collimated. In fact, the laser beam can be focused on the sample to improve the point-spread function. Last, but not least, an OTT scanner can be realized with very low-cost components, as will be discussed in the next section.

9.3.2 Technical Realization of OTT Scanners

To obtain the scan geometry in Figure 9.2, a collimated light source—either a laser or a tightly collimated light emitting diode—is directed onto the sample. At the opposite side, transmitted light is collected by a detector. If a qualitative reconstruction is sufficient and recovery of the absorption and scattering coefficient is not crucial, it is not necessary to take refraction into account. However, the resulting reconstruction will show interior details of the sample poorly. For this reason, the sample is immersed in index-matching fluid, thus requiring a container (sample bath). The use of a sample bath requires that the sample be rotated inside the bath with respect to the acquisition geometry, and rotating the source–detector system around the sample (as in x-ray tomography) is not possible. The basic setup of a pencil-beam OTT system is shown in Figure 9.4. The light source would typically be a red or near-infrared laser, because tissue absorption and scattering is lowest in this wavelength range (tissue window). Because of the relatively high intensity of the incident light, the detector can be realized with an amplified silicon photodiode or an avalanche photodiode. High dynamic range detection is important to capture the full range of intensities, which can be as high as 80 to 100 dB for thin tissue (Gladish et al. 2005). Analog-to-digital conversion (ADC) with 16 bits provides 96 dB and is sufficient for most situations; performing the log transform in hardware increases the dynamic range even further. A block diagram of the acquisition and control circuitry is provided in Figure 9.5.

Environmental light can be attenuated with a narrow-band dichroic filter in the detector path. In addition, modulation of the light source and lock-in detection can further reduce the influence of environmental light. The detector system (dichroic filter, spatial filter, and detector) needs to be housed in a light-proof box, but the remaining parts of the optical system, including the sample, can be used in normal daylight (Gladish et al. 2005).

To obtain a high acquisition speed, fast t motion is crucial. Both DC motor and step motors can be used with the appropriate controllers. For positioning in θ and z, step motors are preferable because of their simpler control. When lock-in demodulation is used, the carrier frequency needs to be much higher than the data acquisition rate. With general-purpose equipment, sample rates of 2000 samples per second can be achieved in a straightforward manner, and one scan in the t-direction with 512 pixels takes less than half a second, including acceleration and deceleration of the motor. The actual resolution depends on the actuator, but the step size would typically not be

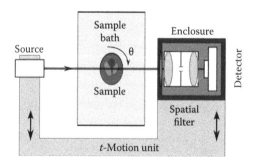

FIGURE 9.4 Basic components for an OTT device (top view). A light source (e.g., a laser) and a detector are mounted on opposite sides of the sample on a linear motion stage. The stage is responsible for scanning along the t-direction. A spatial filter in front of the detector removes off-axis (i.e., scattered or refracted) light, and a dichroic filter removes most of the environmental light. The detector unit is covered in a light-proof enclosure (dark shaded area). The sample is mounted on a rotary stage inside a sample bath. The sample bath contains index-matching fluid, and its transparent side panels are perpendicular to the beam direction to avoid beam refraction. By moving the sample up or down relative to the laser beam, different axial slices can be selected for 3D volumetric scan.

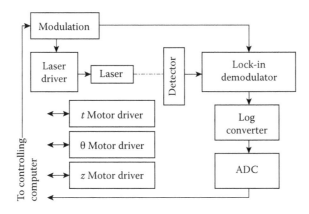

FIGURE 9.5   Block diagram of the control and driver electronics needed for a volumetric scan device (Figure 9.4). Three motors (typically step motors) allow motion in *t*, *z*, and θ directions. Acquisition is synchronized with steps in the *t*-direction. The system proposed in this diagram uses lock-in demodulation to suppress environmental light and an optional log amplifier to perform the log transform of the projection data before ADC.

chosen smaller than the laser beam diameter. Therefore, one slice with 100 projections can be acquired in less than 1 min. A volumetric scan with 60 slices takes almost 1 h to acquire.

More recently, proof of principle was provided that high-speed projection acquisition with large beam diameters is possible (Huang et al. 2008). Large-diameter beam acquisition is the visible-light analog to cone-beam acquisition, but with the key difference that the parallel-beam geometry is maintained. The light source is either a laser with a beam expander or a collimated high-intensity LED combined with a spatial filter. In both cases, the beam diameter is expanded to cover the full width of the sample. Detection takes place with a complementary metal-oxide-semiconductor (CMOS) or charge coupled device (CCD) array, either a line array for single-slice detection or a full 2D chip, which allows the acquisition of a volumetric scan in one single rotation of the sample. Source–detector motion in the *t*-direction is no longer necessary, and positioning in the *z* (axial) direction can be performed in coarse steps. The Gaussian intensity profile of the beam limits the usable area in the *z*-direction. Furthermore, the incident intensity needs to be determined beforehand, without the sample inserted. Unlike the first-generation device in Figure 9.4, the intensity distribution is highly inhomogeneous along both the *t* and *z* axes.

In the pencil-beam device, spatial resolution is limited by the beam diameter. A tightly focused beam allows one to resolve structures of less than 50 μm. Conversely, resolution is limited by the acceptance angle of the spatial filter in the large-diameter beam device. In the study by Huang et al. (2008), crosstalk over 20 adjacent 5 μm pixels was estimated, which corresponds to 100–200 μm resolution. In both cases, higher resolution is theoretically possible, but limited by practical considerations, such as light source intensity, detector sensitivity, or the alignment procedure.

With the large-diameter device, the fast reconstruction software becomes particularly important. Projections can be acquired at the frame rate of the CCD or CMOS

Chapter 9

sensor, which is typically faster than 10 frames per second. Even with small angular increments of 2° or 1° (90 and 180 projections, respectively), projection data for a volumetric section of several millimeters height can be acquired in a few seconds. Highly optimized Fourier-domain reconstruction code is capable of reconstructing one slice in 1–2 s, and with some parallelization, Fourier-based reconstruction has the potential to match the acquisition speed. Unfortunately, Fourier-based reconstruction also provides the lowest image quality. Conversely, ART provides high-quality reconstructions, but its reconstruction performance is one to two orders of magnitude slower than Fourier-domain reconstruction. To make use of the full acquisition speed of the large-diameter beam device, advanced software design approaches need to be taken, such as parallelization with graphics processors or multi-CPU parallelization. With these provisions, OTT with large-beam acquisition is arguably the fastest fully volumetric, high-resolution optical tomography method available.

9.3.3 Differences between OTT and Transmission OPT

As described in the previous sections, OTT creates optical projections through a combination of collimated transillumination and spatially filtered detection. Another approach that can yield a similar optical projection is known as transmission OPT (tOPT). Both OTT and OPT are based on absorption contrast of transmitted illumination; thus, the information that is obtained about the sample is similar in both techniques. Both can be considered equivalent to x-ray CT, implemented in/near the visible range of the electromagnetic spectrum.

The principle difference between OPT and OTT can be thought of as follows: In OTT, the collimated illumination creates the projection through the sample, and the spatial filter in the detection path serves to minimize refractive effects (see Section 9.3.2). In contrast, in OPT, it is the detection optics that provide the projection images of the sample and the illumination that ensures that the image contrast is primarily due to absorption.

Figure 9.6 shows a schematic of a typical implementation of OPT. The basic components required for OPT imaging are the camera, tube lens, objective lens, and the rotation stage for control of the sample orientation. Together, these form a telecentric wide-field telescope that can project an image of the sample on the sensitive surface of the camera. There are two main features that make this core of the OPT machine different from a traditional digital wide-field microscope:

1. The sample, usually embedded in a cylinder of agarose gel as indicated in Figure 9.6, is mounted on a rotation motor such that it can be rotated around an axis perpendicular to the imaging axis of the microscope. This allows a complete set of projections be obtained, as the sample can be viewed from 360°.
2. The detection optics are chosen so as to provide a relatively large depth of field, such that it is roughly half the thickness of the sample. This means that any image captured by the camera will be, to a good approximation, a projection through the proximal half of the sample (the entire sample is imaged by capturing projections over the full 360°, rather than 180° as is typically done for CT).

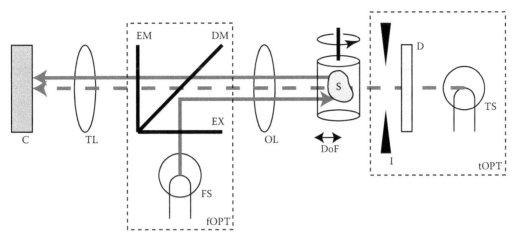

FIGURE 9.6   Schematic of a general OPT system. The sections enclosed in dashed boxes labeled tOPT and fOPT are, respectively, the optics for the transmission and epifluorescence imaging modes. The remaining components are common to both modes. C, camera; TL, tube lens; OL, objective lens; S, sample; DoF, depth of field; tOPT, transmission OPT; TS transmission source (typically an LED or incandescent lamp); D, diffuser; I, iris; fOPT, fluorescence OPT; FS, fluorescence excitation source (a mercury arc lamp); EX excitation filter; EM emission filter; DM dichroic mirror. Thick gray arrows indicate the light propagation paths for tOPT (dashed) and fOPT (solid).

For tOPT, a wide-field incoherent light source is used for illumination of the sample (TS in Figure 9.6). This is typically an incandescent tungsten lamp or a light-emitting diode of the appropriate wavelength. A diffuser between the lamp and the sample ensures uniform incoherent illumination of the sample from a wide range of angles (as opposed to OTT, in which the illumination is collimated and generally at least partially spatially coherent). An iris between the diffuser and sample restricts light that would otherwise pass around the sample and decrease image contrast.

Automated control of an OPT machine is relatively simple, compared to, say, a modern confocal microscope. At a minimum, the camera and rotation stage must be computer-controlled so that the acquisition of the projection images can be synchronized with the sample rotation. For fixed samples in which timing is not critical, other components (e.g., the filters for fOPT and the lamp powers or shutter controls) can be controlled manually. To perform time-lapse recording of live specimens, computer control of the filters and lamps is also beneficial so that experiments lasting hours or days can be undertaken without the need for continuous user supervision.

As discussed in Section 9.3.1 in the context of OTT, refractive index mismatches in the sample cause the acquired optical projections to deviate from the ideal, straight lines that are assumed in the tomographic reconstruction process. This effect can be seen in the pre-reconstruction tOPT projections shown in Figure 9.7. When illuminated with quasi-collimated light (Figure 9.7a), both amplitude and phase features exhibit contrast. Diffuse illumination (Figure 9.7b) reduces the contrast of features that are a result of changes in the (real) refractive index (phase features), leaving an image dominated by the absorbing/scattering objects (amplitude features). Although Figure 9.7a may appear qualitatively more interesting, the presence of phase-contrast features disrupts

Chapter 9

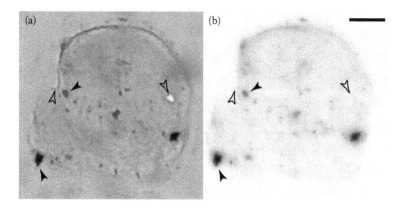

FIGURE 9.7 Single tOPT projections taken with (a) quasi-collimated and (b) diffuse illumination. The sample is a fixed mouse secondary lymphoid organ that has been prepared as described in Section 9.3.6. Note that in (b) phase objects (e.g., features indicated by open arrowheads) have reduced contrast, and the principal signal arises from light attenuated by absorption (solid arrowheads). Scale bar 200 μm. Reconstructions of similar lymphoid organs scanned with fOPT can be seen in Figure 9.14. (Sample courtesy of Jens Stein, Theodor Kocher Institute, U. Bern, Bern.)

the tomographic reconstruction algorithm, meaning that the diffuse illumination of Figure 9.7b will actually yield a reconstruction with fewer artifacts. Note that diffractive tomography techniques (Debailleul et al. 2009; LaRoque et al. 2008) exist that are able to cope with phase samples.

It should be noted that in OPT there is generally a trade-off that must be taken into account when designing the optical system. On the one hand, to generate images with high resolution, the use of a high NA objective is required, because the image resolution varies proportionally with the NA. On the other hand, to image large samples, a large depth of field is needed (as discussed above, the depth of field should be at least about half the sample diameter), and the depth of field varies inversely with the square of the NA. From this, two conclusions can be drawn:

1. For a given sample size, increasing the NA beyond a certain limit (that which yields a depth of field ~1/2 the sample size) can actually decrease the absolute resolution because the images recorded by the camera will not be good approximations to ideal projections through the sample.
2. The relative resolution (the ratio of the sample size to the achieved resolution) will increase with sample size, because of the quadratic dependency of the depth of field on the NA of the imaging system. In practical terms, this means that OPT is better suited to "mesoscopic" samples (with sizes from tens of microns to tens of millimeters) than to single cells or super-resolution applications. It should be noted that other approaches such as the use of annular apertures (Linfoot and Wolf 1953) or axial scanning (Fauver et al. 2005) can alleviate this limitation to some extent and push the resolution of OPT down to the level of subcellular features.

Because it does not employ imaging optics but instead relies on direct optical projection (as is generally done in x-ray CT, where high-quality lens systems are not available), this trade-off is not an issue for OTT. On the other hand, OTT does not receive the resolution benefit that comes from using a diffraction-limited optics to image the sample onto the sensitive surface of the detector.

9.3.4   Fluorescence OPT

Although their implementations differ, both OTT and tOPT are direct analogs of x-ray CT in that they can provide 3D maps of the attenuation of a sample via tomographic reconstruction of measured projections. However, in many applications, it would also be advantageous to be able to utilize the high molecular specificity that can be obtained with fluorescently tagged antibodies or genetically encoded fluorescent proteins.

Fluorescence OPT (fOPT), also known as optical emission CT (optical-ECT; Sakhalkar et al. 2007), is a modification of the basic OPT setup shown in Figure 9.6 by the addition of a lamp for fluorophore excitation and a set of spectral filters (the components in the dashed box labeled fOPT). These effectively add an epifluorescence mode to the transmission contrast given by tOPT. The excitation lamp provides a wide-field source that illuminates the entire 3D volume of the sample, the excitation filter selects the wavelengths from the lamp appropriate to excite fluorescence in the sample, and the emission filter passes the fluorescence to the camera for imaging.

Although the contrast mechanism in fOPT is completely different to that of tOPT, the techniques can be combined in a common platform (as indicated in Figure 9.6), and they share many common features. Principle among these is that the detected signal (the emitted fluorescence in fOPT, or the diffuse illumination of tOPT) is completely incoherent and has a nearly isotropic angular distribution. This means that, in both cases, the resolution and depth of field of the images are primarily determined by the detection objective lens and the wavelength of the detected light. In a system such as that sketched in Figure 9.6, one can switch between imaging in tOPT and fOPT modes "on the fly," and both will have very similar resolutions. Of course, the information obtained in these different channels will in most cases show completely different features of the sample, which will often be complementary. Since they are acquired with the same optics without removing the sample from the instrument, the result is a well-registered multimodal 3D dataset.

An example of the combination of tOPT and fOPT is shown in Figure 9.8, in which a fixed E17 mouse embryo has been imaged using both contrasts. tOPT contrast is primarily from absorption in pigmented (eye), dense (spinal column), or blood-rich (internal organs) tissue. The embryo has not been specifically labeled, so the fOPT signal is due to tissue autofluorescence (excitation 395–455 nm, emission > 480 nm). Note that the image quality is high, isotropic, and uniform throughout the sample in both tOPT and fOPT, and that these channels are coregistered. The specimen was chemically cleared before scanning according to Section 9.3.5.

In addition to imaging the emitted fluorescence intensity (fOPT), OPT has also been recently demonstrated using emission lifetime contrast—fluorescence lifetime imaging OPT (FLIM-OPT; McGinty et al. 2008).

Chapter 9

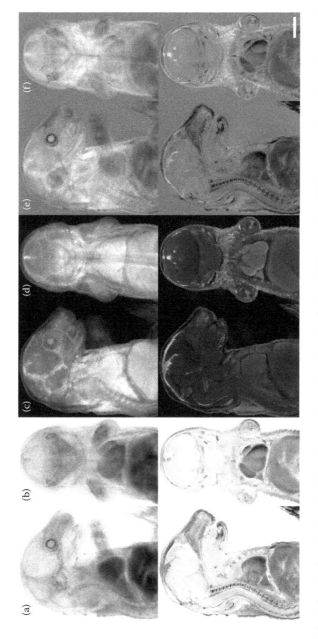

FIGURE 9.8   Selected optical projections (a–f) and reconstructed sections (a–f) of an OPT scan of an E17 mouse embryo. Projections are along, and sections perpendicular to, the right/left axis in (a, c, e) and the dorsal/ventral in (b, d, f). (a, b) tOPT (attenuation). (c, d) fOPT (autofluorescence). (e, f) Overlay. Scale bar 1 mm.

9.3.5   Sample Preparation for OPT

Both tOPT and fOPT were first demonstrated on fixed samples (Sharpe et al. 2002). As discussed in the previous section, OPT is best suited to "mesoscopic"-scale samples, on the order of millimeters. However, typical human soft biological tissues have photon mean free paths (MFPs) on the order of 100 µm in the optical regime (Ntziachristos 2010). Since photon transport in samples with thicknesses >1 MFP cannot be considered ballistic, the assumption that light propagates in straight lines will be invalid in typical mesoscopic samples. In other words, scattering will be significant. Because at most one scattering event per photon is assumed by most CT reconstruction algorithms (see Section 9.3.1), unless measures are taken to alleviate this, high-resolution diffraction-limited reconstructions will not be obtained. Other imaging methods such as time-gated imaging (Bassi et al. 2010), diffuse optical tomography, or photoacoustic tomography (Ntziachristos 2010; see also Chapters 12 and 13 in this book) have their own ways of surmounting this problem; however, in OPT, the method of choice in fixed samples is generally chemical clearing.

In general, chemical clearing of biological tissue is a process by which the water in the sample is replaced by a medium that is closely refractive-index matched to the non-aqueous components of the cell—principally lipids and proteins. This is achieved by a sequence of fixation (in paraformaldehyde), dehydration (in alcohols, typically methanol or ethanol), and clearing [e.g., using methyl salicylate or benzyl-alcohol-benzyl-benzoate (BABB)]. The effect is to increase the MFP in the cleared tissue to the extent that OPT scans of samples such as mouse embryos or adult mouse organs (up to ~15 mm diameter) can be reconstructed with high resolution. Note that this chemical clearing reduces scattering, particularly at water/lipid interfaces (cell and organelle membranes); it does not reduce absorption due to pigmentation, or achieve clearing in dense, calcified tissue such as bone. Details of various clearing protocols can be found in Sakhalkar et al. (2007) and the supplementary material of Sharpe et al. (2002) and Alanentalo et al. (2007).

The advantage of chemical clearing is that it renders many biological tissues almost completely transparent. Unfortunately, there are several associated disadvantages. The first and most obvious of these is that the sample is imaged *ex vivo*, thus eliminating the possibility of studying dynamic processes. Another drawback is that they tend to be time consuming (stretching over the course of days, depending on the sample size), which can be a limitation when high-throughput studies are desired. However, compared to the traditional alternative—physical sectioning of a complete mesoscopic sample—the clearing protocol times are not generally prohibitive. A third disadvantage is that, while small inorganic molecules such as fluorescein- or rhodamine-based dyes generally survive the clearing process, it has a strong tendency to quench the fluorescence of genetically encoded fluorescent proteins (Sakhalkar et al. 2007). This can be overcome by performing fluorescent antibody labeling against the fluorescent protein in question, but this complicates the sample preparation and can lead to penetration issues when performing *in situ* immunolabeling in larger samples. However, there is evidence that by optimizing the preparation and imaging procedures, practical imaging of the intrinsic fluorescence of green fluorescent protein (GFP) in cleared samples is possible (Dodt et al. 2007).

Chapter 9

In addition to imaging cleared samples, OPT can also be used for imaging living tissue, for example, to study the dynamics of morphology and gene expression patterning in mouse development (Boot et al. 2008). Here the sample preparations and imaging conditions are optimized to achieve physiological development of the specific sample of interest, rather than to maximize the imaging resolution. Since clearing is impossible, optical scattering is significant and the resolution achieved is significantly reduced as compared to cleared tissue. However, one gains the advantage of being able to study an individual organism in 3D over time, albeit at reduced spatial resolution (see Section 9.4.2).

In general, CT-based imaging techniques differ from more traditional types of microscopy in two related aspects: (1) one requires optical access (i.e., to be able to take images) from many different orientations of the sample, and (2) the mechanical mounting of the sample must be such that it can be rotated relative to the imaging axis of the system, ideally through a full 360°. This means that traditional microscopy sample-mounting techniques, that is, mounting a thin, flat sample between a glass slide and cover slip, are usually not applicable to OPT. In OPT imaging, to allow all-around optical access, the sample is generally embedded in a block or cylinder of 1% low-melting point agarose gel, which is optically transparent and compatible with chemical clearing protocols (i.e., the sample is embedded before dehydration and clearing). Once solidified, the agarose can be attached to a rotation stage via gluing or mechanical clamping (unlike the case of x-ray CT, it is the sample rather than the scanning apparatus that is rotated in OPT).

9.3.6   Comparison to Other Methods

OTT and OPT generally rely on transmitted light (absorption) or fluorescence for contrast. In this they are similar to several other optical imaging techniques, and a direct comparison can be made with these. In terms of resolution, OTT and OPT lie between confocal and multiphoton microscopies on the one hand and diffuse tomographic and related techniques on the other (Ntziachristos 2010; see also Chapters 12 and 13). Similarly, in terms of size, typical mesoscopic samples for which OTT and OPT are well suited fall between the cell or cell clusters imaged by high-resolution techniques and the whole-animal samples for which diffuse tomography is designed.

There are several other types of contrast that are not applicable to OTT and OPT but have been the basis for very informative and useful optical imaging systems. For such systems, a direct comparison is more difficult because they rely on and image different properties of the sample than do OTT and OPT. Perhaps the most relevant of these is optical coherence tomography (OCT), which is an interferometric technique relying on the collection of low-coherence light back-scattered from a sample (Podoleanu 2005; see also Chapters 10 and 11). OCT can image to depths of about 1 mm, depending on the tissue properties. Because it works in a reflective configuration, complete penetration of photons through the sample is not necessary, and OCT is typically used for imaging the surface layers of larger, intact objects such as human skin or the retina of the eye. However, it relies on the variations in scattering properties between different tissues for contrast and so is not generally the method of choice for studies on the molecular level, for example, of gene expression or protein localization.

Other contrasts in optical imaging include polarization (Chapter 6) and nonlinear effects such as second or third harmonic generation (Chapter 3) or Raman scattering (Chapter 8). Perhaps special mention should be made of phase contrasts (Wang et al. 2011), as quantitative implementations of phase imaging have been incorporated into tomographic instruments (Debailleul et al. 2009; Choi et al. 2007; Jayshree et al. 2000) that are capable of imaging mesoscopic 3D specimens similar to those of interest in OTT and OPT. After determining 2D phase maps from the raw data, these techniques use similar reconstruction algorithms to OTT and OPT to calculate 3D refractive index distributions. All of these contrasts are primarily sensitive to the intrinsic properties of the sample and thus do not require complicated labeling preparation steps. The reliance on intrinsic contrasts also makes these techniques compatible with live imaging, at least in principle. On the other hand, although OPT can image such intrinsic properties as absorption or autofluorescence, it has been primarily used in applications involving specific extrinsic labeling, for example, absorbing stains (Arques et al. 2007; Summerhurst et al. 2008), immunohistochemistry (Alanentalo et al. 2007), or genetically encoded fluorescent labels (Boot et al. 2008).

9.4   Example Applications

9.4.1   OTT: Review of Studies Using OTT

Early OTT systems were designed to demonstrate x-ray tomography principles with low-cost equipment. Schleicher et al. (1998) used 64 pulsed light-emitting diodes, arrayed in an arc, as point sources for an inverse fan-beam geometry with very low resolution. This general principle was improved upon by Bellemann et al. (2002), who used the general principle outlined in Figure 9.4 with a collimated laser source. Scans of simple objects, such as wires and ink-filled tubes, immersed in an index-matching fluid, were presented. Around the same time, Chacko and Singh (2000) examined the information contained in scattered photons. The device proposed by Chacko and Singh uses a fan-beam geometry, in which the laser beams are made divergent with a lens. In addition, the device contains a source–detector configuration that rotates freely around a stationary sample (in contrast to the device sketched in Figure 9.4). Clearly, the detector array collects ballistic, scattered, and refracted photons, and the authors found that only gross sample inhomogeneities were visible in the reconstruction (Chacko and Singh 2000). With a different approach, scattered light can be used to reconstruct the cross-sectional image (e.g., Yuasa et al. 2001; Boas et al. 2001; Siegel et al. 1999). Rather than using the ray-based approach to reconstruct the cross section in conventional CT, diffuse optical tomography examines the propagation of the wave front. Although significant progress has been made in diffuse optical tomography, practical applications still suffer from very low spatial resolution and the need for an index-matching fluid, because refraction cannot be tolerated in the reconstruction of the cross-sectional image.

In early OTT, several approaches were introduced to separate ballistic from scattered photons. One method is ultrafast gating with femtosecond laser pulses (Das et al. 1993; Liu 1993). Multiply-scattered photons have a longer travel path and arrive later at the detector than ballistic photons and those that underwent few scattering events.

Chapter 9

However, the authors found that nearly all detected photons in breast phantoms and breast tissue with thickness of several millimeters are scattered, and consequently, no detectable ballistic photons exist (Liu et al. 1999). For resolutions in the submillimeter range, gating times as low as 10 ps are needed (Marengo et al. 1999). The frequency-domain equivalent of gated detection—coherent detection—was proposed by Yuasa et al. (2002). Coherent detection works analogous to lock-in detection, whereby two light beams, modulated with slightly different modulation frequencies, are merged. A band-pass filter, selective for the difference of the modulation frequencies, allows only the bal-listic component of the light to pass due to the loss of coherence of the scattered photons. According to Yuasa et al. (2002), coherence detection imaging rejects scattered photons and allows beam intensity detection with a dynamic range of 140 dB. Alternatively, multiply-scattered photons can be attenuated at the detector with the help of polariz-ers, because photons lose their polarization orientation upon scattering (Schmitt et al. 1992; Wang et al. 2003). Our own experience, however, shows that polarization acts as a relatively weak filter and does not improve reconstructed image quality in a significant manner.

One popular application of OTT is gel dosimetry [see Oldham et al. (2003) and Oldham and Kim (2004) for a comprehensive introduction into optical dosimetry]. OTT can be used to simulate x-ray irradiation and scattering that occurs during radia-tion treatment. The goal is to provide radiation treatment guidance by estimating the x-ray exposure of the treatment site and surrounding tissue.

Our own studies were driven by the need to image thin-walled vascular grafts (L'Heureux et al. 2006, 2007a,b) *in vivo*, that is, during their growth phase. We exam-ined OTT as one possible alternative to OCT. We chose a spatial filter for the rejection of scattered photons. A spatial filter allows only photons within a very small angular deviation from its optical axis to pass. A simulation study (Yao and Haidekker 2005) confirmed the effectiveness of this approach, but also highlighted the problematic role of refracted photons. In the approach by Chacko and Singh (2000), refracted photons are collected but cannot be distinguished from ballistic and scattered photons. By using a spatial filter, refracted photons are rejected, but in zones affected by refraction, higher apparent absorption is reported. In the worst case, the detector incorrectly reports a light intensity of zero, with a corresponding infinite absorbance. Before reconstruction, absorbance needs to be clamped. However, sections in the projection where the reported intensity is zero can be used to identify zones affected by refraction (Haidekker 2005). Two options to correct this artifact in software advertise themselves. First, any section of the projection where $P(t) = 0$ for $t_1 < P(t) < t_2$ can be interpolated from $P(t_1)$ and $P(t_2)$. The interpolated projection can then be subjected to any reconstruction method. Second, any section of the projection where $P(t) = 0$ can be labeled as invalid and ignored. The second option exists only in ART, because invalid projections can simply be removed from the ART equation system. Both options are associated with some loss of informa-tion. However, other projections still carry the information lost by omitting or inter-polating regions affected by refraction. In a phantom study (Haidekker 2005), it was shown that the reconstruction still carried remarkably detailed information (albeit with some blurring introduced), and that the reconstructed absorption map was accurate to within 25%. Two studies (Gladish et al. 2005; Huang et al. 2008) demonstrated both the remarkable detail that can be obtained from optical transillumination reconstructions and the potential high speed at which projections can be acquired. Two examples are provided in Figures 9.9 and 9.10. Figure 9.9 shows a single cross section of tissue grown on a hollow polymeric tube (mandrel) (Gladish et al. 2005). Defects, such as inhomogeneous tissue thickness and enclosed fluid bubbles, can prominently be seen. Figure 9.10 shows the surface rendering of a volumetric scan, acquired with the first-generation scanner introduced in Figure 9.4. Again, important details can be seen, such as the fixation groove or some fraying of the fixated tissue sample.

OTT advertises itself as a quality control method for vascular tissue engineering. However, some challenges remain for OTT to be adopted in routine imaging, predominantly the need to develop a suitable bioreactor that simultaneously allows the vascular graft to mature and to image it without breaking sterility. As an *in vitro*

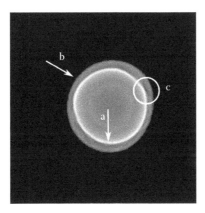

FIGURE 9.9 Reconstruction of a cross-sectional slice of cultured tissue on a mandrel. Reconstruction artifacts outside of the sample have been removed by thresholding. The image clearly reveals the mandrel (a) with a high apparent absorption, mainly due to its refractive properties. Furthermore, inhomogeneous tissue thickness (b) can be seen, as well as an enclosed fluid bubble (c). These types of defects are critical to identify in tissue engineering.

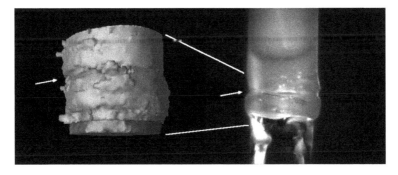

FIGURE 9.10 Volumetric scan (left) of engineered tissue grown on a mandrel related to a photograph (right) of the same object. In the volume rendering, the mandrel is the darker region at the bottom, and the tissue is the remaining upper, lighter region. The fixation groove (arrow) can clearly be seen in the reconstruction. Sections in the tomographic reconstruction where the tissue appears frayed are real and can be observed with the unaided eye when the sample is immersed in water. Tissue sample courtesy of N. L'Heureux (L'Heureux et al. 2006).

Chapter 9

imaging method, OTT has the potential to reach almost real-time acquisition speeds, particularly with the large-diameter beam acquisition method. Such a method would allow examination of, for example, expansion behavior, under pulsatile flow conditions.

9.4.2　Live OPT

For biomedical applications, one would ideally be able to image 3D samples over time periods covering developmental or regenerative processes of interest without harming the sample. In particular, embryonic mouse development is a system of great interest because of (1) its relevance to human development, (2) the relative ease with which it can be genetically manipulated, and (3) the fact that, at least at early stages, anatomical structures are of sizes of the order of the MFP (see Section 9.3.5), which permits the possibility of high-resolution optical imaging in live samples (i.e., without the need for chemical clearing). Despite the challenge of maintaining *in utero* development conditions (e.g., temperature, oxygen and nutrient delivery, protection from phototoxic radiation) *ex utero*, a recent study has demonstrated the potential of OPT to allow dynamic observation of 3D embryonic morphology and gene expression over developmentally interesting time scales (Boot et al. 2008).

Boot et al. were able to quantify the morphology of developing mouse limb buds by culturing them inside their live OPT system. The ectoderm of the embryos was treated with a scattering of fluorescent beads that serve as landmarks on the limb's surface (see Figure 9.11). Thus, by combining tOPT (which yields the overall morphology of the limb) and fOPT (which reveals how individual regions of the ectoderm move relative to each other), they were able to quantify the limb development over time. Note that while the tOPT information could in principle have been obtained by imaging multiple, fixed embryos at different stages of development, the analysis would be complicated by the natural variability of developmental rates between individuals. Additionally, the relative motion of the different parts of the ectoderm supplied by the fluorescent beads would be impossible to measure on fixed samples.

This monitoring of limb ectoderm can be done at high resolution because it is primarily a surface measurement: since the beads are not internalized, they can be imaged clearly, and issues of scattering inside the tissue are not relevant. In contrast, imaging internal structures will of necessity be done with reduced image quality in live, noncleared samples. Nevertheless, Boot et al. have demonstrated that, combined with genetically encoded fluorescent proteins, live OPT is capable of following evolving gene expression patterns such as *Scx* in the limb bud and *Pax6* in the eye for up to 19 h with resolutions on the order of 100 μm or better.

9.4.3　Fixed Sample Imaging

When higher resolution images than those that can be obtained in noncleared biological tissue are needed, one must resort to scanning fixed samples. Although following dynamics is impossible in these cases, an improved 3D resolution is often a more important goal than temporal resolution. Figure 9.8, for example, illustrates that the intrinsic autofluorescence and absorption contrasts in cleared embryonic mouse tissue can be

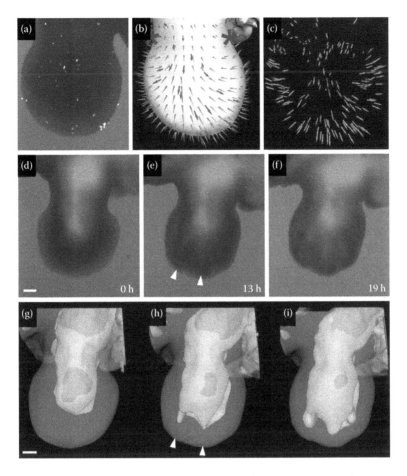

FIGURE 9.11 Live OPT imaging of mouse limb bud development. (a–c) Tracking growth for 6 h (E11.0-E11.25). Imaging fluorescent landmarks on the ectoderm (a) allowed 3D tracking of these points over time (dark arrows, b) and extraction of velocity vector fields over the surface of the limb (light, regularly spaced arrows, b). (c) Overlap of data from two additional experiments with more extensive labeling. (d–i) 4D monitoring Scx-GFP gene expression patterning. (d–f) Raw fluorescence projections at the times indicated. (g–i) 3D rendering of the gene expression pattern (white) and morphology (from the transmission data, gray) at the same time points. Scale bars 200 μm. (Based from Boot MJ et al., *Nature Methods* 5: 609–612, Figures 2 and 3, 2008.)

used to define the organism's morphology with high, isotropic 3D resolution. This type of imaging to reveal morphology has also been used to compile FishNet, a publicly available 3D atlas of zebrafish development (Bryson-Richardson et al. 2007).

Although this 3D morphological information is of interest in itself, OPT is a particularly valuable imaging tool because it can easily capture multiple coregistered channels with different contrasts. This has allowed the combination of mapping 3D gene expression patterns using whole-mount *in situ* hybridization techniques (tOPT) and sample morphology using autofluorescence (fOPT) (Summerhurst et al. 2008) in embryonic mouse tissue. Alternatively, multiple wavelength-multiplexed fluorescence channels can be used to capture morphology and specific labeling using fOPT alone, for example, to visualize vascular development (Walls et al. 2008).

Chapter 9

Figure 9.12 shows an example of the use of tOPT to reconstruct the 3D distribution of Wnt5a-RNA defined by whole-mount *in situ* hybridization. In this case, Wnt5a expression in an E11.5 mouse limb bud is depicted (Summerhurst et al. 2008). This example illustrates the value of 3D imaging techniques such as OPT for quantitative data analysis. In the raw OPT projection (Figure 9.12a), which is essentially a standard wide-field transmission image, even the gross features of the Wnt5a gene expression patterning are difficult to determine, although it is evidently expressed in the limb bud, and perhaps more strongly in distal regions. The 3D reconstruction (Figure 9.12b) gives a better representation of the Wnt5a expression, and it can be seen to form a "cap" covering the anterior, posterior, and distal regions of the limb bud; and by marking selected isosurfaces of WMISH activity (Figure 9.12c), it can be seen that a distal-to-proximal Wnt5a gradient exists.

Figure 9.13 shows an example in which OPT has been used to study a structural feature rather than a gene expression pattern. Several stages (ranging from TS22 to TS25, or about 14 to 17 days postfertilization) of the development of the skeletal structure of a mouse hind limb are visualized using Alcian blue to delineate cartilage (the technique used here is similar to that used to image chick embryos in Nowlan et al. 2007). During this period, the conversion of the skeletal structure from cartilage to bone begins, as can be seen in the disappearance of Alcian blue staining from the central portions of the humerus, radius, and ulna (arrowheads). This type of tOPT imaging can be considered complementary to skeletal development studies, which can be done using μCT (Guldberg et al. 2004); μCT has the advantage of employing intrinsic contrasts from mineralized tissue and can thus be used *in vivo*. However, for this very reason, it is not well suited to imaging soft tissues such as the cartilaginous structures shown in Figure 9.13.

In the past, OPT has been primarily used in developmental biological applications; however, suitability for imaging mesoscopic-scale samples also gives it promise for use for biomedical applications. Although OPT has been employed in studies of early human brain development (Sarma et al. 2005), because of the scales involved in adult human imaging, its most obvious biomedical application is in the use of small animal models for disease studies.

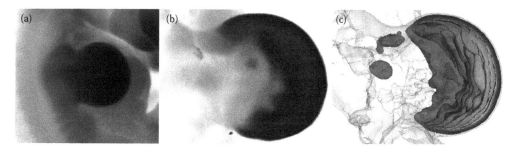

FIGURE 9.12 tOPT imaging of Wnt5a expression in a limb bud of a E11.5 mouse embryo. (a) Single, raw, OPT projection. (b) Volume rendering of the reconstruction. (c) Isosurface rendering of the reconstruction, illustrating the gradient of Wnt5a expression along the proximal/distal axis. Gray isosurfaces: Wnt5a expression; transparent: sample morphology. (Data courtesy of Paula Murphy, Trinity College, Dublin. Reconstruction and visualization courtesy of Henrik Westerberg, CRG, Barcelona.)

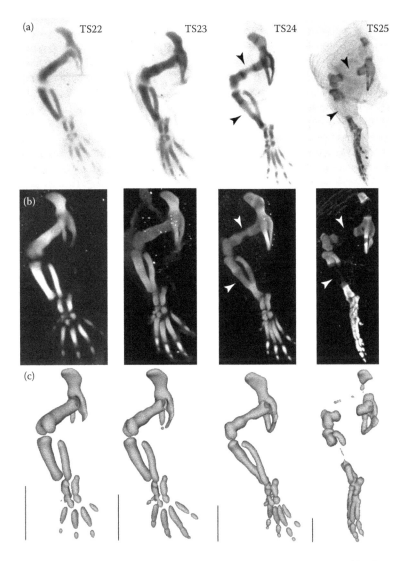

FIGURE 9.13 tOPT scans showing the developing skeletal pattern in mouse hind limbs, at the Theiler stages indicated. (a) Single, raw OPT projections (b) Maximum value projections through reconstructions. (c) Surface renderings. Scale bars 1 mm. (Courtesy of Laura Quintana [embryo harvesting] and Niamh Nowlan [scanning and reconstructions], CRG, Barcelona.)

For instance, non-obese diabetic (NOD) mice are commonly used as a model for type 1 diabetes, an autoimmune response estimated to affect up to 37 of 100,000 people per year in some parts of the world (Soltesz et al. 2007). In several recent studies (Alanentalo et al. 2007, 2010), OPT has been used to quantify the size and distribution of the islets of Langerhans, which contain the insulin-producing β-cells within the pancreata. The combination of specifically labeled insulin-producing cells (through fluorescence immunohistochemistry) and the imaging of the resulting 3D distributions thereof (via fOPT) has quantitatively demonstrated that the smaller, more peripheral islets are the first to be compromised, and that the volume of the larger islets actually increases during the initial stages of insulitis progression (Alanentalo et al. 2010). These observations

Chapter 9

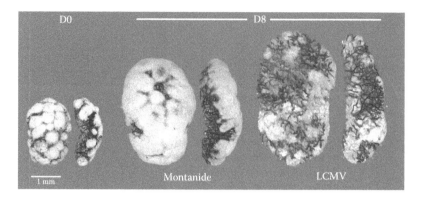

FIGURE 9.14   Mouse inguinal peripheral lymph nodes imaged by fOPT, noninflamed (left, D0) and after 8 days of inflammation by Montadine (center) and LCMV (right). Antibody labeling against Meca79 to visualize the HEV network (dark, filamentous structures) and B220 for B cells (light, amorphous structures). (Courtesy of Renzo Danuser and Jens Stein, Theodor Kocher Institute, U. Bern, Bern.)

rely on the ability to image the islets in 3D throughout the entire pancreatic organs and could not be performed using conventional methods such as serial sectioning. In addition, by colabeling for insulin and CD3 to visualize both the islets and the distribution of the T-cells involved in autoimmune response, Alanentalo et al. were able to demonstrate significant 3D colocalization of these structures.

Another field in which OPT has demonstrated its usefulness for mesoscopic imaging is that of immunology. Although there is a large body of knowledge concerning the biochemistry and molecular interactions between the components of the immune system, less has been known about the quantitative details of the spatial relationships between these components and how they rearrange upon infection. A recent publication has used OPT to examine the remodeling of peripheral lymph nodes (PLNs) following infection (Kumar et al. 2010). In this work, the high endothelial venule (HEV) vasculature and B-cell follicles were imaged using fOPT in intact PLNs at various stages of inflammation (see Figure 9.14). These comprehensive 3D data allowed quantification of the total PLN volume, the volume, number, and distribution of the B-cell follicles, and the length and number of branch points in the HEV. During infection with lymphocytic choriomeningitis virus (LCMV), the PLN volume increases threefold in the first 8 days, with a corresponding increase in the length of the HEV network. As with the above diabetes-related application, this degree of quantification could not be obtained by the classical method of making 2D serial sections of the PLN.

9.5   Conclusions

9.5.1   Summary

Until recently, there has been a dearth of simple, inexpensive, user-friendly methods for imaging mesoscopic-scale biological specimens. At one end of the scale, optical imaging efforts were concentrated on high-resolution microscopy (Pawley 2006) and, more recently, nanoscopy techniques capable of imaging on the subcellular and molecular levels (Klar et al. 2000; Betzig et al. 2006; Rust et al. 2006). Mesoscopic optical techniques have concentrated on low-resolution live imaging in the diffusive-photon range (Ntziachristos 2010), and macroscopic samples have largely been studied using nonoptical methods (Guldberg et al. 2004; Kim et al. 2006; Mariani et al. 2010). However, in the past decade, novel optical techniques have emerged that attempt to image mesoscopic samples at high resolution, that is, in the nondiffusive regime, ideally approaching the diffraction limit. These include OTT and OPT and various light-sheet-based microscopies (Huisken et al. 2004; Dodt et al. 2007).

The need for 3D imaging of mesoscopic biological samples is clear: many embryos, small organs, and tissue samples fit into this broad size category, and many issues of interest concerning them can only be fully understood if their 3D structure can be quantified (Sharpe 2003). The advantages of taking an optical approach to the problem are several. One is practical: visible-wavelength optical sources are generally relatively inexpensive, readily available, and safe; x-ray sources, high-power magnets, or radioactive isotopes are not required. In addition, many decades have been spent developing labeling techniques for light microscopy that allow various features of the sample to be visualized (e.g., immunohistology, *in situ* hybridizations, genetically encoded fluorophores, etc.); many of these can be adapted directly to optical mesoscopy. Thus, although a few optical imaging systems specifically targeting mesoscopic biological samples are currently commercially available, the increasing prevalence of literature references to such techniques leads us to expect them to become increasingly common in the near future.

OTT and OPT are well suited to fill this niche in 3D mesoscopic optical imaging. OTT and tOPT, while differing in implementation, allow visualization of essentially the same sample features: those that are strongly attenuating in the visible region of the spectrum. In addition, fOPT adds the option of sensitivity to fluorescent signals. Considered together, they are able to image samples in the 100 μm to several centimeters range without requiring physical sectioning.

9.5.2 Future Developments

Perhaps the most promising avenue for future development of OTT and OPT for biological applications would be to be able to use the techniques in live tissue at higher resolutions than are currently possible. Although diffraction-limited optical imaging may never be achievable in live mesoscopic samples at optical wavelengths, there is clear room for progress. One possibility is the use of longer wavelengths in the far red or infrared; this would reduce both scattering in the tissue, which limits image quality, and phototoxicity, which can be a serious issue for sensitive living samples (Shu et al. 2009). Another approach would be the development of biocompatible clearing protocols. Although harsh clearing agents such as BABB provide excellent optical clearing, "biofriendly" clearing treatments, even if they do not provide the same level of clearing, would permit more varied 4D experiments.

Another possibility for improvement is in the field of reconstruction algorithms. To date, the standard reconstruction method is the filtered back-projection, which can be implemented easily and efficiently, and has a long history of use in the x-ray CT field. Other algorithms such as ART (see Section 9.3.1), mesoscopic fluorescence tomography (Vinegoni et al. 2009), or maximum likelihood expectation analysis (Darrell et al. 2008) have shown promise for certain applications, although they are generally more complicated and may require additional *a priori* knowledge about the sample and/or imaging apparatus.

A third possible extension of OPT would be to combine tOPT and fOPT at the level of the reconstructions. To date, both imaging modes have been demonstrated to yield coregistered voxel data from a single sample; however, the two datasets have traditionally been treated independently. Given that the information is available, it would be reasonable to combine them using a single algorithm, for example, one that could use absorption data (from tOPT) to correct attenuation-induced artifacts in fOPT data. That such procedures can be successful has been demonstrated in principle (Vinegoni et al. 2009), although in this case, the experimental setup is somewhat different than that described here for fOPT.

Acknowledgments

The authors would like to thank Niamh Nowlan (CRG, Barcelona) for data and discussions regarding the skeletal OPT imaging, Henrik Westerberg (CRG, Barcelona) for his work on the analysis and visualization of gene expression patterns in the mouse limb bud, and the Stein (Theodor Kocher Institute, U. Bern, Bern) and Ludewig (Institute of Immunobiology, Kantonsspital St. Gallen, St. Gallen) groups for fruitful collaborations regarding the mesoscopic imaging of the components of the mouse immune system. J.S. and J.S. thank ICREA and the Swiss Sinergia grant for funding. M.H. thanks the National Institutes of Health for funding through Grant R21 HL081308.

References

Alanentalo T, Asayesh A, Morrison H, Lorén CE, Holmberg D, Sharpe J et al. 2007. Tomographic molecular imaging and 3D quantification within adult mouse organs. *Nature Methods* 4: 31–33.

Alanentalo T, Hörnblad A, Mayans S, Nilsson AK, Sharpe J, Larefalk Å et al. 2010. Quantification and three-dimensional imaging of the insulitis-induced destruction of β-cells in murine type 1 diabetes. *Diabetes* 59: 1756–1764.

Arques CG, Doohan R, Sharpe J & Torres, M. 2007. Cell tracing reveals a dorsoventral lineage restriction plane in the mouse limb bud mesenchyme. *Development* 134: 3713–3722.

Bassi A, Brida D, D'Andrea C, Valentini G, Cubeddu R, De Silvestri S et al. 2010. Time-gated optical projection tomography. *Optics Letters* 35: 2732–2734.

Beckmann EC. 2006. CT scanning the early days. *British Journal of Radiology* 79: 5–8.

Bellemann ME, Baier T, Seitz B & Walther HG. 2002. Development of a laser-optical tomography for demonstration of CT imaging without ionizing radiation. *Biomedizinische Technik/Biomedical Engineering* 47: 457–469.

Betzig E, Patterson GH, Sougrat R, Lindwasser OW, Olenych S et al. 2006. Imaging intracellular fluorescent proteins at nanometer resolution. *Science* 313: 1642–1645.

Boas DA, Brooks DH, Miller EL, DiMarzio CA, Kilmer M, Gaudette JR et al. 2001. Imaging the body with diffuse optical tomography. *IEEE Signal Processing Magazine* 18: 57–75.

Boot MJ, Westerberg CH, Sanz-Ezquerro J, Cotterell J, Schweitzer R, Torres M et al. 2008. *In vitro* whole-organ imaging: 4D quantification of growing mouse limb buds. *Nature Methods* 5: 609–612.

Bryson-Richardson RJ, Berger S, Schilling TF, Hall TE, Cole NJ, Gibson AJ et al. 2007. FishNet: an online database of zebrafish anatomy. *BMC Biology* 5: 34.

Chacko S & Singh M. 2000. Three-dimensional reconstruction of transillumination tomographic images of human breast phantoms by red and infrared lasers. *IEEE Transactions on Biomedical Engineering* 47: 131–135.

Choi W, Fang-Yen C, Badizadegan K, Oh S, Lue N, Dasari RR et al. 2007. Tomographic phase microscopy. *Nature Methods* 4: 717–719.

Darrell A, Swoger J, Quintana L, Sharpe J, Marias K, Brady M et al. 2008. Improved fluorescence optical projection tomography reconstruction. *SPIE Newsroom* DOI: 10.1117/2.1200810.1329.

Das BB, Yoo KM & Alfano RR. 1993. Ultrafast time-gated imaging in thick tissues: a step toward optical mammography. *Optics Letters* 18: 1092–1094.

Debailleul M, Georges V, Simon B, Morin R & Haeberlé O. 2009. High-resolution three-dimensional tomographic diffractive microscopy of transparent inorganic and biological samples. *Optics Letters* 34: 79–81.

Dodt H-U, Leischner U, Schierloh A, Jährling N, Mauch CP, Deininger K et al. 2007. Ultramicroscopy: three-dimensional visualization of neuronal networks in the whole mouse brain. *Nature Methods* 4: 331–336.

Downing KH, Sui H & Auer M. 2007. Electron tomography: A 3D view of the subcellular world. *Analytical Chemistry* 79: 7949–7957.

Fauver M, Seibel EJ, Rahn JR, Meyer GM, Patten FW, Neumann T et al. 2005. Three-dimensional imaging of single isolated cell nuclei using optical projection tomography. *Optics Express* 13: 4210–4223.

Gladish JC, Yao G, Heureux NL & Haidekker MA. 2005. Optical transillumination tomography for imaging of tissue-engineered blood vessels. *Annals of Biomedical Engineering* 33: 323–327.

Guldberg RE, Lin ASP, Coleman R, Robertson G & Duvall C. 2004. Microcomputed tomography imaging of skeletal development and growth. *Birth Defects Research Part C* 72: 250–259.

Haidekker MA. 2005. Optical transillumination tomography with tolerance against refraction mismatch. *Computer Methods and Programs in Biomedicine* 80: 225–235.

Hecht E. 1987. *Optics*, 2nd ed. Reading, MA: Addison-Wesley Publishing Company. Inc.

Huang HM, Xia J & Haidekker MA. 2008. Fast optical transillumination tomography with large-size projection acquisition. *Annals of Biomedical Engineering* 36: 1699–1707.

Huisken J, Swoger J, Del Bene F, Wittbrodt J & Stelzer EHK. 2004. Optical sectioning deep inside live embryos by selective plane illumination microscopy. *Science* 305: 1007–1009.

Jayshree N, Datta GK & Vasu RM. 2000. Optical tomographic microscope for quantitative imaging of phase objects. *Applied Optics* 39: 277–283.

Kak AC & Slaney M. 1988. *Principles of Computerized Tomographic Imaging*. Philadelphia: SIAM.

Kim S-J, Doudet DJ, Studenov AR, Nian C, Ruth TJ, Gambhir SS et al. 2006. Quantitative micro positron emission tomography (PET) imaging for the *in vivo* determination of pancreatic islet graft survival. *Nature Medicine* 12: 1423–1428.

Klar TA, Jakobs S, Dyba M, Egner A & Hell SW. 2000. Fluorescence microscopy with diffraction resolution barrier broken by stimulated emission. *Proceedings of the National Academy of Science* 97: 8206–8210.

Kumar V, Scandella E, Danuser R, Onder L, Nitschké M, Fukui Y et al. 2010. Global lymphoid tissue remodeling during a viral infection is orchestrated by a B cell-lymphotoxin-dependent pathway. *Blood* 115: 4725–4733.

Larabell CA & Le Gros MA. 2004. X-ray tomography generates 3D reconstructions of the yeast Saccharomyces cerevisiae, at 60 nm resolution. *Molecular Biology of the Cell* 15: 957–962.

LaRoque SJ, Sidky EY &Pan X. 2008. Accurate image reconstruction from few-view and limited-angle data in diffraction tomography. *Journal of the Optical Society of America A* 25: 1772–1782.

L'Heureux N, Dusserre N, Konig G, Victor B, Keirre P, Wight TN et al. 2006. Human tissue engineered blood vessel for adult arterial revascularization. *Nature Medicine* 12: 361.

L'Heureux N, McAllister TN & de la Fuente LM. 2007a. Tissue-engineered blood vessel for adult arterial revascularization. *New England Journal of Medicine* 357: 1451–1453.

L'Heureux N, Dusserre N, Marini A, Garrido S, de la Fuente L & McAllister T. 2007b. Technology insight: the evolution of tissue-engineered vascular grafts from research to clinical practice. *Nature Clinical Practice Cardiovascular Medicine* 4: 389–395.

Linfoot EH & Wolf E. 1953. Diffraction images in systems with an annular aperture. *Proceedings of the Physical Society B* 66: 145–149.

Liu F, Yoo KM, Alfano RR. 1999. Ultrafast laser-pulse transmission and imaging through biological tissues. *Applied Optics* 32: 554-558.

Marengo S, Pépin C, Goulct T & Honde D. 1999. Time-gated transillumination of objects in highly scattering media using a subpicosecond optical amplifier. *IEEE Journal of Selected Topics in Quantum Electronics* 5: 895–901.

Mariani G, Bruselli L, Kuwert T, Kim EE, Flotats A, Israel O et al. 2010. A review on the clinical uses of SPECT/CT. *European Journal of Nuclear Medicine and Molecular Imaging* 37: 1959–1985.

McGinty J, Tahir KB, Laine R, Talbot CB, Dunsby C, Neil MAA et al. 2008. Fluorescence lifetime optical projection tomography. *Journal of Biophotonics* 1: 390–394.

Nowlan NC, Murphy P & Prendergast PJ. 2007. Mechanobiology of embryonic limb development. *Annals of the New York Academy of Science* 1101: 389–411.

Chapter 9

Ntziachristos V. 2010. Going deeper than microscopy: the optical imaging frontier in biology. *Nature Methods* 7: 603–614.

Oldham M & Kim L. 2004. Optical-CT gel-dosimetry II: Optical artifacts and geometric distortion. *Medical Physics* 31: 1093.

Oldham M, Siewerdsen JH, Kumar S, Wong J & Jaffray DA. 2003. Optical-CT gel-dosimetry I: Basic investigations. *Medical Physics* 30: 623.

Pawley JB. 2006. *Handbook of Biological Confocal Microscopy*, 3rd ed. New York: Springer Science+Business Media, LLC.

Podoleanu AG. 2005. Optical Coherence Tomography. *British Journal of Radiology* 79: 976–988.

Rust MJ, Bates M & Zhuang X. 2006. Sub-diffraction-limit imaging by stochastic optical reconstruction microscopy (STORM). *Nature Methods* 3: 793–795.

Sakhalkar HS, Dewhirst M, Olver T, Cao Y & Oldham M. 2007. Functional imaging in bulk tissue specimens using optical emission tomography: fluorescence preservation during optical clearing. *Physics in Medicine and Biology* 52: 2035–2054.

Sarma S, Kerwin J, Puelles L, Scott M, Strachan T, Feng G et al. 2005. 3D modelling, gene expression mapping and post-mapping image analysis in the developing human brain. *Brain Research Bulletin* 66: 449–453.

Schleicher E, Jesinghaus M, Hildebrandt G, Liebrecht K, Hampel U & Freyer R. 1998. Optischer Labortomograph für die Lehre und Forschung [Optical laboratory tomography for education and research]. *Biomedizinische Technik (Berlin)* 43: 480–481.

Schmitt JM, Gandjbakhche AH & Bonner RF. 1992. Use of polarized light to discriminate short-path photons in a multiply scattering medium. *Applied Optics* 31: 6535–6546.

Sharpe J. 2003. Optical projection tomography as a new tool for studying embryo anatomy. *Journal of Anatomy* 202: 175–181.

Sharpe J, Ahlgren U, Perry P, Hill B, Ross A, Hecksher-Sørensen J et al. 2002. Optical projection tomography as a tool for 3D microscopy and gene expression studies. *Science* 296: 541–545.

Shu X, Royant A, Lin MZ, Aguilera TA, Lev-Ram V, Steinbach PA et al. 2009. Mammalian expression of infrared fluorescent proteins engineered from a bacterial phytochrome. *Science* 324: 804–807.

Siegel A, Marota JJ & Boas D. 1999. Design and evaluation of a continuous-wave diffuse optical tomography system. *Optics Express* 4: 287–298.

Soltesz G, Patterson CC & Dahlquist G. 2007. Worldwide childhood type 1 diabetes incidence—what can we learn from epidemiology? *Pediatric Diabetes* 8: 6–14.

Summerhurst K, Stark M, Sharpe J, Davidson D & Murphy P. 2008. 3D representation of Wnt and frizzled gene expression patterns in the mouse embryo at embryonic day 11.5 (Ts19). *Gene Expression Patterns* 8: 331–348.

Vinegoni C, Pitsouli C, Razansky D, Perrimon N & Ntziachristos V. 2008. *In vivo* imaging of Drosophila melanogaster pupae with mesoscopic fluorescence tomography. *Nature Methods* 5: 45–47.

Vinegoni C, Razansky D, Figueiredo J-L, Nahrendorf M, Ntziachristos V & Weissleder R. 2009. Normalized Born ratio for fluorescence optical projection tomography. *Optics Letters* 34: 319–321.

Walls JR, Coultas L, Rossant J & Henkelman RM. 2008. Three-dimensional analysis of vascular development in the mouse embryo. *PLoS ONE* 3: e2853.

Wang X, Wang LV, Sun CW & Yang CC. 2003. Polarized light propagation through scattering media: time-resolved Monte Carlo simulations and experiments. *Journal of Biomedical Optics* 8: 608.

Wang Z, Millet L, Mir M, Ding H, Unarunotai S, Rogers J et al. 2011. Spatial light interference microscopy (SLIM). *Optics Express* 19: 1016–1026.

Yao G & Haidekker MA. 2005. Transillumination optical tomography of tissue-engineered blood vessels: a Monte Carlo simulation. *Applied Optics* 44: 4265–4271.

Yuasa T, Tanosaki S, Takagi M, Sasaki Y, Taniguchi H, Devaraj B et al. 2001. Transillumination optical sensing for biomedicine and diagnostics: feasibility of early diagnosis for rheumatoid arthritis. *Analytical Sciences* 17: 515–518.

Yuasa T, Tanosaki S, Sasaki Y, Takagi M, Ishikawa A, Taniguchi H et al. 2002. Fundamental imaging properties of transillumination laser computed tomography based on coherent detection imaging method. *Analytical Sciences* 18: 1329–1333.

10. Optical Coherence Tomography

Overview and Applications

Youbo Zhao, Benedikt W. Graf, and Stephen A. Boppart

Optical Techniques in Regenerative Medicine. Edited by Stephen P. Morgan, Felicity R.A.J. Rose, and Stephen J. Matcher © 2014 CRC Press/Taylor & Francis Group, LLC. ISBN: 978-1-4398-5495-2

Chapter 10

10.1 Introduction

Optical coherence tomography (OCT) is a rapidly emerging, noninvasive biomedical imaging technology (Drexler and Fujimoto 2008a; Huang et al. 1991). By utilizing interferometry-based optical gating to differentiate light signals reflected at different depths, OCT provides 2D cross-sectional and 3D volumetric images of inner structures in scattering biological samples. It generates images based on the spatial variation in refractive index, with no need for exogenous tracers and special sample preparation or tissue excision. In comparison with other prevailing *in vivo* imaging tools, OCT is advantageous in that it has better resolution (up to 1–2 µm) than ultrasound and magnetic resonance imaging (MRI), has higher imaging depth (1–2 mm in highly scattering tissues) than confocal and two-photon microscopes, and is not based on ionizing radiation such as is used in x-ray computed tomography (CT). As a rapidly evolving technology, OCT has the potential to complement or even replace the role in medical diagnosis and regenerative medicine that histological analysis has occupied, wherein complex sample preparation and long processing times are required, and which often rely on the subjective experience of individuals.

Owing to these unique features, OCT research advances have been gaining increasing attention and are finding numerous applications in areas of biological research and medical practice. In the early 1990s, when the technology was first demonstrated, OCT quickly found a niche in ophthalmology, where it provided the first *in situ* high-resolution 3D images of structures in the eye (Drexler and Fujimoto 2008b), including the retina, cornea, and the optic disc. Before long, the advantages of OCT were recognized in many other fields, including cell biology (Boppart et al. 1998; Drexler et al. 1999), dermatology (Gambichler et al. 2005; Weissman et al. 2004), cancer research (Zhou et al. 2010; Zysk et al. 2007), dentistry (Colston et al. 1998; Otis et al. 2000), and tissue engineering (Liang et al. 2009; Matcher 2011), to name a few. The invention and development of endoscopy and catheter OCT devices (Herz et al. 2004; Tearney et al. 1996, 1997) has extended the use of this technology to cardiovascular (Boppart et al. 1997; Jenkins et al. 2006), gastrointestinal (Herz et al. 2004; Rollins et al. 1999), urological (Wang et al. 2011), and gynecological (Jackle et al. 2000) applications. Today, OCT continues to develop rapidly in many areas, with new technologies and variant modalities emerging, which are not only improving the performance of OCT in terms of resolution, contrast, functional analysis capability, and imaging speed but also bringing it to new fields of application. In this chapter, we will introduce the working principle of OCT, review recent advances in OCT technology and applications, and exploit the unique features of OCT for cellular imaging, particularly for the investigation of cell activity and dynamics in *in vivo* skin and engineered tissues.

10.2   Clinical and Biological Measurement Challenges the Method Is Suited to Address

The high 3D resolution and noninvasive imaging of OCT make it a very promising tool for *in vivo* diagnosis of pathologies and real-time investigation of biological activities. OCT has been widely adopted in various clinical and research-based applications to meet particular challenges. For ophthalmic applications, OCT has made significant contributions and has already become a key commercialized diagnostic technology in the areas of retinal diseases and glaucoma (Costa et al. 2006; Drexler and Fujimoto 2008b; Thomas and Duguid 2004). OCT images of the pathology-rich and multilayered structure of the retina have helped to reveal sometimes hidden clinical features associated with disease or treatment, which otherwise are inaccessible with conventional technologies, such as angiography (Antcliff et al. 2000) and ophthalmic ultrasound (Radhakrishnan et al. 2005). OCT-based analysis of retinal structures, along with recently developed algorithms, has allowed for the diagnosis of abnormalities related to glaucoma (Medeiros et al. 2005), diabetic retinopathy (Hee et al. 1998), and other diseases affecting vision (Puliafito et al. 1995; Thomas and Duguid 2004). Ophthalmic applications are continuously enhanced by new advances in OCT technologies, such as high speed imaging, which significantly suppresses motion artifacts and the incorporation of adaptive optics to improve image quality (Miller et al. 2011; Srinivasan et al. 2008). In cardiology, catheter-based endoscopic OCT has enabled high-resolution (compared to intravasculature ultrasound) (Brezinski et al. 1997) *in vivo* imaging of the cardiovasculature (Fujimoto et al. 1999). This allows for visualization of vulnerable plaques in the coronary or carotid arteries and the evaluation of their potential for progression or rupture. Quantification of activated macrophage content and identification of plaque type by OCT has been demonstrated with high sensitivity and specificity (Tearney et al. 2006; Yabushita et al. 2002). In addition, intravascular OCT has been successfully used in the characterization and visualization of cardiac interventions, such as percutaneous coronary intervention and stent implantations. In oncology, OCT applications are targeting the early detection of malignancies associated with a broad spectrum of cancers including those arising in the breast, bladder, gastrointestinal tract, skin, and many others (Adie and Boppart 2010). OCT has recently entered the operating room and is being evaluated for assisting surgeons to assess surgical margins (Nguyen et al. 2009). Portable or handheld OCT devices are under investigation in outpatient clinics for imaging tissue sites that primary care physicians are concerned with, such as the ear, eye, skin, and oral mucosa, for early screening of many types of diseases (Jung et al. 2011). All of these above-mentioned areas are only a small portion of the broad spectrum of areas that OCT has been used to address different challenges. It continues to expand with each rapid development in the field, and readers can refer to other review resources (Drexler and Fujimoto 2008a; Fercher et al. 2003; Schmitt 1999; Zysk et al. 2007) and Chapter 11 in this book.

In regenerative medicine and tissue engineering, OCT is a unique tool for monitoring cellular dynamics in bulk biomaterials and for visualizing the overall structure of scaffolds (Tan et al. 2006; Yang et al. 2006). The fabrication and development of thick-tissue equivalents has been limited by the lack of an efficient imaging tool that has sufficient

Chapter 10

resolution and imaging depth to investigate cell activities deep within the constructs, which is one of the key issues in tissue engineering. With cellular-level spatial resolution and imaging depths up to a few millimeters, OCT is a very promising option to address this challenge. Its noninvasive imaging capabilities allow one to monitor the development of the tissue in real time, and longitudinally over long time periods, without the need to destroy samples at discrete time points for histological evaluation. Moreover, many recent variants of OCT and related technologies have evolved in the past decade, which enables OCT for imaging or probing different parameters of biological tissues. These include flow dynamics by Doppler OCT (Chen et al. 1998; Izatt et al. 1997), birefringence by polarization-sensitive OCT (PS-OCT) (De Boer et al. 1998; Gotzinger et al. 2005), biomechanical properties with optical coherence elastography (OCE) (Ko et al. 2006; Liang et al. 2010), and metabolic states by spectroscopic OCT (SOCT) (Adie et al. 2010; Morgner et al. 2000). New contrast agents (John et al. 2010; Lee et al. 2003; Oldenburg et al. 2006) and computational methods (Adie et al. 2012; Ralston et al. 2007) also have improved the performance of OCT in terms of resolution, imaging depth, and specificity. Operating either in a standard mode for structural OCT imaging or combined with other functions, OCT has the potential to tackle different challenges that arise in tissue engineering. In this chapter, we will outline those technologies and several of the OCT variants with application examples in the field of tissue engineering.

10.3　Technical Implementation of the Method

10.3.1　Optical Principles

In principle, OCT is essentially the optical analogue to ultrasound, except that it uses light rather than sound wave as the probe. In ultrasound, the Axial-scan (A-scan), that is, the measurement of structural changes at different depths within the tissue (distances from the detector), is carried out by directly measuring the sound wave signals with different echo times using time-resolved electronic receivers. Since the time delay difference between the sound waves reflected from two separate points (e.g., separated by 100 μm) is around 100 ns, it can be well distinguished by currently available electronic detectors. In OCT, in which light is used, the corresponding time delay will be tens of femtoseconds (10^{-15} s) for a 100 μm distance change, which is far beyond the response of optical detectors. As a consequence, fast optical gates are required in OCT. While different optical gating approaches have been exploited in the past, such as nonlinear optical gates (Bruckner 1978; Fujimoto et al. 1986), low-coherence optical interferometry (Fercher et al. 1988; Takada et al. 1987) has been widely accepted for the current OCT systems.

Low-coherence interferometry is a well-known optical gating technique that utilizes a low temporal coherence (broadband) light source to realize depth-resolved measurements of objects distributed along the direction of light propagation. A typical Michelson interferometer-based free space time-domain OCT arrangement is shown in Figure 10.1. The low coherence radiation from a broadband light source, which is usually in the near-infrared range, is divided by a beam splitter into two beams, which are separated into the sample and reference arms. Light in the sample arm illuminates the sample and is reflected back along the same path by the inner structures of the sample. In the reference arm, light is incident on and then reflected back from a plane surface mirror

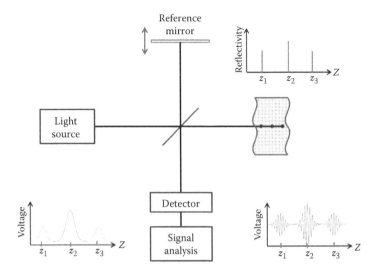

FIGURE 10.1   Schematic drawing of a typical time-domain Michelson interferometer-based OCT system. Three scattering points along the depth direction within the sample (z_1, z_2, and z_3) are schematically shown in the upper-right figure. The points correspond to three interference patterns in the lower right figure and are eventually converted to the shape in the lower-left figure to generate an A-scan. A 2D cross-sectional OCT image is formed by scanning the beam laterally across the sample and assembling adjacent A-scans.

(reference mirror). The back-scattered sample beam and reference beam combine and interfere with each other after the beam splitter and are recorded by a photo sensitive detector at the output of the interferometer. In the case of a broadband light source, interference only happens when the difference of the echo times of the reference and sample beams is within the coherence length of the light source. As the reference mirror is moved along the light propagation direction, the path length (time delay) between the two arms is changed, which results in the alteration of the interference pattern. As shown in Figure 10.1, when multiple scatterers are present along the axial (depth) direction in the sample, their locations are mapped out in a time-domain signal sequence (lower-right insert), if one scans the path length (time delay) of the reference arm by moving the reference mirror. The position and reflectivity can be retrieved by converting the time-resolved interference pattern into space and demodulating to extract the envelope of the interference fringes (lower-left insert). Combined with lateral scanning of the sample beam focus, a 2D or 3D OCT image of the sample is generated.

According to the theory of interferometry, the electric field at the output of the interferometer is the sum of the signal and reference fields, $E_s(t) + E_r(t)$, and a detector measures the intensity of the output, which is proportional to the square of the total fields. For a single reflector in the sample arm, this is given by

$$I_0 \sim |E_r|^2 + |E_s|^2 + 2E_rE_s \times |G(\Delta L)|\cos(2k\Delta L) \tag{10.1}$$

where ΔL is the path length difference between the sample and reference arms of the interferometer. The term $G(\Delta L)$ represents the coherence function of the light source, that is, the autocorrelation function of the electrical field of the light source. As shown in Figure 10.1, the three discrete scatterers along the axial direction, which are infinitely small in space (upper-right insert), correspond to the time-domain (mapped to the z-axis) interference pattern, but are broadened, representing the axial point spread function of the system (lower-right insert). From Equation 10.1, the width of the envelope for each interference pattern is determined by the width of the coherence function, $G(\Delta L)$. In other words, the spatial resolution of the interferometer along the axial dimension, that is, the point spread function, is determined by the coherence length of the light source. The amplitude of the scattered light, which is associated with the reflectivity of the sample structures, is determined by the amplitude of the interference pattern.

Based on the Wiener–Khinchin theorem (Goodman 1985; Schmitt 1999), the temporal coherence is determined by the Fourier transform of the power spectral density of the light source. This relationship reveals that the width and shape of the emission spectrum of the light source are the determining factors for the axial resolution in OCT images. Strictly speaking, the axial resolution of OCT, that is, the full width at half maximum (FWHM) of the axial point spread function, is defined by the round-trip "coherence length" of the light source. Under the assumption of a Gaussian-shaped spectrum of the light source, the axial resolution of OCT is given by

$$\Delta z = \frac{2\ln 2}{\pi}\frac{\lambda_0^2}{\Delta\lambda},\tag{10.2}$$

where λ_0 is the central wavelength and $\Delta\lambda$ is the FWHM spectral bandwidth of the light source spectrum.

In a conventional scanning optical microscope, and also in OCT, the transverse or lateral resolution is determined by the diameter of the spot size of the illumination light beam focused at the sample surface, which is given by

$$\Delta x = \frac{4\lambda_0}{\pi}\cdot\frac{f}{D},\tag{10.3}$$

where D is the beam diameter on the back aperture of the objective, f is the focal length of the objective, and λ_0 is the central wavelength of the light source. Combining Equations 10.2 and 10.3, one can see that the axial and transverse resolutions are determined by independent parameters. This is a unique feature of OCT, which enables one to improve the axial resolution and transverse resolution separately, and advantageously for different applications.

As in any imaging system, another important parameter in OCT is the depth of field (DOF), which is determined by the Rayleigh range of the beam focus. According to the propagation properties of the Gaussian beam, the Rayleigh range is determined by

$$Z_{\mathrm{R}} = \frac{\pi\Delta x^2}{4\lambda_0}\tag{10.4}$$

The DOF is defined as twice the length of Z_{R}, which is also referred to as the confocal parameter. It is apparent from Equations 10.3 and 10.4 that there is a trade-off between the transverse resolution and DOF. Although one can improve the transverse resolution by employing a high numerical aperture (NA) objective, the DOF will be inevitably sacrificed (reduced). As a result, the effective imaging depth of OCT will be restricted by the high NA objective and the short DOF unless other techniques are employed.

10.3.2   Advances in High-Resolution, High-Speed, Volumetric OCT

Thus far, all the discussions have been based on time-domain OCT (TD-OCT), which is usually referred to as the conventional OCT because initial systems were constructed and operated this way. TD-OCT is a simple, robust method to generate high-resolution cross-sectional images of inner structures in highly scattering biological samples. Because a physical scanner must be involved to realize each depth scan, however, the imaging speed of TD-OCT is limited. In spite of many attempts, the fastest imaging speed of TD-OCT is still on the order of a few kilohertz A-scan line rate (Rollins et al. 1998). For this reason, it is challenging to perform *in vivo* imaging of an entire 3D structure of living tissue by TD-OCT due to movement artifacts during the long image acquisition time.

There have been tremendous advances in the performance of OCT, in terms of resolution, imaging speed, detection sensitivity, and 3D information rendering, since the first demonstration of the OCT technology in 1991 (Huang et al. 1991). These advances include the inception of spectral domain OCT (SD-OCT) and the development of new light sources and detectors. The development of these technologies enables one to realize high-resolution, high-speed, and volumetric visualization of intact biological samples.

The invention of SD-OCT could be the most revolutionary advance in the development of the OCT field. SD-OCT was first reported in 1995 (Fercher et al. 1995), but was quickly recognized by the OCT community a few years later (Wojtkowski et al. 2004a; Yun et al. 2003) because of its fast imaging speed and high sensitivity compared to TD-OCT (Choma et al. 2003; Leitgeb et al. 2003). In SD-OCT, the path lengths of both the sample and reference arms are fixed, and wavelength-resolved interference is measured by a spectrometer instead of a single channel detector as in the TD-OCT (Choma et al. 2003; De Boer et al. 2003; Leitgeb et al. 2003). The principle of SD-OCT can be intuitively understood from Equation 10.1, where the interferogram or the spectral domain interference fringes are generated when path length Δl is fixed at a given number, but wavenumber k varies. In this case, through the Fourier transform of the interferogram with respect to wavenumber k, one can map out the depth information of the sample. The depth resolution of SD-OCT, the same as that in TD-OCT, is determined by Equation 10.2.

Compared to TD-OCT, SD-OCT offers a significant improvement in detection sensitivity and imaging speed (Choma et al. 2003; De Boer et al. 2003; Yun et al. 2003). High-speed detection is particularly important for *in vivo* imaging of living samples, in which sample motions introduce image artifacts. Because no moving parts are involved in A-scan acquisition in SD-OCT, its axial scan rate is only limited by the detection or read-out frequency of the spectrometer, which is usually based on high-speed line scan cameras (Oh et al. 2006; Yun et al. 2003). With the advance of high-speed line cameras, A-scan rates of SD-OCT systems have been improved to the level of hundreds of kilohertz at acceptable sensitivity levels (Huber et al. 2007; Potsaid et al. 2008; Srinivasan et al. 2008).

Chapter 10

Another method for performing OCT is swept-source OCT (SS-OCT), which utilizes a wavelength-tunable laser (Chinn et al. 1997; Lexer et al. 1997). Some groups have referred to this method as optical frequency domain imaging (Bouma et al. 2006). The idea of using wavelength-tunable light sources is motivated under certain circumstances by the limited performance of line-scan CCD cameras for spectral measurements (Potsaid et al. 2008). In SD-OCT, the limited number of camera pixels results in a design trade-off between axial resolution and imaging DOF. Furthermore, high-speed cameras are always needed for fast imaging, which in turn results in low detection sensitivity because of the required short camera exposure times at high-speed mode, and the broad bandwidth optical sources for SD-OCT do not have sufficiently high output power to maintain high signal-to-noise ratio at high speeds, compared to the wavelength-swept narrow-frequency lasers. In contrast, SS-OCT employs a fast tunable light source and a single channel or dual-balanced detector along with a high speed A/D convertor, which are straightforward to design and are commercially available. The single detector makes the SS-OCT systems free from these limitations, which promises high-speed, high-resolution, and high-range volumetric reconstruction of the tissues. The imaging performance of SS-OCT is only confined by the properties of the swept source, which have been evolving rapidly in the past years and have been based on a variety of laser tuning mechanisms, ranging from a rotating polygon mirror (Yun et al. 2003), galvo-driven grating filter (Huber et al. 2005), fiber Fabry-Perot tunable filter (Choma et al. 2005; Zhang et al. 2006), and a recently introduced Fourier domain mode locking scheme (Huber et al. 2006).

With the availability of these fast swept sources, the A-scan rates of a few or several tens of megahertz have been demonstrated recently (Klein et al. 2011; Wieser et al. 2010). Using these ultrahigh imaging speeds, for example, operating at a 1 MHz A-scan rate with 1000 depth scans in each B-mode image therefore enables a 3D volume with $1000 \times 1000 \times 1000$ voxels to be generated in 1 s. Such speeds are desirable for *in vivo* imaging of living targets, since the motion artifacts can be significantly suppressed. Furthermore, the high recording speed and thus the high volume data set allows post-data processing algorithms to improve the imaging quality and enable functional analysis. For example, in retinal imaging, a megahertz OCT system has been reported to provide 1900×1900 A-scan data sets over a 70° field-of-view, enabling megapixel *en face* images similar to fundus views. It also enables 2.5 times oversampling for further data processing (Klein et al. 2011).

While the imaging speed of OCT has been significantly improved by virtue of these newly developed fast wavelength tuned lasers, the axial resolution of SD-OCT has been enhanced with the increasing availability of broadband light sources. As stated previously, the axial resolution of OCT is determined by the bandwidth or the coherence length of the light source, independent of the focusing conditions. Improvement of OCT axial resolution has resulted from the advances in broadband light source technology, which has happened in parallel in other fields of optics and photonics. Femtosecond laser pulses generated from solid-state titanium:sapphire lasers were first introduced and are still widely used for OCT systems. Axial resolutions of a few microns have been demonstrated for *in vivo* OCT imaging using few cycle (<10 fs) laser pulses (Drexler et al. 1999; Leitgeb et al. 2004; Wojtkowski et al. 2004b). Superluminescent diodes (SLDs) are compact, low-cost broadband sources for OCT imaging (Baumgartner et al. 1998;

Rao et al. 1993). Multiplex technology-based SLD light sources using two or more SLDs have enabled OCT systems with ultrahigh, 2.5–3.5 µm axial resolutions (Cimalla et al. 2009; Ko et al. 2004a; Leitgeb et al. 2004). Supercontinuum (SC), which can be generated in photonic crystal (PC) fibers with moderate nanojoule ultrashort pulses, is a broad coherent light source and has been used for OCT (Aguirre et al. 2006; Wang et al. 2007a). With hundreds of nanometers of bandwidth, these novel light sources allow for submicron axial resolution in OCT (Hartl et al. 2001; Povazay et al. 2002). The ultrahigh resolutions established with the help of these novel broadband light sources enable much finer structures to be visualized in tissues, such as different layers of the retina (Drexler et al. 2003; Ko et al. 2004b; Srinivasan et al. 2006), human skin (Gasparoni et al. 2005), colon (Hsiung et al. 2005), and blood vessels (Liu et al. 2011), among many others.

Many technologies have recently been reported to improve the transverse resolution of OCT. In particular, optical coherence microscopy (OCM, detailed in Section 10.3.3.1) (Aguirre et al. 2003; Izatt et al. 1994) and full-field OCT (Akiba and Chan 2007; Chang et al. 2005; Dubois et al. 2004) both use high NA objectives and have been extensively exploited to provide high-resolution, high-speed, 3D images of biological samples. Differing from cross-sectional OCT, both of these produce tomographic images in an *en face* orientation (perpendicular to the light propagation direction and parallel to the sample surface) rather than in a cross-sectional depth-resolved orientation. The principle of full-field OCT is similar to TD-OCT except that it does not rely on lateral beam scanning. The tomographic information is extracted from a sequence of interferometric images with respect to a scanned reference path length, which are recorded by an area array sensor, such as a CCD camera. The parallel image recording makes it possible for full-field OCT to realize high-speed rendering of volumetric images of biological tissues. Because high NA objectives are used in both OCM and full-field OCT, these offer higher transverse resolutions compared to conventional OCT systems, up to the order of (or even higher than) 1 µm. These high spatial resolutions enable the two imaging modalities to perform cellular imaging of biological samples *in vivo*. The main disadvantage of the two techniques, however, is the low imaging depth due to the use of the high NA objective and the resulting tight focusing condition. In addition, full-field OCT is also limited by multiple scattering of light and cross-talk between different pixels.

Considering the inherent trade-off between transverse resolution and DOF in OCT, other technologies are emerging to meet this challenge. For example, semi-diffraction free Bessel beams (Durnin 1987) formed through a conical or axicon lens have been proposed to extend the focused depth (Blatter et al. 2011; Lee and Rolland 2008; Leitgeb et al. 2006). Dynamic focusing of the objective lens has been successfully demonstrated to increase imaging depth (Lee et al. 2011; Murali et al. 2009). A multibeam OCT approach that improves lateral resolution without sacrificing the imaging DOF has been employed in a commercial system (Michelson Diagnostics, Inc.). Computed imaging methods, such as interferometric synthetic aperture microscopy (ISAM), are also used to achieve extended DOF imaging (Ralston et al. 2007). Although all of these have specific limitations, these techniques have the potential to further improve the performance of OCT.

All the aforementioned improvements in imaging speed and resolution have both enhanced and extended the functionality of OCT (Fujimoto 2003). While higher spatial resolution allows for visualization of finer structures, high-speed imaging

Chapter 10

improves the performance of OCT in many aspects. In particular, high imaging speed not only helps to suppress motion artifacts and realize real-time volumetric visualization of cellular structures, but also enables functional imaging of transient physiology dynamics of the living biological tissue. These high-speed, high-resolution, volumetric OCT approaches, along with other novel technologies, such as adaptive optics for retinal imaging and catheter probes for endoscopy, have intensified and expanded the applications of OCT in biomedical research and clinical practice in a variety of fields.

10.3.3 OCT Variants

10.3.3.1 Optical Coherence Microscopy

Optical coherence microscopy (OCM) (Aguirre et al. 2003; Hoeling et al. 2000; Izatt et al. 1994) is a high-resolution extension of OCT. In contrast to OCT, in which a low NA objective is usually used to obtain a large DOF, OCM utilizes a high NA objective to realize high lateral resolution. OCM generates *en face* images rather than cross-sectional images in order to avoid the DOF limitation in the cross-sectional plane. In OCM, a coherence gate is generated based on the OCT principle to remove unwanted light scattering from outside the focus, which has the similar effects as the spatial filter (pin hole) induced confocal gate in confocal microscopy. In contrast to confocal microscopy, however, high NA is not necessarily required to achieve fine axial sectioning in OCM, which gives more flexibility in the implementation of OCM. OCM can be implemented in either scanning mode, as in classical confocal microscopy, or in full-field mode, in which the entire field of the image is illuminated and an area array imaging sensor is used to record interferometric images. OCM offers high-resolution, high-speed, and 3D visualization of microstructure within scattering biological tissues. Compared to confocal microscopy, OCM provides higher imaging depth, which is attributed to the increased detection sensitivity due to the heterodyne detection principle. These advantages enable OCM to provide cellular level imaging (Aguirre et al. 2003), which has been extensively exploited with applications in developmental biology (Hoeling et al. 2000), dermatology (Lee et al. 2011), etc.

10.3.3.2 Spectroscopic OCT

Since OCT uses broadband light sources, it is natural to integrate spectroscopic analysis into OCT, which is the technology known as spectroscopic OCT (SOCT) (Leitgeb et al. 2000; Morgner et al. 2000; Schmitt et al. 1998). With SOCT, spatial changes in the absorption and scattering spectra of native or exogenous chromophores in the tissue can be measured, which can be used to enhance image contrast or to extract functional information of the tissue. Sophisticated time–frequency analysis must be utilized for SOCT, though the hardware can be essentially the same as in standard OCT systems (Xu et al. 2005). The algorithms developed have been rather sophisticated, because of the implication applied by the Heisenberg–Gabor uncertainty principle, which implies that one cannot have simultaneously arbitrarily specified frequency and time, that is, the trade-off between spectral and depth resolution specifically in SD-OCT. Applications with contrast enhancement in SD-OCT have been demonstrated in imaging an African frog tadpole (Adler et al. 2004), celery stalk (Xu et al. 2004b), and tissue phantoms (Xu et al. 2004a). An important function of SOCT is to measure localized tissue oxygenation, which is based on the spectral absorption change with oxygen saturation ratio of SO_2 (Adler et al. 2004; Faber et al. 2005). This feature gives SOCT the potential to be utilized in many medical applications, such as monitoring the oxygenation activities of red blood cells.

10.3.3.3   Polarization-Sensitive OCT

During light–tissue interactions, tissues affect not only the amplitude or direction of light but also the polarization state of light. Polarization-sensitive OCT (PS-OCT) is another functional extension of OCT, which takes advantage of the polarization information carried by reflected light to reveal additional tissue information that is not available in conventional OCT images (De Boer et al. 1998; Everett et al. 1998; Yao and Wang 1999). In PS-OCT, specifically controlled polarized light is sent to the sample and reference arms of the interferometer, and the reflected light from both arms is recombined and detected. In addition, rotation-controlled wave retarders are also inserted into both of the arms, and the recombined light can be decomposed into two orthogonal polarization components and detected separately by phase-sensitive detectors. In principle, changes in birefringence, diattenuation, and optical axis can be detected by PS-OCT. Because birefringence is commonly found in tendons, muscle, nerve, collagen, and other highly ordered biological tissues, PS-OCT has found a variety of applications, especially in clinical practice in ophthalmology and dermatology. In particular, PS-OCT has enhanced measurement of the thickness and birefringence of the retinal nerve fiber layer (RNFL) (Cense et al. 2004), which is an important parameter to assess in glaucoma. In dermatology, depth-resolved mapping of the birefringence changes in skin significantly facilitates the assessment of burn depth, which otherwise has to be provided by surgery (Park et al. 2001). With the recent development of techniques that are capable of measuring many other parameters, like the Jones matrix, the Mueller matrix, and Stokes vectors, PS-OCT continues to advance rapidly and find new applications in clinical medicine and biological research.

10.3.3.4   Doppler OCT

Doppler OCT is a functional extension of OCT that combines the Doppler principle with OCT to realize tomographic imaging of tissue structure and measurement of flow dynamics simultaneously (Chen et al. 1999; Izatt et al. 1997). The physical basis of Doppler OCT is that transient motion of the object gives a Doppler frequency shift to OCT interference fringes, which can be retrieved during postprocessing of OCT data to map out the flow distribution. Intuitively, the Doppler frequency shift can be imagined as the change in light wavelength or wave vector in Equation 10.1, which thus changes the frequency of the interference fringes. Doppler OCT has the same optical architecture as in conventional OCT systems, in either the time domain or frequency domain, but signal processing algorithms like the spectrogram method or phase-resolved approaches must be utilized to retrace the Doppler frequency and particle flow velocity. The velocity detection sensitivity in Doppler OCT has reached a level of 10 µm/s in real-time imaging applications. Its high spatial resolution and velocity detection sensitivity make Doppler OCT an exceptional choice for applications associated with measuring blood flow dynamics in the vascular, including the

Chapter 10

capillaries. Successful demonstrations in research or clinical practice have included screening vasoactive drugs (Chen et al. 1998), mapping cortical hemodynamics for brain research (Chen et al. 1999), and imaging ocular blood flow (Wang et al. 2007b). One limitation of Doppler OCT is the fact that it is only sensitive to longitudinal flow, and several other technologies like angle determination have to be utilized to quantify arbitrarily oriented motion.

10.3.3.5 Optical Coherence Elastography

Optical coherence elastography (OCE) (Adie et al. 2010; Ko et al. 2006) is an interesting variant of OCT developed to measure the localized biomechanical properties of tissue. The relationship between OCT and OCE is analogous to ultrasound sonography and elastography. In OCE, the OCT signal is recorded while applying mechanical excitation to the tissue under examination, which can be under static, dynamic, internal, or external conditions (Kennedy et al. 2009; Liang and Boppart 2010). This technique has great potential for detecting mechanical property changes in tissue microstructure, such as tumor invasion (Liang et al. 2010), due to its high-resolution, fast imaging speed and noninvasive features. For example, using a phase-resolved M-mode OCE method and transverse scanning, it was possible to derive an elasticity map of tissue containing both tumor and normal adipose material, which cannot be differentiated with structural OCT alone (Liang et al. 2008). Novel dynamic OCE imaging techniques, including external and internal excitation methods, have been applied quantitatively for measuring and mapping biomechanical properties of tissue phantoms and biological tissues (Adie et al. 2010; Liang et al. 2010; Liang and Boppart 2010). With features of micronscale resolution and noninvasive measurement, these novel OCE technologies have the potential to be used to identify and quantify mechanical tissue properties in various biomedical applications such as early stage tumor detection.

10.3.3.6 Magnetomotive OCT

Contrast of OCT images is formed based on the spatial variation in the refractive index. As a result, OCT is lacking molecular-level discrimination capability, and thus, contrast agents, for example, nanometer-sized metal particles, can be applied to obtain the desired site-specific contrast (Oldenburg et al. 2009; Troutman et al. 2007). Magnetic nanoparticles, which are well known as contrast agents in MRI, can be applied in OCT to enable the molecular imaging capability of OCT. Magnetomotive nanoparticles act as dynamic contrast agents when mechanically actuated by an external alternating magnetic field and subsequently modulate the optical scattering properties of the local tissue microenvironment under observation. The deformations and displacements in the local tissue microenvironment due to magnetomotion can be effectively detected by ultrasensitive phase-resolved OCT systems with a displacement sensitivity of a few tens of nanometers. This technology is referred to as magnetomotive OCT (MM-OCT) (John and Boppart 2011; John et al. 2010), which was first demonstrated with a 3D scaffold of macrophage cells loaded with iron oxide microparticles (Oldenburg et al. 2005). When MM-OCT is performed by using a sinusoidal magnetic excitation field with a frequency matched to the mechanical resonance of the biological specimen, the magnetomotion will be enhanced and thus the mechanical properties of the specimen can be characterized with high sensitivity (Boppart et al. 2005). Specially designed magnetic agents can be specifically attached to target molecules, which enable the molecular imaging capability of MM-OCT. *In vivo* MM-OCT has been demonstrated to investigate a preclinical rat mammary tumor model by targeting specific tumor-cell receptors with magnetic agents (John et al. 2010).

10.3.3.7 Nonlinear Interferometric Vibrational Imaging

Nonlinear interferometric vibrational imaging (NIVI) is another novel variant of OCT with molecular imaging capabilities (Benalcazar and Boppart 2011; Benalcazar et al. 2010; Chowdary et al. 2010; Marks and Boppart 2004). NIVI combines the intrinsic molecular contrast ability of vibration spectroscopy with phase sensitive detection, spectral interferometry, and imaging as in OCT. Without the application of any exogenous contrast agents, vibrational spectroscopy enables intrinsic molecular contrast and provides noninvasive characterization of tissues. In this context, Raman spectroscopy is well suited for biological study, but extremely small Raman scattering cross sections necessitate long acquisition times. Coherent anti-Stokes Raman scattering (CARS) microscopy (Cheng et al. 2002; Zumbusch et al. 1999) is another vibrational contrast-based imaging approach that relies on coherence enhancement to obviate the need for high incident powers and permits video rate imaging compatible with real-time diagnosis. However, CARS is plagued by a nonresonant background. In NIVI, the nonresonant background is suppressed by measuring the time-resolved anti-Stokes signal with the help of spectral interferometry. Furthermore, the interferometer-based measurement permits the use of a broadband excitation source so that multiple resonant bands can be detected simultaneously. This makes NIVI very attractive as a noninvasive molecular imaging technology, and its application in the detection of molecular tumor margins (Chowdary et al. 2010) and skin morphology (Benalcazar and Boppart 2011; Benalcazar et al. 2010), as well as many other biological studies.

10.3.3.8 Interferometric Synthetic Aperture Microscopy

Interferometric synthetic aperture microscopy (ISAM) is a computed imaging technique with a physical framework analogous to synthetic aperture radar (Davis et al. 2008; Ralston et al. 2007). ISAM uses computational imaging to overcome the trade-off between DOF and lateral resolution in OCT. By accurately modeling the scattering process and data collection system, including the defocusing ignored in OCT image formation, the scattering properties of the object can be quantitatively estimated from the collected data (Chen et al. 2010). Combined with interferometric ranging to resolve depth profiles, truly 3D images can be generated with ISAM. More importantly, originally blurred images of out-of-focus regions in traditional OCT images are reconstructed clearly with spatially invariant resolution. As a result, the effective imaging depth of OCT is extended, without the need for any additional hardware. Reconstruction of objects to the depth of several times the Rayleigh range (which is referred to as half the DOF) has been demonstrated (Ralston et al. 2007). The main limitation for further extension of DOF is the low signal-to-noise ratio at deeper planes where the signal from the scattered light is much weaker due to the defocused light illumination. ISAM has been shown to achieve high-speed, cross-sectional, or volumetric imaging of intact specimen with high-resolution and extended imaging depth with no need for axial focus scanning (Ralston et al. 2008).

Chapter 10

10.3.4   Multimodal Integration

10.3.4.1   Optical Coherence and Multiphoton Microscopy

Besides OCT, several optical imaging techniques have emerged in recent years that allow intact tissue to be noninvasively visualized at high resolution. These techniques include confocal microscopy, multiphoton microscopy [MPM, e.g., two-photon excitation fluorescence (TPEF) (Denk et al. 1990; Helmchen and Denk 2005) second harmonic generation (SHG) (Campagnola and Loew 2003; Han et al. 2005), and third harmonic generation (THG) microscopy (Smith et al. 2010; Yelin and Silberberg 1999)], and CARS microscopy (Cheng et al. 2002; Zumbusch et al. 1999). All of these techniques have great potential both as clinical tools and in basic research applications. Since these techniques are based on distinct contrast mechanisms, they have a great potential to provide complementary information about living tissue.

All of the above-mentioned techniques have strengths and weaknesses. For example, TPEF can only function in situations where fluorescence is generated from either exogenous fluorophores, genetically expressed fluorescent proteins, or endogenous proteins with weak fluorescence. SHG signals are only produced at the interfaces with noncentral symmetry. Moreover, all of these have limited imaging depth due to the high NA objectives used to achieve high light intensity, which is needed to induce these nonlinear processes. OCT is a noninvasive imaging modality with high imaging depth that does not need application of contrast agents, but is normally lacking molecular sensitivity. It is thus reasonable that the optimal technique for clinical applications will likely be one that combines multiple modalities into one system. In particular, integrated OCM and MPM is a promising combination of modalities that utilizes both scattering- and fluorescence-based contrast to provide high-resolution views of tissue.

OCM visualizes the structure of biological samples at high resolution based on their scattering properties. It has a higher sensitivity and a deeper penetration depth than confocal microscopy due to the coherence-gated detection (Izatt et al. 1994). The disadvantage of OCM, however, is its relatively weak tissue contrast, which is based on the linear variation of refractive index in space. In contrast, MPM, including TPEF and SHG, has strong molecular discrimination ability. Due to the nonlinear process of fluorescence generation, signal is essentially confined to the focal volume, enabling high resolution without the need for spatial filtering as in confocal microscopy. The combination of these two imaging modalities will provide complementary information simultaneously. MPM, using fluorescently labeled molecules or endogenous proteins, can investigate specific functions of cells at the molecular level, while OCM provides the sample structure based on light scattering properties without the use of contrast agents.

10.3.4.2   Integrated Microscope Hardware

The schematic of a typical integrated OCM–MPM microscope is shown in Figure 10.2. A dual spectrum laser source has been specially designed to optimally perform both modalities. The laser is a high-power, widely tunable titanium:sapphire laser (Mai-Tai HP, Spectra-Physics), which outputs 100 fs pulses with a bandwidth of 10 nm and a center wavelength tunable within the range of 730–1000 nm. The linearly polarized output (with maximum average power of 3 W) from this laser is divided by a 90/10 beam splitter into two portions. The higher power beam is coupled by a 0.4 NA aspheric lens

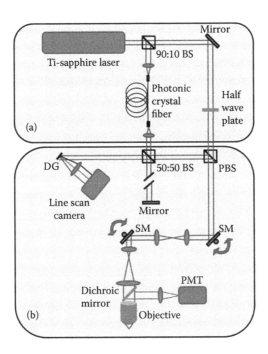

FIGURE 10.2 Schematic of an integrated OCM and MPM microscope. The laser source (a) provides two beams, one for each modality, that are combined and sent to the OCM and MPM microscopes (b). Abbreviations: BS, beam splitter; DG, diffraction grating; PBS; polarizing beam splitter; PMT; photomultiplier tube; SM, scanning mirror.

into a half-meter-long PC fiber with an NA of 0.1 and a mode field diameter of 7 μm (LMA-8, Crystal Fibre A/S). The spectrally broadened OCM beam is collimated by an objective, and the linear polarization is rotated 90° by an achromatic half-wave plate before being recombined with the narrow-band MPM excitation beam at the polarizing beam splitter.

For OCM, the beam is split by a 50/50 beam splitter into the reference and sample arms of the interferometer. In the sample arm, the beam is expanded by a telescope and focused by a microscope objective (20×, 0.95 NA, water immersion, Olympus, Inc.) onto the sample. The sample is positioned on a motorized stage, which can translate the sample in three directions. A pair of galvanometers (Micromax 671, Cambridge Technology) positioned before the telescope scan the beam across the sample. The reference arm light is reflected by a plane mirror mounted on a translation stage. The spectral interference pattern of the reference and sample arm beams is detected for OCM acquisition by a spectrometer, which is based on a diffraction grating and CCD line camera (P2-22-02k40, Dalsa). The frame rate for the line camera depends on the speed and mode of image acquisition (galvanometer) and can be up to 35 kHz. OCM images are generated after several processing steps including compensating for unbalanced dispersion in the sample arm and for nonuniform distribution of the spectrum on the CCD due to nonlinearity of the diffraction grating (Graf and Boppart 2012; Graf et al. 2011).

For MPM, the narrower band excitation laser beam is combined with the OCM sample beam by the polarizing beam splitter. The epi-collected MPM fluorescence signal is diverted by a long-pass dichroic mirror and band-pass filtered. This filter is easily interchanged to detect various fluorescence or SHG signals. The fluorescence signal is detected by a photomultiplier tube (H7421, Hamamatsu). At the output of the interferometer, the polarizer (LPNIR050, Thorlabs) blocks the MPM portion of the spectrum. At 800 nm, the polarizer provides 30 dB attenuation of the MPM beam and 2 dB attenuation of the OCM beam. The power of the MPM and OCM sources can be independently controlled by a set of neutral density filters. The majority of the laser output is used for continuum generation because the degree of spectral broadening is dependent on the input power. The power of the narrow-band MPM reflected from the beam splitter ranges from 30 to 300 mW in the wavelength range 730–1000 nm. Imaging examples with this system are presented in later sections.

There are several other similar OCM–MPM systems reported in the literature. While the essential optical arrangement is similar, the primary difference is with the light source. For example, one MPM/OCT system was based on single 12 fs (100 nm bandwidth at 800 nm center wavelength) titanium:sapphire laser source (Tang et al. 2006). A similar system based on a sub-10-fs laser has been reported to capture coregistered volumetric images of corneal morphology and biochemistry to investigate response of cornea to intraocular pressure (Wu et al. 2011). Because single sources are used, which makes it possible to use one beam to serve as both the MPM excitation beam and OCM sample arm, these systems are simpler in construction. Due to the low wavelength tunability of the broadband femtosecond lasers, however, these single-source systems have to sacrifice some functionality of the imaging modalities. For example, MPM needs a tunable central wavelength to efficiently excite specific fluorophores that have a limited absorption wavelength range. OCM also needs to select different wavelengths to optimize for penetration depth (using longer wavelength) and spatial resolution (using shorter wavelength). Therefore, the previously described dual-band light source with wide tunability has the advantage in optimizing the performance of the two imaging modalities simultaneously.

10.3.4.3 Novel Optical Sources

One important feature of the OCM–MPM system detailed here is the dual-spectrum light source where SC generation in a PC fiber is used. SC generation from ultrashort laser pulses in a PC fiber with spectral bandwidths of up to hundreds of nanometers is an attractive light source for OCT (Wang et al. 2003). SC generation is a well-known nonlinear optical phenomenon that occurs when a high-intensity light pulse propagates in optical waveguides. SC generation in PC fiber is of particular interest, because it provides broader bandwidth with moderate input pulse energy and works over a large wavelength range, compared to silica single mode fibers. A typical PC fiber consists of a small, solid, pure-silica core surrounded by an array of microscopic air holes running along the entire length of the fiber. These holes effectively lower the index of refraction, creating a step-index optical fiber that can be varied according to the application. The large refractive index step between the silica core and the air-silica cladding allows light to be concentrated into a very small area, resulting in enhanced nonlinear effects. The design flexibility of PC fiber is very large, enabling zero dispersion to be varied as desired in different ranges. The nonlinear effects (Shen 2003), such as self-phase modulation, cross-phase modulation, four-wave mixing, and stimulated Raman scattering, which interact and determine the process of SC generation, can be partially controlled by appropriate choice of fiber. This allows one to select different center wavelengths, suppress spiky modulation, and enhance spectrum stability, which are important parameters for applications for OCT.

Ultrahigh-resolution OCT using SC generated in PC fibers has been extensively investigated. Axial resolutions of 2–3 μm or even submicron have been reported with these novel broadband sources (Lim et al. 2005; Povazay et al. 2002). The center wavelengths of these reported SC spectra range from the visible (e.g., 600 nm) to the near-infrared (around 1500 nm) regions, based on different pump sources and fiber designs (Aguirre et al. 2006; Povazay et al. 2002; Wang et al. 2007a). The availability of various wavelength ranges provides another freedom to choose desired wavelength for specific applications, such as shorter wavelengths for better resolution and longer wavelengths for deeper imaging. The drawbacks of SC from PC fiber include power fluctuations, spectral modulations, and additional noise, which need to be carefully considered and controlled when these sources are used for OCT.

10.4 Example Applications

10.4.1 Cellular Imaging Using OCT

Investigating the dynamic functions of cells is one of the fundamental topics in biology. In tissue engineering, cells, along with the scaffold and bioreactor (culture environment), compose the three basic elements required to form functional tissues. Cells play important roles in tissue development, including production of control signals to regulate the production of proteins that comprise the matrix, and directing the location and organization where protein synthesis and deposition must occur. Understanding the functions and responses of cells to the environment is a fundamental task that is crucial to the successful development of engineered tissues. Successful investigation of cell functions is facilitated or limited by the availability of enabling imaging tools (Smith et al. 2010). The desired imaging modality should offer high spatial resolution, noninvasive detection, and functional analysis capability. As a promising imaging technology, OCT has the potential for investigating cell dynamics. In this section, we discuss examples of the application of OCT in tissue engineering to demonstrate the unique features of OCT for cellular imaging.

In the field of tissue engineering, existing imaging techniques such as conventional light microscopy, confocal microscopy (Breuls et al. 2003; Dunkers et al. 2003; Pawley 2006), and micro-CT (Cartmell et al. 2004; Cioffi et al. 2006) are widely used tools for cellular imaging. These techniques, however, have intrinsic limitations when employed to study cell functions in engineered tissues. For example, light microscopy has limited imaging depth and low axial resolution, and thus requires all tissue and samples to be sectioned into thin slices of a few microns to be readily observed. The required cell and tissue fixation procedure for histological analysis with either light or electron microscopy always defines the endpoint of use for the samples and specimens (Rose et al. 2004). Confocal and two-photon microscopy normally need exogenous fluorescent labeling and have limited penetration depth, which can limit their use in *in vivo* investigations. Micro-CT is the micron-scale version of x-ray CT used extensively in medical fields. It provides precise quantitative and qualitative information on the 3D morphology of specimens, is nondestructive, and requires essentially no additional steps to prepare samples for scanning. However, ionizing x-ray radiation is used, which inevitably affects the viability and health of living specimens, and image contrast can be poor, depending on the composition of the sample.

As mentioned earlier, OCT performs noninvasive, label-free, real-time imaging with penetration depths on the order of several millimeters, depending on the optical properties of the sample. OCT has been explored as an alternative tool for the investigation of cell functions in engineered tissues. Not only does OCT have the potential for cellular-level spatial resolution, but it is also capable of monitoring the dynamics of cells in real time. Another reason why OCT is particularly suited for imaging cells in engineered tissues is because artificial scaffolds are usually present to support the growth of the cells and tissue. These scaffolds are commonly nonbiological materials that are normally difficult to label with biological markers and therefore are frequently difficult to observe with fluorescence microscopes. Even worse, in some cases with fluorescence imaging, autofluorescence from the scaffold material may interfere with visualizing the details of the cellular structures, and other contrast mechanisms must be used for better signal separation (Zhao et al. 2012). OCT, however, has the capability to visualize these structures based on the variation of refractive index within the structure, enabling the interaction of cells with the scaffold to be monitored, which is one of the key parameters in the development of a tissue. In addition, recently developed extensions of OCT, for example, Doppler OCT, SOCT, OCE, etc., have extended the functional capability of OCT, enabling OCT to study cellular functions in engineered tissue including cell morphology, proliferation, migration, and viability, among others.

In several early reports, OCT was shown to have sufficient resolution for cellular imaging. For example, high-resolution *in vivo* cellular and subcellular imaging of OCT has been demonstrated in a *Xenopus laevis* tadpole, as shown in Figure 10.3 (Boppart et al. 1998). This imaging ability has been extensively exploited to investigate cellular functions in development biology (Drexler et al. 1999). Three-dimensional volumes of high-resolution OCT data have been acquired from these specimens throughout development and were presented in this paper. From the 3D data set, cells undergoing mitosis were identified and tracked in three dimensions. In a similar manner, 3D data sets were acquired to track single melanocytes (neural crest cells) as they migrated through the living specimens. This result demonstrated the ability of OCT to characterize cellular processes such as mitosis and migration, which are central to many dynamic processes.

Higher spatial resolution on the order of 1–2 μm in three dimensions are enabled by OCM and full-field OCT, which use high NA objectives along with broadband light sources. The ultrahigh resolutions further enhance the cellular imaging performance of OCT. An example of cellular imaging in the *Xenopus laevis* animal model obtained with full-field OCT is shown in Figure 10.4 (Dubois et al. 2004). This figure shows the orthogonal sections extracted from a 3D data set acquired from the tadpole head. Cells are revealed with their membrane and nuclear morphology. Different developmental stages of cell mitosis can be observed in these images. Highly contrasted tissues appear such as the epidermis, an olfactory nerve, and neural crest melanocytes. These images demonstrate that full-field OCT has sufficient resolution to perform 3D cellular-level

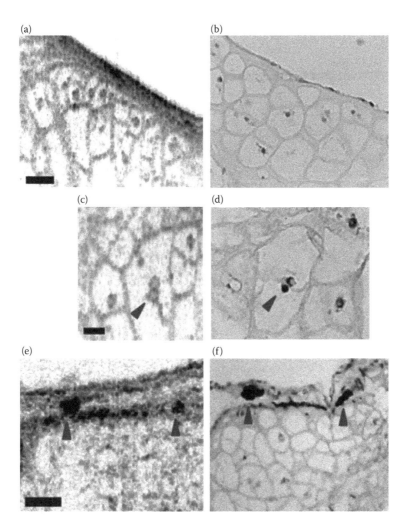

FIGURE 10.3 OCT images (a, c, and e) and corresponding histology (b, d, and f) of *Xenopus* mesenchymal stem cells. Mesenchymal cells of various size and nuclear-to-cytoplasmic ratios are shown in (a) and (b). Cells undergoing mitosis are observed in enlarged images (c and d), with visualization of discrete nuclei following cell division (indicated by arrows). Neural crest cells (melanocytes) presented as pigmented cells in the skin are shown in (e) and (f) (indicated by arrows). The resolutions afforded by OCT are sufficient for imaging and tracking cellular processes such as mitosis and migration in the *Xenopus* model. Scale bars are 50 μm. (Reproduced from Boppart, S. A. et al., *Nature Medicine* 4:861–865, 1998. With permission.)

imaging. More cellular imaging examples with OCM will be presented in Sections 10.4.3 and 10.4.4.

Another important feature of OCT for cellular imaging is the ability to visualize functional features through the use of functional variants of OCT. An example is SOCT analysis, which performs quantitative and qualitative measurements of spectral absorption and/or scattering properties in tissue. A typical study has been done on engineered tissues with fibroblasts and macrophage cells cultured in a 3-mm-thick porous chitosan scaffold. Representative OCT and SOCT images of fibroblast cells within this 3D scaffold are shown in Figure 10.5a and b, respectively, while OCT and SOCT images of macrophage cells are shown in Figure 10.5c and d, respectively. There are a few differences

Chapter 10

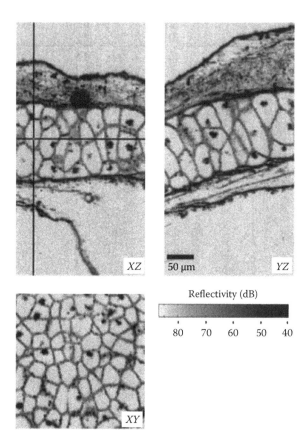

FIGURE 10.4 OCT cross-sectional images of the African frog tadpole *Xenopus laevis*, *ex vivo*, extracted from a stack of 900 *en face* tomographic images representing a volume of 240 × 240 × 450 μm (*X, Y, Z*). The black lines in the *XZ* section indicate the positions of the *XY* and *YZ* sections. (Reprinted from Dubois, A. et al., *Applied Optics* 43:2874–2883, 2004. With permission.)

between the OCT images of these two cell populations. However, there are distinct differences in the SOCT images. From these HSV-color reconstructed images, macrophage cells have more red-shifted centers, which correspond to narrower full-width 80% magnitude values of the spectral autocorrelation of the scattering signal. Fibroblast cells have more blue-shifted centers, which correspond to wider full-width 80% magnitude values of the spectral autocorrelation of the scattering signal. The spectroscopic differences are believed to be due to differences in the cell and organelle morphology between these two types of cells—specifically, differences in cell size, morphological features, nucleus size, and organelle distributions.

10.4.2 OCT of Cell Dynamics in Engineered Tissues

As stated earlier, OCT is a very promising technique for the evaluation of cellular functions and dynamics in engineered tissues (Mason et al. 2004; Matcher 2011; Smith et al. 2010; Xu et al. 2003). It offers noninvasive, high-resolution, 3D volumetric imaging capabilities, which is desirable for investigating cell functions in developing tissues. Cell dynamics, including proliferation, migration, detachment from the scaffold, and

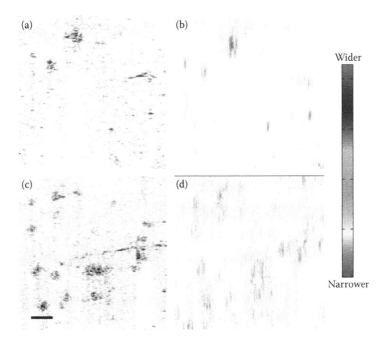

FIGURE 10.5 **(See color insert.)** OCT and SOCT images of two types of cells in engineered tissues. (a) Structural OCT image of fibroblasts in a 3D scaffold. (b) SOCT analysis of (a). (c) Structural OCT image of macrophages in a 3D scaffold. (d) SOCT analysis of (c). The SOCT images are constructed using the HSV color scale. Central color differences are noted for each cell type, representing differing degrees of optical scattering from cell size, shape, nuclei size, and organelle distribution. Scale bar is 50 μm.

interaction with materials, are important parameters indicating the properties of the tissue development and culture environments. Investigation of these cell properties with OCT has been demonstrated in recent years (Mason et al. 2004; Rey et al. 2009; Tan et al. 2006; Xu et al. 2003). These results show the advantages of OCT for the real-time investigation of cell dynamics and activities in developing tissues and enable researchers to gain increasing insight into the developmental process of tissues.

Figure 10.6 shows OCT images of a typical engineered tissue at different developing time points (Tan et al. 2004). The sample is composed of cells, their extracellular matrices, and the 3D scaffolds. Cells in this engineered tissue model are GFP-vinculin-expressing NIH 3T3 cells (American Type Culture Collection, Manassas, VA), and the scaffolds are made of porous chitosan (Tan et al. 2004). The OCT images are acquired from the engineered tissues after 1, 3, 5, and 7 days of culture. A time-domain OCT system with an A-scan rate of 30 Hz was used to obtain these images. The sample arm utilized an achromatic lens and provides 10 μm transverse resolution. The axial resolution of the system was 3 μm, determined by the bandwidth of more than 100 nm generated from an ultrahigh NA fiber pumped by 90 fs laser pulses. Corresponding histology results are also showed in this figure for comparison.

Chitosan scaffolds were prepared according to the following procedure: 2% (wt%) chitosan flakes (Sigma, St. Louis, MO) were dissolved in a 0.2 M acetic acid aqueous solvent. The resulting viscous solution was filtered and transferred to cylindrical molds.

Chapter 10

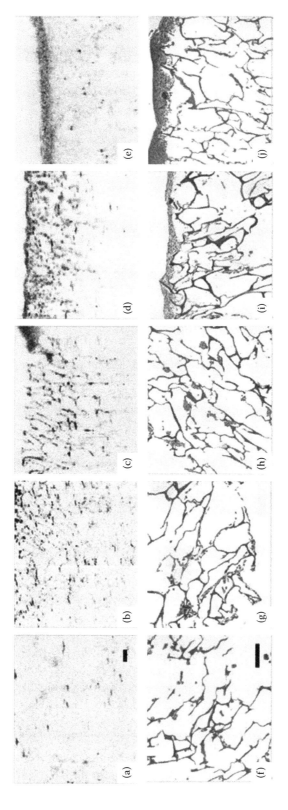

FIGURE 10.6 Two-dimensional OCT images of an engineered tissue (a–e) and their corresponding histological images (f–j) under different stages of development: day 1 (a and f), day 3 (b and g), day 5 (c and h), day 7 (d and i), and day 9 (e and j). Scale bars are 100 μm. (Reprinted from Tan, W. et al., *Tissue Engineering* 10:1747–1756, 2004. With permission.)

Cylinders were frozen at a rate of 0.3°C/min from room temperature to −20°C. Frozen cylinders (8 mm in diameter and 3 mm in height) were lyophilized for 4 days for complete removal of the solvent. The resulting sponges had an interconnected porosity of >80% with an average pore size of 100 μm. Scaffolds were sterilized with 80% ethanol and then rehydrated through a series of ethanol/phosphate-buffered saline (PBS) solutions (70%, 50%, and 0% ethanol), and finally transferred to the final culture medium for overnight hydration. Seeding was done by plating 50 μL of a concentrated cell suspension on the scaffolds at a seeding density of 5×10^6 cells/cm³. Cells were transfected with a green fluorescent protein (GFP)–vinculin plasmid, forming a stable cell line that expressed GFP–vinculin. For transfection, 7×10^5 cells were seeded on each well of a 6-well plate. For each well of cells, 4 μg of DNA and 2 μL of Lipofectamine 2000 reagent were diluted separately with 50 μL of FreeStyle 293 expression medium and then mixed and cultured for 20 min to form DNA–LF2000 complexes. The concentration of cells and DNA was determined in pilot experiments to be optimal for transfection efficiency.

Several stages of engineered tissue development are clearly witnessed from these OCT images, showing the cell proliferation and migration in the tissue. Cell distribution can be seen in the OCT images, although the fine structure of the chitosan is less clear in contrast to conventional histology images, which is attributed to the thin cross-sectional dimensions (typically less than 5 μm) of the scaffold wall and the low backscattering optical property of the material in the absence of cells. As shown in Figure 10.6a, cells initially appear as small, highly backscattering regions, which are attached to the scaffold. Cells and the extracellular matrix are readily apparent compared with the low backscattering chitosan scaffold. Figure 10.6b and c shows that the cells are uniformly distributed throughout the scaffold after 3–5 days in culture (high cell viability deep within the tissue), and that the cell number and matrix density increase with culture time. Figure 10.6d and e shows an uneven distribution in the engineered tissues after 7–9 days in culture. Cells and matrices are dense in the first 100–200 μm, forming a more highly scattering layer near the tissue surface. Cells and deposited extracellular matrix are less abundant deep within the scaffold. It is important to note that all the OCT images were obtained in a noninvasive manner, without any interference with the development of the tissue during the investigation.

An example of 3D volumetric visualization of cell dynamics with OCT is shown in Figure 10.7 (Tan et al. 2006). Figure 10.7 demonstrates the 3D positional changes of the cells in the tissue model over time. Cell positions at each specific time point were labeled with a specific color. Cell migration direction and velocity could be readily obtained from these OCT images. The migration speed in this experiment was approximately 0.67 μm/min (20 μm per 30 min), which correlated with that reported in the literature (Dimilla et al. 1992; Steller 1995). The migration speed was uniform throughout the 2 mm detection depth. By combining the time-lapse 3D images into composite images (Figure 10.7g and h), it was possible to track the paths traveled by individual cells and determine the distribution of phenomenological parameters, including cell speed. The cell migration speed decreased after 2 h, but the cell density in the gel had increased, which may have been a contributing factor.

Tissue models are composed of cells (3T3 fibroblasts or mouse macrophages), matrix scaffolds, or synthetic polymer scaffolds. Gel-based tissue models were prepared by

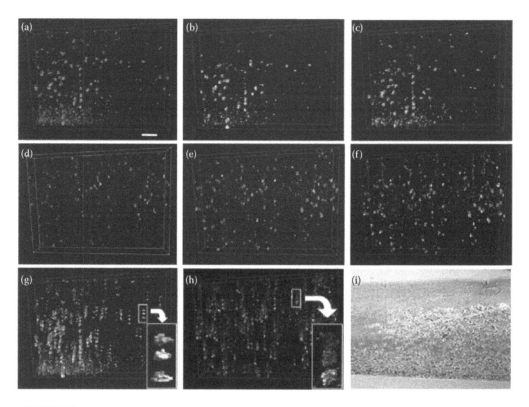

FIGURE 10.7   3D OCT images demonstrate the migration of macrophages (a–f). The interval time is 40 min between (a/b, b/c, d/e, and e/f) and 120 min between (c/d). Individual OCT images are merged to form composite images of individual cell migration in 3D space (g, h). Composite (g) is composed of (a–c) and composite (h) is composed of (d–f). Insets in (g, h) show zoom-in single-cell migration. Corresponding histology after the study is shown in (i), with macrophages collecting at the bottom. Scale bar is 200 μm. (Reprinted from Tan, W. et al., *Optics Express* 14:7159–7171, 2006. With permission.)

mixing cell suspensions with a thawed Matrigel solution (BD Bioscience, Bedford, MA) using 1:1 proportion, and then solidified in a 37°C incubator. Other components such as collagen I (BD Bioscience, Bedford, MA) were added to the gel with appropriate weight proportions to study the effects of matrix components on the OCT images. Cell and tissue cultures were maintained in an incubator at 37°C and with 5% CO_2. For real-time imaging over extended time periods (days), tissue cultures were maintained in a portable microincubator (LU-CPC, Harvard Apparatus, Holliston, MA) that was placed on the microscope stages.

High-resolution OCT is capable of detecting other types of cell dynamics, such as detachment and migration. Figure 10.8 shows the process of cells detaching from a calcium–phosphate scaffold. Time-lapse images demonstrate different detachment processes in an engineered tissue under short-term culture (3 days, Figure 10.8a) from that under long-term culture (10 days, Figure 10.8b). Comparing images at different points of time, the loss of optical scattering, which implies the loss of cells on the scaffold, could be assessed. This tissue exhibited a gradual decline in the optical scattering between 10 and 16 min, which suggests that cells detached from the scaffold slowly over this time.

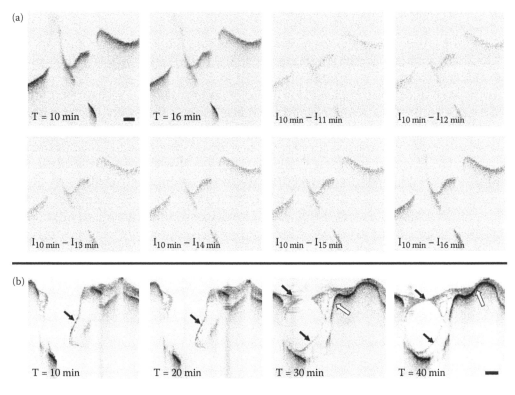

FIGURE 10.8 Time-lapse images showing the process of cell detachment and cell layer movement (black arrows) from a calcium–phosphate scaffold. Cells (3T3 fibroblasts) were cultured on the scaffolds for (a) 3 days and (b) 10 days. Continuous images were taken every minute after engineered tissues were soaked in a trypsin solution. (a) OCT images acquired at 10 and 16 min after placement in trypsin solution. No detectable change was noted before 10 min. Difference images were calculated during cell de-adhesion by subtracting images acquired at later time points. (b) Time-lapse images showing the process of cell layer de-adhesion (black arrows) from the scaffold (white arrows). Detachment of the cell-layer sheet was abrupt between 20 and 30 min. Scale bars represent 200 μm. (Reprinted from Tan, W. et al., *Optics Express* 14:7159–7171, 2006. With permission.)

These images were obtained with a fiber-based OCT system, which use a Nd:YVO$_4$-pumped titanium:sapphire laser as a broad-bandwidth optical source that produced femtosecond pulses with an 80 MHz repetition rate at an 800 nm center wavelength. Laser output was coupled into an ultrahigh-numerical-aperture fiber (UHNA4, Thorlabs, Inc.) to spectrally broaden the light from 20 nm to more than 100 nm, generating an axial resolution of 3 μm for our system (Marks et al. 2002). The transverse resolution was estimated to be 10 μm, which was the full-width at half-maximum of an isolated subresolution scatter.

As shown by the examples above, OCT is appropriate for investigating the cell dynamics, including larger populations of cells, in engineered tissues. Although OCT has less resolution and contrast compared to histology, its noninvasive imaging enables real-time monitoring of cell properties over time during tissue development. Information on the complex dynamics of cells during tissue development will benefit our understanding and allow for better control of the development parameters, facilitating optimized engineered tissue fabrication and function.

Chapter 10

10.4.3 Multimodal Imaging of Cells in Engineered Tissues

Although the previous section demonstrated the potential of OCT for cellular imaging, for many applications, the spatial resolution of OCT is insufficient for visualizing finer cellular structures. To provide higher spatial resolution, OCM, the high-resolution variation of OCT, can be considered. Furthermore, as mentioned in Section 10.3.4, the combination of MPM and OCM can provide different yet complementary information deep in high-scattering tissues. For imaging engineered tissues, in which artificial materials consist of scaffolds, extracellular matrix, and living cells mixed together, the multifunctional imaging modalities have unique advantages. For instance, OCM not only provides structural information about artificial substrates, scaffolds, and the extracellular matrix within which cells develop and interact, but also shows cell morphology and cell populations. Simultaneously, MPM offers complementary information on cell function as well as subcellular structures. All the coregistered information will enhance our understanding of cell physiology and facilitate the optimization of a successful culture of artificial tissue equivalents for regenerative medicine.

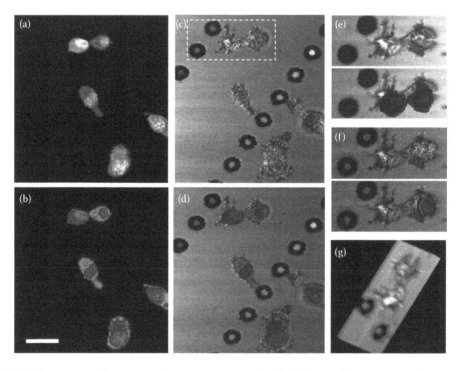

FIGURE 10.9 Functional interactions between GFP-vinculin fibroblasts and a microtextured (micropeg) substrate. Images were acquired under static culture conditions. (a) Combined and (b) separated MPM signals. The fluorescence signal from GFP-vinculin is proportional to the degree of cell adhesions with other cells and the substrate. (c) OCM image showing backscattering signal from both cells and microtextured substrate. (d) Multimodal image of MPM and OCM data showing distinct spatial relationships between cells and substrate features. (e, f) Regions of OCM and combined OCM–MPM data indicated by box in (c) but acquired at different *en face* planes 6 and 12 μm, respectively, above the plane in (c). (g) 3D reconstruction of OCM data set showing cellular morphology on the substrate. Scale bar is 20 μm. (Reprinted from Tan, W. et al., *Microscopy Research and Technique* 70:361–371, 2007. With permission.)

An example of using our integrated MPM/OCM system to image cells in artificial tissue is shown in Figure 10.9 (Tan et al. 2007). The samples under investigation are GFP-vinculin transfected 3T3 fibroblasts seeded on top of a 3D topographically structured polydimethylsiloxane (PDMS) substrate. The microstructured substrate consisted of rows of 10 μm high and 10 μm diameter micropegs, separated by 50 μm between rows. It is seen from these images that OCM is able to resolve both the structures of the microtextured scaffold and the shapes of the cells. Since only back-reflected light is used to generate the OCM images, this obviates the need for fluorescent labeling of the substrate, which is problematic and inevitably affects the long-term viability of the cells. The cell morphology is much clearer in the MPM images, where the fluorescence intensity in cells varies corresponding to the relative degree of substrate adhesion or cell–cell interactions. MPM, however, cannot visualize the position of the cells relative to the scaffold and thus fails to monitor the cell migration when used alone. The problem is well resolved with complementary information provided by OCM, which is shown in the coregistered MPM–OCM images. Using GFP-vinculin-labeled adhesion sites and Hoechst-stained nuclei, the combined MPM–OCM images demonstrate the relative spatial distribution of these signals within individual cells and among these cell populations. It is clear from the combined MPM–OCM images that the fluorescent signals are generated from only portions of the cell, where cell–cell or cell–substrate adhesions are present and where GFP-vinculin is expressed. This method therefore allows the characterization of locally active adhesion sites and provides spatial and temporal variations of cell function.

Complementary structural and functional information provided by the system makes it possible to monitor the cell dynamics and physiology under mechanically stimulated culture conditions, which has been widely accepted as an important control parameter for engineered tissue. Figure 10.10 shows MPM–OCM images of fibroblasts after 18 h of 5% cyclic equibiaxial sinusoidal stretching at a frequency of 1 Hz. Compared to the cells on these microtextured substrates before mechanical stretching, the cells subjected to

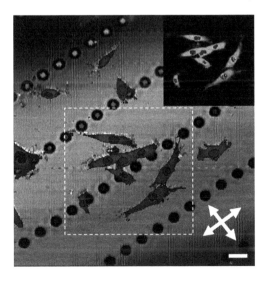

FIGURE 10.10 Integrated OCM–MPM imaging of GFP-vinculin fibroblasts seeded on a microtextured substrate under dynamic culture. The white arrows indicate the directions of stretching. Scale bar is 50 μm.

Chapter 10

mechanical stimuli are more elongated, have a high level of vinculin expression based on the intensity and distribution of fluorescence from the GFP-vinculin, and have an apparent increase in cell–substrate interactions.

Furthermore, the 3D imaging capability enables the system to dynamically study cell–scaffold interactions under more physiological 3D culture conditions. This multimodal technique provides the capability for noninvasive imaging at the molecular, cellular, and tissue level in 3D. It is believed that the mechanical deformation of the biopolymer fibers to which cells attach induced alterations in cell morphology, physiology, and gene expression. The cells cultured under static 3D conditions exhibit elongated morphology and 3D network structure. Cells are oriented and extend isotropically in 3D space, following the random orientation of the matrix structure. After uniaxial mechanical stretching of the 3D constructs, the cells and matrices are densely packed and aligned parallel with the stretched axial direction.

In summary, the combination of OCM and MPM gives complementary information on both the structure and dynamics of cells in engineered constructs, which is not available with either modality alone. This multimodality approach provides another promising option for investigation of tissue development *in situ*.

10.4.4   *In Vivo* Multimodal Imaging of Skin

One potential clinical application of integrated OCM–MPM is in imaging skin for aiding in the diagnosis of various skin diseases and conditions. The multimodal, high-resolution images allow various features of the skin to be assessed. Figure 10.11 is an example of OCM–MPM images acquired from *in vivo* human skin at different depths. In the top layer of the skin, the stratum corneum, the surface topology is visible. In the epidermis, the cell nuclei of keratinocytes appear as dark spots in both the OCM and MPM images. In the superficial dermis layer, the dermal papillae are visible. These structures are projections of the dermis into the epidermis.

One limitation of high-resolution microscopy is that the field of view is inherently limited to several hundred micrometers. For diagnostic applications, it is highly desirable to image wide fields of view to enable direct comparison of different areas and to visualize macroscopic as well as microscopic features of the skin. Using a motorized stage, it is possible to acquire images at various locations of a skin sample and stitch them together into a wide-area mosaic. An example of an OCM–MPM mosaic from the epidermis of *in vivo* human skin is shown in Figure 10.12. While high resolution is maintained, the larger area shown facilitates interpretation of the images.

10.4.5   *In Vivo* Multimodal Imaging and Tracking of Bone-Marrow Derived Cells in Skin

One primary advantage of integrated OCM–MPM is that thick, intact tissue can be noninvasively visualized, enabling repeated imaging of a sample to observe dynamic changes over time. MPM can be utilized to visualize and track fluorescently labeled cells, while OCM visualizes structural features of the tissue environment. The ability to observe single cells and their dynamic interactions within the tissue environment has great potential to improve our understanding of fundamental aspects of cell biology. One

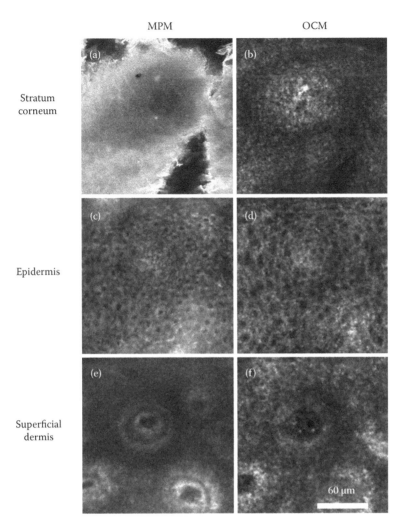

FIGURE 10.11 High-resolution MPM and OCM images from *in vivo* human skin. *En face* MPM and OCM images are from the stratum corneum (a, b), the epidermis (c, d), and the superficial dermis (e, f).

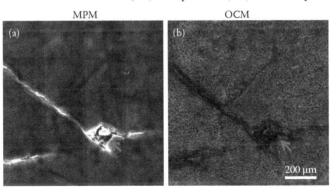

FIGURE 10.12 Wide area (a) MPM and (b) OCM mosaics from *in vivo* human skin from the epidermis. Mosaics increase the field of view and allow larger skin features, such as hair follicles (indicated by arrows), to be visualized. Each mosaic image is composed of 16 stitched individual images consisting of 256 × 256 pixels and obtained in 3 s.

Chapter 10

potential application of integrated OCM–MPM is in studying the role of bone marrow–derived cells in skin. It is well known that these cells play a key role in immune processes in the skin. However, several recent studies have suggested that bone marrow–derived cells found in the skin also contain stem cells that contribute to tissue regeneration after wounding. The role of these bone marrow–derived cells is not fully characterized and is thus an active topic of research. The ability to directly observe the dynamic behavior of bone marrow–derived cells in the steady state as well as in response to wounding of the skin could lead to a better understanding of the role of these cell types. This could, in turn, lead to better treatments for a wide range of medical conditions.

To enable tracking of bone marrow–derived cells in live skin, a chimeric mouse model was utilized. This animal model is based on transplanting bone marrow from transgenic GFP mice into species-matched wild-type hosts. This procedure results in mice whose bone marrow cells (and any cells derived from the bone marrow) are fluorescent. Several weeks following successful engraftment of the GFP bone marrow, fluorescent cells are present in the skin of these mice. To enable imaging of the skin, the mice are anesthetized using isofluorane gas. A specially designed imaging mount is utilized to stabilize the skin and minimize motion artifacts.

Integrated OCM and MPM allow the bone marrow–derived cells to be visualized within their structural environment in the skin. Figure 10.13 shows OCM and MPM images of the skin of a successfully transplanted mouse. The OCM image in this example (Figure 10.13a) shows the location of a hair follicle. The MPM image (Figure 10.13b) shows a cluster of GFP cells present as well as autofluorescence from the hair. Overlaying the MPM and OCM images (Figure 10.13c) allows the cells to be directly visualized in their physical environment. Large areas of skin can be visualized by acquiring OCM–MPM images from different transverse positions and stitching them together into a mosaic.

Dynamic imaging of the bone marrow–derived cells in the skin is challenging because the window of time that a mouse can be repeatedly imaged is limited by practical considerations to several hours. However, the dynamics of wound healing take place over

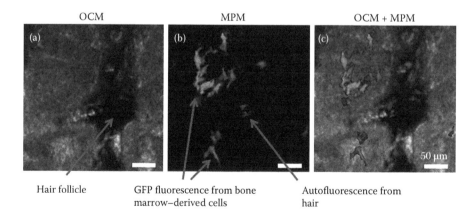

FIGURE 10.13 OCM and MPM images of the skin from a mouse that had received a bone marrow transplant from a GFP donor. (a) OCM image formed based on the scattering signals. (b) MPM image showing GFP fluorescence from bone marrow–derived cells and autofluorescence from hair. (c) Overlay of the OCM and MPM images. (Reproduced from Graf, B. W. et al., *Proceedings of SPIE* 7902:790206, 2011. With permission.)

a period of several weeks. Thus, it is necessary to develop methods for long-term, time-lapse imaging. To image the same location of the skin during different imaging sessions, an area of interest was marked by carefully removing hair using a pair of surgical scissors. This also serves to minimize autofluorescence background. This marking allows approximate alignment of the skin at each imaging session. Since skin is highly flexible, it is also necessary to correct for nonrigid deformations of the skin between different sessions. This is accomplished using a registration algorithm that uses the positions of the hair follicles to correct for the deformations. This software is a plugin for ImageJ entitled bUnwarpJ (Arganda-Carreras 2006). The positions of corresponding hair follicles between two OCM images are manually identified as shown in Figure 10.14. These images were acquired from the same area of mouse skin during different imaging sessions 24 h apart. The second image is warped to match the spatial orientation of the first image. The original and warped OCM images, as well as the grid deformation, are shown in Figure 10.15. Correcting the nonrigid deformation allows accurate comparison of the same area of skin over a long period of time.

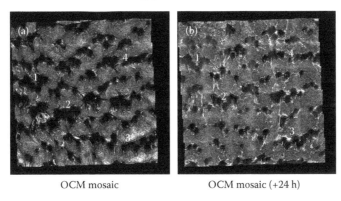

FIGURE 10.14   OCM mosaics of the same region of live mouse skin taken during (a) one imaging session and (b) during a second imaging session 24 h later. Numbers correspond to matching hair follicles. (Reproduced from Graf, B. W. et al., *Proceedings of SPIE* 7902:790206, 2011. With permission.)

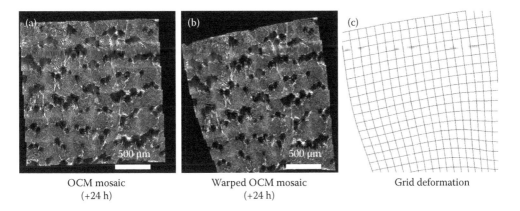

FIGURE 10.15   Warping of the OCM mosaics to enable coregistration of time lapse images. (a) Original OCM mosaic from the second imaging session. (b) The same OCM mosaic after warping based on the landmarks identified in Figure 10.14. (c) Grid deformation between the first and second imaging session. (Reproduced from Graf, B. W. et al., *Proceedings of SPIE* 7902:790206, 2011. With permission.)

Chapter 10

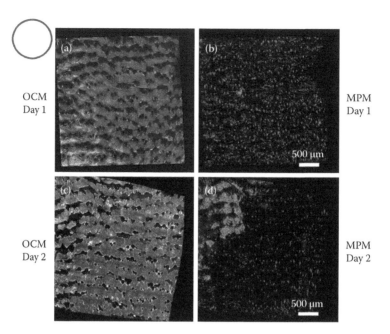

FIGURE 10.16 OCM–MPM mosaic images tracking bone marrow–derived cells in live mouse skin in response to wound healing. (a) OCM and (b) MPM mosaics from the prewound imaging session. Warped (c) OCM and (d) MPM mosaics of the same location 24 h after a superficial skin wound. The circle indicates the approximate region of the skin where the wound was located (which was outside of the area shown in the images) with respect to (a). (Reproduced from Graf, B. W. et al., *Proceedings of SPIE* 7902:790206, 2011. With permission.)

To demonstrate the potential to track bone marrow–derived cells in live mouse skin in response to wound healing, OCM–MPM mosaics were acquired at two time points before and after wounding of the skin. The wound was a superficial laceration that was made outside of the field of view of the images. Figure 10.16 shows the OCM–MPM mosaics taken before and after the wound. Before the wound, the GFP cells are relatively uniformly distributed throughout the area imaged. Following the wound, there is an apparent migration of the cells toward the wound site. This results in a high density of GFP cells in the area close to the wound. The GFP cells that have migrated to the wound site are most likely immune cells. Future work will attempt to track the GFP cell populations over several weeks to determine their role in wound healing and regeneration.

10.5 Conclusions

As the unique capabilities of OCT have been widely recognized, OCT has become an established medical imaging technique. OCT utilizes low-coherence interferometry to differentiate light signals reflected from different depths within the sample and thus generates cross-sectional images of the sample. The high spatial resolution and 3D imaging capability along with the noninvasive imaging nature of OCT enable *in vivo* imaging of intact living tissues without tissue excision or sectioning. While these advantageous features of OCT were quickly appreciated, there has been tremendous progress in the development of OCT in recent years. The working principle and recent advances have been reviewed in this chapter and Chapter 11 in this book.

Intrinsically, the working principles of OCT make this imaging technique advantageous in many aspects. First, the axial or depth resolution is determined by the coherence length of the light source and is not limited by the focusing condition of the imaging system. This feature enables one to improve the axial resolution of OCT by introducing advanced broadband light sources, free from limitations associated with the geometrical setup of the system. With many technological advances to leverage, many types of broadband light sources, including ultrashort solid-state lasers, multiplex technology–based SLDs, and SC generation in nonlinear fibers, have emerged as attractive OCT light sources. These broadband sources, with bandwidths up to several hundreds of nanometers, allow OCT imaging to achieve axial resolutions of about 1 μm, depending on the center wavelength. Second, the introduction of spectral domain OCT significantly improved the imaging speed of OCT because reflections of light from all depths along an axial scan could be detected by the spectrometer, rather than mechanically scanning the reference arm path length delay. While the combination of a broadband source and a spectrometer in SD-OCT can achieve A-scan rates of a few hundred kilohertz, which is limited by the acquisition speed of the available line scan cameras, recently developed swept source OCT systems have an imaging speed up to several megahertz. These advances have enabled OCT to be more widely applied in the clinical fields of ophthalmology, dermatology, and cardiology, among many others.

In the field of tissue engineering, OCT as a noninvasive, high-resolution, real-time imaging modality has the potential to monitor cellular dynamics and functions during the development of various types of engineered tissues. The imaging examples shown in this chapter demonstrate the ability of OCT to differentiate 3D microstructure and morphology during engineered tissue development and to detect and track cell dynamics including migration, proliferation, detachment, and cell–material interactions. Compared to traditional histological processes, OCT imaging provides features for real-time, noninvasive monitoring to investigate engineered tissues. Compared to other imaging modalities such as MRI, ultrasound, microCT, and confocal microscopy, OCT offers important advantages and provides high resolution with deep imaging penetration in highly scattering engineered tissues. Imaging examples of OCM–MPM in *in vivo* skin and stem cell tracking show its potential for clinical applications.

Unique extensions of OCT such as integrated OCM–MPM have also been successfully applied to imaging engineered tissues. OCM integrated with MPM was used to image *en face* 3D structural and functional information from various engineered tissues and cell-seeded substrates. Functional extensions such as SOCT were used to differentiate two cell types in engineered tissues. Functional OCT techniques provide more information on the state of engineered tissue development. These imaging methods and results are likely to provide critical data for tissue-engineering studies and serve as powerful tools for investigating the complex changes that occur during tissue organization and development.

Despite these advantages, further improvements can be made for OCT imaging. First, due to the typical resolutions achieved in OCT (5–10 μm), it is difficult to differentiate individual cell morphology in OCT images. Although OCM offers better resolution by using high NA objectives, its penetration is, in turn, limited to several hundred microns. It is important to improve the spatial resolution of OCT while maintaining acceptable imaging depths for future work. This could be achieved with advances from technologies that extend the DOF, such as ISAM. The second issue is the relatively low contrast of OCT based on only the intrinsic variation of the refractive index. Development and application of high contrast but low toxicity contrast agents could be considered to improve contrast and specificity of OCT detection, such as with MM-OCT. Finally, many of the functional extensions of OCT, such as PS-OCT, NIVI, and OCE, have not been extensively exploited for imaging engineered tissues. Additional effort may be put into in these areas, which will only provide further insight into the complex dynamics and processes of engineered tissue development for regenerative medicine applications.

Acknowledgments

We thank Dr. Wei Tan and Dr. Claudio Vinegoni, former researchers in the Biophotonics Imaging Laboratory, for their significant contributions to the results presented in this chapter. We also thank Dr. Haohua Tu and Eric Chaney for their laboratory assistance with our optical imaging systems and our biological resources. This research was supported in part by grants from the National Science Foundation (CBET 08-52658 ARRA, CBET 10-33906).

Recent updates and extended work on the topic of this chapter can be found at http://biophotonics.illinois.edu.

References

Adie, S. G., and S. A. Boppart. 2010. Optical coherence tomography for cancer detection. In *Optical Imaging of Cancer*. ed. E. Rosenthal and K. R. Zinn, 209–250. New York: Springer.

Adie, S. G., X. Liang, B. F. Kennedy et al. 2010. Spectroscopic optical coherence elastography. *Optics Express* 18:25519–25534.

Adie, S. G., B. W. Graf, A. Ahmad, P. S. Carney, and S. A. Boppart. 2012. Computational adaptive optics for broadband optical interferometric tomography of biological tissue. *Proceedings of the National Academy of Science USA* 109:7175–7180.

Adler, D. C., T. H. Ko, P. R. Herz, and J. G. Fujimoto. 2004. Optical coherence tomography contrast enhancement using spectroscopic analysis with spectral autocorrelation. *Optics Express* 12:5487–5501.

Aguirre, A. D., P. Hsiung, T. H. Ko, I. Hartl, and J. G. Fujimoto. 2003. High-resolution optical coherence microscopy for high-speed, *in vivo* cellular imaging. *Optics Letters* 28:2064–2066.

Aguirre, A. D., N. Nishizawa, J. G. Fujimoto et al. 2006. Continuum generation in a novel photonic crystal fiber for ultrahigh resolution optical coherence tomography at 800 nm and 1300 nm. *Optics Express* 14:1145–1160.

Akiba, M., and K. P. Chan. 2007. *In vivo* video-rate cellular-level full-field optical coherence tomography. *Journal of Biomedical Optics* 12:064024.

Antcliff, R. J., M. R. Stanford, D. S. Chauhan et al. 2000. Comparison between optical coherence tomography and fundus fluorescein angiography for the detection of cystoid macular edema in patients with uveitis. *Ophthalmology* 107:593–599.

Arganda-Carreras, I. 2006. Lecture notes in computer science. *CVAMIA: Computer Vision Approaches to Medical Image Analysis* 4241:85–95.

Baumgartner, A., C. K. Hitzenberger, H. Sattmann, W. Drexler, and A. F. Fercher. 1998. Signal and resolution enhancements in dual beam optical coherence tomography of the human eye. *Journal of Biomedical Optics* 3:45–54.

Benalcazar, W. A., and S. A. Boppart. 2011. Nonlinear interferometric vibrational imaging for fast label-free visualization of molecular domains in skin. *Analytical and Bioanalytical Chemistry* 400: 2817–2825.

Benalcazar, W. A., P. D. Chowdary, Z. Jiang et al. 2010. High-Speed Nonlinear Interferometric Vibrational Imaging of Biological Tissue With Comparison to Raman Microscopy. *IEEE Journal of Selected Topics in Quantum Electronics* 16:824–832.

Blatter, C., B. Grajciar, C. M. Eigenwillig et al. 2011. Extended focus high-speed swept source OCT with self-reconstructive illumination. *Optics Express* 19:12141–12155.

Boppart, S. A., G. J. Tearney, B. E. Bouma et al. 1997. Noninvasive assessment of the developing *Xenopus* cardiovascular system using optical coherence tomography. *Proceedings of the National Academy of Science USA* 94:4256–4261.

Boppart, S. A., B. E. Bouma, C. Pitris et al. 1998. *In vivo* cellular optical coherence tomography imaging. *Nature Medicine* 4:861–865.

Boppart, S. A., A. L. Oldenburg, C. Y. Xu, and D. L. Marks. 2005. Optical probes and techniques for molecular contrast enhancement in coherence imaging. *Journal of Biomedical Optics* 10:41208.

Bouma, B. E., S. H. Yun, G. J. Tearney et al. 2006. Comprehensive volumetric optical microscopy *in vivo*. *Nature Medicine* 12:1429–1433.

Breuls, R. G. M., A. Mol, R. Petterson et al. 2003. Monitoring local cell viability in engineered tissues: A fast, quantitative, and nondestructive approach. *Tissue Engineering* 9:269–281.

Brezinski, M. E., G. J. Tearney, N. J. Weissman et al. 1997. Assessing atherosclerotic plaque morphology: Comparison of optical coherence tomography and high frequency intravascular ultrasound. *Heart* 77:397–403.

Bruckner, A. P. 1978. Picosecond light scattering measurements of cataract microstructure. *Appllied Optics* 17:3177–3183.

Campagnola, P. J., and L. M. Loew. 2003. Second-harmonic imaging microscopy for visualizing biomolecular arrays in cells, tissues and organisms. *Nature Biotechnology* 21:1356–1360.

Cartmell, S., K. Huynh, A. Lin, S. Nagaraja, and R. Guldberg. 2004. Quantitative microcomputed tomography analysis of mineralization within three-dimensional scaffolds *in vitro*. *Journal of Biomedical Materials Research A* 69A:97–104.

Cense, B., T. C. Chen, B. H. Park, M. C. Pierce, and J. F. De Boer. 2004. Thickness and birefringence of healthy retinal nerve fiber layer tissue measured with polarization-sensitive optical coherence tomography. *Investigative Ophthalmology and Visual Science* 45:2606–2612.

Chang, S., X. P. Liu, X. Y. Cai, and C. P. Grover. 2005. Full-field optical coherence tomography and its application to multiple-layer 2D information retrieving. *Optics Communications* 246:579–585.

Chen, X. D., Q. Li, Y. Lei, Y. Wang, and D. Y. Yu. 2010. Approximate wavenumber domain algorithm for interferometric synthetic aperture microscopy. *Optics Communications* 283:1993–1996.

Chen, Z. P., T. E. Milner, X. J. Wang, S. Srinivas, and J. S. Nelson. 1998. Optical Doppler tomography: Imaging *in vivo* blood flow dynamics following pharmacological intervention and photodynamic therapy. *Photochemistry and Photobiology* 67:56–60.

Chen, Z., Y. H. Zhao, S. M. Srinivas et al. 1999. Optical Doppler tomography. *IEEE Journal of Selected Topics in Quantum Electronics* 5:1134–1142.

Cheng, J. X., Y. K. Jia, G. F. Zheng, and X. S. Xie. 2002. Laser-scanning coherent anti-Stokes Raman scattering microscopy and applications to cell biology. *Biophysical Journal* 83:502–509.

Chinn, S. R., E. A. Swanson, and J. G. Fujimoto. 1997. Optical coherence tomography using a frequency-tunable optical source. *Optics Letters* 22:340–342.

Choma, M. A., M. V. Sarunic, C. H. Yang, and J. A. Izatt. 2003. Sensitivity advantage of swept source and Fourier domain optical coherence tomography. *Optics Express* 11:2183–2189.

Choma, M. A., K. Hsu, and J. A. Izatt. 2005. Swept source optical coherence tomography using an all-fiber 1300-nm ring laser source. *Journal of Biomedical Optics* 10:44009.

Chowdary, P. D., Z. Jiang, E. J. Chaney et al. 2010. Molecular histopathology by spectrally reconstructed nonlinear interferometric vibrational imaging. *Cancer Research* 70:9562–9569.

Cimalla, P., J. Walther, M. Mehner, M. Cuevas, and E. Koch. 2009. Simultaneous dual-band optical coherence tomography in the spectral domain for high resolution *in vivo* imaging. *Optics Express* 17:19486–19500.

Cioffi, M., F. Boschetti, M. T. Raimondi, and G. Dubini. 2006. Modeling evaluation of the fluid-dynamic microenvironment in tissue-engineered constructs: A micro-CT based model. *Biotechnology and Bioengineering* 93:500–510.

Colston, B. W., M. J. Everett, L. B. Da Silva et al. 1998. Imaging of hard- and soft-tissue structure in the oral cavity by optical coherence tomography. *Applied Optics* 37:3582–3585.

Costa, R. A., M. Skaf, L. A. S. Melo et al. 2006. Retinal assessment using optical coherence tomography. *Progress in Retinal and Eye Research* 25:325–353.

Davis, B. J., D. L. Marks, T. S. Ralston, P. S. Carney, and S. A. Boppart. 2008. Interferometric synthetic aperture microscopy: Computed imaging for scanned coherent microscopy. *Sensors* 8:3903–3931.

De Boer, J. F., S. M. Srinivas, A. Malekafzali, Z. P. Chen, and J. S. Nelson. 1998. Imaging thermally damaged tissue by polarization sensitive optical coherence tomography. *Optics Express* 3:212–218.

De Boer, J. F., B. Cense, B. H. Park et al. 2003. Improved signal-to-noise ratio in spectral-domain compared with time-domain optical coherence tomography. *Optics Letters* 28:2067–2069.

Denk, W., J. Strickler, and W. Webb. 1990. Two-photon laser scanning fluorescence microscopy. *Science* 248:73–76.

Dimilla, P. A., J. A. Quinn, S. M. Albelda, and D. A. Lauffenburger. 1992. Measurement of individual cell-migration parameters for human tissue-cells. *AIChE Journal* 38:1092–1104.

Drexler, W., and J. G. Fujimoto. 2008a. *Optical Coherence Tomography Technology and Applications.* New York: Springer.

Drexler, W., and J. G. Fujimoto. 2008b. State-of-the-art retinal optical coherence tomography. *Progress in Retinal and Eye Research* 27:45–88.

Drexler, W., U. Morgner, F. X. Kartner et al. 1999. *In vivo* ultrahigh-resolution optical coherence tomography. *Optics Letters* 24:1221–1223.

Drexler, W., H. Sattmarin, B. Hermann et al. 2003. Enhanced visualization of macular pathology with the use of ultrahigh-resolution optical coherence tomography. *Archives of Ophthalmology* 121:695–706.

Dubois, A., K. Grieve, G. Moneron et al. 2004. Ultrahigh-resolution full-field optical coherence tomography. *Applied Optics* 43:2874–2883.

Dunkers, J. P., M. T. Cicerone, and N. R. Washburn. 2003. Collinear optical coherence and confocal fluorescence microscopies for tissue engineering. *Optics Express* 11:3074–3079.

Durnin, J. 1987. Exact-solutions for nondiffracting beams. 1. The scalar theory. *Journal of the Optical Society of America A* 4:651–654.

Everett, M. J., K. Schoenenberger, B. W. Colston, and L. B. Da Silva. 1998. Birefringence characterization of biological tissue by use of optical coherence tomography. *Optics Letters* 23:228–230.

Faber, D. J., E. G. Mik, M. C. G. Aalders, and T. G. Van Leeuwen. 2005. Toward assessment of blood oxygen saturation by spectroscopic optical coherence tomography. *Optics Letters* 30:1015–1017.

Fercher, A. F., K. Mengedoht, and W. Werner. 1988. Eye-length measurement by interferometry with partially coherent light. *Optics Letters* 13:186–188.

Fercher, A. F., C. K. Hitzenberger, G. Kamp, and S. Y. Elzaiat. 1995. Measurement of intraocular distances by backscattering spectral interferometry. *Optics Communications* 117:43–48.

Fercher, A. F., W. Drexler, C. K. Hitzenberger, and T. Lasser. 2003. Optical coherence tomography—Principles and applications. *Reports on Progress in Physics* 66:239–303.

Fujimoto, J. G. 2003. Optical coherence tomography for ultrahigh resolution *in vivo* imaging. *Nature Biotechnology* 21:1361–1367.

Fujimoto, J. G., S. De Silvestri, E. P. Ippen et al. 1986. Femtosecond optical ranging in biological systems. *Optics Letters* 11:150–152.

Fujimoto, J. G., S. A. Boppart, G. J. Tearney et al. 1999. High resolution *in vivo* intra-arterial imaging with optical coherence tomography. *Heart* 82:128–133.

Gambichler, T., G. Moussa, M. Sand et al. 2005. Applications of optical coherence tomography in dermatology. *Journal of Dermatological Science* 40:85–94.

Gasparoni, S., B. Povazay, B. Hermann et al. 2005. Ultrahigh resolution optical coherence tomography of human skin. *Proceedings of SPIE* 5861: TuB3.

Goodman, J. W. 1985. *Statistical Optics.* New York: Wiley.

Gotzinger, E., M. Pircher, and C. K. Hitzenberger. 2005. High speed spectral domain polarization sensitive optical coherence tomography of the human retina. *Optics Express* 13:10217–10229.

Graf, B. W., M. C. Valero, E. J. Chaney et al. 2011. Long-term time-lapse multimodal microscopy for tracking cell dynamics in live tissue. *Proceedings of SPIE* 7902:790206.

Graf, B. W., and S. A. Boppart. 2012. Multimodal *in vivo* skin imaging with integrated optical coherence and multiphoton microscopy. *IEEE Journal of Selected Topics in Quantum Electronics* 18:1280–1286.

Han, M., G. Giese, and J. F. Bille. 2005. Second harmonic generation imaging of collagen fibrils in cornea and sclera. *Optics Express* 13:5791–5797.

Hartl, I., X. D. Li, C. Chudoba et al. 2001. Ultrahigh-resolution optical coherence tomography using continuum generation in an air-silica microstructure optical fiber. *Optics Letters* 26:608–610.

Hee, M. R., C. A. Puliafito, J. S. Duker et al. 1998. Topography of diabetic macular edema with optical coherence tomography. *Ophthalmology* 105:360–370.

Helmchen, F., and W. Denk. 2005. Deep tissue two-photon microscopy. *Nature Methods* 2:932–940.

Herz, P. R., Y. Chen, A. D. Aguirre et al. 2004. Micromotor endoscope catheter for *in vivo*, ultrahigh-resolution optical coherence tomography. *Optics Letters* 29:2261–2263.

Hoeling, B. M., A. D. Fernandez, R. C. Haskell et al. 2000. An optical coherence microscope for 3-dimensional imaging in developmental biology. *Optics Express* 6:136–146.

Hsiung, P. L., L. Pantanowitz, A. D. Aguirre et al. 2005. Ultrahigh-resolution and 3-dimensional optical coherence tomography *ex vivo* imaging of the large and small intestines. *Gastrointestinal Endoscopy* 62:561–574.

Huang, D., E. Swanson, C. Lin et al. 1991. Optical coherence tomography. *Science* 254:1178–1181.

Huber, R., M. Wojtkowski, J. G. Fujimoto, J. Y. Jiang, and A. E. Cable. 2005. Three-dimensional and C-mode OCT imaging with a compact, frequency swept laser source at 1300 nm. *Optics Express* 13:10523–10538.

Huber, R., M. Wojtkowski, and J. G. Fujimoto. 2006. Fourier Domain Mode Locking (FDML): A new laser operating regime and applications for optical coherence tomography. *Optics Express* 14:3225–3237.

Huber, R., D. C. Adler, V. J. Srinivasan, and J. G. Fujimoto. 2007. Fourier domain mode locking at 1050 nm for ultra-high-speed optical coherence tomography of the human retina at 236,000 axial scans per second. *Optics Letters* 32:2049–2051.

Izatt, J. A., M. R. Hee, G. M. Owen, E. A. Swanson, and J. G. Fujimoto. 1994. Optical coherence microscopy in scattering media. *Optics Letters* 19:590–592.

Izatt, J. A., M. D. Kulkarni, S. Yazdanfar, J. K. Barton, and A. J. Welch. 1997. *In vivo* bidirectional color Doppler flow imaging of picoliter blood volumes using optical coherence tomography. *Optics Letters* 22:1439–1441.

Jackle, S., N. Gladkova, F. Feldchtein et al. 2000. *In vivo* endoscopic optical coherence tomography of the human gastrointestinal tract—toward optical biopsy. *Endoscopy* 32:743–749.

Jenkins, M. W., F. Rothenberg, D. Roy et al. 2006. 4D embryonic cardiography using gated optical coherence tomography. *Optics Express* 14:736–748.

John, R., and S. A. Boppart. 2011. Magnetomotive molecular nanoprobes. *Current Medicinal Chemistry* 18:2103–2114.

John, R., R. Rezaeipoor, S. G. Adie et al. 2010. *In vivo* magnetomotive optical molecular imaging using targeted magnetic nanoprobes. *Proceedings of the National Academy of Science USA* 107:8085–8090.

Jung, W., J. Kim, M. Jeon et al. 2011. Handheld optical coherence tomography scanner for primary care diagnostics. *IEEE Transactions on Biomedical Engineering* 58:741–744.

Kennedy, B. F., T. R. Hillman, R. A. McLaughlin, B. C. Quirk, and D. D. Sampson. 2009. *In vivo* dynamic optical coherence elastography using a ring actuator. *Optics Express* 17:21762–21772.

Klein, T., W. Wieser, C. M. Eigenwillig, B. R. Biedermann, and R. Huber. 2011. Megahertz OCT for ultrawide-field retinal imaging with a 1050 nm Fourier domain mode-locked laser. *Optics Express* 19:3044–3062.

Ko, H. J., W. Tan, R. Stack, and S. A. Boppart. 2006. Optical coherence elastography of engineered and developing tissue. *Tissue Engineering* 12:63–73.

Ko, T. H., D. C. Adler, J. G. Fujimoto et al. 2004a. Ultrahigh resolution optical coherence tomography imaging with a broadband superluminescent diode light source. *Optics Express* 12:2112–2119.

Ko, T. H., J. G. Fujimoto, J. S. Duker et al. 2004b. Comparison of ultrahigh- and standard-resolution optical coherence tomography for imaging macular hole pathology and repair. *Ophthalmology* 111:2033–2043.

Lee, K. S., and L. P. Rolland. 2008. Bessel beam spectral-domain high-resolution optical coherence tomography with micro-optic axicon providing extended focusing range. *Optics Letters* 33:1696–1698.

Lee, K. S., K. P. Thompson, P. Meemon, and J. P. Rolland. 2011. Cellular resolution optical coherence microscopy with high acquisition speed for *in vivo* human skin volumetric imaging. *Optics Letters* 36:2221–2223.

Lee, T. M., A. L. Oldenburg, S. Sitafalwalla et al. 2003. Engineered microsphere contrast agents for optical coherence tomography. *Optics Letters* 28:1546–1548.

Leitgeb, R., M. Wojtkowski, A. Kowalczyk et al. 2000. Spectral measurement of absorption by spectroscopic frequency-domain optical coherence tomography. *Optics Lett* 25:820–822.

Leitgeb, R., C. K. Hitzenberger, and A. F. Fercher. 2003. Performance of Fourier domain vs. time domain optical coherence tomography. *Optics Express* 11:889–894.

Leitgeb, R. A., W. Drexler, A. Unterhuber et al. 2004. Ultrahigh resolution Fourier domain optical coherence tomography. *Optics Express* 12:2156–2165.

Leitgeb, R. A., M. Villiger, A. H. Bachmann, L. Steinmann, and T. Lasser. 2006. Extended focus depth for Fourier domain optical coherence microscopy. *Optics Letters* 31:2450–2452.

Lexer, F., C. K. Hitzenberger, A. F. Fercher, and M. Kulhavy. 1997. Wavelength-tuning interferometry of intraocular distances. *Applied Optics* 36:6548–6553.

Chapter 10

Liang, X., and S. A. Boppart. 2010. Biomechanical properties of *in vivo* human skin from dynamic optical coherence elastography. *IEEE Transactions on Biomedical Engineering* 57:953–959.

Liang, X., A. L. Oldenburg, V. Crecea, E. J. Chaney, and S. A. Boppart. 2008. Optical micro-scale mapping of dynamic biomechanical tissue properties. *Optics Express* 16:11052–11065.

Liang, X., B. W. Graf, and S. A. Boppart. 2009. Imaging engineered tissues using structural and functional optical coherence tomography. *Journal of Biophotonics* 2:643–655.

Liang, X., S. G. Adie, R. John, and S. A. Boppart. 2010. Dynamic spectral-domain optical coherence elastography for tissue characterization. *Optics Express* 18:14183–14190.

Lim, H., Y. Jiang, Y. M. Wang et al. 2005. Ultrahigh-resolution optical coherence tomography with a fiber laser source at 1 μm. *Optics Letters* 30:1171–1173.

Liu, L. B., J. A. Gardecki, S. K. Nadkarni et al. 2011. Imaging the subcellular structure of human coronary atherosclerosis using micro-optical coherence tomography. *Nature Medicine* 17:1010–1014.

Marks, D. L., and S. A. Boppart. 2004. Nonlinear interferometric vibrational imaging. *Physical Review Letters* 92:123905.

Marks, D. L., A. L. Oldenburg, J. J. Reynolds, and S. A. Boppart. 2002. Study of an ultrahigh-numerical-aperture fiber continuum generation source for optical coherence tomography. *Optics Letters* 27:2010–2012.

Mason, C., J. F. Markusen, M. A. Town, P. Dunnill, and R. K. Wang. 2004. The potential of optical coherence tomography in the engineering of living tissue. *Physics in Medicine and Biology* 49:1097–1115.

Matcher, S. J. 2011. Practical aspects of OCT imaging in tissue engineering. In *3D Cell Culture: Methods and Protocols.* ed. J. W. Haycock, 261–280. New York: Humana Press.

Medeiros, F. A., L. M. Zangwill, C. Bowd et al. 2005. Evaluation of retinal nerve fiber layer, optic nerve head, and macular thickness measurements for glaucoma detection using optical coherence tomography. *American Journal of Ophthalmology* 139:44–55.

Miller, D. T., O. P. Kocaoglu, Q. Wang, and S. Lee. 2011. Adaptive optics and the eye (super resolution OCT). *Eye* 25:321–330.

Morgner, U., W. Drexler, F. X. Kartner et al. 2000. Spectroscopic optical coherence tomography. *Optics Letters* 25:111–113.

Murali, S., K. P. Thompson, and J. P. Rolland. 2009. Three-dimensional adaptive microscopy using embedded liquid lens. *Optics Letters* 34:145–147.

Nguyen, F. T., A. M. Zysk, E. J. Chaney et al. 2009. Intraoperative evaluation of breast tumor margins with optical coherence tomography. *Cancer Research* 69:8790–8796.

Oh, W. Y., S. H. Yun, B. J. Vakoc, G. J. Tearney, and B. E. Bouma. 2006. Ultrahigh-speed optical frequency domain imaging and application to laser ablation monitoring. *Applied Physics Letters* 88:103902.

Oldenburg, A. L., J. R. Gunther, and S. A. Boppart. 2005. Imaging magnetically labeled cells with magneto-motive optical coherence tomography. *Optics Letters* 30:747–749.

Oldenburg, A. L., M. N. Hansen, D. A. Zweifel, A. Wei, and S. A. Boppart. 2006. Plasmon-resonant gold nanorods as low backscattering albedo contrast agents for optical coherence tomography. *Optics Express* 14:6724–6738.

Oldenburg, A. L., M. N. Hansen, T. S. Ralston, A. Wei, and S. A. Boppart. 2009. Imaging gold nanorods in excised human breast carcinoma by spectroscopic optical coherence tomography. *Journal of Material Chemistry* 19:6407–6411.

Otis, L. L., M. J. Everett, U. S. Sathyam, and B. W. Colston, Jr. 2000. Optical coherence tomography: A new imaging technology for dentistry. *Journal of the American Dental Association* 131:511–514.

Park, B. H., C. Saxer, S. M. Srinivas, J. S. Nelson, and J. F. de Boer. 2001. *In vivo* burn depth determination by high-speed fiber-based polarization sensitive optical coherence tomography. *Journal of Biomedical Optics* 6:474–479.

Pawley, J. B. 2006. *Handbook of Biological Confocal Microscopy*. New York: Springer.

Potsaid, B., I. Gorczynska, V. J. Srinivasan et al. 2008. Ultrahigh speed spectral/Fourier domain OCT ophthalmic imaging at 70,000 to 312,500 axial scans per second. *Optics Express* 16:15149–15169.

Povazay, B., K. Bizheva, A. Unterhuber et al. 2002. Submicrometer axial resolution optical coherence tomography. *Optics Letters* 27:1800–1802.

Puliafito, C. A., M. R. Hee, C. P. Lin et al. 1995. Imaging of macular diseases with optical coherence tomography. *Ophthalmology* 102:217–229.

Radhakrishnan, S., J. Goldsmith, D. Huang et al. 2005. Comparison of optical coherence tomography and ultrasound biomicroscopy for detection of narrow anterior chamber angles. *Archives of Ophthalmology* 123:1053–1059.

Ralston, T. S., D. L. Marks, P. S. Carney, and S. A. Boppart. 2007. Interferometric synthetic aperture microscopy. *Nature Physics* 3:129–134.

Ralston, T. S., D. L. Marks, P. S. Carney, and S. A. Boppart. 2008. Real-time interferometric synthetic aperture microscopy. *Optics Express* 16:2555–2569.

Rao, Y. J., Y. N. Ning, and D. A. Jackson. 1993. Synthesized source for white-light sensing systems. *Optics Letters* 18:462–464.

Rey, S. M., B. Povazay, B. Hofer et al. 2009. Three- and four-dimensional visualization of cell migration using optical coherence tomography. *Journal of Biophotonics* 2:370–379.

Rollins, A. M., M. D. Kulkarni, S. Yazdanfar, R. Ung-arunyawee, and J. A. Izatt. 1998. *In vivo* video rate optical coherence tomography. *Optics Express* 3:219–229.

Rollins, A. M., R. Ung-arunyawee, A. Chak et al. 1999. Real-time *in vivo* imaging of human gastrointestinal ultrastructure by use of endoscopic optical coherence tomography with a novel efficient interferometer design. *Optics Letters* 24:1358–1360.

Rose, F. R., L. A. Cyster, D. M. Grant et al. 2004. *In vitro* assessment of cell penetration into porous hydroxyapatite scaffolds with a central aligned channel. *Biomaterials* 25:5507–5514.

Schmitt, J. M. 1999. Optical coherence tomography (OCT): A review. *IEEE Journal of Selected Topics in Quantum Electronics* 5:1205–1215.

Schmitt, J. M., and A. Knuttel. 1997. Model of optical coherence tomography of heterogeneous tissue. *Journal of the Optics Society of America A* 14:1231–1242.

Schmitt, J. M., S. H. Xiang, and K. M. Yung. 1998. Differential absorption imaging with optical coherence tomography. *Journal of the Optics Society of America A* 15:2288–2296.

Shen, Y. R. 2003. *The Principles of Nonlinear Optics*. Hoboken: Wiley-Interscience.

Smith, L. E., R. Smallwood, and S. Macneil. 2010. A comparison of imaging methodologies for 3D tissue engineering. *Microscopy Research and Technique* 73:1123–1133.

Srinivasan, V. J., T. H. Ko, M. Wojtkowski et al. 2006. Noninvasive volumetric imaging and morphometry of the rodent retina with high-speed, ultrahigh-resolution optical coherence tomography. *Investigative Ophthalmology and Visual Science* 47:5522–5528.

Srinivasan, V. J., D. C. Adler, Y. Chen et al. 2008. Ultrahigh-speed optical coherence tomography for three-dimensional and *en face* imaging of the retina and optic nerve head. *Investigative Ophthalmology and Visual Science* 49:5103–5110.

Steller, H. 1995. Mechanisms and genes of cellular suicide. *Science* 267:1445–1449.

Takada, K., I. Yokohama, K. Chida, and J. Noda. 1987. New measurement system for fault location in optical waveguide devices based on an interferometric technique. *Applied Optics* 26:1603–1606.

Tan, W., A. Sendemir-Urkmez, L. J. Fahrner et al. 2004. Structural and functional optical imaging of three-dimensional engineered tissue development. *Tissue Engineering* 10:1747–1756.

Tan, W., A. L. Oldenburg, J. J. Norman, T. A. Desai, and S. A. Boppart. 2006. Optical coherence tomography of cell dynamics in three-dimensional tissue models. *Optics Express* 14:7159–7171.

Tan, W., C. Vinegoni, J. J. Norman, T. A. Desai, and S. A. Boppart. 2007. Imaging cellular responses to mechanical stimuli within three-dimensional tissue constructs. *Microscopy Research and Technique* 70:361–371.

Tang, S., T. B. Krasieva, Z. Chen, and B. J. Tromberg. 2006. Combined multiphoton microscopy and optical coherence tomography using a 12-fs broadband source. *Journal of Biomedical Optics* 11:020502.

Tearney, G. J., S. A. Boppart, B. E. Bouma et al. 1996. Scanning single-mode fiber optic catheter-endoscope for optical coherence tomography. *Optics Letters* 21:543–545.

Tearney, G. J., M. E. Brezinski, B. E. Bouma et al. 1997. *In vivo* endoscopic optical biopsy with optical coherence tomography. *Science* 276:2037–2039.

Tearney, G. J., I. K. Jang, and B. E. Bouma. 2006. Optical coherence tomography for imaging the vulnerable plaque. *Journal of Biomedical Optics* 11:021002.

Thomas, D., and G. Duguid. 2004. Optical coherence tomography—A review of the principles and contemporary uses in retinal investigation. *Eye* 18:561–570.

Troutman, T. S., J. K. Barton, and M. Romanowski. 2007. Optical coherence tomography with plasmon resonant nanorods of gold. *Optics Letters* 32:1438–1440.

Wang, H., C. P. Fleming, and A. M. Rollins. 2007a. Ultrahigh-resolution optical coherence tomography at 1.15 μm using photonic crystal fiber with no zero-dispersion wavelengths. *Optics Express* 15:3085–3092.

Wang, H., W. Kang, H. Zhu, G. MacLennan, and A. M. Rollins. 2011. Three-dimensional imaging of ureter with endoscopic optical coherence tomography. *Urology* 77:1254–1258.

Chapter 10

Wang, Y. M., Y. H. Zhao, J. S. Nelson, Z. P. Chen, and R. S. Windeler. 2003. Ultrahigh-resolution optical coherence tomography by broadband continuum generation from a photonic crystal fiber. *Optics Letters* 28:182–184.

Wang, Y. M., B. A. Bower, J. A. Izatt, O. Tan, and D. Huang. 2007b. *In vivo* total retinal blood flow measurement by Fourier domain Doppler optical coherence tomography. *Journal of Biomedical Optics* 12:041215.

Weissman, J., T. Hancewicz, and P. Kaplan. 2004. Optical coherence tomography of skin for measurement of epidermal thickness by shapelet-based image analysis. *Optics Express* 12:5760–5769.

Wieser, W., B. R. Biedermann, T. Klein, C. M. Eigenwillig, and R. Huber. 2010. Multi-megahertz OCT: High quality 3D imaging at 20 million A-scans and 4.5 GVoxels per second. *Optics Express* 18:14685–14704.

Wojtkowski, M., T. Bajraszewski, I. Gorczynska et al. 2004a. Ophthalmic imaging by spectral optical coherence tomography. *American Journal of Ophthalmology* 138:412–419.

Wojtkowski, M., V. J. Srinivasan, T. H. Ko et al. 2004b. Ultrahigh-resolution, high-speed, Fourier domain optical coherence tomography and methods for dispersion compensation. *Optics Express* 12:2404–2422.

Wu, Q., B. E. Applegate, and A. T. Yeh. 2011. Cornea microstructure and mechanical responses measured with nonlinear optical and optical coherence microscopy using sub-10-fs pulses. *Biomedical Optics Express* 2:1135–1146.

Xu, C. Y., D. L. Marks, M. N. Do, and S. A. Boppart. 2004a. Separation of absorption and scattering profiles in spectroscopic optical coherence tomography using a least-squares algorithm. *Optics Express* 12:4790–4803.

Xu, C. Y., J. Ye, D. L. Marks, and S. A. Boppart. 2004b. Near-infrared dyes as contrast-enhancing agents for spectroscopic optical coherence tomography. *Optics Letters* 29:1647–1649.

Xu, C. Y., F. Kamalabadi, and S. A. Boppart. 2005. Comparative performance analysis of time-frequency distributions for spectroscopic optical coherence tomography. *Applied Optics* 44:1813–1822.

Xu, X. Q., R. K. K. Wang, and A. El Haj. 2003. Investigation of changes in optical attenuation of bone and neuronal cells in organ culture or three-dimensional constructs *in vitro* with optical coherence tomography: Relevance to cytochrome oxidase monitoring. *European Biophysics Journal* 32:355–362.

Yabushita, H., B. E. Bourna, S. L. Houser et al. 2002. Characterization of human atherosclerosis by optical coherence tomography. *Circulation* 106:1640–1645.

Yang, Y., A. Dubois, X. P. Qin et al. 2006. Investigation of optical coherence tomography as an imaging modality in tissue engineering. *Physics in Medicine and Biology* 51:1649–1659.

Yao, G., and L. V. Wang. 1999. Two-dimensional depth-resolved Mueller matrix characterization of biological tissue by optical coherence tomography. *Optics Letters* 24:537–539.

Yelin, D., and Y. Silberberg. 1999. Laser scanning third-harmonic-generation microscopy in biology. *Optics Express* 5:169–175.

Yun, S. H., G. J. Tearney, J. F. de Boer, N. Iftimia, and B. E. Bouma. 2003. High-speed optical frequency-domain imaging. *Optics Express* 11:2953–2963.

Zhang, J., Q. Wang, B. Rao, Z. P. Chen, and K. Hsu. 2006. Swept laser source at 1 μm for Fourier domain optical coherence tomography. *Applied Physics Letters* 89:073901.

Zhao, Y., B. W. Graf, E. J. Chaney et al. 2012. Integrated multimodal optical microscopy for structural and functional imaging of engineered and natural skin. *Journal of Biophotonics* 5:437–448.

Zhou, C., D. W. Cohen, Y. Wang et al. 2010. Integrated optical coherence tomography and microscopy for *ex vivo* multiscale evaluation of human breast tissues. *Cancer Research* 70:10071–10079.

Zumbusch, A., G. R. Holtom, and X. S. Xie. 1999. Three-dimensional vibrational imaging by coherent anti-Stokes Raman scattering. *Physical Review Letters* 82:4142–4145.

Zysk, A. M., F. T. Nguyen, A. L. Oldenburg, D. L. Marks, and S. A. Boppart. 2007. Optical coherence tomography: A review of clinical development from bench to bedside. *Journal of Biomedical Optics* 12:051403.

11. Application of Polarization-Sensitive OCT and Doppler OCT in Tissue Engineering

Ying Yang, Ian Wimpenny, and Ruikang K. Wang

11.1 Introduction

Within the field of tissue engineering, there is an urgent need to develop high-resolution and nondestructive imaging techniques, which can assess tissue growth within optically dense scaffolds. Such assessment should include the triangular interlinking capacity of characterization, monitoring, and prediction at both the cellular and macroscopic levels as demonstrated in Figure 11.1. In detail, we seek tools that can implement three key actions:

Chapter 11

Optical Techniques in Regenerative Medicine. Edited by Stephen P. Morgan, Felicity R.A.J. Rose, and Stephen J. Matcher © 2014 CRC Press/Taylor & Francis Group, LLC. ISBN: 978-1-4398-5495-2

1. *Characterization:* to obtain the functional property parameters for a specific tissue
2. *Monitoring:* to reveal the evolution of tissue-engineered constructs during *in vitro* culture, microscopically or macroscopically and nondestructively
3. *Prediction:* to establish the correlation of the measurement parameters and the microstructure

In this chapter, we introduce the principle and suitability of polarization-sensitive optical coherence tomography (PS-OCT), Doppler OCT (D-OCT), and Doppler optical microangiography (DOMAG) to meet the assessment requirements aforementioned. Case studies have been used to demonstrate the alteration of the internal structures of tendon constructs or tissues in response to external conditioning or stimulation, illustrating that this can be revealed by PS-OCT in a quantitative and qualitative manner. The local fluid flow and shear stress within three-dimensional porous scaffolds can also be mapped by D-OCT. In addition, DOMAG is highlighted as a technique to produce higher resolution images when compared to D-OCT. The two modalities, therefore, can be useful tools for tracking the perfusion behavior within tissue engineering bioreactors.

Optical coherence tomography utilizes the intrinsic scattering light backscattered from the sample to generate an image. Therefore, it does not require any biochemical labeling or sample processing and is a nondestructive technique, which is ideal for biological samples. OCT has been successfully used in developmental biology (Boppart et al. 2000). Therefore, this technique can be well adapted to dynamically study neo tissue formation, in which the development of the tissue and extracellular matrix (ECM) production on a macroscale is of more interest than specific, individual cellular events.

The OCT technique has been enhanced further through the addition of a polarized light source into the system, becoming PS-OCT. This has attracted considerable interests since a number of biological tissues have a well-organized matrix, mainly consisting of collagen fibers, and the arrangement of these fibers is polarization sensitive. The original intention for the development of PS-OCT was to identify the abnormality in tissues such as in tendon, cartilage, and muscle (Nadkarni et al. 2007). Abnormalities arising from damage/disease are known to cause disruption to the organization of collagen fibers, resulting in alteration of the spatial distribution of polarization. Thus, PS-OCT can detect structural changes in tissues with anisotropic optical properties, for example, tendon, muscle, and cartilage, according to the changes in the birefringence detected by OCT imaging. The back-reflection signal from the tissue is dependent on the polarization state of the incident light. Therefore, organized ECM, for example, oriented collagen fibrils in tendons, can be identified by their birefringent properties; likewise, damage or disruption to this arrangement can be detected.

D-OCT is developed by combining OCT with Doppler velocimetry. It is a functional extension to OCT that can be used to measure localized fluid flows in a highly scattering medium, such as biological tissue. This is based on the principle that the Doppler frequency shift in the light that is backscattered from the moving objects within a sample either adds to or subtracts from the fixed optical frequency, depending on the flow rate (Mason et al. 2004). In comparison to OCT, which utilizes only the amplitude of backscattered light as a function of depth within the tissues, D-OCT additionally employs interferometric phase information to monitor the velocities of moving particles in the backscattered spectrum. Therefore, D-OCT is able to generate simultaneous imaging of both the tissue's architecture and its localized fluid flow of a porous object (Jia et al. 2009a).

We would like to apply these OCT techniques to tissue engineering in order to tackle the inverse challenges, that is, the illustration of microstructural evolution in tissue engineering constructs in response to various external stimulations and culture environments, which is different to the original purpose of developing OCT, that is, to perform early disease diagnosis. In this chapter, we focus on PS-OCT for tendon tissue engineering as an example of the use of this technique and demonstrate the use of birefringence as a marker to correlate the maturation state of tissue-engineered tendon and culture conditions, and investigate the factors that control the birefringence in model tendons. In parallel, D-OCT is used to map the shear strain and fluid flow within porous scaffolds in which mechanical stimulation is a positive regulator for tissue formation. The further comparison of the enhancement effects by DOMAG is also implemented.

11.2   Application of PS-OCT in Tissue-Engineered Tendon

Tendon is a connective tissue linking muscle to bone, permitting the movement of limbs by transmitting the contractile (tensile) force of the muscle to the bone, essential to the movement of limbs. Injury to tendons from sport, accident, or disease has become increasingly common. It is reported that there are 33 million cases of musculoskeletal injuries in the United States alone per year, with nearly 50% of the cases involving injuries to the soft tissue including tendon and ligament (Calve et al. 2004). These types of injuries can greatly reduce mobility and usually require transplantation of tissue or an artificial implant to restore normal functionality. Tissue suitable for transplantation is difficult to obtain while artificial implants have limited life spans because of degradation or loss of coherent connection with surrounding tissues. Tissue engineering could potentially provide an alternative tissue for transplantation. Due to high load bearing demands, tendon possesses a highly organized extracellular environment containing collagen and glycosaminoglycan (GAG) matrix proteins. The major component of the extracellular matrix in tendon tissue is collagen type I (collagen constitutes 75% of the dry tendon weight), which is organized into a highly hierarchical 3D network ranging from thin individual fibrils to large fiber bundles, enabling the tissue to withstand tensile, compressive, or shear stresses. Collagen fibrils are the smallest structural subunit generated from collagen molecules of aggregated triple helices. A bundle of fibrils forms collagen fibers, from which the collagen fiber bundles are bound together into fascicles. The functional nature of tendon and the fiber assembly process cause these subunits to be packed longitudinally and aligned parallel to the long axis of the tendon to ensure appropriate mechanical strength.

Collagen hydrogels have widely been examined as potential scaffolds for engineered tendon tissue *in vitro* due to the similarity of tendon chemical structure (Zeugolis et al. 2009). One of the major limitations of using collagen hydrogels for engineering tendon is the randomly arranged collagen fibrils within the hydrogels. When tenocytes are suspended throughout the hydrogel, over time they re-engineer the structure of the hydrogels through the contraction of collagen fibers and the release of extracellular matrix proteins and enzymes (Noth et al. 2005). However, without specific guidance and stimulation, the tendon constructs cannot evolutes into structurally or mechanically

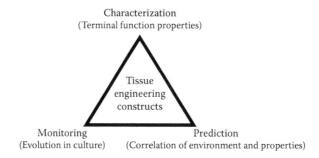

FIGURE 11.1   Triangular relationship of characterization, monitoring, and prediction of tissue-engineered constructs requested for successful tissue engineering.

appropriate tendons due to lack of the specific properties, for example, parallel arranged tendon fascicles.

Various strategies have been implemented to guide the proliferation and differentiation tenocytes, aiming not only to stimulate more matrix production but also replicate the required spatial arrangement of the collagen fibrils. The monitoring of these structural changes in collagen fibrils will provide valuable feedback for better strategies. Traditionally, the monitoring has been performed using either histological staining or electron microscopy, both of which require termination of experiments and destruction of the tissue construct. Therefore, these methods cannot identify the structural evolution of the same engineered constructs over a prolonged and varying culture period. In past years, we have undertaken a suitability study of using PS-OCT as a tool to characterize, monitor, and predict tissue-engineered tendon in order to generate mature tendons. Our results demonstrate that PS-OCT is indeed a powerful tool, able to provide these services.

11.2.1   Characterization of Native and Engineered Tendon: Tissue Structure and Matrix

To develop PS-OCT into a characterization tool for tissue-engineered tendon, it is desirable to study the native tendon under healthy and diseased conditions by PS-OCT. The aligned collagen fibrous structure produces optical anisotropic features. The anisotropic optical properties are manifested as birefringence in polarization light microscope or birefringence, biattenuation, and optical axis orientation in PS-OCT (Kemp et al. 2005). We collected 24 patient specimens (with ethical approval and patient consent) and found good correlation between birefringence and microstructural integrity in ruptured tendon as demonstrated in Figure 11.2 (Bagnaninchi et al. 2010).

Uniquely, we also have undertaken an investigation to examine proteoglycan (PG) influence on the tendon's stability and function by PS-OCT. Collagen fibrils are the smallest structural subunit generated from collagen molecules of aggregated triple helices. However, all the constituent fibrous units are not continuous for the full tendon length. It is logically postulated that there are complex mechanisms holding the individual fibrils, fibers, and fascicles together. Two main mechanisms have been proposed to describe the interfibrillar connections. The first is the cross-linking of collagen fibers during

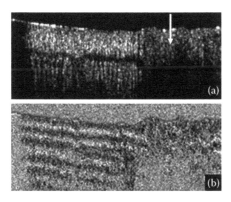

FIGURE 11.2 (a) Structural; (b) birefringence PS-OCT images of a human tendon specimen showing the normal and diseased region within the specimen. The diseased tissue is shown by an arrow in (a).

fibrillogenesis (Provenzano and Vanderby Jr. 2006), and the other important mechanism is by PG binding collagen fiber units (Zhang et al. 2006). Tendons, in addition to having collagen component (75% dry mass), contain 0.2%–5% PG. Together with water and other small portions of noncollagenous macromolecules, PG forms ground substance for tendon (Sharma and Maffulli 2005). It has been proposed that the highly aligned fibers embedded in ground substance allow lateral transference of force between neighboring fibrils (Liao and Vesely 2007). Over the past years, a few models have been proposed to explain the mechanical integrity. Dahners et al. (2000) conducted experiments that led to the conclusion that decorin, a constituent of glycosaminoglycans (GAG), could bind discontinuous collagen fibrils to one another, providing the mechanical integrity to tendon.

We hypothesize that if PG plays a role in forming the interfibrillar connections for collagen fibers in tendon, the organization of such collagen bundles will change when PG is extracted from a tendon. These changes could be observed as an alteration of the birefringence images of PS-OCT and polarizing light microscope. Thus, we have conducted a series of experiments in which time-lapse PS-OCT and polarization light microscopic images were recorded during chemical extraction of PGs in chicken tendons (Yang et al. 2012). The aim of this study was to address the following questions: (1) What is the effect of PG content on collagen fibril organization in terms of birefringence? (2) What is the distribution of PG within tendons?

The fresh chicken tendons showed multiple banded PS-OCT images with an average penetration depth of 700 μm and three to five thick horizontal bands (parallel to tendon axis), while the polarizing light microscopic image exhibited multiple fine colored bands along the long axis of the tendons with less pronounced vertical bands (Figure 11.3). The birefringence bands of the specimen changed rapidly along the reaction time upon application of 4M guanidine hydrochloride (GuHCl) solution which is PG extractor, to the fresh tendons. The number of bands was reduced and the width of the bands increased within a short period (as short as 20 min). The change rate in 4M GuHCl was much faster than in 2M GuHCl solution (Figure 11.4). To identify the distribution of PG in tendons, PS-OCT and polarization light microscopic images of unsheathed tendons (removal of the outermost layer of tendons) were recorded when PGs were extracted. Polarization microscopy demonstrated very subtle changes in the band structure of unsheathed tendons as compared to the sheathed ones after 60 min (Figure 11.4). The corresponding polarization

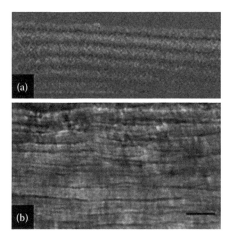

FIGURE 11.3   PS-OCT (a) and polarization light microscopy (b) images of a fresh, intact chicken tendon specimen. The PS-OCT image was the scan of 1 × 4 mm (width × length) area of the specimen. The scale bar is 200 μm. (From Yang, Y. et al., *Journal of Biomedical Optics* 17(8):081417, 2012.)

light microscopy images in Figure 11.5 showed that the unsheathed tendon demonstrated a reduced rate of fiber disorganization in comparison to intact tendon (sheathed tendon).

It is frequently observed that fresh tendon specimens exhibited more and well-defined bands in PS-OCT imaging in comparison to the old specimens (stored in PBS at 4°C for 5 days). To confirm the cause of the difference and the different response of sheathed and unsheathed tendon toward PG extraction, the sheaths removed from freshly dissected and old tendons were stained with Alcian blue to identify sulfated mucins/PGs (appear blue, if present). Darker blue stains appeared in fresh specimens but not

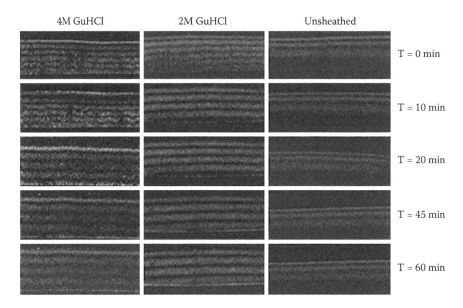

FIGURE 11.4   Time-lapsed PS-OCT images showing the effect of GuHCl concentration and unsheathing on the birefringence bands within tendon during extraction of PG. The size of original OCT image frame is 1 × 4 mm (width × length). (From Yang, Y. et al., *Journal of Biomedical Optics* 17(8):081417, 2012.)

Unsheated   Sheated

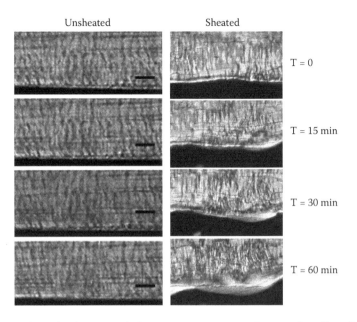

T = 0

T = 15 min

T = 30 min

T = 60 min

FIGURE 11.5 Time-lapsed polarization light microscopic images showing the effect of PG extraction on tendon's structure with and without sheath. The scale bar is 200 μm. (From Yang, Y. et al., *Journal of Biomedical Optics* 17(8):081417, 2012.)

in the old ones (Figure 11.6a). Similarly, the differences in the staining pattern were observed for unsheathed and sheathed tendons. Furthermore, it was demonstrated that the sheath stained homogenously blue while unsheathed tendon stained less intensely (Figure 11.6b), indicating higher concentration of PGs in the sheath region of the tendon than in fascicles.

It was confirmed that extraction of PG from intact tendon caused the disturbance of the collagen arrangement. The reduction in band frequency and increase in the bandwidth after extraction of PG were indicative of lower birefringence, implying a less organized collagen fiber structure. The effect of disrupting the collagen fiber structure by removal of PGs was evident.

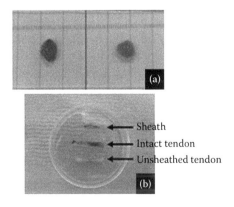

(a)

Sheath
Intact tendon
Unsheathed tendon

(b)

FIGURE 11.6 (a) Alcian blue staining of tendon sheaths: freshly dissected (left) and after 5 days storage in PBS (right); (b) alcian blue staining of a fresh tendon to define the distribution of PG in the tendon sample. PGs stain blue. (From Yang, Y. et al., *Journal of Biomedical Optics* 17(8):081417, 2012.)

Chapter 11

Scott (1988) has successfully visualized the GAG chains of PG molecules in the cornea, sclera, and tendons and the binding site of the PGs along the collagen fibrils by transmission electron microscopy (TEM) and biochemical staining, which provided evidence for the importance of the PG molecules in collagen fibril organization, stabilization, and growth. The proteoglycans are believed to be interwoven with the collagen fibrils; GAG side chains also have multiple interactions with the surface of the fibrils (Canty and Kadler 2005). These interconnections were believed to be formed in the fibril assembly process during tendon development. One theory postulated that dermatan sulfate could be responsible for forming associations between fibrils. When decorin molecules are bound to collagen fibrils, their dermatan sulfate chains may extend and associate with other dermatan sulfate chains on decorin that is bound to separate fibrils, therefore creating interfibrillar bridges and eventually causing parallel alignment of the fibrils (Scott 1996). Our data provide direct evidence to support these theories and visualize the dose responses (time lapse imaging) of PG concentration within collagen fiber organization. When PG was removed from collagen fibers, the stability of the collagen interfibrillar connections reduced and the local collagen organization was disturbed: macroscopically, the original opaque tissue became transparent, manifesting as the degree of birefringence changes at the microscopic level.

Combining the data from the PS-OCT and light microscopy images during the PG extraction confirms two observations: (1) there is a difference in the distribution of PGs within tendon tissue, and (2) the PG extraction rate is different between the outer sheath and the inner collagen fascicles.

The Alcian blue staining (Figure 11.6b) revealed that the PG distribution in the tendon was not homogenous, and high PG density appeared to be in the outer sheath (also known as epitenon). Apparently, extraction of this part of PG led to rapid changes in the PS-OCT images (Figure 11.4). The disorganized sheath layer might further shield the light penetration. Furthermore, our data indicate that PG in the sheath was very vulnerable to decomposition, which may also explain why some stored tendon specimens demonstrate less birefringence bands by PS-OCT, despite the intact organized structure, as determined using polarization microscopy.

In summary, this study establishes a convenient technique to correlate the relationship between PG and collagen fiber organization. PG concentration and the degree of binding with collagen fibers directly control the organization of the hierarchal structure.

The study of native tendon's optical properties in PS-OCT paves a solid foundation for future engineered tendon studies. Here is one pair of images of tissue-engineered tendons with macroscale structure similar to native tendon but less perfect microscopic structure (Figure 11.7). Further characterizations of tissue-engineered tendon will be incorporated into the following sections.

 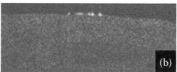 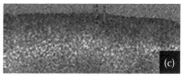

FIGURE 11.7 Image of an engineered tendon. (a) Macroscopic image, (b) PS-OCT structural image; (c) PS-OCT birefringence image showing weak birefringence. The size of the original OCT image frame is 1 × 4 mm (width × length). The scale bar is 15 mm.

11.2.2   Monitoring Tissue-Engineered Tendon

Healthy and mature tendons have demonstrated regular birefringence bands in PS-OCT images. We hypothesize that the alteration of the birefringence bands in a tissue-engineered tendon construct will correlate to the evolution of the microstructure during different culture times. To prove the correlation, we undertook an investigation in which we seeded tenocytes into a collagen gel and then applied tensile stress to the constructs. PS-OCT has been used to visualize the evolution of the alignment of collagen fibers within tissue-engineered tendons nondestructively and online over a 14-day culture period. In addition, the effect of the growth environment on the collagen alignment was examined by altering cell seeding densities and levels of mechanical strain on the constructs. In this study, collagen type I from rat tail (BD Bioscience) was used as the scaffolds for the engineered tendons with a final concentration of 5 mg/mL (Ahearne et al. 2008). Rat tenocytes, which were cultivated from the Achilles tendons of young rats (3-month-old Sprague–Dawley), were used at passage number 2. The tenocytes were suspended throughout the hydrogel solution prior to gelation. Flexcell Tissue Train plates (Flexcell Int.) were placed onto Flexcell trough loaders and a vacuum was applied, resulting in a cavity being created in each well. Hydrogel solution (200 μL) was poured into each cavity on the plate and allowed to set in an incubator at 37°C and 5% CO_2. After 90 min, the plates were removed from the trough and 5 mL of supplemented culture medium was added to each well. The medium was changed every 3 days.

The tendon constructs were uniaxially and cyclically stretched using the Flexcell Tissue Train Culture system. There were nine groups in total with two samples in each group. Groups had a cell seeding densities of either 1, 2.5, or 5×10^5 cells/mL and had a strain applied of 1.0% or 1.5% at 1 Hz for 1 h every day for 14 days. Controls were not subjected to mechanical conditioning. A PS-OCT system, built in-house, was used to examine the alignment of collagen fibers within the engineered tendons at the four culture time points and several different locations along each specimen at each time point (Ahearne et al. 2008). The PS-OCT system used a superluminescent diode with a central wavelength of 1310 nm and bandwidth of 52 nm. The beam from the diode is sent through a polarizer, a polarizer modulator, and then split 50/50 by a 2×2 all-fiber beam splitter into a sample and reference arm. After recombination, the polarized light is sent through a polarization beam splitter. The interference fringes are detected with two balanced detectors, for horizontal and vertical polarization. The system has an axial resolution of 14 μm in free space and penetration depth up to 2 mm. The birefringence Δn, the indicator of fiber alignment, was calculated over several A-scans by the following equation 11.1 (de Boer and Milner 2002):

$$\Delta n = \frac{\lambda o}{2\pi} \frac{d\delta}{dz} \tag{11.1}$$

where z is the depth, λo is the source wavelength, and δ is the phase retardation, which can be calculated from the Stokes vectors from the backscattered light as described by Park et al. (2003). Phase retardation noise was calculated to be 0.025 rad/mm.

PS-OCT images were recorded parallel to the elongation axis. It can be seen that there was increased birefringence manifested by the appearance of a banded structure

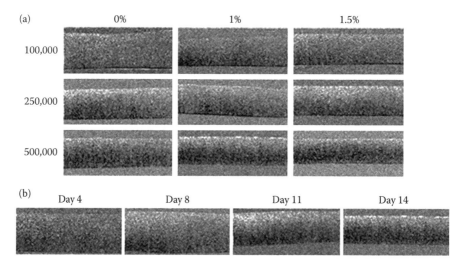

FIGURE 11.8 PS-OCT images demonstrating the effect of cell density and applied strain (a) and culture time (b) on the birefringence manifested by the appearance of a banded structure in tissue-engineered tendons. (From Ahearne, M. et al., *Journal of Tissue Engineering and Regenerative Medicine* 2:521–524, 2008.)

in culture over time. Birefringence also increased in response to an increase in initial cell seeding concentration (Figure 11.8). It was also noticeable that the mean diameter of samples decreased over time. These results suggest that the cell seeding density and culture time affected the organization of collagen within the tissue-engineered tendons potentially by increasing the amount of aligned collagen (Ahearne et al. 2008). Each image obtained by PS-OCT was analyzed to quantify the associated birefringence by phase retardation (Figure 11.9). Statistical analysis of the data was performed using ANOVA with a 95% CI Tukey's post-hoc test. It was found that the degree of birefringence increased significantly with culture time between day 4 and day 14 for all samples. These results suggest that the cells increased the density of aligned collagen in the tissue-engineered tendons. Increasing the cell number and culture time influenced the tissue thickness by increasing the rate and degree of collagen contraction, which led to a thinner construct and higher density of collagen fiber orientation. The increased organization of collagen fiber in response to the mechanical load and culture were cross-validated by polarization light microscopy. When viewed under polarized light microscopy, certain degrees of birefringence were observed in these specimens (Figure 11.10), which supported the observation from PS-OCT and showed collagen running parallel to the direction of strain. Samples without cells did not display birefringence. The close relationship of banded pattern images in PS-OCT and birefringence images under polarization light microscopy has been reported previously (Drexler et al. 2001).

Thus, PS-OCT has been demonstrated as a valuable technique for nondestructively examining the alignment of collagen within tissue-engineered tendons, potentially providing real-time analysis. The evolution of the tissue organization has successfully been examined at several different culture time points in the same sample. This approach will be of great benefit in the development of functional tissue-engineered tendons.

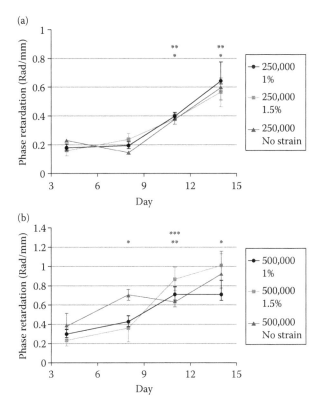

(a)

(b)

FIGURE 11.9 Quantification of the associated birefringence of the tissue-engineered tendons: (a) 250,000 cells per specimen; (b) 500,000 cells per specimen versus strain level and culture time (as in Figure 11.8) by phase retardation calculated from corresponding PS-OCT images. (From Ahearne, M. et al., *Journal of Tissue Engineering and Regenerative Medicine* 2:521–524, 2008.)

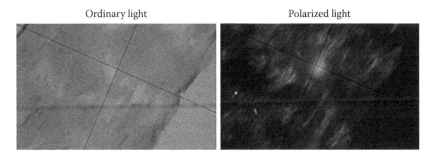

Ordinary light   Polarized light

FIGURE 11.10   Normal and polarized light microscopic images of the representative tissue-engineered tendons after 14 day loading shown in Figure 11.8.

11.2.3   Prediction of Engineered Tendon Constructs

Intensive studies by histology, SEM, and TEM have revealed the macroscopic and microscopic structures of native tendon in healthy and diseased states, especially the fibrous architecture (Franchi et al. 2007). However, there has been no detailed investigation that correlates the birefringence and microstructure in tissue-engineered constructs. In recent years, we have established new techniques that can introduce nanoscale features

Chapter 11

into tissue engineering constructs including aligning collagen fibrils during the fibrillogenesis. With these techniques, we have generated a series of model tendons. Their corresponding PS-OCT images were collected and analyzed, attempting to establish the correlation of birefringence and microstructure of tendons, which can be used to predict the microstructure in real tissue-engineered tendon during culture and assist for better culture condition design and execution strategy.

11.2.3.1 Tendon Model with Aligned Collagen Fibrils Formed by Magnetic Field

Application of a high magnetic field to align collagen fibers during the fibrillogenesis is a proven technique (Torbet and Ronziere 1984). Collagen molecules are diamagnetically anisotropic, and when a strong magnetic field is present during fibrillogenesis, alignment is induced, creating organized architecture. The process is rapid and samples can be processed and maintained in a sterile state. Using this magnetic alignment technique, scaffolds with orthogonally aligned collagen lamellae for corneal stroma reconstruction have recently been created (Torbet et al. 2007). We investigated how the aligned collagen fiber, the collagen density, and the formation of high density–oriented collagen-based matrices affected the birefringence and its suitability for tendon tissue engineering applications.

To achieve the required collagen fiber alignment, collagen solutions (1.5–4 mg/mL) were placed in a 12 T superconducting magnet at room temperature before being heated to 30°C for gelation. The total gelation time under the magnetic field was 30 min. The sample temperature was controlled by circulating water through the system using a temperature-controlled water bath. The high magnetic field resulted in collagen fiber alignment perpendicular to the magnetic field direction. Once the samples had cured, magnetically induced alignment of the collagen fibers became permanent. The clear

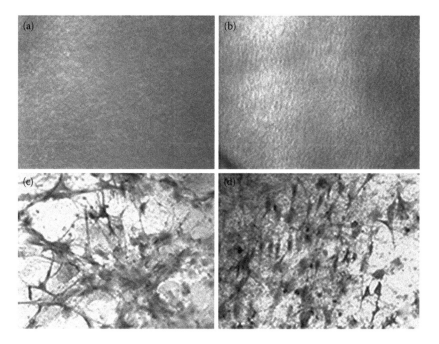

FIGURE 11.11 Light microscopic images of randomly formed (control) (a, c) and magnetically aligned collagen scaffolds (c, d) with and without (a, b) tenocyte culture (cultured for 2 days and with H & E staining).

collagen solution became opaque as the hydrogel formed. Examination by light micros-copy revealed that the collagen specimens subjected to a strong magnetic field during fibrillogenesis exhibited an aligned morphology. Its counterpart formed in the absence of a magnetic field consisted of randomly oriented collagen fibrils, as demonstrated in Figure 11.11. Samples were seeded with tenocytes. After 2 days of culture, cells on the aligned specimens showed orientation in the direction of the collagen alignment. This was not observed with the nonaligned specimens, which shows the effect of cell con-tact guidance determined by the alignment of collagen. However, these aligned collagen specimens, with or without the presence of tenocytes, did not exhibit banded birefrin-gence in PS-OCT images.

11.2.3.2   Tendon Model Incorporating Nanofiber Bundles and Higher Collagen Density

The aforementioned studies on native and tissue-engineered tendon confirm that bire-fringence images in PS-OCT are very sensitive to alterations in the internal structure of tendons and collagen density. Inversely, the impact of artificial alteration of structural components on tendon's physical properties could be revealed by PS-OCT. In this study, we generated tendon models with controllable internal structure and composition in order to investigate how collagen concentration, the density of aligned collagen fibrils, and the dimension of artificial fibril bundles (made from polyester nanofibers) influence the organization of tendon tissue in terms of birefringence in PS-OCT.

To define which parameters and to what extent these parameters control the spatial organization of tendon tissues, especially the cause of the appearance of birefringence banding images, we have produced four types of collagen hydrogel–based tendon mod-els as follows:

1. Spiral of plastic compressed collagen gel
2. Spiral of plastic compressed collagen gel with aligned fibrils induced by a magnetic field
3. Spiral of plastic compressed collagen gel incorporated low density of poly(lactic acid) (PLA) nanofiber meshes
4. Spiral of plastic compressed collagen gel incorporated high density of PLA nano-fiber meshes

The production of a spiral of plastic compressed collagen gels in models 1 and 2 fol-lowed the similar protocol described previously (Brown et al. 2005) in order to increase collagen density. To construct models 3 and 4, PLA polyester nanofiber meshes were produced by electrospinning as described elsewhere (Yang et al. 2011). To incorporate the nanofiber meshes into a gel, a collagen gel around 1.5 mm thick (3.5 mg/mL) was formed, then the nanofiber mesh was carefully placed on top of it, followed by the addi-tion of a further 1.5 mm thick collagen layer on that to stabilize the composite. The polyester nanofiber meshes had a highly aligned morphology with an average diameter of around 500 nm, as determined by SEM (Yang et al. 2011). The line density of low-density nanofiber meshes in model 3 was around 45 nanofiber/100 μm, while line den-sity in model 4 was around 180 nanofiber/100 μm. Nanofiber-hydrogel composites were plastically compressed.

Chapter 11

Prior to the plastic compression, none of the four models showed birefringence in PS-OCT images. After plastic compression, the gel specimens were transformed into very thin sheets. They were too thin to show birefringence by PS-OCT either. However, all spirals rolled from the thin sheets showed birefringence to different degrees. Randomly formed collagen gel or in combination with a low-density nanofiber mesh showed broad and poorly defined banding images in PS-OCT (Figure 11.12a′ and c′). The samples containing magnetically oriented collagen, however, showed clear birefringence, which was distributed homogenously throughout the spiral (Figure 11.12b′). The spiral containing a dense nanofiber mesh exhibited strong birefringence. The bands were distinct and narrow in shape and greater in number in these specimens (Figure 11.12d′).

The studies of the model tendon specimens further confirm that birefringence is predominately determined by two important factors: collagen concentration and the density of aligned fibril bundles. Using the magnetic induction technique and the incorporation of nanofiber, we can easily alter one or two parameters in the tendon model and investigate the effect of these parameters on the overall organization separately and independently by assessing the degree of birefringence. The preliminary data imply that if the collagen concentration in the tendon model is low, for instance, lower than 1%, the high fibril organization and alignment do not generate birefringence, or the birefringence cannot be detected by the current setting of the PS-OCT instrument. On the other hand, the plastic compression process of collagen gels can

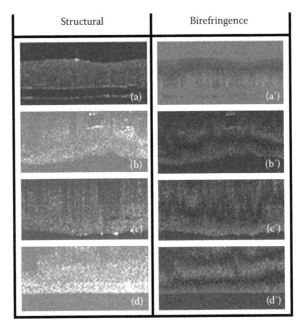

FIGURE 11.12   PS-OCT images of the four tendon models. (a–d) Structural images; (a′–d′) birefringence images. (a) Spiral of plastic compressed random organized collagen gel; (b) spiral of plastic compressed collagen gel with aligned fibrils induced by magnetic field; (c) spiral of plastic compressed collagen gel incorporating low density of PLA nanofiber bundles; (d) spiral of plastic compressed collagen gel incorporating high density of PLA nanofibers bundles. The size of original OCT image frame is 1 × 4 mm (width × length).

generate a certain degree of birefringence, even using randomly arranged collagen gels, when collagen concentration was increased to around 10% to 20%. Aligned polyester nanofiber meshes can act as artificial collagen bundles due to their size and orientation. The preliminary data in this study demonstrated that it is feasible to construct a model tendon by distributing nanofibers in an appropriate pattern within a high-density collagen matrix.

In summary, the study of model tendon constructs through analysis of birefringence patterns in PS-OCT images will lead to a better understanding of the relationship between individual components, their internal structure, and physiological functions in tendon. This can facilitate the optimization of the culture conditions required for successful tissue engineering of tendon.

11.3   Application of D-OCT and DOMAG in Tissue Engineering

11.3.1   D-OCT and Porous Scaffold

Cell-seeded scaffold constructs are frequently subjected to perfusion in various bioreactors in order to have better nutrient supply, exerting mechanical stimulation and accelerating the maturation of the constructs. Establishing a relationship between perfusion rate and fluid shear stress in a 3D cell culture environment is still a challenging task. Thus, 3D porous scaffolds not only play a role in the initial cell attachment and proliferation but also influence the cells in the way that they form their surrounding extracellular environment. The main parameters of scaffolds that will affect fluid flow will be pore density, pore size, and shape. The introduction of appropriate pores (i.e., mimetic of cell niches, which encourage attachment, growth, and development of cells) by manipulation of porogens or pore-generating materials will greatly control cell growth behavior. Porosity of scaffolds links to permeability and is important for the mass transport of fluids in and around scaffolds. It is also a requirement for tissue engineering scaffolds to provide enough space for cells to attach and proliferate to ensure growth into 3D. In particular, when cell–scaffold constructs are cultured in dynamic conditions, such as perfusion bioreactors, the fluid flow pattern is of importance. The fluid flow pattern will determine the shear stress by the relations of $\tau = -\gamma\mu$, where τ is the shear stress (tangential stress), γ is the shear rate, and μ is the dynamic viscosity of the fluid (Gerhart et al. 1992).

We explored the application of D-OCT in the measurement of fluid flow and shear stress in porous scaffolds. Unlike conventional OCT, D-OCT uses additional phase information of the complex OCT signals to monitor the velocities of moving particles in the backscattered spectrum, providing information regarding the architecture and local fluid flow through the scaffold simultaneously. We used chitosan as model scaffolds because chitosan can be fabricated with a simple technique and the pore parameters can be tailored easily. We specifically investigated the effect of pore size and pore shape on the fluid flow rate and shear stress distribution.

Porous chitosan scaffolds were fabricated by two techniques. Three sets of scaffolds were prepared for the current study:

1. Low porosity chitosan scaffolds (LPCSs) with a pore size ranging from 30 to 100 μm (Jia et al. 2009b)
2. High porosity chitosan scaffolds (HPCSs) with a pore size between 100 and 200 μm (Jia et al. 2009b)
3. Chitosan scaffold with regular, spherical pores and pore size between 106 and 250 μm

To produce the elongating pores in chitosan scaffolds, a freeze drying technique was used. 2% chitosan solution in acetic acid was cast in tubular molds (1.5 mm inner diameter) which were frozen at –20°C overnight and subsequently freeze dried. The resulting scaffolds were re-hydrated in a gradient of ethanol (100%, 70% and 50%) and stored in PBS. The scaffolds with different porosity and pore size were obtained by adjusting the freezing rate (Bagnaninchi et al. 2007). To produce chitosan scaffolds with regular round pores, a combination of porogen and alkaline gelation techniques has been developed. Paraffin spheres with 106–250 μm were used as porogens and were introduced to a tube with an internal diameter of 2 mm and external diameter of 3 mm. One end of the tube had been plugged with molten paraffin and allowed to cool, forming an air tight seal. Chitosan solution was drawn over the spheres by creating a vacuum over the spheres. The chitosan-sphere complexes were frozen at –20°C for overnight and then placed in 3M NaOH in ethanol to gelate and neutralize. After 24 hours, they were removed and rinsed in ethanol and then placed in a hexane solvent and stirred very gently to leach out the paraffin spheres for up to eight hours. Scaffolds were then transferred for storage in PBS.

The chitosan scaffolds were imaged using a laboratory-built spectral domain D-OCT. The system used a superluminescent diode with a central wavelength of 842 nm, which yields a measured axial resolution of ~7 μm in air. The sample light was coupled into a probe, consisting of a pair of *X–Y* galvanometer scanners and the optics to deliver the probing light onto and collect the light backscattered from the sample. The detection system was a custom-built high-speed spectrometer, consisting of a 30 mm focal length collimator, a 1200 lines/mm diffracting grating, and an achromatic focusing lens with 150 mm focal length. The focused light spectrum was captured in parallel by a line scan charge coupled device (CCD). To obtain Doppler shift images, the scaffolds were perfused by a precision DC pump. The perfusion system consisted of 1.5 mm inner diameter tubing connected to the pump. The sample chamber, made of a portion of the tubing, was fixed on a goniometer (used to rotate an object around a fixed axis—the precise angle and position can be controlled by the operator) under the D-OCT probe at an angle of 8° from the horizontal to achieve a Doppler angle of 82°. The system was set up to deliver a constant input flow rate of 0.5 mL/min. A solution of 0.5% latex microspheres (0.3 μm in diameter) suspension was used as the light scattering medium to monitor the fluid flow through the scaffolds situated within the tube (Jia et al. 2009b).

Figure 11.13a shows a typical cross-sectional image of LPCS acquired by the D-OCT system. The bidirectional flow velocity map obtained by D-OCT is presented in Figure 11.13b. Both the distribution and the magnitude of the flow can be seen. Figure 11.13b clearly reveals the heterogeneous distribution of the flow velocity in the porous structures. Although the input flow rate was constant, the local fluid flow in this complex construct varied greatly in both magnitude and direction. Furthermore, the flow in

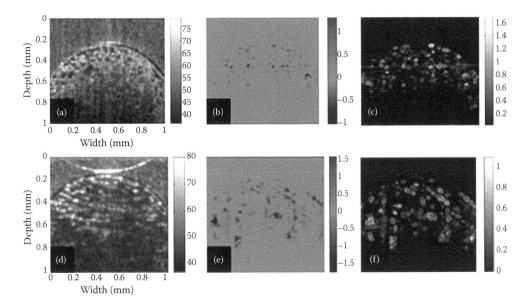

FIGURE 11.13 D-OCT images of fluid flow in LPCS (a–c) and HPCS (d–f). (a, d) Structural images; (b, e) bidirectional local fluid flow images; (c, f) local shear stress distribution images. (From Jia, Y. et al., *Journal of Biomedical Optics* 14:034014–034019, 2009.)

the micropores did not show parabolic distributions. Consequently, the fluid shear stress, as shown in Figure 11.13c, differed between individual pores, with values ranging from 0 to 0.165 N m^{-2}. The structural image, the flow velocity map, the fluid shear stress, and the flow magnitude of a typical cross section of HPCS are shown in Figure 11.13d through f with the same imaging and perfusion conditions as for LPCS. From these figures, we can see that fluid flow existed mainly in the big pores. The magnitude of the flow was inhomogeneously distributed within the pores, resulting in shear stresses ranging from 0 to 0.107 N m^{-2}. The 3D reconstruction of the flow, and its associated isosurface representation, displayed ellipsoidal and elongated pores interconnected in a unidirectional way. The fluid flowed preferentially along sequentially aligned pores (Jia et al. 2009b).

Figure 11.14a shows the OCT structural image of the chitosan scaffold with the round-shaped pore scaffold. When perfusing fluid in such scaffolds, it is found that the flow pattern in the pores (Figure 11.14b) was quite distinct from LPCS and HPCS (Figure 11.13b and e). In general, the flow rates in elongated pores were more homogenous, although the edge of the pores had a low flow rate. The flow rate in round shape pores exhibited a heterogeneous flow pattern. Interestingly, it is found that in a symmetric flow pattern, a wrapping flow pattern has been generated in a round shape pore (Figure 11.14b). Although most pores were round in the scaffold, not all round pores have such defined flow pattern, which might link to the adjacent pore interconnectivity. The considerably higher liquid volume in the round pores might be another factor causing the wrapping flow pattern. This study has clearly delineated the flow pattern in a scaffold with round pores with detailed flow rate distribution.

Chapter 11

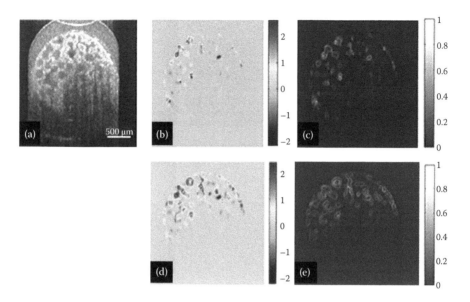

FIGURE 11.14 (a–c) D-OCT images for chitosan scaffolds with round-shaped pore structures. (a) Structural image; (b) flow rate; and (c) shear stress. (d–e) DOMAG images of the same scaffolds in (a), showing the better resolution of the flow rate (d) and shear stress (e).

In summary, D-OCT can reveal localized fluid flow within porous scaffolds. This study has gained new insight on how a process of designing and engineering a scaffold can make impacts on fluid flow pattern and therefore shear stress distribution within the different structural pores when the scaffolds are subject to a constant input flow rate. This knowledge can enhance our understanding for controlling fluid flow within the scaffolds.

11.3.2   DOMAG and Porous Scaffold

In recent years, a more powerful novel imaging technique, called DOMAG, has been developed (Wang and An 2009). DOMAG combines the phase resolved technique in D-OCT and flow signals from optical microangiography (OMAG), which generates low noise flow images, reflecting the real flow rate more precisely. In this study, we undertook a preliminary study of porous scaffolds in DOMAG in order to compare the resulted flow patterns from D-OCT imaging technique.

Figure 11.14d shows the flow pattern images of the same scaffold as in Figure 11.14a by DOMAG. It is very clearly seen that there are two improvements for these images. First is the higher permeation depth of the image, and second is the much clearer flow pattern in each pore using the DOMAG technique, which resulted in more accurate shear stress images (Figure 11.14e). It is reported that the background noise from a non-flow region of the scaffold and random noise were imposed onto D-OCT flow images, resulting in a difficulty with precisely measuring small flow velocity (Jia et al. 2009a; Wang and An 2009). DOMAG is able to battle this issue and separate the background flow signals from the background signals, which leads to low noise images.

FIGURE 11.15   Diagram indicating the future applications of OCT in tissue engineering, highlighting ability to characterize, monitor, and predict features and tissue development in constructs.

11.4   Conclusions

In summary, techniques such as PS-OCT, DOMAG, and D-OCT have been used to illustrate the alterations of the internal structures/properties of tendon tissues, engineering constructs, and scaffolds in response to external conditioning or stimulation or fabrication. These techniques have provided information or parameter changes in a quantitative and qualitative manner, and are used as tools for tracking the perfusion in tissue engineering bioreactors. Our data demonstrate that OCT is one of the most powerful tools for the tissue engineering field, capable of characterization, monitoring, and prediction of construct properties as summarized in Figure 11.15 if the right contract mechanisms are selected. The case studies in this chapter demonstrate the importance and impact of online, nondestructive optical characterization techniques in the development of new strategies for regenerative medicine. Therefore, extensive application and standardization of these techniques will provide beneficial information to researchers and clinicians for the application of tissue-engineered constructs permitting real-time predictions of feedback and characterization of novel therapies.

References

Ahearne, M., P. O. Bagnaninchi, Y. Yang, and A. J. El Haj. 2008. Online monitoring of collagen fibre alignment in tissue-engineered tendon by PS OCT. *Journal of Tissue Engineering and Regenerative Medicine* 2:521–524.

Bagnaninchi, P. O., Y. Yang, N. Zghoul, N. Maffulli, R. K. Wang, and A. J. El Haj. 2007. Chitosan microchannel scaffolds for tendon tissue engineering characterized using optical coherence tomography. *Tissue Engineering* 13:323–331.

Bagnaninchi, P. O., Y. Yang, M. Bonesi, G. Maffulli, C. Phelan, I. Meglinski, A. El Haj, and N. Maffulli. 2010. In-depth imaging and quantification of degenerative changes associated with Achilles ruptured tendons by polarization-sensitive optical coherence tomography. *Physics in Medicine and Biology* 55:3777–3787.

Boppart S. A., M. E. Brezinski, and J. G. Fujimoto. 2000. Optical coherence tomography imaging in developmental biology. *Methods Molecular Biology* 135:217–233.

Brown, R. A., M. Wiseman, C. B. Chuo, U. Cheema, and S. N. Nazhat. 2005. Ultrarapid engineering of biomimetic materials and tissues: Fabrication of nano- and microstructures by plastic compression. *Advanced Functional Materials* 15:1762–1770.

Calve, S., R. G. Dennis, P. E. Kosnik, K. Baar, K. Grosh, and E. M. Arruda. 2004. Engineering of functional tendon. *Tissue Engineering* 10:755–761.

Canty, E. G., and K. E. Kadler. 2005. Procollagen trafficking, processing and fibrillogenesis. *Journal of Cell Science* 118:1341–1353.

Dahners, L. E., G. E. Lester, and P. Caprise. 2000. The pentapeptide NKISK affects collagen fibril interactions in a vertebrate tissue. *Journal of Orthopaedic Research* 18:532–536.

de Boer, J. F., and T. E. Milner. 2002. Review of polarization sensitive optical coherence tomography and Stokes vector determination. *Journal of Biomedical Optics* 7:359–371.

Drexler, W., D. Stamper, C. Jesser, X. D. Li, C. Pitris, K. Saunders, S. Martin, M. B. Lodge, J. G. Fujimoto, and M. E. Brezinski. 2001. Correlation of collagen organization with polarization sensitive imaging of *in vitro* cartilage: Implications for osteoarthritis. *Journal of Rheumatology* 28:1311–1318.

Franchi, M., A. Trire, M. Quaranta, E. Orsini, and V. Ottani. 2007. Collagen structure of tendon relates to function. *The Scientific World Journal* 7:404–420.

Gerhart, P. M., R. J. Gross, and J. I. Hochstein. 1992. *Fundamentals of Fluid Mechanics.* Addison-Wesley Pub. Co. Boston, USA.

Jia, Y., L. An, and R. K. Wang. 2009a. Doppler optical microangiography improves the quantification of local fluid flow and shear stress within 3-D porous constructs. *Journal of Biomedical Optics* 14(5):050504.

Jia, Y., P. O. Bagnaninchi, Y. Yang, A. El Haj, M. T. Hinds, S. J. Kirkpatrick, and R. K. Wang. 2009b. Doppler optical coherence tomography imaging of local fluid flow and shear stress within microporous scaffolds. *Journal of Biomedical Optics* 14:034014–034019.

Kemp, N. J., J. Park, H. N. Zaatari, H. G. Rylander, and T. E. Milner. 2005. High-sensitivity determination of birefringence in turbid media with enhanced polarization-sensitive optical coherence tomography. *Journal of the Optical Society of America a-Optics Image Science and Vision* 22:552–560.

Liao, J., and I. Vesely. 2007. Skewness angle of interfibrillar proteoglycans increases with applied load on mitral valve chordae tendineae. *Journal of Biomechanics* 40:390–398.

Mason, C., J. F. Markusen, M. A. Town, P. Dunnill, and R. K. Wang. 2004. Doppler optical coherence tomography for measuring flow in engineered tissue. *Biosensors and Bioelectronics* 20:414–423.

Nadkarni, S. K., M. C. Pierce, B. H. Park, J. F. de Boer, P. Whittaker, B. E. Bouma, J. E. Bressner, E. Halpern, S. L. Houser, and G. J. Tearney. 2007. Measurement of collagen and smooth muscle cell content in atherosclerotic plaques using polarization-sensitive optical coherence tomography. *Journal of the American College of Cardiology* 49:1474–1481.

Noth, U., K. Schupp, A. Heymer, S. Kall, F. Jakob, N. Schutze, B. Baumann, T. Barthel, J. Eulert, and C. Hendrich. 2005. Anterior cruciate ligament constructs fabricated from human mesenchymal stem cells in a collagen type I hydrogel. *Cytotherapy* 7:447–455.

Park, B. H., M. C. Pierce, B. Cense, and J. F. de Boer. 2003. Real-time multi-functional optical coherence tomography. *Optics Express* 11:782–793.

Provenzano, P. P., and R. Vanderby Jr. 2006. Collagen fibril morphology and organization: Implications for force transmission in ligament and tendon. *Matrix Biology* 25:71–84.

Scott, J. E. 1988. Proteoglycan fibrillar collagen interactions. *Biochemical Journal* 252:313–323.

Scott, J. E. 1996. Proteodermatan and proteokeratan sulfate (decorin, lumican/fibromodulin) proteins are horseshoe shaped. Implications for their interactions with collagen. *Biochemistry* 35:8795–8799.

Sharma, P., and N. Maffulli. 2005. Basic biology of tendon injury and healing. *The Surgeon* 3:309–316.

Torbet, J., and M. C. Ronziere. 1984. Magnetic alignment of collagen during self-assembly. *Biochemical Journal* 219:1057–1059.

Torbet, J., M. Malbouyres, N. Builles, V. Justin, M. Roulet, O. Damour, A. Oldberg, F. Ruggiero, and D. J. S. Hulmes. 2007. Orthogonal scaffold of magnetically aligned collagen lamellae for corneal stroma reconstruction. *Biomaterials* 28:4268–4276.

Wang, R. K., and L. An. 2009. Doppler optical micro-angiography for volumetric imaging of vascular perfusion *in vivo*. *Optics Express* 17:8926–8940.

Yang Y., A. Rupani, P. Bagnaninchi, I. Wimpennya, and A. Weightmana. 2012. The study of optical properties and proteoglycan content of tendons by PS-OCT. *J Biomedical Optics* 17(8):081417.

Yang, Y., I. Wimpenny, and M. Ahearne. 2011. Portable nanofiber meshes dictate cell orientation throughout three-dimensional hydrogels. *Nanomedicine-Nanotechnology Biology and Medicine* 7:131–136.

Zeugolis, D. I., G. R. Paul, and G. Attenburrow. 2009. Cross-linking of extruded collagen fibers—A biomimetic three-dimensional scaffold for tissue engineering applications. *Journal of Biomedical Materials Research Part A* 89A:895–908.

Zhang, G., Y. Ezura, I. Chervoneva, P. S. Robinson, D. P. Beason, E. T. Carine, L. J. Soslowsky, R. V. Iozzo, and D. E. Birk. 2006. Decorin regulates assembly of collagen fibrils and acquisition of biomechanical properties during tendon development. *Journal of Cellular Biochemistry* 98:1436–1449.

Chapter 11

12. Photoacoustic Tomography

Wiendelt Steenbergen

12.1 Introduction

The technology highlighted in this chapter is photoacoustic tomography (PAT). In the context of this book, this includes techniques that in the literature also are being referred to as photoacoustic imaging or optoacoustic imaging. PAT refers to ultrasound imaging technology in which the ultrasound is created by light. The conversion from light to sound is due to absorption by light inside the tissue. Absorption of light by tissue chromophores such as hemoglobin provides a large contrast, while the ultrasound radiation provides a high spatial image resolution and imaging depth. This gives photoacoustics a major advantage compared to purely optical methods to image tissues at a depth larger than 0.5–1 mm.

Optical Techniques in Regenerative Medicine. Edited by Stephen Morgan, Felicity R.A.J. Rose, and Stephen J. Matcher © 2014 CRC Press/Taylor & Francis Group, LLC. ISBN: 978-1-4398-5495-2

Chapter 12

In this chapter, first the measurement challenges in the application field of regenerative medicine will be outlined that photoacoustic imaging is likely to overcome. Next the principles of photoacoustics will be discussed, including the algorithms used for image reconstruction, the current status of the technique (with a distinction between deep imaging and superficial microscopy), and its current applications. Finally, its potential for application in regenerative medicine will be discussed.

12.2   Imaging Challenges in Regenerative Medicine to Be Addressed by PAT

The ultimate aim of regenerative medicine is the restoration of organs and tissues. Throughout the process of regeneration, the tissues undergo a transition in both their structure and function. For instance, the starting point of the therapy may be a porous scaffold of biodegradable polymer with seeded cells and loaded with growth factor releasing vehicles, while the end point is a tissue performing its intended function. A common aspect of many of these therapies will be the development of a functional vascular bed that supports the metabolic processes. The presence of a fully functioning microcirculatory system and of a proper connection to the macrocirculation system are vital conditions for tissues and organs. Therefore, an important aspect of monitoring therapy and assessing its outcome is to observe and follow in time the structure and function of the vascular bed. Keeping in mind that regenerative therapies will become routine in health care, the imaging technology must have the following features:

- It must be noninvasive, nontoxic, and non-ionizing and preferentially work without contrast agents.
- It should have sufficient penetration depth and spatial imaging resolution to provide local information at relevant body sites.
- It should provide quantitative information regarding the functioning of the tissue as far as witnessed by the blood, represented by quantities such as blood concentration, blood flow, and oxygenation.
- It should be applicable in an outpatient setting, therefore leading to requirements regarding the total costs, the level of staff training, and the speed of the procedure.

This book provides a broad picture of the use of light in regenerative medicine and highlights the unique features of optical techniques. In principle, the large absorption contrast of blood compared to bloodless tissue, and the unique absorption spectra of hemoglobin and oxyhemoglobin, make light the radiation of choice for imaging blood and its oxygenation. However, the disadvantage of purely optical techniques when used *in vivo* is the dominance of light scattering. Scattering of light makes optical imaging of the vascular system and the blood oxygenation based on ballistic (= unscattered) light impossible at depths larger than several hundreds of micrometers. Even in the presence of strong scattering, the use of mathematical inversion techniques may result in obtaining local quantitative information from scattering tissues; however, light diffusion puts a fundamental limit on the spatial resolution and therefore on the sensitivity of all-optical techniques for absorption imaging. This fundamental limit associated with all-optical techniques is overcome by PAT, which enables imaging of the spatial distribution of optical absorption at resolutions defined by the ultrasound waves. In short, PAT allows for high-resolution imaging of optical absorption. In the context of regenerative medicine, its primary virtue is to visualize the development of the vascular bed and the blood oxygenation level until a depth of various centimeters, at a resolution ranging from 10 to 50 μm at a depth of 0–5 mm, up to 200–500 μm at a larger depth. The maximum depth that can be imaged depends on many factors, such as the laser pulse energy, the noise equivalent pressure (NEP) of the ultrasound transducer, and the scattering and absorption properties of the tissue at the used laser wavelength. The signal-to-noise ratio, and therefore the imaged depth, can be improved by averaging signals generated by several laser pulses, but this will increase the image acquisition time. Other applications will be speculated on at the end of this chapter. An extensive textbook description of PAT and its theory, implementations, and applications for biomedical purposes can be found in Wang (2009). Recent reviews of photoacoustic imaging are given in Xu and Wang (2006), Tang et al. (2010), and Mallidi et al. (2011).

12.3 Principles and Practice of Photoacoustics

In this section, the principles of PAT will be described. The process steps (light propagation, absorption, thermal conversion, stress relaxation, ultrasound propagation, and detection) will be elaborated. Attention will be given to the specific features of photoacoustic signals: their shape for ideal absorbers, stress and thermal confinement aspects, and the large bandwidth. Furthermore, we will describe methods to reconstruct images from ultrasound signals. Finally, the most important technical implementations of the technology will be discussed.

12.3.1 Ultrasound Generation by Light

Light or generally electromagnetic radiation can generate sound through various processes, each working at different optical energy density regimes. The process of specific importance in biomedical photoacoustics is the thermoelastic effect: the formation of mechanical stress upon local absorption of light. While the pressure (or stress) distribution in a tissue as a function of time in response to light is generally described by an inhomogeneous partial differential equation (see Chapter 1 in Wang 2009), here we take a simpler point of view for the sake of conveying the basic principles. The situation is depicted in Figure 12.1, where a turbid medium is illuminated by a short light pulse (Figure 12.1, left). Inside the tissue, this leads to an optical fluence distribution $\Phi(x, y, z)$ (in J/m^2) and a distribution of absorbed optical energy density $E_a(x, y, z) = \mu_a\Phi$ (in J/m^3) with $\mu_a(x, y, z)$ the local optical absorption coefficient (1/m). Absorption of light will have two effects: (1) light will be reemitted by fluorescence, and (2) the local temperature will increase. The energy density associated with fluorescence will depend on the quantum yield, which for naturally occurring chromophores such as hemoglobin and melanin will be relatively low (<0.01), so in the remaining text, it will be neglected. The amount of temperature increase will depend on the time scale in which the light has been deposited, hence the pulse duration τ_p, compared to the time scale τ_d, in which heat is lost by (mainly) diffusive transport. Here we will assume that $\tau_p \ll \tau_d$, which means that within

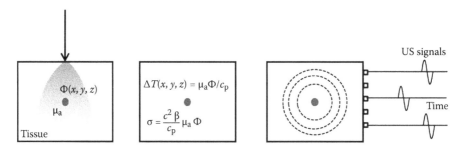

FIGURE 12.1 Ultrasound generation in tissue by light pulses. Left: pulsed light leads to a local fluence distribution $\Phi(x, y, z)$ as a result of scattering and absorption and absorption in the entire medium. Center: local optical absorption leads to temperature increase ΔT and thermoelastic stress σ (fluorescence is neglected here). Right: relaxation of stress leads to emission of (ultra)sound, which can be measured at the tissue surface.

the laser pulse duration, no heat is lost by diffusion, a situation that is usually referred to as thermal confinement. Even the smallest absorbing structures in the body, which are the blood capillaries with a diameter of 9 μm, will have $\tau_d > 20$ μs (see Welch and van Gemert 1995 for an overview of work on heat flow in laser-irradiated tissue).

As we will see, this time scale by far exceeds the pulse duration of lasers used for photoacoustics. Hence, after the laser pulse, we obtain a temperature distribution $\Delta T(x, y, z) = \mu_a \Phi/c_p$ with c_p being the specific heat capacity of the tissue (Figure 12.1, center). Due to the temperature rise, the tissue will locally expand. On the other hand, in the usual case of an inhomogeneous temperature rise, such as by blood compared to the relatively cooler surrounding medium, the expansion will be frustrated. The thermal expansion interacting with the surrounding medium will lead to a mechanical stress. This stress is conveyed to the neighboring matter, and in this way, we see the formation of a stress wave: the locally higher stress relaxes by the emission of ultrasound. The actual stress level reached on light absorption will depend on the time scale τ_s at which the stress relaxes. For an absorbing object of size d, we have $\tau_s = d/c$ with c being the speed of sound within the object. For maximum sound creation, we require the so-called stress confinement, which implies that the pulse duration is much shorter than the transit time of the ultrasound through the absorbing object; hence $\tau_p \ll \tau_s$. For a typical speed of sound $c = 1540$ m/s in soft biological tissues, this means that for blood capillaries of 9 μm diameter, we require laser pulses with a duration $\tau_p \ll 6$ ns. This means that for typical Nd:YAG lasers giving pulse durations of 5–10 ns, which are often used for photoacoustics, the condition of stress confinement may not be obeyed for the smallest blood vessels. If we nevertheless assume stress confinement, we obtain a local stress distribution at the end of the laser pulse:

$$\sigma(x, y, z, t = 0) = \Gamma \mu_a \Phi = \frac{c^2 \beta}{c_p} \mu_a \Phi \tag{12.1}$$

with β as the thermal expansion coefficient. The parameter Γ is called the Grueneisen parameter, which describes the efficiency of conversion of thermal into mechanical energy. The resulting situation is a medium in which regions of higher absorption possess an elevated mechanical stress level. The medium will return to its equilibrium through internal pressure waves, which are actually sound waves. These waves will travel through the medium in a relatively undisturbed manner to reach the tissue surface, where they can be detected with ultrasound transducers (Figure 12.1, right). In principle, PAT signals can be measured with a standard ultrasound transducer as used in normal medical ultrasound imaging devices. However, it is important to keep in mind the difference between normal ultrasound imaging and PAT. Normal ultrasound imaging works on the basis of ultrasound waves internally reflected by the tissue, while PAT images the tissue with ultrasound generated inside the tissue. This means that the spectral content of normal ultrasound imaging is mainly determined by the ultrasound injected in the medium. In contrast, the spectral content of photoacoustic signals is determined by the size of the absorbers. For a spherical absorber with a Gaussian absorption distribution (assumed for mathematical ease), Sigrist and Kneubuhl (1978) proved the photoacoustic signal to be bipolar. Based on their solution, Kolkman et al. (2004a) proved that a blood vessel with diameter d generates a bipolar signal with a peak-to-peak time of $\tau_{pp} = d/2c$. Elaborating on this model, they calculated the ultrasound frequency range as a function of blood vessel diameter, as shown in Figure 12.2 (Kolkman et al. 2008a). This figure shows that for an absorbing object with a size smaller than 1 mm, the sound frequencies all exceed the audible range and in fact are in the ultrasound range. What does this mean for photoacoustic imaging? Firstly, either the ultrasound spectrum or the temporal signals allow for determining the size of regions that absorb light, provided (1) that an ultrasound transducer is used, which has sufficient measurement bandwidth and (2) that all frequencies will reach the detector. Due to frequency-dependent ultrasound attenuation, the imaging resolution of PAT will generally deteriorate with depth. Secondly, for a known speed of sound c, we can localize optical absorbers based on combining signals measured by multiple transducers as shown in Figure 12.1 (right). Hence, PAT can provide three-dimensional images of optical absorption, with the imaging resolution governed by the ultrasound. Since scattering and attenuation of ultrasound by biological tissues is orders of magnitude lower than that of light, the imaging quality of PAT compared to diffuse optical tomography is superior for tissue regions where diffuse light dominates (hence regions larger than 0.5–1 mm).

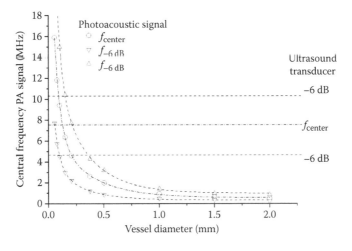

FIGURE 12.2 Frequency range of photoacoustic signals vs. blood vessel diameter. (Reprinted from Kolkman, R. G. M. et al., *Journal of Biomedical Optics* 13, 2008. With permission by SPIE.)

12.3.2   Photoacoustic Systems

12.3.2.1   Light Sources

Photoacoustic signals can be obtained with both modulated continuous-wave light sources and pulsed light sources. Modulated continuous wave (CW) light sources combined with lock-in detection methods might have advantages in terms of signal-to-noise ratio (Fan et al. 2004). In between CW and pulsed mode is the use of a femtosecond laser with 80 MHz pulses, combined with lock-in detection techniques (van Raaij et al. 2010). However, the vast majority of PAT systems are based on pulsed light sources. Pulsed light sources have the advantage of being more widely available than CW light sources with sufficient power and tuning range of the modulation frequency. While for CW light sources a large modulation frequency range is required, pulsed sources with sufficiently short pulse durations intrinsically provide all these frequencies. Furthermore, pulsed mode photoacoustics is more compatible with the available ultrasound detection array technology and will allow for integration of PAT and traditional medical echography, as will be described in Section 12.3.2.2.

Choosing a light source for PAT is guided by the following technical considerations:

- The pulse duration should be sufficiently short to comply with the condition of stress confinement as given in Section 12.3.1.
- The pulse energy should allow for generating photoacoustic signals at the required depth in the tissue, while observing the maximum exposure limits as imposed by safety standards such as the IEC laser safety standard IEC 60825-1.
- The wavelength should provide sufficient penetration depth in the tissue and sufficient absorption contrast between the tissue and the chromophore.

Of primary interest is that the wavelength is chosen appropriately. This is mainly guided by the absorption properties of the tissue and the main chromophores to be imaged. Absorption spectra of three main tissue chromophores are given in Figure 12.3. While the absorption spectrum of the hemoglobins would suggest a wavelength in the

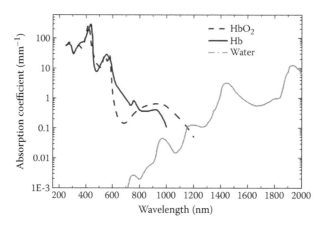

FIGURE 12.3   Absorption spectra of the main tissue chromophores hemoglobin, oxyhemoglobin, and water. (From Kolkman, R. G. M. et al., *Physics in Medicine and Biology* 49:4745–4756, 2004.)

blue-green region (450–550 nm) where the absorption is high, the high scattering at these wavelengths strongly limits the penetration depth of light.

The local excitation is low beyond a certain depth, reducing the quantity $\mu_a\Phi$, which determines the level of photoacoustic signals. For imaging at small depths, a frequency-doubled ND:YAG laser may be used at 532 nm or a frequency-tripled Nd:YAG with an optical parametric oscillator (OPO) for wavelength tuning in the range 480–650 nm (Ma et al. 2009). However, more often wavelengths in the red to near-infrared range are used, such as Nd:YAG lasers at 1064 nm, frequency-doubled ND:YAG lasers with an OPO for the range 690–950 nm, alexandrite lasers, mainly at 755 nm (Brecht et al. 2009), or tunable dye lasers pumped by an Nd:YAG laser (Kim et al. 2010) with a broad tuning range in the visible and near-infrared, and Ti:Sapph oscillators pumped by a frequency doubled Nd:YAG laser (Ku and Wang 2005) giving a wavelength range of 650–1000 nm.

Using the criterion for stress confinement given in Section 12.3.1, imaging the smallest blood capillaries with a size of 6–8 μm requires lasers with a pulse duration shorter than 5 ns. The additional requirement of sufficient pulse energy in many cases leads to the use of Nd:YAG lasers (as stand-alone or pumping other lasers) with a pulse duration of 5–15 ns as a trade-off. Also, larger pulse durations can be used, for instance, when it is not necessary to image the smallest possible objects such as blood capillaries, but rather bigger vessels such as arteries and veins.

As for the pulse energy: the general approach will be (apart from budget considerations) to apply the maximum pulse energy that can be practically used within the safety limits. This implies that photoacoustic microscopy with focused illumination will use pulses with nanojoule energy (Zhang et al. 2010a). At the other extreme, pulses with energy of tens or hundreds of millijoules will be used for imaging objects at depths of centimeters (Kim et al. 2010; Ku and Wang 2005; Manohar et al. 2007). In order not to exceed exposure limits, in this case, the laser beam has to be expanded to spread the energy over a large tissue area. For deep tissue imaging, this is not disadvantageous since at several centimeters depth, the light beam will expand by diffusion anyway.

All lasers mentioned above have the disadvantage of being bulky: they need an optical table and a remote station for driving and water cooling. A recent development of high interest is the advent of quasi-continuous wave diode lasers with ever-increasing powers. These light sources, including their drivers, have the size of a cigarette box. Currently, they are mainly available within the wavelength range of 700–900 nm, which is suitable for photoacoustics. Even if their pulse energies as yet do not exceed 20 μJ for a pulse duration of 100 ns, their use for biomedical photoacoustics has been demonstrated (Allen and Beard 2006; Kolkman et al. 2006). The low pulse energy is partly compensated by the high pulse rate of more than 1000 pulses/s. Since the performance of high power diode lasers is ever increasing, it is likely that these very compact and low-cost light sources will replace Nd:YAG lasers with pulse energies within the millijoule range, in those situations in which the pulse duration range of 100–200 ns is sufficient from the application point of view. These pulse durations will allow for imaging resolutions of 150–300 μm.

Other new lasers currently being used for photoacoustics are fiber lasers and microchip lasers. In Shi et al. (2010), a diode-pumped fiber laser was evaluated with a pulse repetition rate of 100 kHz, a pulse duration of 250 ns, a wavelength of 537 nm, and a pulse energy of 6.5 μJ. The microchip laser was a diode-pumped Nd:Cr:YAG chip giving 1064 nm, with a frequency doubling fiber yielding 532 nm. Pulse rates of 10 kHz and

Chapter 12

pulse energies of 60 μJ for a 1 ns pump were realized. The above pulse energies exceed those needed for photoacoustic microscopy. In terms of compactness, these lasers must be considered in between standard Nd:YAG lasers and diode lasers.

The above illustrates the wide choice of light sources available today, and it is to be expected that in the near future, major steps will be made toward more compact and affordable high-performance light sources for photoacoustics.

12.3.2.2 Measurement Configurations

Since photoacoustics is a highly dynamic field of research, the amount of system configurations has increased tremendously in the last few years. Giving a complete overview is beyond the scope of this book, and some of the early arrangements will soon cease to be relevant. Therefore, the discussion will be restricted to those configurations that are likely to be relevant for research and (pre)clinical application of therapies in regenerative medicine.

In general, we currently see a rapid change from PAT systems based on single ultrasound transducers to systems using ultrasound detection arrays. Since single ultrasound transducers have to be scanned for obtaining an image, this development implies a drastic reduction of the image acquisition times. Although not giving full coverage, we distinguish between reflection mode imaging systems and systems for PAT in the true sense. From the clinical point of view, reflection mode systems have the obvious advantage of only requiring one-sided access to the tissue. True tomography, in contrast, requires detection of photoacoustic signals all around the object. Therefore, reflection mode systems are more suitable for clinical application and for high-resolution imaging of blood vessels close to the tissue surface where the ultrasound is not deteriorated by attenuation. True PAT systems, on the other hand, have the potential of whole body imaging of small animals and of imaging extremities of the human body, such as breasts, fingers, and limbs.

12.3.2.2.1 Reflection Mode Systems

Given the feasibility of using nanosecond pulses, it must be possible to achieve imaging resolutions of several micrometers for small tissue depths where optical attenuation is less relevant. From our earlier discussion, it is clear that the use of ultrasound transducers with a bandwidth of 50 MHz or more is a prerequisite. However, for such high-resolution photoacoustic imaging, more challenges have to be met. The maximum amount of acoustic energy emitted per voxel (volumetric pixel) will be proportional to the voxel volume and therefore will rapidly decrease with voxel size. Furthermore, a good resolution is required both longitudinally and laterally. The problem is illustrated in Figure 12.4a, showing a photoacoustic B-scan of the skin of a human volunteer. The image has been taken with a double-ring sensor (Kolkman et al. 2004b) scanning parallel to the skin surface and providing focused detection of the photoacoustic (PA) signals. Since in this configuration each blood vessel gives a bipolar signal, the blood vessels appear as pairs of lobes representing the upper and lower sides of the lumen. Fitting the signals composing the B-scan with bipolar model functions and subsequent integration along vertical lines gives more "natural" images of the blood vessels, as shown in Figure 12.4b. While the longitudinal or axial resolution is secured by a laser pulse duration of 8 ns and an ultrasound detection bandwidth of 50 MHz, the vessel image is elongated in the lateral direction. The reason is that the image is based on ultrasound collected only

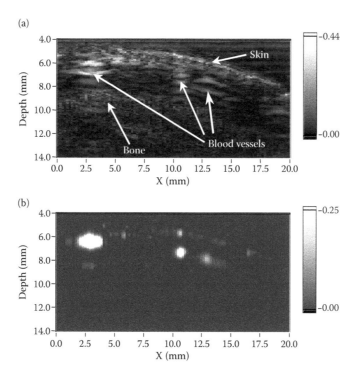

FIGURE 12.4 Photoacoustic B-scan obtained with a double-ring sensor. The lateral resolution is impaired by the limited NA of the acoustic detection. (a) B-scan of rectified line signals. (b) Result of fitting with bipolar model function and subsequent integration. (From Kolkman, R. G. M. et al., *Journal of Biomedical Optics* 9:1327–1335, 2004. With permission by SPIE.)

within a limited range of angles, hence with a limited numerical aperture (NA) for detection. In a manner completely analogous to optical microscopy, the lateral resolution will improve when the angular range over which ultrasound is detected is increased.

Figure 12.5 illustrates three strategies to image in reflection mode a photoacoustic point source (= a point absorber) with both high sensitivity and lateral (or transverse) resolution by collecting light ultrasound from the photoacoustic point source with a large NA. These strategies apply to all reflection mode PA systems, both for high-resolution

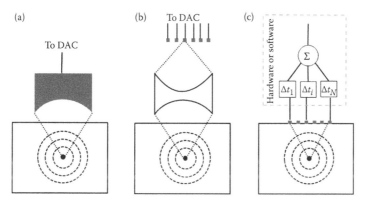

FIGURE 12.5 Various configurations of photoacoustic imaging in reflection mode. (a) Focused ultrasound transducer. (b) Acoustic lens. (c) Array detector with electronic/digital focusing.

Chapter 12

microscopy and for lower-resolution imaging of deeper objects. In Figure 12.5a, the ultrasound is collected using a single focused ultrasound transducer with focusing achieved by the concave shape of the detector surface. This approach is often used in photoacoustic microscopes (Shi et al. 2010; Rao et al. 2010). Another form of focusing is achieved by using an array of concentric ring detectors. For PA, imaging this approach has been adopted by Kolkman et al. (2004b). They used a double-ring sensor that provides dynamic focusing by varying the delay between the signals of both rings. While focused transducers can achieve a high signal-to-noise ratio, in particular, for focused light beams, their disadvantage is the need to scan the tissue mechanically in the lateral direction to obtain a 2D or 3D image. Although rapid scanning techniques are available, mechanical scanning will add to the complexity of such systems. This may be overcome when using acoustic lenses such as illustrated in Figure 12.5b.

While this concept was first used for imaging the photoacoustic stress in a water tank that was scanned using a Schlieren system (Niederhauser et al. 2004), it has been used later in photoacoustic systems using a single transducer (which makes scanning necessary) or an ultrasound array (Valluru et al. 2011; Chen et al. 2007; Balogun et al. 2010; Chen et al. 2010; Wei et al. 2008). While a lens provides a planar image, still some sort of depth scanning is required. The most compact solution that does not need mechanical scanning for obtaining images at various depth is the use of ultrasound detection arrays, which are placed directly on the tissue surface (see Figure 12.5c). In these systems, image reconstruction is performed by giving the signal of each detector element a proper delay after which all signals are summed, using the so-called "delay-and-sum" algorithm (Hoelen and de Mul 2000). This procedure can be performed in software or in hardware. Hardware implementation is present in many standard echography devices and will become obsolete with the expected complete digitization of ultrasound imaging. We can regard this process as electronic focusing. Each detector element is a fragment of a tunable "lens," and the focusing quality will increase with the number of elements. This suggests the use of many small elements. However, we still must require that each element provides sufficient signal-to-noise ratio, which imposes a lower limit on the element size. This challenge may be overcome by optical detection of ultrasound, for instance, on the basis of a polymer film that acts as a Fabry–Perot interferometer (Zhang et al. 2008). Ultrasound arrays with electronic focusing for image reconstruction have been used for high-resolution photoacoustic imaging, providing real-time images of subcutaneous vasculature and its dynamics (Song et al. 2010). An image of the vasculature in the human hand obtained with this system is shown in Figure 12.6. Blood vessels are visible down to a depth of 2 mm. In the vertical cross sections through the blood vessels, the difference in axial and lateral resolution can be observed. Array transducers with software-based image reconstruction also have been used for lower-resolution imaging of large tissue volumes, such as in photoacoustic mammography (albeit in transmission mode) (Manohar et al. 2007). Transmission mode photoacoustic imaging with synthetically focusing ultrasound arrays enables the integration of photoacoustics in a normal clinical echography system. Steps toward such a hybrid imaging modality have been made by various groups (Kolkman et al. 2008a; Kim et al. 2010; Niederhauser et al. 2005; Zeng et al. 2004; Haisch et al. 2010). In all cases, fiber optic light delivery was performed by an optical element added to the ultrasound transducer, rather than by integration of optics inside the probe. Combining photoacoustic and

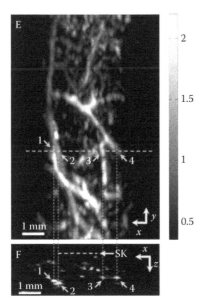

FIGURE 12.6 Photoacoustic image of vasculature in the human hand, using a photoacoustic microscope with an ultrasound detection array. (From Song, L. A. et al., *Journal of Biomedical Optics* 15, 2010. With permission by SPIE.)

ultrasound imaging will result in a versatile imaging modality in which the echographic ultrasound images, which are mainly structural, will be enriched with the functional information of the vascular bed provided by photoacoustics.

12.3.2.2.2 True PAT Systems

The contrast between reflection mode PAT and true PAT is that in the latter, ultrasound signals are collected from many positions around the object. This will give better photoacoustic images than reflection mode systems. This can be understood from the idea that image reconstruction basically involves reconstruction of the initial pressure distribution that was obtained immediately after the laser pulse. This reconstruction will be exact when the complete wave is recorded over a closed area surrounding the object, provided that the acoustic properties of the medium are known.

True PAT systems are ideally suited for research on small animal models. The early systems (Kruger et al. 1995; Wang et al. 2002) collected ultrasound signals using a single detector rotating around the objects. This resulted in long scanning times, so imaging was mostly reduced to single tissue slices. A first realization of a PAT system with multiple detectors is by Kruger et al. (2003). They used a linear ultrasound array, observing the ultrasound emitted by a phantom or animal rotating along its axis in 16 steps to realize 360° observation. A slice image was obtained with a thickness of 1.5 mm and an in-plane resolution of 200 μm. In recently developed systems, ultrasound detection arrays are placed in a curved arrangement, while the systems mainly differ in the configuration of the illumination. An impression of the various arrangements of illumination and detection published recently is given in Figures 12.7 through 12.8.

The system described by Brecht et al. (2009) contains a curved array with 64 detectors, with the arc spanning 152° (see Figure 12.7a). Each detector is square, given a similar

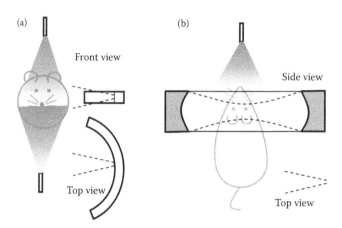

FIGURE 12.7 (a) PAT systems by Brecht et al. (2009) comprising 64 detection elements with identical divergent angular apertures (dashed lines) in two directions, arranged on an arc. Illumination is performed from two sides. (b) PAT system by Gamelin et al. (2009) with a completely closed array ring containing 512 curved transducers, focused in one direction and divergent in the other direction.

angular detection aperture both in the plane of the arc and perpendicular to it. By rotating the animal, detection along an almost complete spherical surface is realized. The resolution reported is 0.5 mm, and the acquisition time is 8 min for 2.4° steps. Illumination is performed from the sides using a fiber bundle, with the beam illuminating the entire object.

Gamelin et al. (2009) describe a system with a closed ring array with 512 elements, as illustrated in Figure 12.7b. The radius of the ring is 25 mm. The concave shape of the elements provides focusing in one direction, limiting detection to a slice with a thickness of 0.6–2.5 mm. The spatial resolution in the other two directions is 200 μm. Since all 512 detectors are read out simultaneously, a rate of 8 frames/s is obtained. This system is mainly used for imaging mouse brain, where the object is illuminated from the top, in a direction perpendicular to the plane of the detection ring.

The system by Taruttis et al. (2010) contains 64 elements placed on an arc that spans 172° (Figure 12.8a). The concave elements provide cylindrical focusing that limits observation to the plane of the transducer. Scanning the array along the axis of the animal allows for 3D imaging. Single slice imaging at 10 fps is possible, limited by the pulse repetition rate of the laser. A set of 10 fibers provide illumination all around the animal, in a plane parallel to the plane of the slice.

Jose et al. (2011a) describe a system with 32 elements over an angle of 85° and with a radius of 40 mm (see Figure 12.8b). The elements are relatively long in one direction, which provides some focusing, giving a slice thickness of 5 mm. Illumination is performed by a fiber-delivered beam opposite to the detection array. By rotating the animal in steps of 8°, 360° detection can be realized. A particular feature is the use of a passive optical absorber (horsetail hair or graphite rod) in front of the optical fiber. This creates a cylindrical ultrasound wave that is recorded simultaneously with the photoacoustic wave from the animal. Based on the arrival times of the cylindrical ultrasound wave, images of the distributions of the acoustic speed of sound and acoustic attenuation can be constructed.

An example of the potential of true PAT is the whole body image shown in Figure 12.9, taken from Brecht et al. (2009). The relative visibility of arteries, veins, and organs

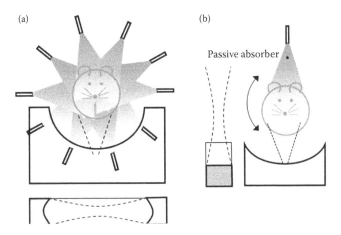

FIGURE 12.8 (a) The system by Taruttis et al. (2010) contains 64 elements placed on an arc. The concave elements provide focusing (dashed lines) that limits observation to the plane of the transducer. A set of 10 fibers provide illumination all around the animal. (b) System by Jose et al. (2011a) containing a 32-element curved array with rectangular elements providing some focusing in one plane. The passive absorber enables speed-of-sound and acoustic attenuation tomography.

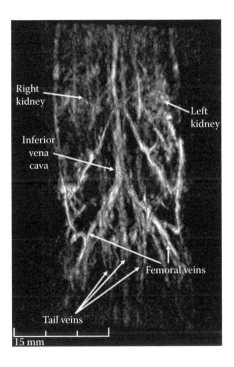

FIGURE 12.9 Photoacoustic whole body image obtained with the system described in Figure 12.7a of a nude mouse illuminated at 1064 nm. (From Brecht, H. P. et al., *Journal of Biomedical Optics* 14, 2009. With permission of SPIE.)

Chapter 12

such as kidneys will depend on the wavelength used, due to the dependence of penetration depth on the wavelength, and the wavelength-dependent response of optical absorption on oxygenation.

The configurations of the various true PA tomographers can also be used for imaging extremities of the human body such as breast, fingers, and limbs. Examples of these applications will be discussed in Section 12.4.3.

12.4   Image Reconstruction

Since a photoacoustic measurement results in a collection of pressure signals recorded outside the object, reconstruction of the interior of the object will result in some pressure fields, too. Actually, a perfect reconstruction of the object's interior encompasses reconstruction of the initial thermoelastic stress distribution $\sigma(x, y, z, t = 0)$, where $t = 0$ denotes the end of the laser pulse.

In a medium without absorption or scattering of ultrasound, the propagation of sound waves is governed by a wave equation, which is time invariant. The consequence is that if we record a collection of waves (by definition obeying the wave equation) moving through a medium, playing the recorded waves backward will result in a wave field that also obeys the wave equations. Since photoacoustic waves are of a broadband nature and therefore have a limited time duration, if we "freeze" the photoacoustic wave when it is completely outside the medium and we apply the wave equation in the reverse direction, eventually the situation will return to the initial stress distribution $\sigma(x, y, z, t = 0)$. This procedure represents an ideal case that cannot be realized: in photoacoustic imaging, we cannot freeze the wave, but at best we can record the wave by obtaining ultrasound signals in a limited number of transducers with finite size and aperture, placed on a surface. Furthermore, the acoustic properties of the medium are not exactly known, and acoustic attenuation will destroy a certain part of the information, in particular, regarding small structures that generate high ultrasound frequencies. Nevertheless, most image reconstruction methods are more or less based on the above idea of "time reversal," a concept that has been elaborated for reversing the propagation of ultrasound waves (Fink 1992).

The most straightforward image reconstruction method is the backprojection algorithm. From the wave equation, it can be derived that for a homogenous lossless medium, the local pressure as a function of time $p(r,t)$ in point r is given by the Poisson integral

$$p(\vec{r},t) = \frac{\partial}{\partial t}\left(\frac{t}{4\pi} \iint_{|\vec{r}'-\vec{r}|=ct} \sigma(\vec{r}',t=0) \right) d\Omega \qquad (12.2)$$

(see also Figure 12.10 as illustration) which implies that the pressure in a point at time t is completely determined by the integrated initial pressure (or stress) on a sphere around that point with radius ct, c being the speed of sound.

The concept that the pressure transient at a certain point provides information on the average pressure history on a sphere expanding in time leads to the most simple and straightforward image reconstruction algorithm, which is radial backprojection of the recorded signals. The method is illustrated in two dimensions in Figure 12.11, where the

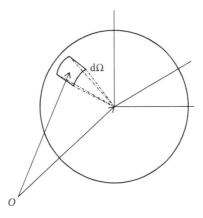

FIGURE 12.10 As a solution of the wave equation, the pressure transient in $\vec{r}'$ at time t is determined by the pressure at time 0 on a sphere centered around $\vec{r}'$, with radius ct.

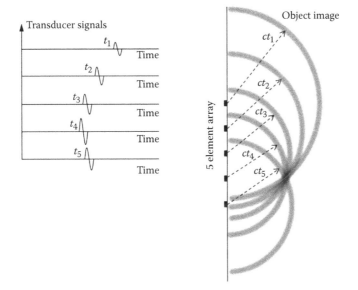

FIGURE 12.11 Image reconstruction using radial backprojection, illustrated for an object containing a single spherical absorber, imaged with a linear array of five detectors. In the backprojection, no angular weighting function has been assumed. The bipolar signal has been transformed into a monopole. The limited number of detectors leads to arc-shaped artifacts. An angular aperture of 2π has been assumed.

signals of a linear array of ultrasound detectors are shown when measuring on an object containing one spherical absorber. Backprojection of the signals in all directions leads to spherically symmetric pressure distributions, which intersect at the location of the assumed spherical absorber. The bipolar signals have been transformed to monopoles by integration and $1/r$ attenuation has been corrected for. This type of backprojection leads to large arc-shaped artifacts, which can be reduced by increasing the number of detectors. In practice, detectors will have a limited angular aperture. In the configuration of Figure 12.11, this will lead to an attenuated contribution of the outer detectors and consequent blurring of the image in the lateral direction. Furthermore, since the aperture function is symmetrical, off-axis absorbers will lead to imaging artifacts situated on-axis. Other reconstruction methods used for planar detection geometries are the "delay-and-sum" algorithm (Hoelen and de Mul 2000) and the synthetic aperture focusing technique (Liao et al. 2004). Although they will differ in detail and involve refinements that reduce imaging artifacts and improve lateral resolution, they do not fundamentally differ from the backprojection approach described above. The above reconstruction algorithms work in the time domain and may be relatively time consuming. Furthermore, the physics of wave propagation, such as the deformation of monopolar initial pressure distributions into bipolar signals, is not present implicitly. This is partly overcome by the introduction of Fourier transform methods such as found in Kostli and Beard (2003), and Kostli et al. (2001).

While the above reconstruction methods are attractive because of their simplicity, they are not quantitative in the sense that, in most cases, essential aspects of the physics of ultrasound propagation are ignored. The consequence is that while the size and shape of absorbing structures may be retrieved, the distribution of the absorbed energy density is not. Hence, the eventual determination of absorption coefficients or concentrations of chromophores is not possible. For that reason, in the past decade, many other, more exact image reconstruction algorithms have been developed, for cylindrical, spherical, and planar geometries. These approaches have in common that they are based on an analytic solution of the inverse problem.

Here we mention the approximation of the Radon transform developed by Kruger et al. (1999). While the usual Radon transform involves projections on planar surfaces, here integration is performed on cylindrical or spherical surfaces. However, if we assume the object to be placed in the center of the sphere, while the size of the object is much smaller than the radius of the sphere, the exact Radon transform is obtained, and the reconstruction equation becomes

$$\sigma(\vec{r}', t=0) \propto \iint_{\Omega} \left(t \frac{\partial p(\vec{r}_0, t)}{\partial t} + 2p(\vec{r}_0, t) \right)_{t=|\vec{r}'-\vec{r}_0|/c} \mathrm{d}\Omega \qquad (12.3)$$

with $\vec{r}_0$ being the collection of position vectors of the cylinder or sphere at which ultrasound detection takes place, $p(\vec{r}_0, t)$ being the detected pressure transients, and $\sigma(\vec{r}', t=0)$ being the reconstructed initial pressures. It must be stressed that this reconstruction method will be less accurate for objects that are too large compared to the distance from the center of the object to the ultrasound detectors. Other reconstruction algorithms using a form of backprojection while having a wider validity are given in Wang et al. (2002) and Xu and Wang (2005). A review of more sophisticated image reconstruction algorithms can be found in Xu and Wang (2006). In view of the analytic solutions underlying these methods, these algorithms may be more or less time consuming, and often some form of regularization is needed to keep the solutions within the realm of physical reality, such as smoothing. A recent approach that is more computationally efficient is the "interpolated-matrix-model inversion" approach (Rosenthal et al. 2010). Here "model-based inversion" means that rather than an underlying analytic solution, the photoacoustic image is obtained by searching for a minimum error between the measured signals and signals expected from the modeled initial stress distribution.

12.5   Applications of Photoacoustics

12.5.1   Challenges in Quantitative Detection of Tissue Chromophores

Since PAT is a modality that images optical absorption, it would be reasonable to expect that eventually we can obtain the concentration of chromophores. This, however, is impossible for the state-of-the-art PAT methodology. The background of this limitation is demonstrated by Equation 12.1, in which we may substitute for the absorption coefficient (assuming a single dominant chromophore) $\mu_a = c\varepsilon_a$, with c the chromophore concentration and ε_a its molar absorption coefficient. Suppose that we succeed in reconstructing the initial stress distribution $\sigma(x, y, z, t = 0)$; from Equation 12.1, we observe that we additionally need to know the local fluence Φ and the Grueneisen parameter Γ. While the latter may be obtained from an *in vitro* experiment, the fluence distribution $\Phi(x, y, z)$ is essentially unknown. The local fluence in a certain tissue location will depend on the scattering and absorption properties in those regions of the tissue that were addressed by the photons prior to absorption. Therefore, one of the current challenges in photoacoustics is to achieve quantification by correcting the photoacoustic image for the local fluence Φ. Here we mainly distinguish six approaches:

(1) In specific cases, it is possible to translate the temporal shape of the photoacoustic signal into an optical extinction coefficient of the blood in one specific blood vessel. While making assumptions on the influence of scattering, the extinction coefficient can then be related to the absorption coefficient μ_a, which in turn may give the hemoglobin concentration c. This approach has been applied to the carotid artery (Esenaliev et al. 2004), where the photoacoustic signals have been fitted with solutions inspired by optical diffusion theory. In general, however, such a solution is not available.

(2) The local fluence can be measured by obtaining PA signals of a dye, which is injected in the blood stream with a known concentration (Rajian et al. 2009). However, this is an invasive procedure and dye dilution will complicate its application.

(3) Another approach is the iterative combination of photoacoustic signals with a model for light transport (Yuan et al. 2007; Cox et al. 2006, 2009; Laufer et al. 2007, 2010), while in some cases, measurements are added on diffuse optical reflection or transmission (Ranasinghesagara and Zemp 2010; Yin et al. 2007).

(4) Fluence variation corrections were realized by making use of the different spatial variations of the variations in fluence and absorption, and describing the latter in terms of wavelets (Rosenthal et al. 2009).

(5) An until now numerically investigated approach is the application of multiple points for pulsed illumination (Shao et al. 2011; Zemp 2010). Most of the above methods know inherent limitations in terms of the prior knowledge required, for example, of the scattering coefficient of the surrounding tissues. Furthermore, inherent limitations are involved in the use of the diffusion approximation to calculate fluence variations.

(6) A purely experimental approach is to combine photoacoustics with acousto-optic modulation (Daoudi et al. 2012)—also see Chapter 13 on ultrasound modulated optical tomography in this book. Acousto-optic modulation encompasses a focused ultrasound beam, which is used to modulate the light that traverses the ultrasound focus. Part of this modulated light can be observed externally by speckle pattern

Chapter 12

modulation. The biggest challenge in this method is the low signal-to-noise ratio and speckle decorrelation in living tissues.

12.5.2   General Functionalities of Photoacoustics

While quantitatively imaging the concentration of chromophores with photoacoustics is still being pursued, photoacoustic technology and the exploration and demonstration of applications have seen a tremendous development in the last decade.

In general, photoacoustics is able to map optical absorption with resolutions ranging from 15 μm at depths until 1–2 mm, to 1–2 mm at depths until 50 mm. Depending on the wavelength range, sensitivity can be obtained to substances such as hemoglobin, oxyhemoglobin, melanin (Viator et al. 2004), water (Xu et al. 2010), and lipids (Wang et al. 2010a). While for familiar wavelengths such as 532 and 1064 nm (Nd:YAG laser with and without frequency doubling) and 755 nm (Alexandrite laser) the hemoglobins form the dominant absorber, for separating between Hb and HbO_2, spectroscopy is required, often in the range 690–900 nm (frequency doubled Nd:YAG laser plus optical parametric oscillator). Also for identification of melanin, water and fat spectroscopic approaches are required, which allows separating these substances from each other (water and lipids) and the hemoglobins. Apart from the above endogenous chromophores, photoacoustics has been used for imaging contrast agents such as Evans Blue (Yao et al. 2009), indocyanine green (Ku and Wang 2005), and gold nanoparticles (Li et al. 2007). In a multispectral approach using contrast agents with steep slopes in their absorption coefficients, combining photoacoustic images obtained at nearby wavelengths can yield sensitivities to very low contrast agent concentrations (Razansky et al. 2009) by suppressing the background absorption.

While such absorption mapping can be regarded as structural imaging, photoacoustics can also reveal tissue function by measuring oxygenation and flow. Measuring absolute blood oxygenation is challenging for reasons given in the previous section. A recent single wavelength approach is to use the absorption saturation effect of hemoglobin, which is related to the oxygen-dependent relaxation times. This can be achieved by exciting the tissue with various pulse energy levels (Danielli et al. 2010). It is even possible to measure tissue-dissolved oxygen using a pump-probe technique, where the pump pulse excites a methylene blue dye and the probe pulse gives a photoacoustic signal, the amplitude of which depends on the oxygen-related relaxation time of this dye (Ashkenazi 2010).

But in general, the quantity measured is relative blood oxygenation, using multiwavelength photoacoustics. Some references will be given in Section 12.5.3. Another functional parameter is blood flow, which can be probed using the acoustic Doppler effect. Since the Doppler effect to be expected is small compared to the normal bandwidth of pulsed photoacoustic ultrasound, photoacoustic signals obtained with a CW light sources were used in which the Doppler shift can be directly obtained from the ultrasound spectrum (Fang et al. 2007). An approach that allows position-sensitive measurement of both absorption and Doppler shift is tone burst or pulse train excitation (Sheinfeld et al. 2010). This method is similar to pulsed Doppler ultrasound. *In vivo* measurements on blood flow were performed by studying the broadening of the photoacoustic ultrasound spectrum (Yao et al. 2010). This method is sensitive to transverse flow, for which the Doppler method loses its sensitivity.

12.5.3 Specific Application Fields of PAT

In this section, we will first discuss already proven specific applications of photoacoustics outside regenerative medicine and then highlight its use in regenerative medicine in Section 12.5.4.

12.5.3.1 Oncology

Photoacoustics can be used to visualize the process of neovascularization or tumor angiogenesis (Siphanto et al. 2005), and the distribution of blood oxygenation in and around a tumor can be obtained (Lungu et al. 2007). The potential of photoacoustics to image deep breast tissue and breast tumors has been shown (Manohar et al. 2007; Kruger et al. 2010; Ermilov et al. 2009). Using spectroscopy in melanoma, the blood and melanin content of the melanoma can be identified (Zhang et al. 2006). Photoacoustics can provide support in the identification of sentinel lymph nodes such as in breast surgery, since lymph vessels can be colored using an optical contrast agent such as indocyanine green (Kim et al. 2010). Furthermore, developments are underway toward *in vivo* detection of tumor metastases in lymph nodes (Galanzha et al. 2009a). For this application, usually a contrast agent is required, although for melanoma, intrinsic melanin contrast may be used (Jose et al. 2011b). Finally, photoacoustics has been used for *in vivo* flow cytometry for the detection of circulating tumor cells (Galanzha et al. 2009b; Zharov et al. 2006; Weight et al. 2006). A more complete overview of the applications of photoacoustics in oncology can be found in Mallidi et al. (2011).

12.5.3.2 Neurophysiology

Photoacoustics is capable of functional imaging of the brain of small animals, since certain stimuli affect both the hemoglobin content and blood oxygenation in the brain. From the point of view of light penetration depth, benefit can be taken from the fact that functional brain activity is mostly prominent in the cortex of the brain, hence the outer layer. While initially brain imaging was limited to mice with a thin skull, it can now be applied also to monkeys with a skull 2–4 mm thick (Nie et al. 2011). This suggests that human neonatal brain imaging will be possible. The fastest imaging of a large volume of a small animal brain is with the real-time imager presented in Gamelin et al. (2009), while the highest spatial resolution can be obtained with photoacoustic microscopy. Examples of photoacoustic functional imaging can be found in Liao et al. (2010) and Hu et al. (2009).

12.5.3.3 Dermatology (Portwine Stains)

Photoacoustics is particularly promising for imaging vascular malformations and melanin-related diseases; however, until now, the number of studies in the field of dermatology is very limited (except for skin cancer). Photoacoustic estimation of the depth of portwine stains has been demonstrated using a one-point measurement (Viator et al. 2002, 2003), while 2D photoacoustic imaging of portwine stains was realized using a double-ring probe (Kolkman et al. 2008b). A basic study has been performed into estimation of epidermal melanin content (Viator et al. 2004), but no publications are known regarding pigmentation diseases studied by photoacoustics.

12.5.3.4   Joint Disease

In view of their size, fingers seem ideal objects for PAT. Therefore, PAT might have potential for the diagnosis and treatment monitoring of joint diseases such as rheumatoid arthritis and osteoarthritis. Photoacoustics might also reveal both the mechanical properties of cartilage like in Ishihara et al. (2005), Sato et al. (2011), and Sato et al. (2005). The same group developed an arthroscope-based photoacoustic system for measuring the viscoelasticity of articular cartilage in joints (Ishihara et al. 2006). But in most cases, the target of study is optical absorption, which might reveal modifications in hemoglobin and water content of the joint cavity, as associated with inflammation. Another quantity of interest revealed by PAT may be the interphalangeal joint gap, which is decreased in affected joints. Photoacoustic images have been published of both unaffected joints in animals (Wang et al. 2006), human cadavers (Wang et al. 2007), and living humans (Sun and Jiang 2009) and human joints affected by osteoarthritis (Sun et al. 2011; Xiao and He 2010; Xiao et al. 2010). In general, affected joints had an elevated absorption coefficient compared to healthy joints. Photoacoustic spectroscopy (Xiao and He 2010) could reveal the elevated water content and decreased hemoglobin oxygen saturation in osteoarthritic joints. Studies on joint disease published so far are very limited in size, precluding conclusions regarding the clinical value of PAT for this field of medicine.

12.5.3.5   Cardiovascular Medicine

In recent years, several systems have been developed for intravascular photoacoustic (IVPA) imaging. The effort is to obtain a complete impression of the mechanical, chemical, and biological properties by combining intravascular ultrasound imaging (IVUS) with photoacoustic spectroscopy. Substances of interest might be hemoglobin in the *vasa vasorum* and lipids that form a major risk factor in plaque rupture. A recent overview of IVPA imaging is given in Wang et al. (2010b). Two systems for IVPA–IVUS were realized for *ex vivo* use: one with external light delivery through the vessel wall and one with delivery through a side-emitting optical fiber. In both cases, a commercial single-element IVUS-catheter was used for normal IVUS and IVPA measurements. Measurements are until now limited to rabbit arteries with plaque, and spectroscopy has enabled the identification of lipid regions. Use of plasmonic gold nanoparticles has enabled the identification of macrophages. This is based on the increase in absorption at 680 nm by plasmonic coupling when gold particles accumulate by phagocytosis. More recently, the development of a combined IVUS–IVPA system containing a side-looking polymer micro ring resonator for optical detection of ultrasound has been reported (Hsieh et al. 2010). The first measurements on human coronary arteries *ex vivo* were recently performed, in which wavelength scanning around 1200 nm revealed the presence of lipids. In this study, a "home-built" IVUS–IVPA was used with a side-emitting fiber and a single side-facing piezoelement for ultrasound generation and detection. As yet, no *in vivo* measurements have been reported.

12.5.4   Photoacoustics Applied to Regenerative Medicine

As yet, most applications of photoacoustics are in oncology and neurophysiology, while work on regenerative medicine is very sparse. It also depends on the definition of "regenerative medicine" and the type of photoacoustic technology to be included. Photoacoustic technologies in the wider sense have been used for research of materials. The type of photoacoustics that is most remote from tomography is photoacoustic-Fourier transform infrared spectroscopy (PA-FTIR), which was developed in the 1970s. Unlike the PAT technology described here, in PA-FTIR spectroscopy, a sample is illuminated by an amplitude-modulated light source that can be wavelength-tuned. The ultrasound signal is of a narrow-band nature, allowing for phase-sensitive detection. The photoacoustic spectra give insight into the chemical composition of the sample; PA-FTIR has been used, for instance, for assessing mineralization of hydroxyapatite under various circumstances (Verma et al. 2006). An example is the *in vivo* assessment of cartilage's mechanical properties using the photoelastic effect (Sato et al. 2011). Using laser pulses rather than an amplitude modulated light gives ultrasound waves with a wide frequency range, enabling a more versatile mechanical tissue analysis. The viscoelasticity of engineered cartilage has been studied using the photoelastic effect induced by nanosecond laser pulses by judging the relaxation times of photoacoustically induced stresses by measuring the exponential decay rate of the photoacoustically induced ultrasound wave (Ishihara et al. 2005). This method has been used *in vitro* but can be applied to grown cartilage *in vivo* as well. The same method has been used to analyze the mechanical properties of sheets and plates of tissue-engineered chondrocytes (Sato et al. 2008). The photoelastic effect is also being investigated for its therapeutic use, for instance, in the *in vitro* stimulation of stem cells in bone tissue engineering (Sitharaman et al. 2011), promoting differentiation toward osteoblasts. Application of PAT in regenerative medicine is much more limited. Photoacoustic microscopy at a resolution of 15 and 45 μm in longitudinal and lateral directions has been performed to image and quantify the proliferation of melanoma cells in poly(lactic-co-glycolic acid) scaffolds of 1.5 mm thickness (Zhang et al. 2010b). Melanoma cells were used for their natural pigmentation and served as model cells in this study to demonstrate the potential of photoacoustic microscopy for quantifying cell growth in scaffolds.

A branch of medicine that can be viewed as part of regenerative medicine is wound healing, in particular, diagnosis and healing of burns and the use of skin grafts. Skin grafts have been studied with photoacoustics for visualizing both the vascularization process (Yamazaki et al. 2006) and the formation of granulation tissue (Hatanaka et al. 2010). These studies involve photoacoustic monitoring with a single ring-shaped ultrasound transducer, which due to its narrow angular detection aperture, enables depth profiling of the absorption in the tissue. A similar sensor was used to study burns at various degrees of severity (Sato et al. 2005), and burns of different depth could be distinguished based on the photoacoustic signal. The same group performed photoacoustic spectroscopy on burns (Yamazaki et al. 2005), within the wavelength range 500–650 nm, and 2D images of burns were obtained by lateral scanning of the ring sensor. By spectroscopic analysis, hemoglobin was identified as the dominant absorber, and in burns, a zone with low photoacoustic response was present compared to healthy tissue. A next step was to photoacoustically monitor the burn healing process by making photoacoustic depth profiles at various wavelengths. A distinct behavior was found in the first and the second period of 24 h postburn. In the period 24 to 48 h postburn, the wavelength dependency of the photoacoustic signals in the range 532–600 nm changed, indicating a change in the tissue metabolism. It furthermore appeared possible to spectroscopically distinguish methemoglobin, characterized by signal maxima for 500 and 633 nm (Aizawa et al. 2009).

12.6   Concluding Remarks

This chapter illustrates the huge boost that photoacoustic technology has experienced in the last decade, and is still undergoing. Within the foreseeable future, systems will be available for real-time photoacoustic imaging ranging from almost cellular resolution at a shallow depth to angiography with a resolution of 0.5–1 mm at centimeters depth. Ultrasound systems with handheld probes will be available for hybrid ultrasound-photoacoustic imaging, along with tomographic systems for whole body small animal imaging, and for imaging human limbs and individual tissues. Spectroscopy will allow for functional imaging, while application of targeted contrast agents will allow for selective molecular-based disease detection and molecular imaging of biological processes. In view of the current status, it seems that the limiting factor in photoacoustics is the laser technology, since real-time functional photoacoustics will require high repetition rate lasers with rapid wavelength tunability.

The PAT platforms will, in general, be applicable for regenerative medicine as long as the site of therapy is in reach of the light delivery. Those situations will be most challenging in which scaffolds and other biomaterials are used with acoustic properties that do not match those of the surrounding tissues: this will require further developments in imaging reconstruction algorithms, taking full account of acoustic heterogeneities. Similarly, further steps have to be made in quantitative imaging of tissue chromophores to give a full description of the functional state of the tissue and the progress of therapy.

References

Aizawa, K., S. Sato, D. Saitoh, H. Ashida, and M. Obara. 2009. *In vivo* photoacoustic spectroscopic imaging of hemoglobin derivatives in thermally damaged tissue. *Japanese Journal of Applied Physics* 48:062302.

Allen, T. J., and P. C. Beard. 2006. Pulsed near-infrared laser diode excitation system for biomedical photoacoustic imaging. *Optics Letters* 31:3462–3464.

Ashkenazi, S. 2010. Photoacoustic lifetime imaging of dissolved oxygen using methylene blue. *Journal of Biomedical Optics* 15:040501.

Balogun, O., B. Regez, H. F. Zhang, and S. Krishnaswamy. 2010. Real-time full-field photoacoustic imaging using an ultrasonic camera. *Journal of Biomedical Optics* 15:021318.

Brecht, H. P., R. Su, M. Fronheiser, S. A. Ermilov, A. Conjusteau, and A. A. Oraevsky. 2009. Whole-body three-dimensional optoacoustic tomography system for small animals. *Journal of Biomedical Optics* 14:064007.

Chen, X. A., Z. L. Tang, Y. H. He, H. F. Liu, and Y. B. Wu. 2010. A simultaneous multiple-section photoacoustic imaging technique based on acoustic lens. *Journal of Applied Physics* 108:073116.

Chen, Z. X., Z. L. Tang, and W. Wan. 2007. Photoacoustic tomography imaging based on a 4f acoustic lens imaging system. *Optics Express* 15:4966–4976.

Cox, B. T., S. R. Arridge, K. P. Kostli, and P. C. Beard. 2006. Two-dimensional quantitative photoacoustic image reconstruction of absorption distributions in scattering media by use of a simple iterative method. *Applied Optics* 45:1866–1875.

Cox, B. T., S. R. Arridge, and P. C. Beard. 2009. Estimating chromophore distributions from multiwavelength photoacoustic images. *Journal of the Optical Society of America A: Optics, Image Science, and Vision* 26:443–455.

Danielli, A., C. P. Favazza, K. Maslov, and L. V. Wang. 2010. Picosecond absorption relaxation measured with nanosecond laser photoacoustics. *Applied Physics Letters* 97:163701.

Daoudi, K., A. Hussain, E. Hondebrink, and W. Steenregen. 2012. Correcting photoacoustic signals for fluence variations using acousto-optic modulation. *Optics Express* 20:14117–14129.

Ermilov, S. A., T. Khamapirad, A. Conjusteau et al. 2009. Laser optoacoustic imaging system for detection of breast cancer. *Journal of Biomedical Optics* 14:024007.

Esenaliev, R. O., Y. Y. Petrov, O. Hartrumpf, D. J. Deyo, and D. S. Prough. 2004. Continuous, noninvasive monitoring of total hemoglobin concentration by an optoacoustic technique. *Applied Optics* 43:3401–3407.

Fan, Y., A. Mandelis, G. Spirou, and I. A. Vitkin. 2004. Development of a laser photothermoacoustic frequency-swept system for subsurface imaging: Theory and experiment. *Journal of the Acoustical Society of America* 116:3523–3533.

Fang, H., K. Maslov, and L. V. Wang. 2007. Photoacoustic Doppler flow measurement in optically scattering media. *Applied Physics Letters* 91:264103.

Fink, M. 1992. Time-reversal of ultrasonic fields. 1. Basic principles. *IEEE Transactions on Ultrasonics Ferroelectrics and Frequency Control* 39:555–566.

Galanzha, E. I., M. S. Kokoska, E. V. Shashkov, J. W. Kim, V. V. Tuchin, and V. P. Zharov. 2009a. *In vivo* fiber-based multicolor photoacoustic detection and photothermal purging of metastasis in sentinel lymph nodes targeted by nanoparticles. *Journal of Biophotonics* 2:528–539.

Galanzha, E. I., E. V. Shashkov, P. M. Spring, J. Y. Suen, and V. P. Zharov. 2009b. *In vivo*, noninvasive, label-free detection and eradication of circulating metastatic melanoma cells using two-color photoacoustic flow cytometry with a diode laser. *Cancer Research* 69:7926–7934.

Gamelin, J., A. Maurudis, A. Aguirre et al. 2009. A real-time photoacoustic tomography system for small animals. *Optics Express* 17:10489–10498.

Haisch, C., K. Eilert-Zell, M. M. Vogel, P. Menzenbach, and R. Niessner. 2010. Combined optoacoustic/ultrasound system for tomographic absorption measurements: possibilities and limitations. *Analytical and Bioanalytical Chemistry* 397:1503–1510.

Hatanaka, K., S. Sato, D. Saitoh, H. Ashida, and T. Sakamoto. 2010. Photoacoustic monitoring of granulation tissue grown in a grafted artificial dermis on rat skin. *Wound Repair and Regeneration* 18:284–290.

Hoelen, C. G. A., and F. F. M. de Mul. 2000. Image reconstruction for photoacoustic scanning of tissue structures. *Applied Optics* 39:5872–5883.

Hsieh, B. Y., S. L. Chen, T. Ling, L. J. Guo, and P. C. Li. 2010. Integrated intravascular ultrasound and photoacoustic imaging scan head. *Optics Letters* 35:2892–2894.

Hu, S., K. Maslov, V. Tsytsarev, and L. V. Wang. 2009. Functional transcranial brain imaging by optical-resolution photoacoustic microscopy. *Journal of Biomedical Optics* 14:040503.

Ishihara, M., M. Sato, S. Sato, T. Kikuchi, J. Mochida, and M. Kikuchi. 2005. Usefulness of photoacoustic measurements for evaluation of biomechanical properties of tissue-engineered cartilage. *Tissue Engineering* 11:1234–1243.

Ishihara, M., M. Sato, N. Kaneshiro et al. 2006. Development of a diagnostic system for osteoarthritis using a photoacoustic measurement method. *Lasers in Surgery and Medicine* 38:249–255.

Jose, J., R. G. H. Willemink, S. Resink et al. 2011a. Passive element enriched photoacoustic computed tomography (PER PACT) for simultaneous imaging of acoustic propagation properties and light absorption. *Optics Express* 19:2093–2104.

Jose, J., D. J. Grootendorst, T. W. Vijn et al. 2011b. Initial results of imaging melanoma metastasis in resected human lymph nodes using photoacoustic computed tomography. *Journal of Biomedical Optics* 16:096021.

Kim, C., T. N. Erpelding, K. Maslov et al. 2010. Handheld array-based photoacoustic probe for guiding needle biopsy of sentinel lymph nodes. *Journal of Biomedical Optics* 15:046010.

Kolkman, R. G. M., J. Klaessens, E. Hondebrink et al. 2004a. Photoacoustic determination of blood vessel diameter. *Physics in Medicine and Biology* 49:4745–4756.

Kolkman, R. G. M., E. Hondebrink, W. Steenbergen, T. G. van Leeuwen, and F. F. M. de Mul. 2004b. Photoacoustic imaging of blood vessels with a double-ring sensor featuring a narrow angular aperture. *Journal of Biomedical Optics* 9:1327–1335.

Kolkman, R. G. M., W. Steenbergen, and T. G. van Leeuwen. 2006. *In vivo* photoacoustic imaging of blood vessels with a pulsed laser diode. *Lasers in Medical Science* 21:134–139.

Kolkman, R. G. M., P. J. Brands, W. Steenbergen, and T. G. van Leeuwen. 2008a. Real-time *in vivo* photoacoustic and ultrasound imaging. *Journal of Biomedical Optics* 13:050510.

Kolkman, R. G. M., M. J. Mulder, C. P. Glade, W. Steenbergen, and T. G. van Leeuwen. 2008b. Photoacoustic imaging of port-wine stains. *Lasers in Surgery and Medicine* 40:178–182.

Chapter 12

Kostli, K. P., and P. C. Beard. 2003. Two-dimensional photoacoustic imaging by use of Fourier-transform image reconstruction and a detector with an anisotropic response. *Applied Optics* 42:1899–1908.

Kostli, K. P., M. Frenz, H. Bebie, and H. P. Weber. 2001. Temporal backward projection of optoacoustic pressure transients using Fourier transform methods. *Physics in Medicine and Biology* 46:1863–1872.

Kruger, R. A., P. Y. Liu, Y. R. Fang, and C. R. Appledorn. 1995. Photoacoustic ultrasound (PAUS)—reconstruction tomography. *Medical Physics* 22:1605–1609.

Kruger, R. A., D. R. Reinecke, and G. A. Kruger. 1999. Thermoacoustic computed tomography-technical considerations. *Medical Physics* 26:1832–1837.

Kruger, R. A., W. L. Kiser, D. R. Reinecke, and G. A. Kruger. 2003. Thermoacoustic computed tomography using a conventional linear transducer array. *Medical Physics* 30:856–860.

Kruger, R. A., R. B. Lam, D. R. Reinecke, S. P. Del Rio, and R. P. Doyle. 2010. Photoacoustic angiography of the breast. *Medical Physics* 37:6096–6100.

Ku, G., and L. H. V. Wang. 2005. Deeply penetrating photoacoustic tomography in biological tissues enhanced with an optical contrast agent. *Optics Letters* 30:507–509.

Laufer, J., D. Delpy, C. Elwell, and P. Beard. 2007. Quantitative spatially resolved measurement of tissue chromophore concentrations using photoacoustic spectroscopy: application to the measurement of blood oxygenation and haemoglobin concentration. *Physics in Medicine and Biology* 52:141–168.

Laufer, J., B. Cox, E. Zhang, and P. Beard. 2010. Quantitative determination of chromophore concentrations from 2D photoacoustic images using a nonlinear model-based inversion scheme. *Applied Optics* 49:1219–1233.

Li, P. C., C. W. Wei, C. K. Liao et al. 2007. Photoacoustic imaging of multiple targets using gold nanorods. *IEEE Transactions on Ultrasonics Ferroelectrics and Frequency Control* 54:1642–1647.

Liao, C. K., M. L. Li, and P. C. Li. 2004. Optoacoustic imaging with synthetic aperture focusing and coherence weighting. *Optics Letters* 29:2506–2508.

Liao, L. D., M. L. Li, H. Y. Lai et al. 2010. Imaging brain hemodynamic changes during rat forepaw electrical stimulation using functional photoacoustic microscopy. *Neuroimage* 52:562–570.

Lungu, G. F., M. L. Li, X. Y. Xie, L. H. V. Wang, and G. Stoica. 2007. *In vivo* imaging and characterization of hypoxia-induced neovascularization and tumor invasion. *International Journal of Oncology* 30:45–54.

Ma, R., A. Taruttis, V. Ntziachristos, and D. Razansky. 2009. Multispectral optoacoustic tomography (MSOT) scanner for whole-body small animal imaging. *Optics Express* 17:21414–21426.

Mallidi, S., G. P. Luke, and S. Emelianov. 2011. Photoacoustic imaging in cancer detection, diagnosis, and treatment guidance. *Trends in Biotechnology* 29:213–221.

Manohar, S., S. E. Vaartjes, J. C. G. van Hespen et al. 2007. Initial results of *in vivo* non-invasive cancer imaging in the human breast using near-infrared photoacoustics. *Optics Express* 15:12277–12285.

Nie, L., Z. Guo, and L. V. Wang. 2011. Photoacoustic tomography of monkey brain using virtual point ultrasonic transducers. *Journal of Biomedical Optics* 16:076005.

Niederhauser, J. J., M. Jaeger, and M. Frenz. 2004. Real-time three-dimensional optoacoustic imaging using an acoustic lens system. *Applied Physics Letters* 85:846–848.

Niederhauser, J. J., M. Jaeger, R. Lemor, P. Weber, and M. Frenz. 2005. Combined ultrasound and optoacoustic system for real-time high-contrast vascular imaging *in vivo*. *IEEE Transactions on Medical Imaging* 24:436–440.

Rajian, J. R., P. L. Carson, and X. D. Wang. 2009. Quantitative photoacoustic measurement of tissue optical absorption spectrum aided by an optical contrast agent. *Optics Express* 17:4879–4889.

Ranasinghesagara, J. C., and R. J. Zemp. 2010. Combined photoacoustic and oblique-incidence diffuse reflectance system for quantitative photoacoustic imaging in turbid media. *Journal of Biomedical Optics* 15:046016.

Rao, B., L. Li, K. Maslov, and L. H. Wang. 2010. Hybrid-scanning optical-resolution photoacoustic microscopy for *in vivo* vasculature imaging. *Optics Letters* 35:1521–1523.

Razansky, D., J. Baeten, and V. Ntziachristos. 2009. Sensitivity of molecular target detection by multispectral optoacoustic tomography (MSOT). *Medical Physics* 36:939–945.

Rosenthal, A., D. Razansky, and V. Ntziachristos. 2009. Quantitative optoacoustic signal extraction using sparse signal representation. *IEEE Transactions on Medical Imaging* 28:1997–2006.

Rosenthal, A., D. Razansky, and V. Ntziachristos. 2010. Fast semi-analytical model-based acoustic inversion for quantitative optoacoustic tomography. *IEEE Transactions on Medical Imaging* 29:1275–1285.

Sato, M., M. Ishihara, K. Furukawa et al. 2008. Recent technological advancements related to articular cartilage regeneration. *Medical and Biological Engineering and Computing* 46:735–743.

Sato, M., M. Ishihara, M. Kikuchi, and J. Mochida. 2011. A diagnostic system for articular cartilage using non-destructive pulsed laser irradiation. *Lasers in Surgery and Medicine* 43:421–432.

Sato, S., M. Yamazaki, D. Saitoh et al. 2005. Photoacoustic diagnosis of burns in rats. *Journal of Trauma-Injury Infection and Critical Care* 59:1450–1455.

Shao, P., B. Cox, and R. J. Zemp. 2011. Estimating optical absorption, scattering, and Grueneisen distributions with multiple-illumination photoacoustic tomography. *Applied Optics* 50:3145–3154.

Sheinfeld, A., S. Gilead, and A. Eyal. 2010. Photoacoustic Doppler measurement of flow using tone burst excitation. *Optics Express* 18:4212–4221.

Shi, W., S. Kerr, I. Utkin et al. 2010. Optical resolution photoacoustic microscopy using novel high-repetition-rate passively Q-switched microchip and fiber lasers. *Journal of Biomedical Optics* 15:056017.

Sigrist, M. W., and F. K. Kneubuhl. 1978. Laser-generated stress waves in liquids. *Journal of the Acoustical Society of America* 64:1652–1663.

Siphanto, R. I., K. K. Thumma, R. G. M. Kolkman et al. 2005. Serial noninvasive photoacoustic imaging of neovascularization in tumor angiogenesis. *Optics Express* 13:89–95.

Sitharaman, B., P. K. Avti, K. Schaefer, Y. Talukdar, and J. P. Longtin. 2011. A novel nanoparticle-enhanced photoacoustic stimulus for bone tissue engineering. *Tissue Engineering Part A* 17:1851–1858.

Song, L. A., K. Maslov, K. K. Shung, and L. H. V. Wang. 2010. Ultrasound-array-based real-time photoacoustic microscopy of human pulsatile dynamics *in vivo*. *Journal of Biomedical Optics* 15:021303.

Sun, Y., and H. B. Jiang. 2009. Quantitative three-dimensional photoacoustic tomography of the finger joints: phantom studies in a spherical scanning geometry. *Physics in Medicine and Biology* 54:5457–5467.

Sun, Y., E. S. Sobel, and H. B. Jiang. 2011. First assessment of three-dimensional quantitative photoacoustic tomography for *in vivo* detection of osteoarthritis in the finger joints. *Medical Physics* 38:4009–4017.

Tang, M. X., D. S. Elson, R. Li, C. Dunsby, and R. J. Eckersley. 2010. Photoacoustics, thermoacoustics, and acousto-optics for biomedical imaging. *Proceedings of the Institution of Mechanical Engineers Part H-Journal of Engineering in Medicine* 224:291–306.

Taruttis, A., E. Herzog, D. Razansky, and V. Ntziachristos. 2010. Real-time imaging of cardiovascular dynamics and circulating gold nanorods with multispectral optoacoustic tomography. *Optics Express* 18:19592–19602.

Valluru, K. S., B. K. Chinni, and S. A. Rao. 2011. Photoacoustic imaging: opening new frontiers in medical imaging. *Journal of Clinical Imaging Science* 1:24.

van Raaij, M. E., M. Lee, E. Cherin, B. Stefanovic, and F. S. Foster. 2010. Femtosecond photoacoustics: integrated two-photon fluorescence and photoacoustic microscopy. In *Photons Plus Ultrasound: Imaging and Sensing 2010*. A. A. Oraevsky, and L. V. Wang, editors. Spie-Int Soc Optical Engineering, Bellingham.

Viator, J. A., G. Au, G. Paltauf et al. 2002. Clinical testing of a photoacoustic probe for port wine stain depth determination. *Lasers in Surgery and Medicine* 30:141–148.

Viator, J. A., B. Choi, M. Ambrose, J. Spanier, and J. S. Nelson. 2003. *In vivo* port-wine stain depth determination with a photoacoustic probe. *Applied Optics* 42:3215–3224.

Viator, J. A., J. Komadina, L. O. Svaasand, G. Aguilar, B. Choi, and J. S. Nelson. 2004. A comparative study of photoacoustic and reflectance methods for determination of epidermal melanin content. *Journal of Investigative Dermatology* 122:1432–1439.

Verma, D., K. Katti, and D. Katti. 2006. Experimental investigation of interfaces in hydroxyapatite/polyacrylic acid/polycaprolactone composites using photoacoustic FTIR spectroscopy. *Journal of Biomedical Materials Research Part A* 77A:59–66.

Wang, B., J. L. Su, J. Amirian, S. H. Litovsky, R. Smalling, and S. Emelianov. 2010a. Detection of lipid in atherosclerotic vessels using ultrasound-guided spectroscopic intravascular photoacoustic imaging. *Optics Express* 18:4889–4897.

Wang, B., J. L. Su, A. B. Karpiouk, K. V. Sokolov, R. W. Smalling, and S. Y. Emelianov. 2010b. Intravascular photoacoustic imaging. *IEEE Journal of Selected Topics in Quantum Electronics* 16:588–599.

Wang, L. H. V., editor. 2009. *Photoacoustic Imaging and Spectroscopy*. CRC Press, Boca Raton, FL.

Wang, X. D., Y. Xu, M. H. Xu, S. Yokoo, E. S. Fry, and L. H. V. Wang. 2002. Photoacoustic tomography of biological tissues with high cross-section resolution: reconstruction and experiment. *Medical Physics* 29:2799–2805.

Wang, X. D., D. L. Chamberland, P. L. Carson et al. 2006. Imaging of joints with laser-based photoacoustic tomography: an animal study. *Medical Physics* 33:2691–2697.

Wang, X. D., D. L. Chamberland, and D. A. Jamadar. 2007. Noninvasive photoacoustic tomography of human peripheral joints toward diagnosis of inflammatory arthritis. *Optics Letters* 32:3002–3004.

Wei, Y. D., Z. L. Tang, H. C. Zhang, Y. H. He, and H. F. Liu. 2008. Photoacoustic tomography imaging using a 4f acoustic lens and peak-hold technology. *Optics Express* 16:5314–5319.

Chapter 12

Weight, R. M., J. A. Viator, P. S. Dale, C. W. Caldwell, and A. E. Lisle. 2006. Photoacoustic detection of metastatic melanoma cells in the human circulatory system. *Optics Letters* 31:2998–3000.

Welch, A. J., and M. J. C. Van Gemert, editors. 1995. *Optical-Thermal Response of Laser-Irradiated Tissue.* Plenum Press, New York.

Xiao, J. Y., and J. S. He. 2010. Multispectral quantitative photoacoustic imaging of osteoarthritis in finger joints. *Applied Optics* 49:5721–5727.

Xiao, J. Y., L. Yao, Y. Sun, E. S. Sobel, J. S. He, and H. B. Jiang. 2010. Quantitative two-dimensional photoacoustic tomography of osteoarthritis in the finger joints. *Optics Express* 18:14359–14365.

Xu, M. H., and L. H. V. Wang. 2005. Universal back-projection algorithm for photoacoustic computed tomography. *Physical Review E* 71:016706.

Xu, M. H., and L. H. V. Wang. 2006. Photoacoustic imaging in biomedicine. *Review of Scientific Instruments* 77:041101.

Xu, Z., C. H. Li, and L. V. Wang. 2010. Photoacoustic tomography of water in phantoms and tissue. *Journal of Biomedical Optics* 15:036019.

Yamazaki, M., S. Sato, H. Ashida, D. Saito, Y. Okada, and M. Obara. 2005. Measurement of burn depths in rats using multiwavelength photoacoustic depth profiling. *Journal of Biomedical Optics* 10:064011.

Yamazaki, M., S. Sato, D. Saitoh, Y. Okada, H. Ashida, and M. Obara. 2006. Photoacoustic monitoring of neovascularities in grafted skin. *Lasers in Surgery and Medicine* 38:235–239.

Yao, J. J., K. Maslov, S. Hu, and L. H. V. Wang. 2009. Evans blue dye-enhanced capillary-resolution photoacoustic microscopy *in vivo. Journal of Biomedical Optics* 14:054049.

Yao, J. J., K. I. Maslov, Y. F. Shi, L. A. Taber, and L. H. V. Wang. 2010. *In vivo* photoacoustic imaging of transverse blood flow by using Doppler broadening of bandwidth. *Optics Letters* 35:1419–1421.

Yin, L., Q. Wang, Q. Z. Zhang, and H. B. Jiang. 2007. Tomographic imaging of absolute optical absorption coefficient in turbid media using combined photoacoustic and diffusing light measurements. *Optics Letters* 32:2556–2558.

Yuan, Z., Q. Wang, and H. B. Jiang. 2007. Reconstruction of optical absorption coefficient maps of heterogeneous media by photoacoustic tomography coupled with diffusion equation based regularized Newton method. *Optics Express* 15:18076–18081.

Zemp, R. J. 2010. Quantitative photoacoustic tomography with multiple optical sources. *Applied Optics* 49:3566–3572.

Zeng, Y. G., D. Xing, Y. Wang, B. Z. Yin, and Q. Chen. 2004. Photoacoustic and ultrasonic coimage with a linear transducer array. *Optics Letters* 29:1760–1762.

Zhang, C., K. Maslov, and L. H. V. Wang. 2010a. Subwavelength-resolution label-free photoacoustic microscopy of optical absorption *in vivo. Optics Letters* 35:3195–3197.

Zhang, E., J. Laufer, and P. Beard. 2008. Backward-mode multiwavelength photoacoustic scanner using a planar Fabry-Perot polymer film ultrasound sensor for high-resolution three-dimensional imaging of biological tissues. *Applied Optics* 47:561–577.

Zhang, H. F., K. Maslov, G. Stoica, and L. H. V. Wang. 2006. Functional photoacoustic microscopy for high-resolution and noninvasive *in vivo* imaging. *Nature Biotechnology* 24:848–851.

Zhang, Y., X. Cai, S. W. Choi, C. Kim, L. H. V. Wang, and Y. N. Xia. 2010b. Chronic label-free volumetric photoacoustic microscopy of melanoma cells in three-dimensional porous Scaffolds. *Biomaterials* 31:8651–8658.

Zharov, V. P., E. I. Galanzha, E. V. Shashkov, N. G. Khlebtsov, and V. V. Tuchin. 2006. *In vivo* photoacoustic flow cytometry for monitoring of circulating single cancer cells and contrast agents. *Optics Letters* 31:3623–3625.

13. Ultrasound–Modulated Optical Tomography

Stephen P. Morgan, Nam T. Huynh, Haowen Ruan, and Felicity R.A.J. Rose

13.1 Introduction

Ultrasound modulated optical tomography (USMOT) is an imaging technique that combines light and ultrasound (US) to overcome the effects of light scattering in thick tissue. In this context, "thick" tissue is defined as beyond the penetration depth of techniques such as confocal or nonlinear microscopy, optical projection tomography, or optical coherence tomography (typically >2 mm). The aim of USMOT is to obtain the imaging resolution

Chapter 13

Optical Techniques in Regenerative Medicine. Edited by Stephen P. Morgan, Felicity R.A.J. Rose, and Stephen J. Matcher © 2014 CRC Press/Taylor & Francis Group, LLC. ISBN: 978-1-4398-5495-2

of US with the functional information that can be provided by optical techniques. The main difference between USMOT and photoacoustic tomography (PAT; described in Chapter 12) is that in PAT, light is used to generate sound within the tissue via the absorption of light (usually by blood). The sound waves are then detected and are used to form an image. In USMOT, light illuminates the tissue and US (delivered via an US transducer) is used to modulate the light within the tissue. The modulated optical signal is then detected and is used to form an image. It can be used to image absorption, scattering, or fluorescence within tissue or 3D cell cultures.

The next section summarizes the main challenges in regenerative medicine that USMOT has the potential to address. This is followed by a description of the basic principles of USMOT, including the mechanisms of modulating light in tissue using US and image formation. A range of different optical detection systems have been developed to improve signal-to-noise ratio (SNR) and these are summarized. Finally examples of the application of USMOT, including imaging blood vessels, necrotic tissue, and fluorescence, are provided.

13.2   Challenges to Be Addressed by US-Modulated Optical Tomography

USMOT has the potential to map the optical properties of thick tissue in 3D at the imaging resolution of US. There are therefore several applications in regenerative medicine that may benefit from USMOT as scaffolds are often opaque with thicknesses in the length scale of millimeters to centimeters. Opacity continues to be a problem, or becomes a problem even if the initial scaffold is transparent, as extracellular matrix deposition and tissue formation occur. USMOT would therefore be extremely valuable for the imaging of 3D cell cultures within thick scaffolds *in vitro*, for any regenerative medicine construct implanted in an animal model, for monitoring tissue development within a bioreactor, and for assessing the impact of implanting 3D constructs into humans. The latter is absolutely crucial as flexibility in monitoring graft viability and survivability is limited in these cases, and yet it is absolutely paramount that we confirm functionality.

Similar to PAT, if the local absorption properties in tissue are mapped, then it is possible to obtain images of blood vessels. Extending the optical illumination to more than one wavelength allows blood oxygen saturation to be obtained due to the different absorption spectra of oxyhemoglobin and deoxyhemoglobin (Tuchin 2007). Determining the extent of vascularization of an implanted tissue-engineered construct is essential for graft survival (reviewed in Laschke and Menger 2012). To date, USMOT has not been able to replicate the high-quality images that have been observed in PAT for imaging blood vessels. However, the instrumentation can be relatively easily incorporated into a commercial US scanner.

There may also be other applications in which USMOT can be useful and provide additional or complementary information to PAT. For example, USMOT is sensitive to variations in the scattering properties of tissue. This can be used to image necrotic tissue; for example, in 3D tissue culture in bioreactors, a necrotic core often occurs due to insufficient nutrient supply to the tissue (Silva et al. 2006). The presence of a necrotic core is often only established by terminating the experiment and carrying out histological analysis. In addition, USMOT would allow spatial information about cell ingress into a scaffold and the extent to which this occurs (indicating restricted nutrient supply prevascularization). This would inform scaffold design or bioreactor parameters to ensure that cell penetration was achieved throughout the construct. Once *in vivo*, it would allow monitoring of the construct, as mentioned above, to ensure graft survivability. In addition, as the initial tissue construct matures, extracellular matrix deposition changes (e.g., calcification and mineralization that occurs during bone formation) could potentially be monitored.

Although challenging in terms of SNR, USMOT can be used to map fluorescence in 3D, which opens up a wide range of applications in imaging 3D tissue culture and in preclinical imaging. Monitoring cell cultures for the expression of key functional markers is a key technique in cell biology as it indicates the differentiation state of cells and/or can be used to identify extracellular matrix production during cell/tissue formation (see Chapter 2). Fluorescence imaging including confocal microscopy allows visualization of these markers, but penetration depths do not exceed 200–300 µm, limiting the utility of such methods for 3D construct imaging. Imaging of fluorescence in PAT is carried out indirectly through imaging the loss of excitation light due to absorption by a fluorophore. However, if multiple absorbers or multiple fluorophores are present, this may present a problem. USMOT detects fluorescence at the emission wavelength, providing a direct measurement. An intriguing possibility is to use USMOT to map chemiluminescence or bioluminescence (see Chapter 15), providing a higher spatial resolution alternative to track stem cell location following injection, for example, although it is likely to be extremely challenging in terms of SNR. In this case, PAT cannot be applied as the light source is generated within the tissue and an external source is not required.

13.3   Technical Implementation of US-Modulated Optical Tomography

In this section, the principles of USMOT are described, starting with the simplest system configuration and a description of image formation. This is followed by an explanation of the three mechanisms by which light can be modulated by US. Several variants of the basic system have been proposed, mainly based on different optical detectors to improve SNR, and the most prevalent of these are reviewed.

13.3.1   Image Formation

The simplest USMOT system is shown in Figure 13.1. Light (usually from a laser) illuminates an optically scattering medium such as tissue. US is applied to the tissue and the light that passes through the US column becomes modulated by the change in material properties caused by the US. If the US is focused, then light that passes through the US focus produces a stronger modulation effect than anywhere else in the medium. This means that light that has passed through the US focus has effectively become "tagged" by the US, creating a modulated light "beacon" within the tissue, which can be distinguished from light that has taken other random paths. The detected signal is most affected by the optical properties of the sample within the focal zone of the US. An image can be

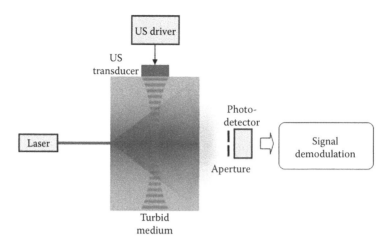

FIGURE 13.1　Basic USMOT configuration. Laser light enters the scattering medium. Light passes through the US column, becomes modulated by the US, and is detected by a photodetector, for example, photomultiplier tube (PMT). An image is built up by scanning the US focus and recording the modulated light at each position.

built up by scanning the US focus (i.e., moving the modulated light beacon) to different positions in the sample and recording the received optical signal at each US position. In continuous wave (CW) US, the image resolution is determined by the size of the focal zone. Alternatively, if pulsed (Lev and Sfez 2003; Hisaka et al. 2001) or frequency-swept US (Forget et al. 2003) is used, then scanning along the US column can be achieved without moving the transducer, with the resolution being determined by the length of the US pulse. Axial scanning can also be achieved using a CW system (Lesaffre et al. 2009) by introducing identical but random phase jumps on both the optical and ultrasonic radiation applied to the sample. As these phase jumps in the light and US are correlated with one another, the strongest contribution to the signal is due to a single layer within the sample where the US and optical signal are highly correlated. The width of this layer depends upon the length over which the phase remains constant in the random sequence. Recently improved imaging resolution has been demonstrated using time-reversed ultrasonic encoding (TRUE) (Xu et al. 2011; Lai et al. 2012) in which the wave front of the US-modulated light emerging from the sample is detected by writing a hologram to a photorefractive crystal (Section 13.3.3.4). A time-reversed (phase conjugate) version of this wave front is then transmitted back through the hologram and into the medium. The time-reversed wave front compensates for the effect of scattering within the medium and focuses the light to a small focal zone, allowing high-resolution imaging to be achieved.

Optical illumination is usually applied at a single wavelength either at a single point or with an expanded beam. The majority of configurations involve the light source and detector being coaxial. Reflection mode imaging has been carried out applying ring illumination to help remove the effects of surface and subsurface backscatter (Kim et al. 2009). Multiple wavelengths have also been applied in order to obtain spectroscopic information with the main aim of measuring oxygen saturation within a scattering medium (Kim and Wang 2007).

Due to the weak US modulation of the optical signals, the majority of the research has been devoted to developing more sensitive detection methods, and this is described in more detail in Section 13.3.3. There are relatively few examples of reconstruction algorithms being applied to USMOT imaging, and so imaging is largely qualitative at present. Algorithms applied to diffuse optical tomography (Gibson et al. 2005) are applicable in USMOT, and so there is scope for rapid improvement in image reconstruction. This approach has been taken by Bratchenia et al. (2011a,b), where quantitative mapping of the local absorption has been achieved. A maximum likelihood algorithm has also been applied to image reconstruction (Huynh et al. 2012a).

13.3.2 Modulation Mechanisms

Three mechanisms (Figure 13.2) have been identified for the ultrasonic modulation of light in a scattering medium (Wang 2001a,b): (1) US-induced variation of the optical properties (absorption and scattering) due to compression and rarefaction of the medium, (2) variation of the optical phase due to US-induced displacement of scatterers (Leutz and Maret 1995), and (3) modulation of the optical phase due to US modulation of the index of refraction. The mechanism described in item 1 is a weak effect that can only be observed with low SNR when incoherent light is used. This has been observed in fluorescence experiments and has also been described theoretically and modeled (Mahan et al. 1998). It has been shown (Liu et al. 2008) that the contribution of the nonphase mechanism 1 to the modulated signal is typically two to three orders of magnitude lower than for the phase mechanisms 2 and 3. The contribution of mechanisms 2 and 3 to the modulation depth of the detected signal is comparable when the acoustic wave number $k_a \leq 0.559\mu_s$, where μ_s is the optical scattering coefficient. As k_a increases above this value, mechanism 3 dominates and the modulation depth increases significantly (Wang 2001b). This theory has been extended to take into account anisotropic scattering (Sakadzic and Wang 2002) and pulsed US (Sakadzic and Wang 2005).

The US modulated signals manifest themselves as a fluctuating intensity signal and a very small change in the frequency (color) of the illumination due to a Doppler frequency shift. If the detected light is coherent (usually from a laser), the fluctuating intensity signal occurs because scattered light takes many random paths through the tissue and emerges having a random phase at the detector plane (due to mechanisms 2 and 3). As the light is

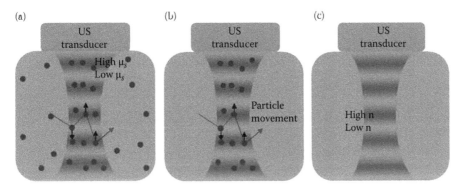

FIGURE 13.2 Mechanisms of ultrasonic modulation of scattered light: (a) mechanism 1, change in scattering and absorption coefficients; (b) mechanism 2, motion of scatterers producing phase modulation; and (c) mechanism 3, change in refractive index producing phase modulation.

coherent, an interference pattern can be observed. For the case of two plane waves interfering, a distinctive fringe pattern can be observed due to constructive and destructive interference as obtained in the well-known Young's slits experiment. For multiple scattered waves of random phase emerging from a scattering medium, the interference is a granular interference pattern known as a speckle pattern (Figure 13.3). As described in the previous section, the two dominant modulation mechanisms (2 and 3) both cause a change in the optical path length in the light emerging from the tissue, which can be observed as a phase change in the detected scattered light and a change in the detected speckle pattern. The interference of modulated and unmodulated electric fields emerging from the medium produces a beating effect, which is observed as a fluctuating speckle pattern.

The form of the detected speckle signals due to mechanisms 2 and 3 can be derived from theory originally developed in dynamic multiple light scattering. The autocorrelation of the detected scattered light field is (Leutz and Maret 1995)

$$G_1(t) = \langle E(0)E^*(t) \rangle = \int_l^\infty p(s) \langle E_s(0)E_s^*(t) \rangle ds \tag{13.1}$$

where <> denotes averaging over time, $p(s)$ is the distribution of path length s within the scattering medium, l is the scattering mean free path, and E and E_s are the electric fields of the scattered light and electric field of light scattered along a path of length s, respectively. This equation can be related to the power spectrum of the detected light using the Wiener–Khinchin theorem:

$$S(\omega) = \int_{-\infty}^\infty G_1(t)e^{i\omega t} dt \tag{13.2}$$

where ω is relative to the optical frequency ω_0, that is, the beat frequency between the modulated and unmodulated electric fields. The US modulation of the optical field can be observed as a sideband in the spectrum described by Equation 13.2.

FIGURE 13.3   Speckle pattern due to multiple plane waves interfering.

It should be noted that other effects can cause the speckle pattern to change, for example, Brownian motion of particles in a scattering medium or moving red blood cells. This causes the speckle to decorrelate between measurements and can be a hindrance for speckle-based detection systems as changes in the speckle pattern cannot be solely attributed to the effect of the applied US. The speckle decorrelation for tissue is typically 0.1–1 ms for blood perfused tissue (Elson et al. 2011).

Motivated by the need to increase the modulation depth of the detected signal or to provide higher spatial resolution, other modulation mechanisms have been exploited. As in conventional US, microbubbles can be used to produce a greater change in the scattering properties of the tissue than could be achieved in tissue alone (Honeysett et al. 2011, 2012; Hall et al. 2009). Intense acoustic bursts (Elson et al. 2011; Li et al. 2011; Zemp et al. 2007; Kim et al. 2006) have been applied in order to generate an acoustic radiation force (ARF), which produces large scatterer displacements in the tissue and increases the modulation depth of the signal. The ARF is obtained when US is absorbed, scattered, or reflected within a medium. It is a small unidirectional force that produces large tissue displacements and hence has the opportunity to increase the modulation depth of USMOT signals. Along with the ARF, shear waves are also produced in a perpendicular direction to the ARF. The effects of these can reduce lateral resolution, but time-gating the detection before the shear wave propagates significantly can reduce this effect (Elson et al. 2011; Li et al. 2011).

In addition, at high acoustic pressures, the US propagation becomes nonlinear. As the nonlinear effect occurs in a localized region, this can be used to improve image contrast and resolution in both conventional US and USMOT (Selb et al. 2002; Ruan et al. 2012). Figure 13.4 shows simulated fundamental and second harmonic US focal zones obtained using freely available software (the "Bergen code"; Aanonsen et al. 1984;

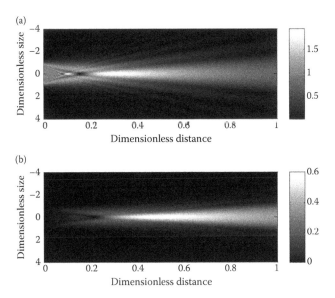

FIGURE 13.4 Simulated comparison of fundamental (a) and second harmonic (b) focal zones. Note the more localized pressure distribution in the lower figure, which results in higher contrast and better spatial resolution.

Chapter 13

Bernsten 1990) demonstrating a narrower focal region with reduced sidelobes for the second harmonic case.

13.3.3   Detection of US Modulated Signals

The modulation mechanisms described in the previous section are relatively weak, and so much research has been directed into the development of sensitive detection methods. These can be broadly classified into those that demodulate the fluctuating speckle pattern and those that directly detect the small frequency shift (i.e., the small color change) in the US modulated optical signal (Table 13.1). It should be noted that if incoherent light is used, for example, for fluorescence imaging, then an interference pattern cannot be observed, but a modulated output signal can still be detected (albeit two to three orders of magnitude smaller) due to mechanism 1 described in the previous section. A change in the scattering and absorption coefficient in the focal region results in light taking a different path within the tissue, which changes the spatial distribution of the emerging light.

Table 13.1   Comparison of Detection Techniques in USMOT

	SNR	Immunity to Speckle Decorrelation	Imaging Speed	Detector	Speckle-Based Detection
Single photodetector direct detection	Poor	Good	Slow as averaging required	PMT/PD	Yes (single speckle)
Speckle contrast detection	Average	Average	Average	Camera	Yes
Lock-in parallel speckle detection	Good	Poor	Slow	Camera	Yes
Holography-based detection	Excellent	Good	Average (digital holography) Good (crystal-based holography)	Camera (digital holography) PMT/PD (crystal-based holography)	Yes
Optical filter-based detection	Good	Excellent	Very fast	PMT/PD	No

Note: PD, photodiode; PMT, photomultiplier tube.

13.3.3.1   Single Photodetector

The simplest detection scheme (Figure 13.1) uses a single, fast photodetector, usually a PMT, followed by a transimpedance amplifier to convert the photocurrent into a voltage (Lev and Sfez 2003). The speckle amplitude can then be extracted, for example, by calculating the autocorrelation function and taking the Fourier transform (Equations 13.1 and 13.2). The disadvantage of this approach is that in order to obtain the maximum modulation depth of the detected signal, it is necessary to restrict the detector size to the order of a single speckle or smaller, which results in a relatively small amount of scattered light being collected. The detection has a low etendue (the product of the detector area and the collection angle), which is undesirable. For the case of incoherent light, such as fluorescence, it is not necessary to use a small detector as there are no observable speckles because the light is incoherent.

13.3.3.2   Speckle Difference

In order to match the speckle size to the detector area and increase the detection etendue, a camera can be used. The best SNR can be obtained when four pixels sample a single speckle (Kirkpatrick et al. 2008), with the speckle size being determined by (Boas and Dunn 2010)

$$\text{speckle size} = \frac{2.44\, f\, (1 + M)}{D} \tag{13.3}$$

where λ is the wavelength of the light, f is the focal length of the lens, D is the diameter of the lens, and M is the magnification of the imaging system.

The simplest example of this type of detection involves acquiring an image with the US on and off (Hisaka et al. 2001; Li et al. 2002a). Figure 13.5a shows an example of such a system with the numerical aperture (and hence speckle size) controlled by the size of the aperture. The application of US increases the motion of the scatterers within the US focal zone, which results in the speckle pattern fluctuating more rapidly. Time integration of the speckle pattern (over the image acquisition time of the camera) means that there will be more averaging of the fluctuating speckle pattern when US is applied, which results in a lower contrast speckle pattern. Measuring the difference in the speckle contrast, defined as the ratio of the standard deviation of the speckles to the mean, over the entire speckle image for the US on and US off cases indicates the magnitude of the US modulation. Figure 13.5b and c shows an example of a speckle pattern obtained from a scattering gel in the absence (Figure 13.5b) and presence (Figure 13.5c) of US. There is a measurable change in the speckle contrast, but it is interesting to note that the effect of applying the US is small as the images qualitatively do not change significantly. This is an example of where speckle decorrelation (Section 13.3.2) is not observable as the scatterers in the gel are relatively static compared to *in vivo* tissue. For *in vivo* work, scatterers such as red blood cells move and the speckle pattern changes even in the absence of US. However, the advantage of this approach is that a global measure of speckle contrast in a single frame is made (US on) and then subtracted from the global measure in the next frame (US off). It is not essential that the speckle patterns remain correlated between frames and so it is relatively immune to speckle decorrelation.

Chapter 13

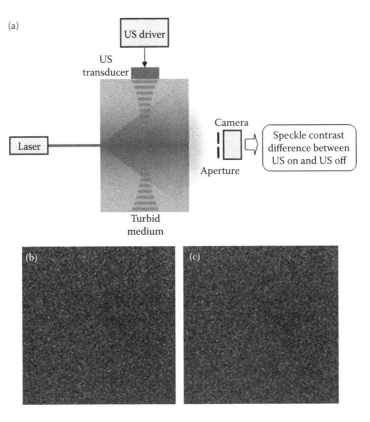

FIGURE 13.5   (a) Speckle difference imaging system; speckle pattern obtained from a light scattering gel with (b) US off, contrast = 0.52, and with (c) US on, contrast = 0.49.

13.3.3.3   Parallel Lock-in Detection

A camera-based system (Figure 13.6a) can also be used to obtain the amplitude and phase of an individual modulated speckle by implementing a parallel lock-in detection scheme (Leveque-Fort 2000; Leveque-Fort et al. 2001). Summing the modulated magnitude over the entire detector array provides an $N^{0.5}$ improvement in SNR compared to the single pixel case (N = number of pixels). The optical illumination is strobed at four different phases in the acoustic modulation cycle (Figure 13.6b) over four consecutive camera frames. For example in frame 1, strobe position S1 is applied many times within the camera frame; in frame 2, the strobe position switches to strobe S2; and so on. The signal obtained in each frame is given by (Resink et al. 2012)

$$S_1 = NT_0 \left[\frac{I_{dc}}{4} + \frac{\sqrt{2}}{2\pi} I_{ac} \cos(\Phi) \right] \tag{13.4}$$

$$S_2 = NT_0 \left[\frac{I_{dc}}{4} + \frac{\sqrt{2}}{2\pi} I_{ac} \sin(\Phi) \right] \tag{13.5}$$

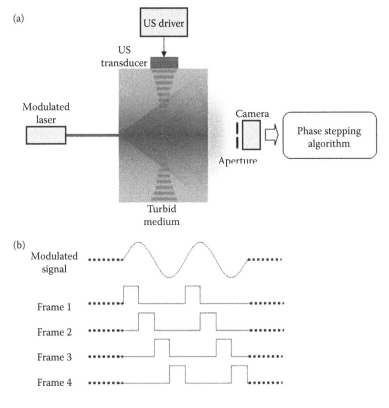

FIGURE 13.6   Lock-in detection algorithm implemented using strobed laser illumination and charge coupled device (CCD) camera detection: (a) system; (b) US modulated optical signal showing strobe position in different camera frames.

$$S_3 = NT_0\left[\frac{I_{dc}}{4} - \frac{\sqrt{2}}{2\pi}I_{ac}\cos(\Phi)\right] \tag{13.6}$$

$$S_4 = NT_0\left[\frac{I_{dc}}{4} - \frac{\sqrt{2}}{2\pi}I_{ac}\sin(\Phi)\right] \tag{13.7}$$

Combining the measurements made in the four frames, the amplitude and phase of each speckle can be obtained:

$$I_{ac} = \frac{\pi}{NT_0\sqrt{2}}\sqrt{(S_1 - S_3)^2 + (S_2 - S_4)^2} \tag{13.8}$$

$$\Phi = \arctan\left(\frac{S_2 - S_4}{S_1 - S_3}\right) \tag{13.9}$$

Chapter 13

However, as four frames (typically of 20 ms duration, frame rate 50 Hz) are required to produce a signal, SNR can be reduced due to speckle decorrelation. This is a problem for imaging tissue for which speckle decorrelation times are of the order of 1 ms, but it can be applied for imaging gel scaffolds for which speckle decorrelation times are longer due to the absence of moving red blood cells (Morgan et al. 2011). A method using two phase steps has been demonstrated to provide comparable performance (Li and Wang 2002b) and can reduce image acquisition times. Variations of this algorithm have been developed for second harmonic imaging (Selb et al. 2002; Ruan et al. 2012). Smart CMOS sensors have also been developed (Patel et al. 2011), which can perform lock-in detection of the incident light at each pixel and removes the requirement to strobe the illumination light.

13.3.3.4 Interferometric/Holographic Methods

A useful approach for detecting the optical signal due to US modulation is based on holography. A proportion of the light illuminates the sample and is modulated at the frequency of the US (Figure 13.7a) (Gross et al. 2003). The remainder is used as a reference beam, which is frequency-shifted at the US frequency by two acousto-optic modulators. This reference beam interferes with the light emerging from the sample at a CCD detector. This process is analogous to recording a hologram of the aperture as illustrated in Figure 13.7a. As the light modulated within the medium and the reference beam are at the same optical frequency, the fringes due to interference between these two sources at the detection plane will be static (in the absence of speckle decorrelation).

On the other hand, the interference between unmodulated light and the reference beam results in time-varying fringes, which are filtered out by the camera, which acts as a low-pass filter. The amplitude and phase of the static fringes can then be extracted by applying numerical heterodyne holography reconstruction to this hologram. In this case, phase shifts (usually 4) are applied to the reference arm, and a phase stepping algorithm is applied (Gross et al. 2003). The intensity of the reconstructed aperture image represents the US modulation depth. An advantage of this approach over the lock-in detection method described in the previous section is that the amplitude of the detected signals is dependent on both the magnitude of the light emerging from the medium and the reference beam. Therefore, the intensity of the reference beam can be set to be greater than the thermal or readout noise of the camera, thus improving SNR. Applying the phase stepping algorithm requires four images to be recorded in series and so speckle decorrelation could present a problem. However, this is reduced by tilting the reference beam at an angle to the beam emerging from the sample (Figure 13.7a), which means that the static fringes due to the modulated light from the sample is shifted in spatial frequency and can be separated from speckle decorrelation in the spatial frequency domain.

The CCD-based holography method has been developed further through the application of photorefractive crystals (Ramaz et al. 2004; Murray et al. 2004). Photorefractive crystals are materials whose refractive index or absorption changes with illuminating light and as such allows a hologram to be dynamically written to it. One method of detecting US modulated light using a photorefractive crystal is shown in Figure 13.7b (Ramaz et al. 2004). As with the CCD-based holography approach, a static speckle pattern is obtained when the US-modulated light interferes with the reference beam

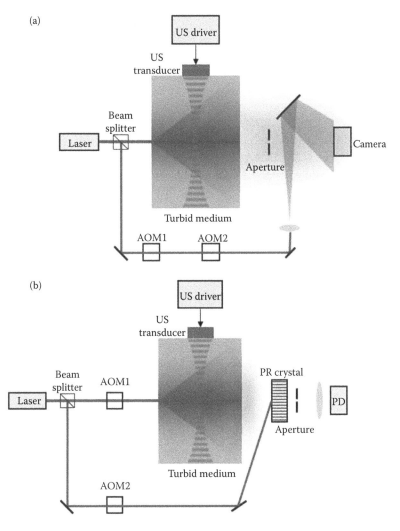

FIGURE 13.7 Holographic detection (a) using a CCD and (b) using a photorefractive crystal (PR). AOM: acousto-optic modulator.

because they have the same frequency. This static speckle pattern writes a volume holo-gram within the crystal. All other frequency components produce a fluctuating speckle pattern because they are at different frequency to the reference and are therefore aver-aged out. The reference beam is then diffracted by the volume hologram into a field, which has the same wave front as the US-modulated field emerging from the scattering medium. The multiplication of the US-modulated wave front by the matched reference wave front at the detector provides a significant improvement in SNR. The elegance of this approach is that as these wave fronts are matched, the random phases that cause a speckle pattern are not present in the detected interference pattern, and so a large-area single detector such as a PMT can be used. Again the modulated signal amplitude is extracted using a phase stepping algorithm (Ramaz et al. 2004). Holographic meth-ods are not immune to speckle decorrelation, as the response of the crystals should be shorter than the speckle decorrelation times. The response time is typically ~100 ms,

Chapter 13

which is much longer than the speckle decorrelation time in tissue, although certain types of photorefractive crystals such GaAs could achieve a response time of 0.25 ms (Elson et al. 2011).

13.3.3.5 Fabry–Perot Interferometer

The previously described detection methods all rely on measurement of the effect of the US modulation on laser speckles. As such, they are affected to a greater or lesser extent by speckle decorrelation, which is caused by the fluctuations in the speckle pattern not caused by US modulation, such as speckle fluctuations caused by moving red blood cells. Both the Fabry–Perot interferometer described here and the spectral hole burning technique described in the next section involve detecting the small frequency shift (color change) that occurs at the optical frequency due to the application of the US. As such, they are immune to the effects of laser speckle, and so in this respect, they are an attractive option for detecting USMOT signals. Another advantage is that detection using a narrow band-pass filter blocks the DC background light and so removes the problems associated with detecting a small, modulated light signal on a large DC light background. A confocal Fabry–Perot interferometer (Figure 13.8) (Sakadzic and Wang 2004; Rousseau et al. 2009; Kothapalli and Wang 2008) has been used as a narrow band-pass filter for the detection of US-modulated light. By changing the separation of the mirrors, the cavity can be tuned to different optical frequencies. For example, the cavity has been tuned to the optical frequency plus 15 MHz (Sakadzic and Wang 2004) and the optical frequency plus 75 MHz for high spatial resolution USMOT imaging (Kothapalli and Wang 2008).

13.3.3.6 Spectral Hole Burning

Spectral hole burning has recently received attention (Li et al. 2008) as it can provide a narrow band-pass filter with a high etendue. Like the Fabry–Perot interferometer described in the previous section, spectral hole burning is a technique that detects the small frequency shift (color change) in the light due to US modulation. This is achieved through using spectral hole burning as a narrow band spectral filter. Detection of US-modulated light occurs as one of the US-modulated side bands is

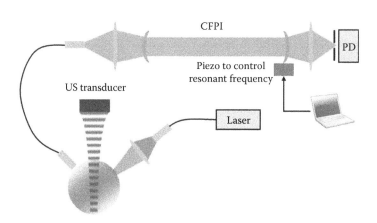

FIGURE 13.8 Confocal Fabry–Perot interferometer detection.

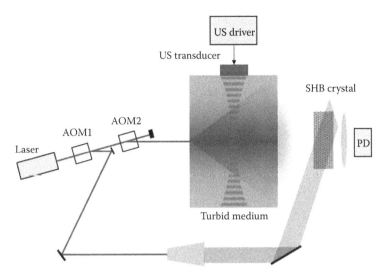

FIGURE 13.9 Spectral hole burning system. AOM: acousto-optic modulator; PD: photodetector; SHB: spectral hole burning.

centered within the passband of this filter and unmodulated light and other side-bands lie outside the passband. The main element of the spectral hole burning system is a cryogenically cooled spectral hole burning crystal (e.g., a Tm^{3+}:YAG crystal cooled to 4 K was used by Li et al. 2008). Without the application of a pump beam, the crystal is a wideband optical absorber. Upon application of a pump beam, resonant ions are excited from their ground state to a higher energy state. If the pump beam is sufficiently intense to excite all resonant ions, then the crystal cannot absorb further light at this frequency (color). A spectral hole is thus burned in the crystal at this frequency, allowing light within this narrow band to pass. Figure 13.9 shows how spectral hole burning has been implemented in US-modulated optical tomography. The pump beam, which was at the same frequency as the US-modulated signal, passes outside the medium and burns a spectral hole in the crystal, which has a lifetime of ~10 ms. The pump beam is then switched off and detection of the US-modulated light occurs.

13.4 Example Applications

US-modulated optical tomography has potential in a range of applications in regenerative medicine both for *in vivo* imaging and monitoring in bioreactors. As US-modulated light is more heavily affected by the optical properties of tissue within the US focal zone, the technique has the potential to map the optical absorption, scattering, and fluorescence in 3D, which can all be related to the tissue properties.

13.4.1 Absorption Imaging

Due to the absorption of light by oxyhemoglobin and deoxyhemoglobin, USMOT offers the potential for 3D imaging of blood vessels, for example, for monitoring new vessel formation in regenerative medicine. Using two or more wavelengths will

Chapter 13

also allow oxygen saturation to be obtained. Kothapalli and Wang (2009) applied US-modulated optical tomography to image blood vessels *ex vivo* utilizing 30 MHz focused US and confocal Fabry–Perot interferometer detection. The system has been demonstrated to have an axial resolution of 60 μm and lateral resolution of 70 μm at a depth of 2.5 mm in tissue. Results obtained using the system are shown in Figure 13.10. Figure 13.10a shows the photograph of the *ex vivo* tissue, highlighting the region of interest. The A-scans (Figure 13.10b) clearly show the difference between

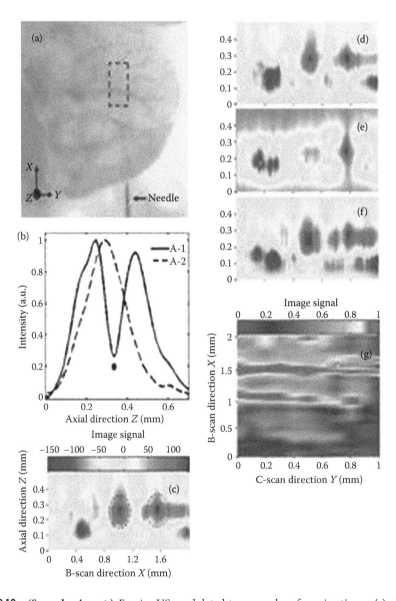

FIGURE 13.10 **(See color insert.)** *Ex vivo* US modulated tomography of *ex vivo* tissue: (a) photograph of blood vasculature in rat ear highlight region of interest; (b) A-1 is an US-modulated optical A-scan through a blood vessel and A-2 away from a vessel; (c–f) B-scans obtained 2 mm apart; (g) image obtained using maximum image projection processing.

a line when a vessel is present (A-1) and away from a vessel (A-2). A selection of B-scans obtained at different positions is shown in Figure 13.10c through f, and an image processed using maximum image projection is shown in Figure 13.10g. Such a technique would be valuable in determining vascularization *in vivo*, in particular, in response to scaffolds or microparticles with tailored growth factor release to encourage neovascularization (e.g., vascular endothelial growth factor [VEGF]) and support subsequent vessel maturation (e.g., platelet derived growth factor [PDGF]) (Bible et al. 2012; Saif et al. 2010).

Gunadi and Leung (2011) have investigated the potential for carrying out oxygen saturation measurements using US-modulated optical tomography. Monte Carlo simulations were used to demonstrate that US-modulated images are more sensitive to the optical properties of the region of interest compared to near-infrared spectroscopy, which is more sensitive to superficial tissue. Microbubbles have also been suggested as a method of improving contrast in this technique (Honeysett et al. 2012). Development of accurate reconstruction algorithms will be an important part of the development of a quantitative spectroscopy tool (Bratchenia et al. 2011b).

From Figure 13.10, vessels can be clearly identified, but at present, the image quality is significantly poorer than that achieved using PAT. Without significant improvement, for example, by using TRUE, it is therefore questionable whether this technique will be routinely used for blood vessel imaging.

13.4.2 Scattering Imaging

In addition to being sensitive to the absorption properties of the tissue, USMOT is sensitive to the scattering coefficient of the tissue. The scattering coefficient is different between healthy and necrotic tissue and so USMOT offers potential for monitoring tissue necrosis. This can be useful in monitoring 3D cell culture in bioreactors in tissue engineering. For example, with cartilage constructs, one barrier to successful *in vitro* 3D culture is restricted nutrient exchange in thicker tissues, which results in nonuniform tissue growth (Freed et al. 1999; Martin et al. 2004). The rapid formation of tissue at the outer regions of the scaffold limits the access of cells and nutrients into the deeper regions, leading to a necrotic core. To an extent, this can be addressed by scaffold design (Silva et al. 2006), but as tissue forms within the scaffold and diffusion routes are restricted, this again can be an issue. The ability to monitor tissue necrosis within a scaffold noninvasively would be an important advance in this field, allowing the tailoring of scaffold and bioreactors design strategies.

USMOT has been applied to monitoring high-intensity focused US (HIFU) of soft tissue tumors (Lai et al. 2010). In this case, thermal damage is induced in soft tissue tumors, and it is of interest to monitor tissue necrosis during treatment in order to improve efficacy and safety. Lai et al. (2010) implemented a system based on a HIFU transducer and detection using a photorefractive crystal. Figure 13.11 shows the acousto-optic signal obtained from a chicken breast sample for different pulse durations of HIFU for a fixed peak US intensity of 3300 W/cm^2. A 5 s exposure time had relatively little effect on the acousto-optic signal (Figure 13.11a and d, top image), whereas with longer exposure times, the acousto-optic signal changed (Figure 13.11b and c) along with the volume of tissue necrosis (Figure 13.11d, II and III images).

Chapter 13

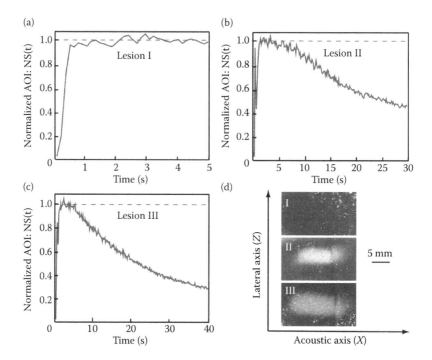

FIGURE 13.11 Normalized acousto-optic signals from lesions with different exposure times: (a) 5 s, (b) 30 s, and (c) 40 s. (d) Photographs of the lesions for different exposure times: (I) 5 s, (II) 30 s, and (III) 40 s.

13.4.3 Fluorescence Imaging

Potentially, one of the most useful applications of USMOT in regenerative medicine is in imaging fluorescence in thick tissue. PAT has, to date, outperformed USMOT in imaging optically absorbing features such as blood vessels. However, as PAT is sensitive to the absorption of light in tissue, techniques to monitor fluorescence using PAT are based on relating the absorption of light to fluorescence (Razansky et al. 2009). Interpreting the results may be challenging in the presence of multiple absorbers or multiple fluorophores. US-modulated fluorescence tomography (USMFT) has an advantage in this case, as fluorescence can be monitored directly because the fluorescence emission light is modulated. The drawback is that the detected signal is very weak because fluorescent light is incoherent and modulation is due to a change in intensity of the detected light due to a change in the optical absorption and scattering coefficients (mechanism 1, Section 13.3.2). As discussed in Section 13.3.2, the signal is typically two to three orders of magnitude lower than when coherent light is imaged. As the light is incoherent, speckle is not present, and so speckle-based detection techniques do not offer an improvement in SNR. Relatively simple photomultiplier-based detection systems (Section 13.3.3.1) with a large area detector are therefore applied.

Much of the research carried out so far has been fundamental, investigating the mechanisms for modulation (Yuan et al. 2008) and concentrating on the position of the US focus relative to the fluorescent target. CW US has generally been applied in all USMFT, although pulsed US has recently been applied (Huynh et al. 2012b) to enable time-gated fluorescence imaging along the acoustic axis. Potential applications in regenerative medicine would include tracking the movement of cells that have been transformed to express a fluorescent protein within a scaffold or expression of a key marker of interest whose expression is linked to the expression of a fluorescent protein (reviewed in Stepanenko et al. 2011). This would also have application in a wider context of *in vitro* testing, for which the cell could be used as a biosensor (Hofmann et al. 2013). For example, fluorescence-based fiber-optic oxygen sensors have been used to monitor oxygen levels in collagen gels; such sensors only allow measurements within a particular area of the sample, and they are invasive (Cheema et al. 2012).

Contrast agents offer potential for improving SNR. For example, microbubbles can be used to produce a greater change in the optical properties than could be achieved with US alone (Hall et al. 2009; Honeysett et al. 2011). One proposition is the use of bubbles comprising fluorophore and quencher (Yuan 2009; Schutt et al. 2011), which produce fluorescent signals in the presence of US but are quenched in its absence. Outside the fluorescence focal zone, fluorescence is significantly quenched due to the close proximity of fluorophore and quencher of the bubble in its base state, producing a low background signal. Within the US focal zone, the bubbles are expanded to separate the fluorophore and quencher, allowing the fluorophore to probe the properties of the medium within this region. An alternative approach is to use thermosensitive polymers (Lin et al. 2012; Yuan et al. 2012), which produce a different fluorescence intensity and lifetime signal at different temperature. As a change of temperature occurs within the US focal zone, this effect can be used to produce localized fluorescent probes in a region of interest. Indocyanine green–loaded pluronic nanocapsules have been used (Lin et al. 2012), which demonstrate a change in fluorescence lifetime and intensity with change in signal. A HIFU transducer is scanned across a tissue phantom ($4 \times 10 \times 10$ cm agar containing Intralipid and India ink, reduced scattering coefficient = 0.005 mm^{-1} and absorption coefficient = 0.6 mm^{-1}) containing a 3 mm inclusion of the contrast agent. Although the reduced scattering coefficient is relatively low compared to that of tissue (typically 1 mm^{-1}), the results demonstrate proof of concept. As the HIFU is scanned across the medium (Figure 13.12a and b), a change in both fluorescence intensity (Figure 13.12c and e) and lifetime (Figure 13.12d and f) can be observed. A similar approach has also been demonstrated using a DBD-AA–labeled thermosensitive polymer in which the change in lifetime of the fluorophore is used to select between background fluorescence and fluorescence from the US focal zone (Yuan et al. 2012).

13.4.4 Luminescence Imaging

Another application for which USMOT may have important applications is when the light source is generated within the tissue itself without the need for an external (excitation) source; US modulation of sonoluminescence (Wang and Shen 1998), chemiluminescence, and bioluminescence imaging (Huynh et al. 2013) have all been proposed. In these cases, there are clear advantages over PAT as an external light source is essential to generate sound in photoacoustic imaging.

Sonoluminescence involves the production of light when microbubbles collapse (cavitate). As this occurs in regions of high pressure, localized sonoluminescence imaging can be achieved using a focused US transducer. There are concerns that the cavitation of bubbles can produce tissue damage, although sonoluminescence

Chapter 13

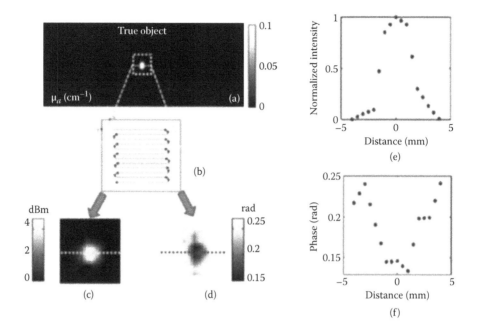

FIGURE 13.12   Temperature-modulated fluorescence imaging: (a) 3 mm inclusion filled with pluronic-ICG embedded within a 4 × 10 × 10 cm agar phantom; (b) HIFU is scanned over an 8 × 8 mm area; (c) intensity; (d) lifetime; (e and f) normalized profiles of (c) and (d), respectively.

has been observed *in vivo* at lower pressures than the FDA safety limit (He et al. 2002). US has been used to modulate light emitted by a chemiluminescent target embedded within a scattering medium as a means of localizing the region of interest (Huynh et al. 2013). Figure 13.13a shows the configuration of an experiment in which a chemiluminescent target is placed within a scattering medium and focused US is used to modulate the light emerging from the target. Figure 13.13b shows that the imaging resolution is clearly superior compared to the unmodulated light case. If SNR challenges can be overcome, the technique will have applications in biolumines-cence imaging (see Chapter 15). Bioluminescence is widely used within regenerative medicine to track the migration of cells following injection (there are many papers that describe this, but an example is given in Han et al. 2012) and has, for example, been used in conjunction with fluorescence labeling (Luc-GFP-labeled mesenchymal stem cells) to determine that ischemia is the prime cause of stem cell death in tissue-engineered constructs following implantation *in vivo* (Becquart et al. 2012). Again, there are many more examples in the literature illustrating the utility of such imaging techniques in regenerative medicine.

13.4.5   Elastography

Tissue mechanical properties can be measured by applying sound and measuring the change in the detected speckle pattern at the surface of the tissue. For example, Kirkpatrick et al. (2006) applied low-frequency acoustic waves (1–5 Hz) to tissue and measured the dynamic shift in the speckle pattern reflected from the tissue. A

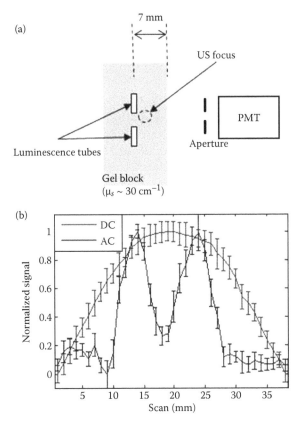

FIGURE 13.13 US-modulated chemiluminescence imaging: (a) experiment configuration; (b) modulated and unmodulated line scans.

mechanically stiffer tissue will produce a more slowly changing speckle pattern upon the application of sound. In a small sample, different types of skin lesions were demonstrated to have different speckle properties, which appeared to correlate with whether any part of the lesion lays within the dermis.

An approach termed transient elastography has been used to measure the elastic properties at depth within a medium (Daoudi et al. 2009). An ARF (Section 13.3.2) occurs when sound propagates in an acoustically attenuating medium, and when pulsed focused US is used, a transient shear stress is induced in the focal region. Within the medium, this produces shear waves that affect the measured speckle pattern at the surface, allowing the elastic properties of the tissue to be monitored. Li et al. (2011) imaged surface acoustic waves in tissue phantoms using speckle contrast and in tissue phantoms demonstrated that surface waves could be used to indicate mechanical stiffness. The mechanical properties of a matrix are well known to influence cell behavior, including stem cell differentiation (Kshitiz et al. 2012), and matrix stiffness will change during initial population of a scaffold with cells, during subsequent matrix deposition, and finally during tissue maturation. Being able to monitor such matrix stiffness changes during the regenerative process would provide valuable spatial information in a rapid format to inform tissue engineering strategies.

Chapter 13

13.5   Conclusions and Future Prospects

US-modulated optical tomography offers the potential for overcoming the effects of light scattering and enabling optical imaging of thick tissue and 3D tissue cultures at the resolution of US. One of the major problems is that the modulation of light by sound within tissue is a very weak effect, and so much effort has gone into optimizing the detection systems. To date, the image quality has been inferior to that of PAT when imaging optically absorbing structures such as blood vessels. Time-reversal techniques show great promise in improving the imaging resolution of US-modulated optical tomography systems. The technique may also find application in high-resolution imaging of fluorescence, sonoluminescence, chemiluminescence, and bioluminescence. In these cases, SNR is a significant challenge, but it is likely that novel acoustically activated contrast agents will significantly improve performance.

Acknowledgments

This work was supported by the *Biotechnology and Biological Sciences Research Council (BBSRC) UK* (BB/F004826/1 and BB/F004923/1). We thank ML Mather and BR Hayes-Gill for useful discussions.

References

Aanonsen, SI, Barkve, T, Tjotta, JN, Tjotta, S. 1984. Distortion and harmonic generation in the nearfield of a finite amplitude sound beam. *Journal of the Acoustical Society of America* 75:749–768.

Becquart, P, Cambon-Binder, A, Monfoulet, LE, Bourguignon, M, Vandamme, K, Bensidhoum, M, Petite, H, Logeart-Avramoglou, D. 2012. Ischemia is the prime but not the only cause of human multipotent stromal cell death in tissue-engineered constructs *in vivo*. *Tissue Engineering Part A* 18(19–20):2084–2094.

Bernsten, J. 1990. Numerical calculations of finite amplitude sound beams. *Frontiers of Nonlinear Acoustics: Proceedings of 12th ISNA*, edited by M. F. Hamilton and D. T. Blackstock. London: Elsevier.

Bible, E, Qutachi, O, Chau, DYS, Alexander, MR, Shakesheff, KM, Modo, M. 2012. Neo-vascularization of the stroke cavity by implantation of human neural stem cells on VEGF-releasing PLGA microparticles. *Biomaterials* 33(30):7435.

Boas, DA, Dunn, AK. 2010. Laser speckle imaging in biomedical optics. *Journal of Biomedical Optics* 15:01109.

Bratchenia, A, Molenaar, R, van Leeuwen, TG, Kooyman, RPH. 2011a. Acousto-optic-assisted diffuse optical tomography. *Optics Letters* 36:1539–1541.

Bratchenia, A, Molenaar, R, Kooyman, RPH. 2011b. Towards quantitative acousto-optic imaging in tissue. *Laser Physics* 21:601–607.

Cheema, U, Rong, ZM, Kirresh, O, MacRobert, AJ, Vadgama, P, Brown, RA. 2012. Oxygen diffusion through collagen scaffolds at defined densities: implications for cell survival in tissue models. *Journal of Tissue Engineering and Regenerative Medicine* 6(1):77–84.

Daoudi, K, Boccara, AC, Bossy, E. 2009. Detection and discrimination of optical absorption and shear stiffness at depth in tissue-mimicking phantoms by transient optoelastography. *Applied Physics Letters* 94(15):154103.

Elson, DS, Li, R, Dunsby, C, Eckersley, R, Tang, MX. 2011. Ultrasound-mediated optical tomography: a review of current methods. *Interface Focus* 1:632–648.

Forget, B, Ramaz, F, Atlan, M, Selb, J, Boccara, AC. 2003. High-contrast fast Fourier transform acousto-optical tomography of phantom tissues with a frequency-chirp modulation of the ultrasound. *Applied Optics* 42:1379–1383.

Freed, L, Martin, I, Vunjak-Novakovic, G. 1999. Frontiers in tissue engineering—in vitro modulation of chondrogenesis. *Clinical Orthopedics and Related Research* 367:S46–S58.

Gibson, AP, Hebden, JC, Arridge SR. 2005. Recent advances in diffuse optical imaging. *Physics in Medicine and Biology* 50:R1–R43.

Gross, M, Goy, P, Al-Koussa, M. 2003. Shot-noise detection of ultrasound-tagged photons in ultrasound modulated imaging. *Optics Letters* 28:2482–2484.

Gunadi, S, Leung TS. 2011. Spatial sensitivity of acousto-optic and optical near-infrared spectroscopy sensing measurements. *Journal of Biomedical Optics* 16:127005.

Hall, DJ, Hsu, MJ, Esener, S, Mattrey, RF. 2009. Detection of US-modulated photons and enhancement with US microbubbles. *Proceedings of SPIE* 7177:71771L.

Han, XG, Yang, N, Cui, YY, Xu, YS, Dang, GT, Song, CL. 2012. Simvastatin mobilizes bone marrow stromal cells migrating to injured areas and promotes functional recovery after spinal cord injury in the rat. *Neuroscience Letters* 521(2):136–141.

He, YH, Xing, D, Tan, SC, Tang, YH, Ueda, K. 2002. *In vivo* sonoluminescence imaging with the assistance of FCLA. *Physics in Medicine and Biology* 47:1535–1541.

Hisaka, M, Sugiura, T, Kawata, S. 2001. Optical cross-sectional imaging with pulsed ultrasound wave assistance. *Journal of the Optical Society of America A* 18:1531–1534.

Hofmann, U, Michaelis, S, Winckler, T, Wegener, J, Feller, KH. 2013. A whole-cell biosensor as in vitro alternative to skin irritation tests. *Biosensors and Bioelectronics* 39(1):156–162.

Honeysett, JE, Stride, E, Leung, TS. 2011. Microbubble enhancement of US-modulated optical sensing with incoherent light. *Proceedings of SPIE* 7899:789919.

Honeysett, JE, Stride, E, Leung, TS. 2012. Feasibility study of non-invasive oxygenation measurement in a deep blood vessel using acousto-optics and microbubbles. *Advances in Experimental Medicine and Biology* 737:277–283.

Huynh, NT, He, D, Hayes-Gill, BR, Crowe, JA, Walker, JG, Mather, ML, Rose, FRAJ, Parker, NG, Povey, MJW, Morgan, SP. 2012a. Application of a maximum likelihood algorithm to ultrasound modulated optical tomography. *Journal of Biomedical Optics* 17:026014.

Huynh, NT, Ruan, H, He, D, Hayes-Gill, BR, Morgan, SP. 2012b. Effect of object size and acoustic wavelength on pulsed ultrasound modulated fluorescence signals. *Journal of Biomedical Optics* 17:076008.

Huynh, NT, Hayes-Gill, BR, Zhang, F, Morgan, SP. 2013. Ultrasound modulated imaging of luminescence generated within a scattering medium. *Journal of Biomedical Optics* 18(2):020505.

Kim, C, Wang, LV. 2007. Multi-optical-wavelength ultrasound-modulated optical tomography: a phantom study. *Optics Letters* 32:2285–2287.

Kim, C, Zemp, RJ, Wang, LV. 2006. Intense acoustic bursts as a signal-enhancement mechanism in ultrasound-modulated optical tomography. *Optics Letters* 31:2423–2425.

Kim, C, Song, KH, Maslov, K, Wang LV. 2009. Ultrasound-modulated optical tomography in reflection-mode with ring-shaped light illumination. *Journal of Biomedical Optics* 14:024015.

Kirkpatrick, SJ, Wang, RK, Duncan, DD, Kulesz-Martin, M, Lee, K. 2006. Imaging the mechanical stiffness of skin lesions by *in vivo* acousto-optical elastography. *Optics Express* 14:9770–9779.

Kirkpatrick, SJ, Duncan, DD, Wells-Gray, EM. 2008. Detrimental effects of speckle-pixel size matching in laser speckle contrast imaging. *Optics Letters* 33:2886–2888.

Kothapalli, S, Wang, LV. 2008. Ultrasound modulated optical microscopy. *Journal of Biomedical Optics* 13:054046.

Kothapalli, S, Wang, LV. 2009. Ex vivo blood vessel imaging using ultrasound modulated optical microscopy, *Journal of Biomedical Optics* 14:014015.

Kshitiz, Park, J, Kim, P, Helen, W, Engler, AJ, Levchenko, DH, Kim, A. 2012. Control of stem cell fate and function by engineering physical microenvironments. *Integrative Biology* 4(9):1008–1018.

Lai, PX, McLaughlan, JR, Draudt, AB, Murray, TW, Cleveland, RO, Roy, RA. 2010. Real-time monitoring of high-intensity focused ultrasound lesion formation using acousto-optic sensing. *Ultrasound in Medicine and Biology* 37:239–252.

Lai, PX, Xu, X, Liu, HL, Wang, LHV. 2012. Time-reversed ultrasonically encoded optical focusing in biological tissue. *Journal of Biomedical Optics* 17:030506.

Laschke, MW, Menger, MD. 2012. Vascularization in tissue engineering: angiogenesis versus inosculation. *European Surgical Research* 48(2):85–92.

Lesaffre, M, Farahi, S, Gross, M, Delaye, P, Boccara AC, Ramaz, F. 2009. Acousto-optical coherence tomography using random phase jumps on ultrasound and light. *Optics Express* 17:18211–18218.

Leutz, W, Maret, G. 1995. Ultrasonic modulation of multiply scattered light. *Physica B* 204:14–19.

Lev, A, Sfez, BG. 2003. Pulsed ultrasound-modulated light tomography. *Optics Letters* 28:1549–1551.

Chapter 13

Leveque-Fort, S. 2000. Three dimensional acousto-optic imaging in biological tissues with parallel signal processing. *Applied Optics* 40:1029–1036.

Leveque-Fort, S, Selb, J, Pottier, L, Boccara, AC. 2001. In situ local tissue characterization and imaging by backscattering acousto-optic tomography. *Optics Communications* 196:127–131.

Li, J, Ku, G, Wang, LV. 2002a. Ultrasound-modulated optical tomography of biological tissue by use of contrast of laser speckles. *Applied Optics* 41:6030–6035.

Li, J, Wang, LV. 2002b. Methods for parallel detection based ultrasound modulated optical tomography. *Applied Optics* 41:2079–2084.

Li, R, Elson, DS, Dunsby, C, Eckersley, R, Tang, MX. 2011. Effects of acoustic radiation force and shear waves for absorption and stiffness sensing in ultrasound modulated optical tomography. *Optics Express* 19:7299–7311.

Li, Y, Hemmer, P, Kim, C, Zhang, H, Wang, LV. 2008. Detection of ultrasound-modulated diffuse photons using spectral-hole burning, *Optics Express* 16:14862–14874.

Lin, Y, Kwong, TC, Bolisay, L, Gulsen, G. 2012. Temperature-modulated fluorescence tomography based on both concentration and lifetime contrast. *Journal of Biomedical Optics* 17:056007.

Liu, Q, Norton, S, Vo-Dinh, T. 2008. Modeling of nonphase mechanisms in ultrasonic modulation of light propagation. *Applied Optics* 47:3619–3630.

Mahan, GD, Engler, WE, Tieman, JJ, Uzgiris, E. 1998. Ultrasonic tagging of light: theory. *Proceedings about the National Academy of Science* 95:14015–14019.

Martin, I, Wendt, D, Heberer, M. 2004. The role of bioreactors in tissue engineering. *Trends in Biotechnology* 22(2):80–86.

Morgan, SP, Huynh, NT, Hayes-Gill, BR, Crowe, JA, Mather, ML, Rose, FRAJ, Parker, NG, Povey, MJW. 2011. Characterizing tissue scaffolds using optics and ultrasound. *Proceedings of SPIE* 7897:789719.

Murray, TW, Sui, L, Maguluri, G, Roy, RA, Nieva, A, Blonigen, F, DiMarzio, CA. 2004. Detection of ultrasound-modulated photons in diffuse media using the photorefractive effect. *Optics Letters* 29:2509–2511.

Patel, R, Achamfuo-Yeboah, S, Light, R, Clark, M. 2011. Widefield heterodyne interferometry using a custom CMOS modulated light camera. *Optics Express* 19:24546–24556.

Ramaz, F, Forget, BC, Atlan, M, Boccara, AC, Gross, M, Delaye, P, Roosen, G. 2004. Photorefractive detection of tagged photons in ultrasound modulated optical tomography of thick biological tissues. *Optics Express* 12:5469–74.

Razansky, D, Distel, M, Vinegoni, C, Ma, R, Perrimon, N, Koster, RW, Ntziachristos, V. 2009. Multispectral opto-acoustic tomography of deep-seated fluorescent proteins *in vivo*. *Nature Photonics* 3:412–417.

Resink, SJ, Steenbergen, W, Boccara, AC. 2012. State-of-the art of acousto-optic sensing and imaging of turbid media. *Journal of Biomedical Optics* 17(4):040901.

Rousseau, G, Blouin, A, Monchalin, JP. 2009. Ultrasound-modulated optical imaging using a high-power pulsed laser and a double-pass confocal Fabry-Perot interferometer. *Optics Letters* 34:3445–3447.

Ruan, H, Mather, ML, Morgan, SP. 2012. Pulse inversion ultrasound modulated optical tomography. *Optics Letters* 37:1658–1660.

Saif, J, Schwarz, TM, Chau, DYS, Henstock, J, Sami, P, Leicht, SF, Hermann, PC, Alcala, S, Mulero, F, Shakesheff, KM, Heeschen, C, Aicher, A. 2010. Combination of injectable multiple growth factor-releasing scaffolds and cell therapy as an advanced modality to enhance tissue neovascularization. *Arteriosclerosis Thrombosis and Vascular Biology* 30(10):1897–1904.

Sakadzic, S, Wang, LV. 2002. Ultrasonic modulation of multiply scattered coherent light: an analytical model for anisotropically scattering media. *Physical Review East* 66:026603.

Sakadzic, S, Wang, LV. 2004. High-resolution ultrasound-modulated optical tomography in biological tissues. *Optics Letters* 29:2770–2772.

Sakadzic, S, Wang, LV. 2005. Modulation of multiply scattered coherent light by ultrasonic pulses: an analytic model. *Physical Review East* 72:036620.

Schutt, CE, Benchimol, MJ, Hsu, MJ, Esener, SC. 2011. Ultrasound-modulated fluorescent contrast agent for optical imaging through turbid media. *Proceedings of SPIE* 8165:81650B.

Selb, J, Pottier, L, Boccara, AC. 2002. Non-linear effects in acousto-optic imaging, *Optics Letters* 27:918–920.

Silva, MMCG, Cyster, LA, Barry, JJA, Yang, XB, Oreffo, ROC, Grant, DM, Scotchford, CA, Howdle, SM, Shakesheff, KM, Rose, FRAJ. 2006. The effect of anisotropic architecture on cell and tissue infiltration into tissue engineering scaffolds. *Biomaterials* 27:5909–5917.

Stepanenko, OV, Stepanenko, OV, Shcherbakova, DM, Kuznetsova, IM, Turoverov, KK, Verkhusha, VV. 2011. Modern fluorescent proteins: from chromophore formation to novel intracellular applications. *Biotechniques* 51(5):313.

Tuchin, VV. 2007. *Tissue Optics*. Bellingham: SPIE Press.

Wang, LV. 2001a. Mechanisms of ultrasonic modulation of multiply scattered coherent light: an analytic model. *Physics Review Letters* 87:043903.

Wang, LV. 2001b. Mechanisms of ultrasonic modulation of multiply scattered coherent light: a Monte Carlo model. *Optics Letters* 26:1191–1193.

Wang, LV, Shen, Q. 1998. Sonoluminescent tomography of strongly scattering media. *Optics Letters* 23:561–563.

Xu, X, Liu, H, Wang, LV. 2011. Time-reversed ultrasonically encoded optical focusing into scattering media. *Nature Photonics* 5:154–157.

Yuan, B. 2009. US-modulated fluorescence based on a fluorophore-quencher-labeled microbubble system. *Journal of Biomedical Optics* 14:024043.

Yuan, B, Gamelin, J, Zhu, Q. 2008. Mechanisms of the ultrasonic modulation of fluorescence in turbid media. *Applied Physics* 104:103102.

Yuan, B, Uchiyama, S, Liu, Y, Nguyen, KT, Alexandrakis, G. 2012. High-resolution imaging in a deep turbid medium based on an ultrasound switchable fluorescence technique. *Applied Physics Letters* 101:033703.

Zemp, RJ, Kim, C, Wang, LV. 2007. Ultrasound-modulated optical tomography with intense acoustic bursts. *Applied Optics* 46:1615–1623.

Chapter 13

Optical Macroscopic Imaging in Regenerative Medicine

14. Macroscopic Imaging in Regenerative Medicine

Sean J. Kirkpatrick, Stephen J. Matcher, and Stephen P. Morgan

14.1 Introduction

The techniques described within this book highlight a wide range of techniques that can be applied in regenerative medicine. Depending on the spatial resolution, penetration depth, and functional information required, different approaches should be utilized. Many of the techniques are relatively sophisticated; for example, nonlinear microscopy requires the application of a pulsed laser, optical coherence tomography requires interferometry, and photoacoustic imaging requires a pulsed laser and an array of ultrasound transducers.

Here we discuss what can be achieved with relatively simple imaging systems at the macroscopic scale in regenerative medicine. Three different techniques are investigated: (1) time-lapsed imaging, (2) laser speckle imaging, and (3) spectral imaging. In the next chapter, an additional macroscopic technique, bioluminescence imaging, is described. The techniques are common as they require a simple system involving illumination of an area of a sample and forming an image via a lens on a camera. In time-lapsed imaging, a movie is recorded of an event (e.g., tissue growth in a bioreactor) and image processing is applied to extract properties of the

Optical Techniques in Regenerative Medicine. Edited by Stephen P. Morgan, Felicity R.A.J. Rose, and Stephen J. Matcher © 2014 CRC Press/Taylor & Francis Group, LLC. ISBN: 978-1-4398-5495-2

Chapter 14

tissue, for example, cell aggregate size. Additional information can be obtained if coherent illumination is applied (usually from a laser) and the properties of the detected speckle are analyzed. The contrast of the detected speckle pattern can be related to the biomechanical properties of the tissue or the flow of blood cells within vessels. If illumination is performed using different colors, then tissue properties such as blood oxygen saturation can be imaged. The advantages of such approaches are that they are very robust and relatively simple, and so are well suited for long-term monitoring and automation of tissue-engineered processes. The drawback is that such macroscopic imaging is at low spatial resolution and is often only carried out at the surface of the tissue, with no (or very little) depth discrimination.

The factors affecting image quality in a macroscopic imaging system are described here, followed by the modifications required to perform laser speckle and spectroscopic imaging. The theory of laser speckle imaging and how these measurements can be related to the biomechanical properties of tissue and blood flow within tissue is also described. This is followed by fundamental theory on spectroscopy of highly scattering tissue. Light scattering is a major problem in this case and needs to be taken into account when relating the images to the oxygen saturation of the tissue. Examples then follow of how such techniques can be applied in regenerative medicine.

14.2   Clinical and Biological Measurement Challenges

The techniques described are relatively simple and robust and so can be easily applied to macroscopic imaging of tissue within the biology laboratory or the clinic. With appropriate image processing, the techniques are well suited to monitoring the manufacture of regenerative medicine products and subsequent automation of this process. Tissue growth in bioreactors can be monitored using time-lapsed imaging, and in Section 14.4.1, an example is provided in which the number and size of aggregates in a rotating bioreactor were monitored over several hours. An example is also provided of the use of time-lapsed imaging in the production of tissue scaffolds in a supercritical fluid process. Tissue biomechanics can be measured using laser speckle methods by applying a force to the tissue and then monitoring the properties of the speckle pattern as the force increases or is released. Laser speckle imaging can also be applied to monitor flows and is well suited to monitoring blood flow in tissue. This can be used to identify regions where a tissue-engineered product, such as a 3D scaffold and/or cells, should be implanted. It can also be used to monitor tissue regeneration after the application of the product. For example, in wound healing, high blood flow has been demonstrated to correlate with wounds that will heal (La Hei et al. 2006). Spectroscopic imaging can be used to perform a similar role in identifying ischemic tissue, which requires intervention, and also in monitoring the subsequent regeneration of the tissue.

14.3   Technical Implementation

14.3.1   Time–Lapsed Imaging

The basic system for time-lapsed macroscopic imaging is very simple. Figure 14.1 shows an example in which a time-lapsed imaging system is used to monitor a rotating bioreactor. The system consists of a light source, a camera, imaging optics, optical filters, and a computer for system control and image processing. It is important to achieve the best

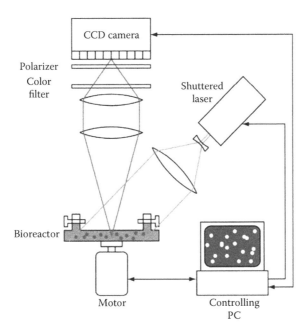

FIGURE 14.1 Example of a macroscopic imaging system. The use of a coherent light source and appropriate signal processing enables measurement of flow or tissue biomechanics through laser speckle imaging. Spectral imaging with appropriate signal processing enables measurement of tissue properties such as blood oxygen saturation.

image quality possible in order to provide the best quality data for subsequent image processing algorithms.

Illumination is a key component of the system. The four cornerstones of lighting have been described as follows (National Instruments 2012):

1. Geometry: relative positions of the camera, light source, and sample
2. Structure: the shape of the light projected onto the sample
3. Wavelength: depends on the properties of the source, sample, and detector
4. Filters: preferentially allow different wavelengths or polarization states to pass

In establishing the optimum system geometry, it is important to reduce the ambient light to a level that is insignificant compared to the direct illumination, for example, via light proof boxes. The position of the light source relative to the sample and detector produces different effects in the image. For example, illumination from behind the sample can be useful in determining the shape of an object, but it is relatively insensitive to surface features. Back-illumination is utilized in one of the examples in Section 14.4.1 when monitoring the shape of foamed polymeric scaffolds during their production. Diffuse illumination is useful for illuminating curved surfaces that have high specular (surface) reflections (e.g., this can be useful when imaging cylindrical bioreactors). Directional lighting (as shown in Figure 14.1) is frequently used for direct illumination of a sample, although it can be susceptible to hot spots due to direct reflections from surfaces. Dark field lighting, which involves illumination at low angles of incidence, is useful in reducing the effects of surface reflections. A form of dark field illumination is utilized in an

Chapter 14

example of imaging a rotating bioreactor in Section 14.4.1, for which sheet illumination from the side of the bioreactor (perpendicular to the camera) is employed. A spatially uniform intensity light source is ideal, although this is often difficult to achieve. Controlled angle holographic diffusers can be employed or beam expanders can be used to produce broad Gaussian-shaped intensity distributions, which are relatively uniform across the field of view of the imaging system. At the very least, if uniform illumination cannot be achieved, then the nonuniformity should be measured and compensation made in subsequent image processing.

The camera usually employed is a charge coupled device (CCD), but sometimes for low-power, high-frame-rate, low-cost applications, complementary metal oxide semiconductor (CMOS) sensors are often preferred and are improving in performance (Litwiller 2005). For low light level applications, a low-temperature sensor (e.g., Peltier cooled) should be used. For low contrast images, a camera with a high well capacity and a high-resolution analogue-to-digital converter is more capable for discriminating small-intensity changes on a high-intensity background.

The light source, the sample, and the detector all affect the spectral response of the system. In regenerative medicine, the sample spectral response (Section 14.3.3) is important in determining parameters such as the oxygen saturation of the tissue. However, "white" light sources such as LEDs or halogen lamps do not produce outputs that are uniform across the wavelength range. Also, the quantum efficiency of CCD and CMOS cameras varies with wavelength. In order to measure the true spectral response of the sample, it is necessary to calibrate the effects of the nonuniform illumination and detection of spectral response using a test target, such as Spectralon, which has uniform reflectivity across a wide wavelength range. Laser illumination can be useful as it produces monochromatic, intense, unidirectional illumination. However, the images obtained under laser illumination are susceptible to laser speckle, which can reduce image quality. A useful property of laser speckle is that it is sensitive to motion of the sample, and this can be utilized to measure blood flow or tissue biomechanics (Section 14.3.2).

Filters can be used to select different wavelength bands for spectroscopic and fluorescence imaging. Sophisticated multispectral and hyperspectral cameras that incorporate wavelength scanning are described in Section 14.3.3. Polarization filters can also be used to reduce the effects of specular reflections. Illuminating the sample with a given polarization state (e.g., vertical) and detecting in the orthogonal polarization (horizontal in this example) can reduce hot spots in the image due to surface reflections, as these maintain their original polarization state upon reflection from the surface.

The image processing algorithms applied will benefit greatly from careful design of the optical system, which will provide high-quality images. The range of image processing algorithms that can be applied in regenerative medicine is beyond the scope of this chapter; however, useful textbooks are available (e.g., Sonka et al. 2007). Examples of machine vision approaches applied in regenerative medicine are given in Section 14.4.1, and examples on a microscopic scale are provided in Chapter 5.

14.3.2　Laser Speckle Imaging

Laser speckle is the granular intensity pattern observed when an optically rough surface or a scattering volume is illuminated by coherent laser light (Figure 14.2).

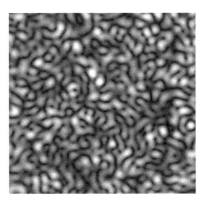

FIGURE 14.2   Magnified image of a speckle pattern. The minimum speckle size is 8 pixels.

Speckle, as a phenomenon, is not restricted to laser illumination, but is observed whenever an appropriately rough surface or volume is illuminated from any coherent source. This includes acoustic and coherent x-ray sources. In addition to laser speckle–based techniques, ultrasound speckle–based techniques, in particular, hold a great deal of potential for monitoring the development and properties of engineered tissue constructs.

Goodman (2007) provides a sound introduction to the theory of laser speckle as do several other publications (e.g., Goodman 1984). Instead of repeating these earlier works, we will instead discuss some of the more salient features of the first- and second-order statistics of laser speckle and how these statistical properties of speckle influence the results that can be inferred from laser speckle imaging experiments.

In a very general sense, speckle can be classified based upon the manner in which it is observed. Imaged or "subjective" speckle (Goodman 1984), as the name implies, involves the use of an imaging system to observe the speckle, while nonimaged or "objective" speckle does not. A clear distinction can be made between the two when one views speckle as arising from the interference of randomly phased secondary sources (Huygen's wavelets). In the case of nonimaged speckle, the source wavelets arise from a real source, a coherently illuminated object. In the case of imaged speckle, the source is the exit pupil of the imaging system, which is a virtual one. Thus, it becomes apparent that the physics of speckle formation is identical for the two cases, and thus both cases display identical statistics.

First-order intensity statistics of laser speckle patterns describe the statistical properties of the speckle pattern at a particular point in the observation plane. Frequently, the statistic of interest is the probability distribution function (PDF) of the intensity $p_I(i)$. For polarized speckle patterns, it can easily be demonstrated that the intensity is distributed in a negative exponential fashion (Goodman 2007). The PDF of intensity takes the form

$$p_I(i) = \frac{1}{2\sigma^2} e^{-\frac{i}{2\sigma^2}} \; ; i \geq 0 \tag{14.1}$$

where σ is the standard deviation of the underlying circular complex Gaussian process. Alternatively, by examining the moments of this distribution, it is straightforward to demonstrate that the PDF can be rewritten as

$$p_I(i) = \frac{1}{\mu} e^{-\frac{i}{\mu}}; i \geq 0. \tag{14.2}$$

It must be noted that the form of the PDF given above specifically refers to the case in which there is only a single population of randomly distributed phasors giving rise to the total intensity. Such would be the case when viewing the speckle pattern through a polarizer or when single scattering dominates the process. However, it is also of interest to examine the case of the incoherent sum of two statistically independent speckle patterns. In physical terms, this can be considered as a simple heuristic for describing the first-order intensity statistics of a speckle pattern arising from a scattering volume, such as biological tissue, where multiple scattering dominates and the speckle pattern is observed without a polarizer placed in front of the detector. In this case, the intensity PDF follows a Rayleigh distribution of the form

$$p_I(i) = \frac{1}{\mu_0^2} e^{-\frac{i}{\mu_0}}; \mu_1 = \mu_2 = \mu_0. \tag{14.3}$$

If the summed speckle patterns are correlated, the intensity PDF takes on different forms depending upon the degree of correlation. A discussion of these cases is beyond the scope of this chapter, and interested readers are referred to Goodman (1984) for a detailed description.

Higher-order intensity statistics yield information regarding how the intensity at different locations in the speckle pattern relates to each other. Second-order statistics provide a convenient measure of the minimum speckle size, or coarseness, of the speckle pattern. In other words, the second-order intensity statistics describe the dependence of the speckle intensity at two points in space. Again, we will rely upon a simple heuristic, the Young's double slit experiment, to gain an intuitive understanding of speckle size.

Consider the fringe spacing in a double slit experiment. Because of the Fourier relationship between the separation of the slits and the spacing of the fringes, a small slit separation yields large fringe spacing and vice versa. Extended to include an infinite number of scatterers (i.e., slits), then it is clear that the (objective) speckle pattern is due to all possible pairwise scatterers within the object plane and that the smallest speckle is due to the largest separation in the object plane. For most practical applications, this separation is the diameter of the illuminated spot. Thus, from the math typically associated with the Young's double slit experiment, the smallest speckle in the speckle pattern will be

$$\sigma_i = \frac{L}{S} \tag{14.4}$$

where L is the observation distance, and S is the maximum separation between scatterers (i.e., the diameter of the illuminated spot).

Recalling that in subjective (imaged) speckle, the source is virtual, that is, the exit pupil of the imaging system, then the smallest speckle is given by

$$\sigma_i = \frac{2.44 d_i}{S} \tag{14.5}$$

where d_i is the image distance, and S is the maximum separation of the secondary Huygen's wavelets within the exit pupil, that is, the diameter of the aperture.

A widely used first-order statistic of speckle is the spatial (or temporal) contrast. Speckle contrast c is typically defined as the ratio of the standard deviation of the intensity and the mean intensity usually calculated over a small window (e.g., 7×7) of pixels:

$$c = \frac{\sigma_i}{\mu_i}. \tag{14.6}$$

For a polarized speckle pattern, the maximum value of c is identical unity. For unpolarized speckle, the maximum value of c is lower at $1/\sqrt{2}$. It must be noted, however, that the speckle contrast itself has a statistical distribution. Duncan et al. (2008) demonstrated that the local contrast is modeled closely by a lognormal distribution and that values substantially higher and lower than 1.0 are possible for polarized speckle patterns. The exact distribution is determined by the size of the local window over which contrast is calculated and the relative size of the speckles compared to the pixel size. Furthermore, it has been demonstrated that spatially undersampling the speckle pattern also leads to an artificial reduction in speckle contrast. This reduction can be avoided by ensuring that the minimum speckle size is at least 2 pixels. That is, the speckle pattern must be spatially sampled at or above the Nyquist rate (i.e., twice the highest spatial frequency component present in the speckle pattern) (Kirkpatrick et al. 2008). It is possible to account for incorrect spatial sampling by invoking the so-called coherence factor, β (Boas and Dunn 2010), which in effect is simply the inverse of Goodman's M parameter (the number of speckles per pixel). When possible, however, it is advisable to sample the speckle pattern adequately and avoid such correction factors.

Speckle imaging for assessing fluid (blood) flow, particle dynamics, or tissue mechanical properties can be implemented with a deceivingly simple system comparable to that shown in Figure 14.1. In the simplest scenario, all that is required is a coherent or partially coherent source such as a superluminescent diode and a camera, although typically a polarizer and an analyzer are employed so as to evaluate only a single component of the scattered field. The techniques are full field, thus promising the ability to compare and contrast spatially separated regions of the tissue in the same image. Speckle approaches are highly attractive in that they always yield results, that is, a moving speckle pattern, and it is relatively easy to demonstrate a correlation between speckle motion and fluid motion for blood flow imaging and tissue stiffness for the measurement of mechanical properties.

Chapter 14

14.3.3 Spectroscopic Imaging

Multispectral and hyperspectral imaging refers to a class of techniques in which each pixel in a 2D image field contains a wavelength-resolved spectrum rather than a single intensity value. The enormous power of this technique stems from the fact that this 3D spectral "data cube" can be spectrally analyzed pixel-by-pixel in order to provide 2D maps of the distributions of many distinct light-absorbing compounds. The number of unique compound images that can be extracted is limited by the number of effective measurement wavelengths at each pixel and by the number of compounds that possess measurable and separable absorption spectra in this wavelength range. The terms "multispectral" and "hyperspectral" do not describe fundamentally different techniques but reflect differences of degree: systems employing a "small" (20 or less) number of wavelengths tend to be described as multispectral, whereas if there are tens to hundreds of wavelengths, then the term "hyperspectral" tends to be used. Consequently, there are significant differences in the types of instrumentation that are used to implement each technique. Both techniques have been extensively developed in recent years, not purely in biomedicine but in the general area of remote sensing. Aerial and satellite imagery have perhaps been the main driving areas. The last decade has seen an explosion in interest in the technique, driven by factors such as improved light sources, cameras, and filter components but also by the continued improvements in computer power. This makes it feasible to reduce the enormous volumes of raw data that can be generated into useful chemical maps in a reasonable time: potentially in real time.

Many schemes exist for collecting the raw data. Multispectral imagers have tended to employ conceptually straightforward schemes in which a 2D area-scan camera is simply placed behind an interchangeable band-pass filter and the sample illuminated with a broadband white light source. Different slices through the spectral data cube are collected by switching in different band-pass filters, for example, by mounting them in a filter wheel controlled by a stepper motor.

The obvious disadvantage of such schemes arises when imaging dynamic phenomena. If the distribution of chemical species changes between sequential filtered image acquisitions, then the spectral unmixing algorithms will fail in ways that are hard to predict. To reduce this problem, we can speed up the switching between different filters, or to eliminate it completely, we can use multiple cameras to acquire different wave bands simultaneously. The former approach has the disadvantage that the wavelength switching time limits the speed at which dynamic changes can be followed, while the latter approach obviously leads to more cumbersome instruments and effectively precludes more than a handful of wavelength bands.

Hyperspectral imaging cameras combine very large numbers of effective bands (hundreds typically) with very fast switching times (10 ms or better) by using electronically controllable band-pass filters. A popular voltage-controlled filter is the Lyot filter. This device is basically a stack of repeating structures, sequentially arranged, in which each structure is itself a consecutive arrangement of a liquid crystal retarder and a fixed retarder sandwiched between two fixed polarizers. The liquid crystal is a voltage-controlled retarder and a detailed analysis shows that a stack of such elements transmits a narrow range of wavelengths centered on a value that varies with the voltage applied to the liquid crystals. The passband is thus, in principle, continuously variable. A commercial embodiment of this design is the Nuance series of instruments from Caliper Life Sciences Inc. While millisecond switching is much faster than can be achieved by mechanical designs involving filter wheels, etc., even this can be improved to a few hundred microseconds using filter elements such as the acousto-optic tunable filter (AOTF). The AOTF basically uses an acoustic traveling wave in a crystal to effectively form a diffraction grating with which to dispersively filter radiation. The center wavelength is determined by the acoustic wavelength, which in turn is determined electronically by the applied driving frequency. As with the Lyot filter, the center wavelength is continuously adjustable. A commercial example of this design is, for example, the Chromodynamics Inc HSi-300.

Truly parallel acquisition of different spectral bands can be achieved at the expense of spectral coverage. An elegantly simple scheme, which is of particular value in imaging oximetry (Section 14.4.3), is to use a single red, green, and blue (RGB) color video camera as the detector. Since each pixel is divided into three separate light sensing elements, each with a broad band-pass filter in front of it centered on the RGB parts of the visible spectrum, two or, at most, three separate compounds can be separated by using the three separate monochrome images output by the camera. Serendipitously, it transpires that oxygenated and deoxygenated hemoglobin possesses distinct absorption spectra that are separable using measurements of mean absorbance, averaged over the broad filter passbands of the RGB filters (Steimers et al. 2011).

Finally, we should mention the concept of raster scanning a point of illumination across the sample and spectrally resolving the output using a dispersive element and multiple light detectors. This approach has been commercially embodied in fluorescence microscopy by systems such as the Zeiss LSM 510 META. For absorption spectroscopy, a broad-band but spatially coherent source of illumination is required. An obvious candidate technology is the supercontinuum laser (Kaminski et al. 2008).

The goal of spectral imaging is to determine spatial maps of the concentration of n distinct absorbing compounds. The basic mathematical framework for this is the Beer–Lambert law, which relates the attenuation of light by a medium to the distance the light has traveled and the absorbing species within the medium:

$$OD(\lambda) = \log_{10}\left(\frac{I_0(\lambda)}{I(\lambda)}\right) = \alpha(\lambda)l = \sum_{i=1}^{n} C_i \varepsilon_i(\lambda)l \tag{14.7}$$

Here OD is the light absorption in optical densities, and l is the optical path length through the absorbing compound. I_0 and I are the spectrum of the light in a reference condition and after traversing the sample, respectively. This equation embodies a linear relationship between the measured variable (OD) and the variables to be determined (the n concentrations C_i). The coefficients $\varepsilon_i(\lambda)$ are the specific absorption coefficient spectra of the n known absorbing species. Measurements of $OD(\lambda)$ at m different wavelengths ($m \geq n$) then allow this equation to be recast in matrix–vector form:

$$\mathbf{OD}/l = \varepsilon \cdot \mathbf{C} \tag{14.8}$$

The $m \times n$ matrix ε is known as the "design set." If $m = n$, the design set is square and invertible (provided the n compound spectra are distinct) and hence the concentrations

C can easily be determined. If $m > n$, then the pseudo-inverse of ε is calculated and multiplied onto **OD**, which, as can be shown, is equivalent to finding a least-squares solution to the overdetermined problem of fitting the measured spectrum with a linear combination of the compound spectra. This technique is commonly termed "multilinear regression."

A particular example of this technique that is of great value in medicine is "oximetry," that is, the optical or near-IR determination of the oxygenation state of blood. The technique is based on the fact that the oxygen-carrying molecule hemoglobin undergoes a conformational change when bound to oxygen, and this causes a pronounced change in the absorption spectrum. Measurements of OD at a minimum of two wavelengths then in principle allows one to measure the concentrations of oxygenated $[O_2Hb]$ and deoxygenated [HHb] hemoglobin independently. From these two measurements, other related quantities can be derived, mainly the total hemoglobin concentration $[O_2Hb]$ + [HHb] and the percent of saturation = 100% × $[O_2Hb]/([O_2Hb]$ + [HHb]). The percent of saturation is particularly important as, along with the total blood volume flow rate, it directly determines the tissue oxygen supply. The typical arterial saturation is around 90%, whereas in the venous circulation, it drops to 50%–60% typically. Abnormally low measured values of saturation are generally associated with hypoxia and/or ischemia and can be an important early indicator of lack of perfusion, which will quickly lead to a loss of tissue viability.

A major complication arises when the Beer–Lambert law is applied to noninvasively probe biological tissues. When optical and near-infrared light interacts with tissue, the predominant interaction is elastic scattering rather than absorption. Hence, the measured OD contains a contribution from scattering that is unrelated to the concentration of blood. In fact, scattering is far and away the dominant interaction mechanism (by up to 100× in the 700–1000 nm range), which profoundly alters the form of the Beer–Lambert law. Firstly, the overall measured OD will be much higher than that expected on the basis of absorption alone because scattering will remove photons from the beam that would otherwise have reached the detector. Secondly, the effect of an absorber will be greatly increased because strong multiple scattering leads to a "path lengthening" effect. Photons essentially undergo a random walk between scattering centers, which means that the effective value of l is now much larger than the physical separation between the light source and the detector. Other, subtler effects also exist; however, to a first approximation, these two confounding effects can be described by a comparatively simple modification of the original Beer–Lambert law, which leads to the so-called "modified Beer–Lambert law":

$$OD(\lambda) = DPF \cdot l \cdot \sum_{i=1}^{n} C_i \cdot \varepsilon_i(\lambda) + G \qquad (14.9)$$

This modified equation describes both the path lengthening effect (l is multiplied by a factor termed the "differential path length factor" or DPF) and beam attenuation due to scattering of photons out of the beam (additive factor G). Refinements of this basic equation include making both G and DPF functions of wavelength, and even more sophisticated approaches based on accurate physical modeling of light transport in tissue can be developed (Matcher 2002). These are of great value in the near-IR; however, in the visible (400–700 nm), the specific absorption coefficient of hemoglobin is typically 50× higher than in the near-IR and so scattering is less dominant. To utilize the modified Beer–Lambert law for optical oximetry, a major simplification arises by noting that the saturation is independent of any arbitrary scaling factor that is applied equally to [O$_2$Hb] and [HHb]. Hence, the unknown overall photon path length DPF × l need not be determined. In order to account for G, various approaches have been suggested. Shonat et al. (1997), using an AOTF-based hyperspectral microscope, simply assumed that G was a constant and included this as another term in the multilinear regression. This method was adopted by Sorg et al. (2005) to obtain spatial maps of oxygen saturation using a Lyot-filter hyperspectral camera. Styp-Rekowska et al. (2007) also used this idea but extended it by approximating G with a first-order polynomial rather than a constant. An alternative approach to eliminating G is to make dynamic measurements, that is, to measure changes in hemoglobin concentrations relative to a baseline image. If ΔOD is defined as the change in OD between two time points, then if it can be assumed that the scattering term G remains constant, the modified Beer–Lambert law becomes

$$\Delta OD(\lambda) = DPF \cdot l \cdot \sum_{i=1}^{n} \Delta C_i \cdot \varepsilon_i(\lambda) \tag{14.10}$$

Hence, multilinear regression can be used to calculate changes in oxygenation, that is, Δ[O$_2$Hb] and Δ[HHb], without making assumptions about the spectrum of G. Prakash et al. (2007) used this approach to produce maps of oxygenation changes in the mouse and rat somatosensory cortex during sensory stimulation. The major disadvantage of this approach is that relative changes in hemoglobin concentration are not sufficient to determine the absolute saturation unless other information is also available (Matcher 2002).

14.4   Example Applications

14.4.1   Time–Lapsed Imaging: Monitoring Tissue Growth in Bioreactors

Time-lapsed imaging has been used as a method of monitoring tissue growth within bioreactors with the objective of ultimately automating the cell culture process and understanding cell aggregation. For example, a 3D culture of MCF-7 breast carcinoma cells has been monitored in a rotating cell culture bioreactor system (Sawyer et al. 2008). The bioreactor (43.5 mm internal diameter, 6.7 mm depth) was redesigned so that the vessel ports were situated at the side and did not obscure the field of view of the imaging system. Under direct illumination using the system shown in Figure 14.1, the spheroids caused by cell aggregation are visible (dark circles in the image shown in Figure 14.3a); but there is a strong contribution from surface reflections (hot spots), even using crossed polarizers, which provide a significant reduction in their intensity, and the background membrane of the bioreactor is clearly visible. The presence of these artifacts makes applying image processing methods for automation extremely challenging. In an early design of this system, the cells were fluorescently labeled, and a high-pass filter was used

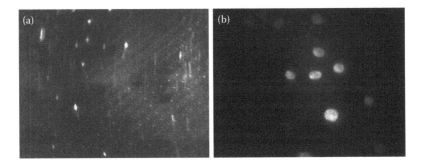

FIGURE 14.3 Image of 3D tissue constructs (spheroids) in a rotating bioreactor (a) under direct illumination and (b) at the fluorescence emission wavelength. The spheroidal aggregates shown are ~2 mm in diameter.

to block light at the excitation wavelength. Figure 14.3b shows that this is an effective approach for eliminating surface reflections and reducing the effects of the background membrane on the image quality. However, it was desirable that imaging was carried out label-free as it was found that the fluorescence molecule interfered with cell aggregation and so the illumination of the system was modified (Figure 14.4). Laser light was expanded and then focused using a cylindrical lens so that the central plane of the rotating bioreactor was illuminated with a sheet of light. The cell spheroids interacting with the light sheet scattered light in all directions and allowed a proportion of the scattered light to be imaged by the camera.

The full data set comprised 15 images, taken at three different exposures (1/30, 1/60, and 1/125 s) every 10 min. Figure 14.5 shows an example of a time-lapsed image sequence. Initially the vessel was seeded with a cell suspension, which, when illuminated, gives a cloudy appearance to the image (Figure 14.5, top left image). The scattering of light

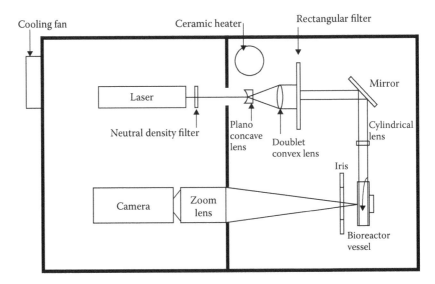

FIGURE 14.4 Imaging system utilizing sheet illumination from the side of the vessel to enable label-free imaging while reducing the effects of surface reflections.

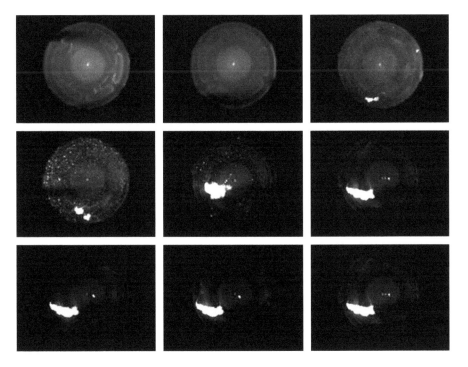

FIGURE 14.5   Time-lapsed, label-free images of 3D tissue construct within a rotating bioreactor. From top left, $t = 0, 1, 2, 3, 4, 5, 6, 7, 8$ h. The diameter of the bioreactor is 43.5 mm.

from the cell suspension also made the rear surface of the bioreactor visible. This effect reduced with aggregation of the cells. With time, the cells began to aggregate into spheroids and larger aggregates. Two hours after seeding (Figure 14.5, top right image), one large aggregate and a large number of smaller aggregates had formed and rotated with the vessel (i.e., performed solid body wall rotation). Once the large aggregate had started to form, it merged with smaller ones with which it came into contact. Between hours 3 and 4 (Figure 14.5, middle left and center image), the aggregate ceased to become part of the solid wall rotations and settled at approximately 8 o'clock in the image (Figure 14.5, middle right and bottom row).

Image processing enabled information such as number and size of aggregates to be automatically detected over time. This identified the aggregates, counted the number present, and determined their cross-sectional areas. Large aggregates appeared as bright objects and were easily identified against the background by setting a high amplitude threshold. Smaller aggregates were more difficult to detect against the residual background. A second stage of detection removed the background and large aggregates from the images. Using MATLAB®'s morphological opening function (Bashtanov et al. 2004), an estimation of the background was obtained, which could be subtracted from the original image. Setting a lower amplitude detection threshold allowed the less bright, smaller aggregates to be detected. The processed images were used to obtain aggregate population and cross-sectional area. Values of each of these parameters were calculated for every image in the subset, giving 15 values per parameter at each time interval.

Results (Sawyer et al. 2008) demonstrate that the number of aggregates reached a peak after ~3 h and then gradually reduced as a single large aggregate is formed. The processing also tracked the size of the largest aggregate, which was demonstrated to reach a maximum at ~4 h due to loose aggregation of the cells, and then it becomes more compact due to cell–cell adhesion. This demonstrates the potential for quantitatively monitoring 3D cell culture, allowing process parameters to be dynamically altered, reducing the need to manually sample the bioreactor, and ultimately, generating robust *in situ* data.

Other notable work using this approach has involved the design of a bioreactor that contains an integrated camera that eliminates the effects of surface reflections from the bioreactor surface (Ziegelmueller et al. 2010). The system has been used to qualitatively monitor the growth of tissue-engineered heart valves under mechanical stimulation. A machine vision system has also been used to monitor the growth of tissue-engineered blood vessels within a perfusion bioreactor (Xu et al. 2005). A series of image processing steps was used to extract the diameter of the vascular construct; the region of interest on the image was manually located, and then the edges in the image were identified using a binary gradient mask; any gaps in the edges were removed by applying a dilated gradient mask. The region that represented the vessel was then filled in, and the area and diameter of the vessel calculated. The pressure of the peristaltic pump was set to maintain a deformation of the diameter of 10% with the pressure automatically gradually increasing as the vessel walls become tougher and thicker. This was compared with a control group in which a step increase in the pressure was made every week. The burst strengths of the constructs were demonstrated to be 31.6% higher using the automated method compared to the control.

Time-lapsed macroscopic imaging has also been used to monitor scaffold production of foamed polymeric scaffolds within a supercritical fluid process (Mather et al. 2009). Two sapphire windows coaxially located on opposite sides of a high-pressure chamber were utilized. One allowed both imaging and illumination from the front of the vessel; the other window was used for back-illumination of the sample. Illumination from the same side of the vessel was useful in extracting features on the surface of the scaffold, whereas illumination from behind was more useful in extracting the overall shape of the scaffold. The latter case was used to monitor the change in the position of the scaffold surface with time during venting of the chamber over 10, 30, and 60 min. The change in the position of the scaffold surface corresponded to bubble coalescence and the onset of the glass transition locking in the scaffold structure.

14.4.2 Laser Speckle Imaging

In regenerative medicine applications, laser speckle imaging can be used to monitor both tissue biomechanics and blood flow. Ensuring that engineered tissues have suitable and appropriate mechanical properties, such as strength and stiffness, as well as appropriate dynamic, viscoelastic properties is critical for the long-term success of the tissue. A variant of laser speckle imaging, optical elastography (Kirkpatrick and Duncan 2002), provides avenues for the assessment of static and dynamic mechanical properties of biological tissues and engineered tissue constructs.

Optical elastography, in the broadest sense, encompasses a variety of techniques mostly involving inferring mechanical strain by tracking the motion of an imaged laser speckle pattern that results from coherently illuminating a mechanically stressed tissue. The techniques are highly attractive for tissue engineering, largely because of their versatility, ease of experimental configuration, and high strain resolution. Optical elastography methods excel at evaluating strains in the microstrain to several hundred microstrain regimes.

An early use of optical elastography techniques in tissue engineering was the application of a differential laser speckle strain gauge to evaluate the viscoelastic behavior of an elastin-based tissue scaffold (Kirkpatrick et al. 2003). The concept behind the differential laser speckle strain gauge is quite straightforward. A dynamically loaded sample is sequentially illuminated from equal but opposite off-axis angles, and the speckle motion from each illumination is recorded and subtracted from each other to eliminate unwanted motion terms. With careful experimental arrangement, including well-collimated beams (parallel rays), the in-plane deformation term can be isolated, and from this term, the in-plane strain can be inferred. Viscoelastic properties, specifically the mechanical loss factor tan δ, can be determined if the phase difference between the imposed dynamic mechanical load and the motion of the speckle is calculated.

Accurately quantifying the speckle motion is the key to any elastography technique. Simple cross-correlation techniques can be applied; however, these are susceptible to error due to noise or a poor reference image. Without interpolation techniques, cross-correlation techniques are also limited to large speckle motions; the speckles need to translate >0.5 pixel per frame for motion to be detected. Alternative approaches to tracking speckle have been proposed that have a much better displacement resolution and are also less susceptible to noise (Kirkpatrick et al. 2003, 2006). These approaches are based on a maximum likelihood estimator for calculating the shifts in the speckle patterns. Phase-based techniques have also been employed in optical coherence tomography-based elastography applications (Wang et al. 2007).

One particular advantage of optical elastography techniques in tissue engineering is the simplicity of the experimental configuration. Optical elastography lends itself readily to implementation within tissue culture incubators as the optical signals can be carried via optical fibers and thus keep the sensitive electronics outside of the incubator and safe from the humid environment.

Blood flow imaging is the other important application of laser speckle imaging in regenerative medicine. Over 30 years ago, Fercher and Briers (1981) introduced the concept of estimating blood flow based on the contrast of a time-integrated laser speckle pattern. Since that time, numerous minor variations to this theme have been proposed and implemented. Yet, despite all the different approaches, the underlying concept has remained consistent. Namely, the idea is to infer a temporal correlation time constant from the observed speckle contrast c and relate this time constant to flow velocity. An example laser speckle image of a mouse stroke model using a commercial laser speckle imager is shown in Figure 14.6.

While laser speckle contrast imaging (LSCI), or laser speckle contrast analysis (LSCA), as these approaches are called, is highly attractive, the real challenge is to provide a truly quantitative relationship between the observed spatial contrast, the temporal correlation time constant, and the flow velocity. This has yet to be accomplished in a convincing fashion. Boas and Dunn (2010) summarized the development of LSCI using the formalisms of diffusing wave spectroscopy and quasi-elastic light scattering as an initial starting point. It must be noted that while these techniques are closely related to

Chapter 14

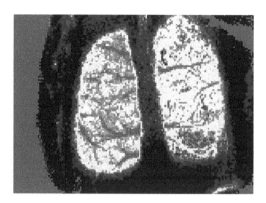

FIGURE 14.6 **(See color insert.)** Middle cerebral artery occlusion imaged with a moorFLPI through a thinned rat skull. The image was obtained as part of a sequence taken at 25 Hz, at a spatial resolution of 152 × 113 pixels to study fast dynamic responses. Higher-resolution images, 760 × 565 pixels, can be obtained at 1 Hz. (Courtesy of Dr. Thaddeus Nowak, University of Tennessee, USA.)

LSCI, their data collection requirements and data processing demands are quite different. Duncan and Kirkpatrick (2008) enumerated a variety of issues that have impeded the progress of LSCI from a correlative technique to a truly quantitative technique.

LSCI has not been used extensively or in a convincing fashion yet in the field of tissue engineering. However, it is an extremely promising tool for understanding and quantifying relative blood flow in neo-tissues. A significant feature of LSCI is that the spatial resolution of the LSCI system can readily be altered from the size of very small vessels up to large areas of tissue. Perhaps the greatest potential for LSCI in the field of tissue engineering lies in assessing the flow of blood (and therefore nutrients) to the developing tissues. The technique lends itself to being used within the confines of an incubator and can provide a temporal history of changes in the relative blood flow.

14.4.3 Spectroscopic Imaging: Oxygen Saturation Imaging

Chin et al. (2010) used a CombiVu-R hyperspectral camera (Hypermed Inc.) to study perfusion changes in rat skin *in vivo* during periods of applied mechanical stretch. Applied mechanical forces are emerging as an important mechanism by which to upregulate a range of cellular responses in tissue-engineered constructs. A key finding was that the applied stretch produced a marked increase in total hemoglobin content 1 h after cessation of a 4-h stretching protocol, but this returned to baseline over the next few days. Interestingly, reverse-transcription polymerase chain reaction (RT-PCR) showed expression of HIF-1alpha, a known marker of local hypoxia; marked increases in cell proliferation and blood vessel density were also noted. The authors interpret these results as suggesting that mechanical stretch induces transient hypoxia, followed later by reperfusion, and further that transient hypoxia might have a stimulatory effect on cell proliferation.

Nighswander-Rempel et al. (2006) used a hyperspectral camera based around a liquid-crystal tunable filter to study regional variations in oxygenation in the isolated, perfused porcine heart during periods of induced ischemia reperfusion and global hypoxia. Stacks of around 40 spectral images with a 10 nm bandwidth were collected over the range of 650 to 1050 nm in around 5 min and converted to hemoglobin concentration maps using the modified Beer–Lambert and G assumed constant. The oxygenation images produced average oxygen saturations that correlated with increasing grades of ischemia induced over 15 min periods and verified that 15 min of reperfusion is sufficient to restore oxygenation back to baseline in all cases.

The ability to directly image microvessels via their hemoglobin absorption is likely to increase in importance in tissue engineering because of the recognized need to promote angiogenesis in order to improve the viability of tissue-engineered constructs *in vivo* (Laschke et al. 2006). Although a number of approaches are currently under development to promote vascularization of implanted constructs, none are as yet capable of regenerating a vascular bed that can meet the tissue demand for oxygen. This is emerging as a key limitation in the successful translation of tissue-engineered constructs into the clinic. New approaches such as the incorporation of vascular progenitor cells or stem cells into the construct show promise. Imaging tools that can quantify both the physical extent of perfusion and also give an insight into the local balance between O_2 supply and consumption (reflected in the hemoglobin saturation) may have a very direct role in evaluating the efficacy of these approaches, either in isolation or in combination.

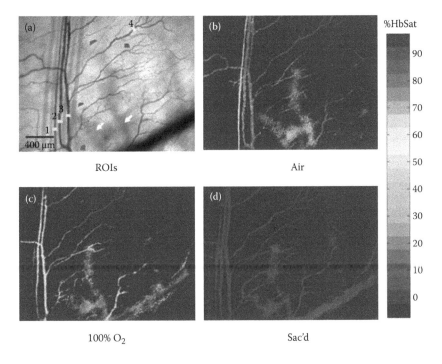

FIGURE 14.7   (**See color insert.**) Example of transmission hyperspectral imaging on a nude mouse dorsal skinfold chamber model. (a) Conventional bright-field image, showing blood vessels as dark shadows due to strong light absorption by hemoglobin. (b) Hemoglobin saturation image extracted from a hyperspectral data set collected in parallel with the bright-field image, while the animal was breathing air. (c) Identical image field while breathing 100% oxygen. Note the increase in hemoglobin saturation, as expected under conditions of unaltered cellular metabolism. (d) Following pentobarbital-induced sacrifice. Continued cellular respiration and O_2 extraction following failure of the circulation is consistent with the observed very low saturation values. (Adapted from Sorg, B. S. et al., *Journal of Biomedical Optics* 10(4):044004, 2005.)

Chapter 14

A key recent innovation in this area has been the introduction of the established dorsal skin-fold chamber model into tissue engineering. This model allows the direct visualization of the microvasculature in living rodent skin longitudinally over several weeks. It also allows the incorporation of small diameter constructs into this environment, for example, to study integration of the construct into the host vasculature. Figure 14.7 shows the results obtainable using a Lyot-filter hyperspectral camera. The increased use of such technology in studying angiogenesis models is likely to prove very fruitful.

14.5 Conclusions

Macroscopic imaging can be performed using very simple instrumentation, including a light source, camera, computer, and optics elements such as lenses and filters. It is important to obtain the best image quality prior to image processing, and techniques for achieving this have been applied. Using coherent illumination (e.g., from a laser) produces a speckle pattern at the detector plane, and the fluctuations in this speckle pattern can be related to tissue biomechanics and blood flow. The color reflected by a sample can be analyzed, and, providing the effects of light scattering are taken into account, the spectral images can be related to oxygen saturation of the blood. The ability to image microvessels is likely to increase in importance in regenerative medicine because of the need to promote angiogenesis for construct survival and integration following implantation.

Acknowledgment

SPM acknowledges the support of the Medical Research Council (United Kingdom) for the rotating bioreactor imaging research described in this chapter.

References

Bashtanov, M.E., Goodyear, R.J., Richardson, G.P., and Russell, I.J. 2004. The mechanical properties of chick (Gallus domesticus) sensory hair bundles: Relative contributions of structures sensitive to calcium chelation and subtilisin treatment. *Journal of Physiology* (London) 559(1):287–299.

Boas, D.A. and Dunn, A.K. 2010. Laser speckle imaging in biomedical optics. *Journal of Biomedical Optics* 15:011109.

Chin, M.S., Ogawa, R., Lancerotto, L. et al. 2010. *In vivo* acceleration of skin growth using a servo-controlled stretching device. *Tissue Engineering Part C* 16(3):397–405.

Duncan, D.D. and Kirkpatrick, S.J. 2008. Can laser speckle flowmetry be made a quantitative tool? *Journal of the Optical Society of America A* 25:2088–2094.

Duncan, D.D., Kirkpatrick, S.J., and Wang, R.K. 2008. Statistics of local speckle contrast. *Journal of the Optical Society of America A.* 25:9–15.

Fercher, A.R. and Briers, D.A. 1981. Flow visualization by means of single-exposure speckle photography. *Optics Communications* 37:326–330.

Goodman, J.W. 1984. Statistical properties of laser speckle patterns. In *Topics in Applied Physics, Vol. 9: Laser Speckle and Related Phenomena*, ed. J.C. Dainty, 9–75. Berlin: Springer-Verlag.

Goodman, J.W. 2007. *Speckle Phenomena in Optics*. Englewood, CO: Roberts & Company.

Kaminski, C.F., Watt, R.S., Elder, A.D. et al. 2008. Supercontinuum radiation for applications in chemical sensing and microscopy. *Applied Physics B* 92: 367–378.

Kirkpatrick, S.J. and Duncan, D.D. 2002. Optical assessment of tissue mechanics. In *Handbook of Optical Biomedical Diagnostics*, ed. V.V. Tuchin, 1037–1084. Bellingham, WA: SPIE Press.

Kirkpatrick, S.J., Hinds, M.T., and Duncan, D.D. 2003. Acousto-optical characterization of the viscoelastic nature of a nuchal elastin tissue scaffold. *Tissue Engineering* 9:645–656.

Kirkpatrick, S.J., Wang, R.K., and Duncan, D.D. 2006. OCT-based elastography for large and small deformations. *Optics Express* 14:11585–11597.

Kirkpatrick, S.J., Duncan, D.D., and Wells-Gray, E.M. 2008. Detrimental effects of speckle-pixel size matching in laser speckle contrast imaging. *Optics Letters* 33:2886–2888.

La Hei, E.R. Holland, A.J.A., and Martin, H.C.O. 2006. Laser Doppler imaging of paediatric burns: Burn wound outcome can be predicted independent of clinical examination. *Burns* 32:550–553.

Laschke, M.W., Harder, Y., Amon, M. et al. 2006. Angiogenesis in tissue engineering: Breathing life into constructed tissue substitutes. *Tissue Engineering* 12(8):2093–2104.

Litwiller, D. 2005. CMOS v CCD, maturing technologies, maturing markets. *Photonics Spectra*.

Matcher, S.J. 2002. Signal quantification and localization in tissue near-infrared spectroscopy. In *Handbook of Optical Biomedical Diagnostics*, ed. V. Tuchin, 487–584. Bellingham: SPIE Press.

Mather, M.L., Morgan, S.P., White, L.J. et al. 2009. Time-lapsed imaging for in-process evaluation of supercritical fluid processing of tissue engineering scaffolds. *Biotechnology Progress* 25:1176–1183.

National Instruments. 2012. A Practical Guide to Machine Vision Lighting, Parts 1–3. Available at www.ni.com/white-paper/6901/en, accessed November 2012.

Nighswander-Rempel, S.P., Kupriyanov, V.V., and Shaw, R.A. 2006. Regional cardiac tissue oxygenation as a function of blood flow and pO2: A near-infrared spectroscopic imaging study. *Journal of Biomedical Optics* 11(5):054004.

Prakash, N., Biag, J.D., Sheth, S.A. et al. 2007. Temporal profiles and 2-dimensional oxy-, deoxy-, and total-hemoglobin somatosensory maps in rat versus mouse cortex. *Neuroimage* 37:S27–S36.

Sawyer, N.B.E., Worrall, L.K., Crowe, J.A. et al. 2008. In situ monitoring of 3D *in vitro* cell aggregation using an optical imaging system. *Biotechnology and Bioengineering* 100:159–167.

Shonat, R.D., Wachman, E.S., Niu, W. et al. 1997. Near-simultaneous hemoglobin saturation and oxygen tension maps in mouse brain using an AOTF microscope. *Biophysical Journal* 73:1223–1231.

Sonka, M., Hlavac, V., and Boyle, R. 2007. Image processing, analysis and machine vision. CL Engineering.

Sorg, B.S., Moeller, B.J., Donovan, O. et al. 2005. Hyperspectral imaging of hemoglobin saturation in tumor microvasculature and tumor hypoxia development. *Journal of Biomedical Optics* 10(4):044004.

Steimers, A., Gramer, M., Takagaki, M. et al. 2011. Simultaneous imaging of haemoglobin oxygenation and blood flow with RGB reflectometry and LASCA during stroke in rats. *Proceedings of SPIE* 8088:808808.

Styp-Rekowska, B., Mecha Disassa, N., Reglin, B. et al. 2007. An imaging spectroscopy approach for measurement of oxygen saturation and hematocrit during intravital microscopy. *Microcirculation* 14:207–221.

Wang, R.K., Kirkpatrick, S.J., and Hinds, M.T. 2007. Phase-sensitive optical coherence elastography for mapping tissue microstrains in real time. *Applied Physics Letters* 90:164105.

Xu, J., Ge, H., Zhou, X. et al. 2005. Tissue-engineered vessel strengthens quickly under physiological deformation: Application of a new perfusion bioreactor with machine vision. *Journal of Vascular Research* 42:503–508.

Ziegelmueller, J.A., Zaenkert, E.K., Schams, R. et al. 2010. Optical monitoring during bioreactor conditioning of tissue-engineered heart valves. *ASAIO Journal* 56:228–231.

Chapter 14

15. Bioluminescence Imaging

Juliaan R.M. van Rappard,
Preston Lavinghousez, and Joseph C. Wu

15.1   Introduction

Recent advances in cellular and molecular biology have enabled the utilization of convenient and sensitive *in vivo* imaging techniques. Bioluminescence is the enzymatic generation of visible light by a living organism that occurs in nature (Paroo et al. 2004). *In vivo* bioluminescence imaging (BLI) provides images of biologic components genetically tagged with a reporter gene that encodes a light-generating enzyme. BLI allows for serial monitoring and is a preferable alternative to analyzing a large number of experimental animals at multiple time points by histological sections. It enables relatively simple yet extremely sensitive *in vivo* detection of biological processes during the course of an experiment in a variety of areas.

BLI requires the transfection of cells or tissues with a gene cassette consisting of a selected promoter that controls the expression of the bioluminescence reporter gene. Once transfected, the expression cassette produces the luciferase enzyme, and after the specific

Optical Techniques in Regenerative Medicine. Edited by Stephen P. Morgan, Felicity R.A.J. Rose, and Stephen J. Matcher © 2014 CRC Press/Taylor & Francis Group, LLC. ISBN: 978-1-4398-5495-2

Chapter 15

reporter probe is introduced (luciferin or coelenterazine), an oxidation reaction occurs in the presence of magnesium, oxygen, and adenosine triphosphate (ATP; for luciferase-luciferin), emitting visible light photons that can be detected by a charge coupled device (CCD) camera. This reaction allows *in vivo* tracking of the injected cells within the subject via detecting the emitted photon light signal (Figure 15.1). In planar BLI, there is a 10-fold loss of photon density for each centimeter of tissue depth (Contag et al. 1995). Thus far, repeated exposure of animals to reporter probes appears to be safe and nontoxic. Due to tissue attenuation properties, the cells emitting light closer to the surface appear brighter compared to deeper sources in creating surface-weighted images (Weissleder 2001; Wu et al. 2001).

Currently, the most commonly studied light-generating enzyme is luciferase. Luciferase enzymes from the North American firefly *Photinus pyralis* catalyze the oxidation of reporter probe D-luciferin into inactive oxyluciferin, resulting in photons of light emitted at wavelengths 530–640 nm that peaks at 562 nm (green) (Gould and Subramani 1988). Unlike fluorescence imaging, BLI requires no exogenous excitatory light source, but instead is a byproduct of the oxidative chemical reaction when chemical energy is released as visible light. A significant advantage of BLI is that the background autofluorescence *in vivo* is very low due to the paucity of intrinsic bioluminescence from mammalian cells, which makes this type of imaging highly sensitive. Therefore, images can be generated with remarkably high signal-to-noise ratios (de Wet et al. 1987).

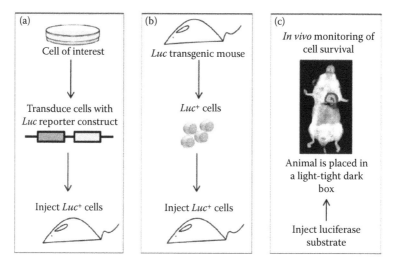

FIGURE 15.1 Diagram illustrating common BLI steps. (a) Cells of interest undergo transduction with a luciferase reporter gene construct and are injected into an animal. (b) Alternatively, luciferase-expressing cells can be isolated from a luciferase transgenic animal and injected into the experimental subject. (c) After systemic injection or reporter substrate (D-luciferin or coelenterazine), a CCD camera can be used to localize the luciferase photon signals *in vivo*. Pseudo-colored images that represent signal intensity are overlaid with gray-scale reference images of the animal to facilitate localization of the signal. Cell survival, proliferation, and homing, as well as the expression of therapeutic genes, can be monitored using this strategy. (Adapted from de Almeida, P. E. et al., *Am J Physiol Heart Circ Physiol*, 301(3), H663–71, 2011. With permission.)

Of the library of luciferase enzymes, only a small subset is utilized as reporter genes. The most commonly used gene for *in vivo* bioluminescence as described is the firefly luciferase (FLuc) gene. Other commonly used bioluminescence reporter proteins are click beetle luciferase (*Pyrophorus plagiophthalamus*), which emits green to orange light (546–593 nm); renilla luciferase from sea pansy (*Renilla reniformis*), which emits blue light (480 nm); and Gaussia luciferase (marine copepod), which emits blue light (480 nm). The differences in emission spectra are actually quite beneficial because they allow dual imaging when FLuc is paired with either RLuc and GLuc, as they are in different spectra (Luker and Luker 2008). The reporter probes are different as well, with the substrate for FLuc (D-luciferin) having lower background signals than the substrate for RLuc and GLuc luciferases (coelenterazine).

At present, it is unlikely BLI can be applied to humans due to limited tissue penetration. Compared to radiotracers used for clinical positron emission tomography that emit photon energy of 511,000 eV, BLI emits photon energy of only 2–3 eV. However, the utility of BLI as a preclinical tool to investigate human disease processes makes it one of the most valuable technologies at the present time. In this chapter, bioluminescence reporter genes will be covered in detail as well as the applicability of BLI to the fields of oncology, immunology, cardiovascular disease, infectious disease, neurology, and regenerative medicine.

15.2   Clinical and Biological Measurement Challenges the Method is Suited to Address

The use of BLI requires exogenous expression of a luciferase enzyme, followed by systemic delivery of its reporter substrate. In nature, a large number of organisms are known for their ability to emit light. This biological production of light, or "bioluminescence," is used as a disguise for fleeing prey, as signals for courtship and mating, and as a lure for prey. Luciferases occur in bacteria, fungi, radiolarians, dinoflagellates, and about 17 metazoan phyla. Of all these luminous species from more than 700 genera, 80% are of marine origin (Widder 2010). In the last few decades, bioluminescent visualization has been used in the biomedical field. In these applications, luciferases have been isolated, sequenced, and used to build DNA vectors. In 1985, de Wet et al. (1985) cloned the cDNA encoding the luciferase (Luc) of the firefly *Photinus pyralis* and expressed it in *Escherichia coli*. Since then, other methods like viral and RNA transcription have been developed, and various techniques have been optimized so that the Luc gene can be expressed at high levels and its product localized in the cells' cytoplasm. The light intensity measured by the CCD cameras is dependent on the optical properties as well as the thickness of the tissue through which it travels. Within the tissue, hemoglobin and melanin in pigmented animals are the main endogenous chromophores within the body. Hemoglobin absorbs primarily the blue to green part of the visible light, but the absorption of wavelengths higher than 600 nm is reduced, allowing transmission of red light through several centimeters of tissue (Zhao et al. 2005).

For an ideal reporter system to track cells *in vivo*, it must not alter the biological processes within the cell, it should be highly sensitive to small changes in cell distribution and function, and it must be able to reveal molecular and cellular processes throughout an entire study period. Of all the possible luciferases, only a few have been developed and applied as reporter genes.

15.3 Technical Implementation of the Method

15.3.1 Bioluminescence Reporter Genes

The luciferases that are found to be the most useful for molecular imaging are firefly luciferase (*Photinus pyralis* [FLuc]), green or red click beetle luciferase (*Pyrophorus plagiophthalamus* [CBR, CBG]), sea pansy luciferase (*Renilla reniformis* [RLuc]), and copepod luciferase (*Gaussia princeps* [GLuc]). Each of these luciferases has its unique characteristics. Of all these luciferases, FLuc (550 aa, 62 kDa) is one of the best studied and most commonly used luciferases. Its high quantum yield (>88%) is generated by oxidation of reporter substrate D-luciferin and requires the cofactors Mg^{2+}, ATP, and molecular O_2. The substrate D-luciferin is poorly catalyzed in mammalian tissues and thus remains in circulation longer compared to other substrates (Zhao et al. 2004). FLuc is the only known luciferase to be temperature dependent, with a peak emission at 578 nm at 25°C to 612 nm at 37°C (Zhao et al. 2005). However, more thermostable mutant forms of FLuc have been developed (Baggett et al. 2004). The relatively red-shifted emission spectrum of FLuc thus has greater applicability for measuring biological processes in deeper tissue. Recently, a red-shifted mutant form of FLuc was generated by changing a single amino acid. This mutant, called Ppy RE-TS reporter, was described to have an emission maximum of 612 nm, a narrow emission bandwidth, better thermostability, and a half-life of 8.8 h at 37°C versus 0.26 h for the wild type. This reporter has a demonstrated superior imaging performance when compared to the wild-type FLuc and has been used successfully in studies to visualize cancer progression in small animals (Branchini et al. 2007, 2010).

Similar to FLuc, luciferases from the click beetle *Pyrophorus plagiophalamus* also use D-luciferin as a reporter substrate. The light they produce varies from green (530 nm) to red wavelengths (635 nm). Mixed click beetle red and green luciferase fragments enable dual-colored imaging of interacting proteins simultaneously due to the combination of the coding sequences CBG and CBR. These dual imaging techniques may prove useful in providing information on the signal depth and localization in tissue, as well as allowing a detailed study of multiprotein interactions (Zhao et al. 2005; Villalobos et al. 2010).

Distinct from the terrestrial luciferases Fluc are marine luciferases *Renilla reniformis* and *Gaussia princeps*, which use coelenterazine as the reporter substrate. Both RLuc and GLuc catalyze the oxidative decarboxylation of coelenterazine. The reaction only requires O_2 and no ATP or Mg^{2+} is needed, in contrast to FLuc. Hence, these luciferases can be used to image cells independent of their metabolic state (Bhaumik and Gambhir 2002; Raty et al. 2007). RLuc and GLuc have much shorter coding sequences, 936 and 558 bp, than FLuc (1653 bp). The marine luciferases are therefore well suited for applications for which short sequences are required, such as gene transfer and gene expression studies (Kimura et al. 2010). Both RLuc and GLuc catalyze the oxidation of coelenterazine to produce a broad spectrum of light with a blue emission peak at 480 nm, and the emitted light is highly absorbed by tissue chromophores such as hemoglobin. The utility of RLuc and GLuc as reporter genes for *in vivo* studies could therefore benefit from improvements in signal intensity. This has been accomplished by raising the level of light emitted in the longer wavelengths of their spectra. Previous studies have enhanced mutant forms of RLuc that are more stable and produce brighter red-shifted light, making it more resistant to tissue absorption (Loening et al. 2006, 2007). GLuc is different from RLuc in that the former is naturally secreted. Genetically engineered GLuc, through the addition of a CD8 transmembrane domain to the carboxy terminus of the enzyme (extGLuc$^+$), may overcome this limitation, allowing retention of the luciferase construct at the cell surface (Santos et al. 2009). The humanized form of GLuc is more highly expressed in mammalian cells and has a 1000-fold higher bioluminescent signal intensity in mammalian cells compared to humanized FLuc and humanized RLuc, and a 200-fold higher signal *in vivo*. The short coding sequence of humanized GLuc makes it suitable for small viral vectors, such as AAV (Tannous et al. 2005).

The pharmacokinetics of coelenterazine is somewhat different compared to D-luciferin. Coelenterazine binds to serum proteins (Campbell et al. 1989), is prone to quick inactivation due to degradation through autoxidation, and is cleared rapidly from the bloodstream (Zhao et al. 2004). It uses flash kinetics versus glow kinetics for D-luciferin, which rapidly decays over time (Bhaumik and Gambhir 2002). Acquiring a signal immediately after substrate admission *in vivo* is necessary (Zhao et al. 2004). The short half-life of coelenterazine, however, can be advantageous when using both coelenterazine and D-luciferin for dual imaging within the same animal (Zhao et al. 2004; Bhaumik and Gambhir 2002).

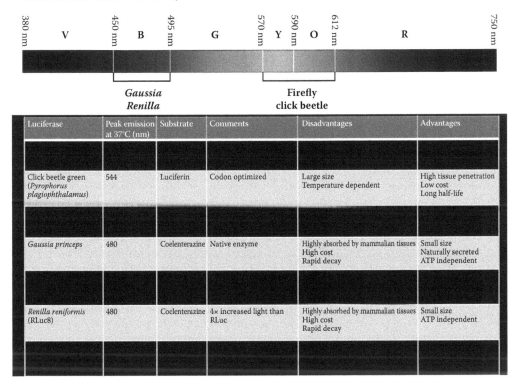

Luciferase	Peak emission at 37°C (nm)	Substrate	Comments	Disadvantages	Advantages
Click beetle green (*Pyrophorus plagiophthalamus*)	544	Luciferin	Codon optimized	Large size Temperature dependent	High tissue penetration Low cost Long half-life
Gaussia princeps	480	Coelenterazine	Native enzyme	Highly absorbed by mammalian tissues High cost Rapid decay	Small size Naturally secreted ATP independent
Renilla reniformis (RLuc8)	480	Coelenterazine	4× increased light than RLuc	Highly absorbed by mammalian tissues High cost Rapid decay	Small size ATP independent

FIGURE 15.2   Emission wavelengths for the most commonly used reporter luciferase enzymes in BLI, peak emission wavelengths at normal body temperature, necessary reporter probes, and their advantages and disadvantages. V = violet, B = blue, Y = yellow, O = orange, R = red.

Chapter 15

A unique luciferase is the bacterial luciferase (Lux) derived from the bacteria-like *Photorhabdus luminescens*. These luciferase reporters emit blue light with a wavelength of 490 nm and are unique in that no substrate addition is needed for whole animal BLI. To fully exploit the advantages of bacterial luciferase, all five genes (lux CDABE) of the lux operon must be expressed simultaneously (Close et al. 2010). The luciferase component is a heterodimer formed from the products of the luxAB genes. The luxCDE genes encode for the enzymatic complex responsible for the synthesis of the reaction's substrate, a long-chain aldehyde. Besides this long-chain aldehyde, molecular oxygen and reduced riboflavin phosphate (FMNH2) are needed; both are naturally occurring products within the cell. To produce light, the luciferase protein first binds FMNH2, followed by O_2 and the synthesized aldehyde.

In summary, many luciferases are known and they vary in the emission spectra, brightness, sensitivity, and stability. Creating mutant forms of known luciferases, and cloning new luciferases from different organisms like *Luciola italica* (Branchini et al. 2006) and *Cratomorphus distinctus* (Viviani et al. 2004), can certainly improve current applications and make novel ones possible. Some of the most utilized luciferases used in BLI are outlined in Figure 15.2.

15.3.2 Instrumentation

To track cells *in vivo* by BLI, the cells of interest need to be genetically modified to express luciferase. After anesthetic induction with 2% isoflurane, the animal recipient of the cells receives the luciferase substrate intravenously or intraperitoneally. The animals (to a maximum of five at a time) are placed in a light-tight chamber, where luminescence is detected by a cooled CCD camera that can detect very low levels of visible light emitted from internal organs. The sensitivity of detection depends on the expression levels of the enzyme in the target cells, the wavelength of light expression, the location of the source of bioluminescence in the animal, the efficiency of the collection optics, and the sensitivity of the detector (Wu et al. 2001).

As far as the sensitivity of the detector is concerned, several optical instrumentations are commercially available, including the IVIS 200 series (Caliper Life Sciences, Hopkinton, MA, USA) and the ORCA-2BT Imaging system (Hamamatsu Photonics, Hamamatsu City, Japan). Region-of-interest (ROI) analysis using Living image (Xenogen) or Wasabi imaging software (Hamamatsu) can be performed for each scan to derive the peak signal in terms of maximum radiance. BLI signals are quantified by the maximum photons per second per centimeter squared per steradian (p/s/cm^2/sr).

15.4 Cardiovascular Disease

Coronary artery disease remains the leading cause of death in the Western world. The human heart is not able to significantly regenerate damaged cells after myocardial infarction, and impaired cardiac function is a common result of ischemic disease. Currently, the only cure for end-stage chronic heart failure is orthotopic heart transplantation (OHT), which is limited by organ shortage and high costs. For this reason, new therapies have been studied extensively over the past decade as a novel therapy to reverse or minimize myocardial injury. Gene therapy is one of these new therapies. However, at present, gene therapy is plagued by suboptimal delivery of genes, poor transfection efficiencies *in vivo*, and inadequate expression of potential therapeutic genes (Hiona and Wu 2008).

BLI is a powerful tool that allows for noninvasive visualization of a variety of biological processes in real time, both *in vitro* and *in vivo*. The information gathered in small animals can be used to design clinical trials that will help advance gene therapy for cardiac repair (de Almeida et al. 2011). Until now, BLI has been used to determine the effects of transgene expression on cardiomyoblast survival and cardiac repair (Kutschka et al. 2006). BLI also has been used to demonstrate that gene transfection efficiency can be improved by treating mice with vascular endothelial growth factor (VEGF) prior to gene delivery (Sato et al. 2009). Delivery of a reporter gene using a novel nonviral minicircle vector significantly improves gene transfection efficiency to the heart, as well as the duration of the transgene expression (Huang et al. 2009). BLI can also visualize the biological role of short hairpin RNA (shRNA) therapy for improving cardiac function. For example, inhibition of prolyl hydroxylase-2 (PHD2) protein by shRNA led to significant improvement in angiogenesis and contractility in *in vitro* and *in vivo* experiments (Huang et al. 2008).

Besides gene therapy, stem cell–based therapies also have generated much interest as a novel potential treatment for ischemic heart diseases. The majority of stem cell therapy studies in small animals have shown that implantation of various cell types (e.g., bone marrow–derived mononuclear cells, mesenchymal stem cells (MSCs), embryonic stem cell–derived cardiac or endothelial cells, circulating progenitor cells, cardiac resident stem cells, and skeletal myoblasts) can provide varying degrees of improvement in terms of reduced infarct size, reversed ventricular remodeling, and improved left ventricular systolic function (Segers and Lee 2008; Hansson et al. 2009). The exact mechanisms by which these stem cells exert their effect remain poorly characterized, but generally have been linked to cells stimulating angiogenesis, secreting growth factors, or recruiting endogenous stem cells for enhanced cardiac repair. However, despite the potential of these cells, many fundamental questions remain unanswered. For instance, do these cells migrate, proliferate, and differentiate? Do the injected cells survive long term? Direct imaging of the cells *in vivo* after implantation into living subjects can offer great insights into the mechanisms of action as well as the survival, distribution, and pharmacokinetics of these cells.

Cell imaging requires the use of a cell marker that ideally (1) generates signal only when the marker is associated with viable cells (high imaging specificity), (2) emits robust signal for detection (high imaging sensitivity), (3) minimally perturbs cellular function (low cytotoxicity), and (4) causes minimal toxicity to the subject when it (or its metabolite) is released into the circulation during excretion (low systemic toxicity) (Chen and Wu 2011). BLI, although not clinically applicable due to the lack of deeper tissue imaging, is an important tool to answer some of the remaining pressing questions by uncovering the dynamics of cell migration, survival, and expansion in small animal studies (Wang et al. 2003; Zhou et al. 2006) For example, several studies have indicated that embryonic stem cell–derived cardiomyocytes (ESC-CMs) can benefit cardiac function following myocardial ischemia (van Laake et al. 2009; Laflamme et al. 2007; Caspi et al. 2007; Cao et al. 2008). However, there are several drawbacks of pluripotent stem cell therapy, including tumorigenicity of undifferentiated cells, immunogenicity from

Chapter 15

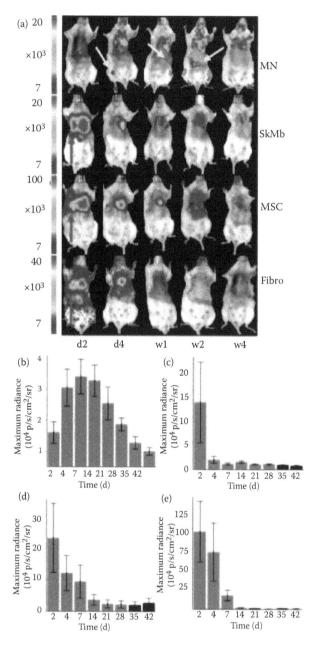

FIGURE 15.3 Representative examples of longitudinal *in vivo* BLI of transplanted cell survival. (a) Images from the same representative animal from each group reveal cell proliferation, death, and migration. Mononuclear cells (MNs) show retention in the heart and can home in on the femur, spleen, and liver (arrows in d4, MN; w1, MN and w2, MN images). BLI images from animals 2 days after injection of skeletal myoblasts (SkMbs), MSCs, and fibroblasts (Fibro) show retention not only in the heart but also in the lungs (arrows in d2, SkMb; d2, MSC and d2, Fibro images). Decreasing signal intensity over time is indicative of acute donor cell death in these groups. Scale bars represent BLI signal in photons/s/cm²/sr. (b) Quantification of BLI signals on fixed regions of interest (ROIs) over the heart reveals an early increase in signal from day 2 until day 7 in the MN group, while signal intensity in the SkMb (c), MSC (d), and Fibro (e) groups clearly decreases to background signal (black bars) at week 3 to 5. Bars represent mean ± SEM. (Reprinted from van der Bogt, K. E. et al., *Circulation,* 118, S121–9, 2008. With permission.)

allogeneic source, and poor survival of differentiated derivatives. BLI has demonstrated that a minimum of 10,000 to 100,000 undifferentiated ESCs are necessary to form a tumor in the heart (Lee et al. 2009). Moreover, >90% of adult stem cells or differentiated pluripotent stem cells will die within the first 4 weeks following delivery (van der Bogt et al. 2008; Li et al. 2007, 2009; Sheikh et al. 2010). BLI demonstrating a similar cell death pattern involving adult stem cells is displayed in Figure 15.3. The lack of long-term survival of transplanted adult stem cells or differentiated pluripotent stem cells might be a reason for the short duration of cardiac function improvement observed in these studies. Overall, the use of BLI in small animal studies has provided significant insights into the biology of stem cells *in vivo*.

15.5 Cancer

Traditionally, the use of BLI in cancer research allows for repetitive, sensitive, real-time monitoring of neoplastic cells and the assessment of *in vivo* effectiveness of antitumor interventions. It has been used to image the mass and location of xenografted cells expressing a luciferase gene in mice (Runnels et al. 2011), spontaneous tumor growth in transgenic mice (Vooijs et al. 2002; Lyons et al. 2006), and assessment of temporal tumor growth (Edinger et al. 2002). BLI has also been used to assess the tumorigenicity of cell lines and to monitor metastasis and response to chemotherapy (Jenkins et al. 2003). The ability to monitor tumors pretreatment and posttreatment over time in the same experimental subject significantly improves the overall quality of data obtained and reduces the number of subjects necessary to generate them. BLI can be used to track human multiple myeloma cells expressing both bioluminescent and fluorescent reporters while assessing tumor growth and chemotherapy response in a mouse xenograft model (Runnels et al. 2011). Here the BLI data demonstrated tumor growth in mice with characteristics of human multiple myeloma disease, showing a distribution of cancer cells in the hips, femurs, vertebrae, ribs, and skull without the need to sacrifice multiple animals at multiple time points (Runnels et al. 2011). In another study, BLI was used to demonstrate metastasis of engineered human tumor cell lines in subcutaneous, intravenous, and spontaneous metastasis models (Jenkins et al. 2003). Bioluminescent PC-3M-luc-C6 human prostate cancer cells were subcutaneously injected into SCID-beige mice and were monitored for tumor growth and response to treatment. The BLI data correlated well with total lung weight at necropsy. Overall, these studies show that BLI provides rapid, noninvasive monitoring of tumor growth and regression in animals, and its application to previously used oncology animal models offers quantitative and sensitive analysis of tumor growth and metastasis (Runnels et al. 2011; Jenkins et al. 2003).

The role of angiogenesis is indisputable in relation to tumor growth. Various imaging modalities have been used to assess tumor angiogenesis. Although currently impractical for use in patients, BLI has great utility for monitoring tumorigenesis in animal models of disease (Sun et al. 2010). Many studies have focused on antiangiogenic strategies as a way to inhibit tumor growth and formation, including long-term serial analysis of the molecular events involved in the angiogenic process and monitoring of responses to antiangiogenic agents (Sanz et al. 2008). Here BLI was used to quantify angiogenesis via coimplantation of human primary endothelial cells expressing FLuc and with human

Chapter 15

bone marrow MSCs in immunodeficient mice. The investigators found that FLuc activity correlated with the formation of a network of functional, mature blood vessels of human nature inside the implant that critically depends on the presence of MSCs (Sanz et al. 2008).

Anticancer therapies have been developed in the field of regenerative medicine. In 2002, a group of investigators showed that the tumor microenvironment preferentially promotes the engraftment of MSCs, and MSCs expressing IFN-β can inhibit tumor growth by releasing their therapeutic payload locally (Studeny et al. 2002). MSCs are resistant to the cytokine tumor necrosis factor apoptosis ligand (TRAIL), and when engineered to express secreted recombinant TRAIL, they can induce caspase-mediated apoptosis in established glioma cell lines *in vitro*. BLI has been used to visualize tumor load before, during, and after MSC-TRAIL cell treatment, which resulted in killing of both residual and invasive tumor cells (Sasportas et al. 2009). In addition, using a murine lung metastasis model, MSCs expressing TRAIL have the ability to seek out metastases at sites far removed from the primary tumor, which resulted in complete clearance of metastases in 38% of the mice compared to 0% controls (Loebinger et al. 2009).

15.6 Transplant Immunology

In the area of immunology, BLI has emerged as a valuable imaging strategy with the potential to elucidate the temporal patterns of immune responses and the spatial distribution of lymphocytes within the body, and to probe endogenous immune response signaling pathways. BLI has repeatedly demonstrated effectiveness in murine models of graft versus host disease (GVHD) to monitor the location and activity of engrafted T cells *in vivo*. Conventional techniques such as flow cytometry and immunofluorescence have been used to monitor location and activity of cell populations in tissues at a given time point, but these methods cannot assess dynamic cell migration patterns *in vivo*. Monitoring the localization and timing of critical immunological events *in vivo* has revealed basic pathogenic mechanisms. For example, luciferase-labeled allogeneic splenocytes have been transplanted and their tissue distribution monitored by BLI (Beilhack et al. 2005). High-resolution analyses showed initial proliferation of donor CD4$^+$ T cells followed by CD8$^+$ T cells in secondary lymphoid organs with subsequent homing to the intestines, liver, and skin. Transplantation of purified naive T cells caused GVHD that was initiated in secondary lymphoid organs followed by target organ manifestation in gut, liver, and skin. By contrast, the investigators found that transplanted CD4$^+$ effector memory T cells did not proliferate in secondary lymphoid organs *in vivo* and, despite their *in vitro* alloreactivity in mixed leukocyte reaction assays, did not cause acute GVHD. These results demonstrated the potential of BLI for visualizing trafficking patterns of immunologic cells and responses *in vivo* (Beilhack et al. 2005).

The continuing progress in regenerative medicine has provided novel transplant therapies for disease. Diabetes research has made recent headway in its pursuit of a cure with promising pancreatic islet cell transplantation. The difficulty in assessing pancreatic islet cells after transplantation in preclinical models can be alleviated by the introduction of BLI to the field. With the ability to follow islet cell transplantations *in vivo*, interventions to improve islet survival can emerge. Fowler et al. (2005) transplanted luciferase-expressing murine or human islets into intrahepatic or renal subcapsular of immunodeficient mice and quantified using BLI. Cell signal of injected pancreatic islet cells was stable for greater than 8 weeks posttransplantation. The authors concluded that BLI allows for quantitative, serial measurements of pancreatic islet mass after transplantation and should be useful in assessing interventions to sustain or increase islet survival of transplanted islets (Fowler et al. 2005). Cao et al. (2005) showed that pancreatic islet cell transplants could be monitored via BLI and elucidated the dynamic range of the entire process of transplantation. BLI was used to noninvasively visualize engraftment, survival, and rejection of transplanted tissues from a transgenic donor mouse that constitutively expressed luciferase. Dynamic early events of hematopoietic reconstitution were accessible, and engraftment from as few as 200 transplanted whole bone marrow cells resulted in bioluminescent foci in lethally irradiated, syngeneic recipients. The transplantation of autologous pancreatic Langerhans islets and of allogeneic heart revealed the tempo of transplant degeneration or immune rejection over time (Cao et al. 2005).

BLI has also proven useful in the study of nuclear factor-kappa B (NF-kB) pathway signaling. NF-kB is a transcription factor that plays a critical upstream role in coordinating the inflammatory and wound healing cascades by initiating the transcription of certain cytokines, chemokines, adhesion molecules, and proinflammatory genes (Roth et al. 2006). Monitoring expression of NF-kB provides insight into the immune response. Roth et al. (2006) derived transgenic mice expressing the proximal 5′ human immunodeficiency virus long terminal repeat (referred to as HIV-LTR/Luciferase, HLL mice), a known NF-kB–dependent promoter sequence. In this transgenic mouse model, when NF-kB binds to its promoter sequence to initiate transcription of immune response proteins, luciferase enzyme is also produced. Therefore, the amount of signal absorbed by BLI is indicative of NF-kB activity. Imaging showed statistically significant differences between preoperative and postoperative data points per mouse, as well as differences in NF-kB activity between transplant groups over a 6-week period (Roth et al. 2006). Increasing our understanding of NF-kB may lead to future therapeutic gains and a better understanding of the immune response. Future studies showing how to downgrade NF-kB may lead to a decreased inflammatory response, which in turn may aid other studies within the transplantation field of regenerative medicine research.

15.7   Central Nervous System

BLI has also been used in central nervous system research such as in studying gene expression in the brain. Pike et al. (2006) described a herpes simplex virus (HSV) amplicon vector-based system that allows for fast, noninvasive, semiquantitative analysis of gene expression in the brain. An HSV amplicon vector expressing FLuc cDNA from an inducible promoter was inserted into cells, then the vectors were stereotactically injected into the mouse brain, and FLuc expression was measured noninvasively using BLI. Rapamycin-mediated induction of FLuc from an HSV amplicon vector in culture resulted in dose-dependent expression of FLuc as measured by a luminometer and digital analysis. In the mouse cortex, a single injection of an HSV amplicon vector [2 µL, 1×10^8 transducing units (t.u.)/mL] expressing FLuc from a viral cytomegalovirus promoter was sufficient to detect robust luciferase activity for at least 1 week. Similarly, an HSV amplicon vector expressing FLuc under an inducible promoter was also detectable in the mouse cortex after a single dose (2 μL, 1×10^8 t.u./mL) for up to 5 days, with no detectable signal in the uninduced state (Pike et al. 2006).

BLI has also been used to examine the transcriptional activity of estrogen receptors in the brain of male and female mice in order to help answer the question of estrogen receptor differences between sexes and sites of activity. Exploiting the estrogen response element-luciferase reporter mouse, Stell et al. (2008) used BLI to study the brain's estrogen receptor transcriptional activity. The authors were able to demonstrate that estrogen receptors are similarly active in male and female brains and that the estrous cycle affects estrogen receptor activity in regions of the central nervous system not known to be associated with reproductive functions, such as the limbic area (Stell et al. 2008).

The multipotent C17.2 neuroprogenitor cells (NPCs) have been widely studied in different brain diseases. For tracking of neuronal cell migration, Pineda et al. (2007) used BLI to demonstrate that these cells were able to migrate from their injection site in the frontal lobe of the brain in quinolinate-lesioned nude mouse striata toward an orthotopic glioblastoma tumor in the contralateral frontal lobe within 2 weeks of transplantation in a Huntington's disease mouse model. Injection of the glial cell line-derived neurotrophic factor and the firefly luciferase gene (GDNF/Luc) cells into the ventricle contralateral to the glioblastoma yielded even more cells migrating into the tumor, whereas intravenous and intraperitoneal injections yielded markedly less and no tumor infiltration at all, respectively. Implanted NPCs showed migratory behavior, homing in on the granular and glomerular cell layers of the olfactory bulb when they were grafted in the rostral migratory stream, which is the natural migratory pathway for NPCs of the subventricular zone (Pineda et al. 2007). Shah et al. (2005) have used C17.2 derived mouse NPCs expressing TRAIL to treat gliomas in mice brains. BLI has shown that the NPCs expressing TRAIL can migrate into the gliomas and result in considerable reduction in growth of highly malignant gliomas. Using such a cell line with a greater apoptosis-inducing ability in combination with real-time BLI will bring therapeutic advances in treating brain tumors in the future.

15.8 Conclusions

In summary, BLI is an efficient, cost-effective, and highly sensitive technology that has made possible the *in vivo* study of a plethora of disease processes across myriad biomedical specialties. Recent advancements in three-dimensional BLI are further improving this exciting imaging technique (Zinn et al. 2008). The capability to monitor a process of interest longitudinally in the same experimental animal is a tremendous advantage of BLI over many other imaging modalities. The longitudinal analysis of disease progression or cellular activity may help to explain the pathophysiology of disease. The use of BLI in cellular and small animal studies has provided valuable results, paving the way for preclinical and translational research in the field of regenerative medicine. The use of BLI has also helped researchers better understand crucial biological processes and transplanted cell behavior within different fields of regenerative medicine. It has greatly aided our understanding of and ability to visualize immune responses in the transplant immunology field, enabling the development of new anticancer therapies based on anti-angiogenic strategies and transplanted MSCs and providing vital information about cell migration, survival, and expansion in regenerative medicine research.

Acknowledgments

We thank Amy Morris for preparing the illustrations. Due to space limitations, we are unable to include all of the important papers relevant to BLI and its applications; we apologize to those investigators whose work we omitted here. This work was supported by NIH HL099117 and NIH EB009689 (JCW).

References

Baggett, B., Roy, R., Momen, S., Morgan, S., Tisi, L., Morse, D. & Gillies, R. J. 2004. Thermostability of firefly luciferases affects efficiency of detection by *in vivo* bioluminescence. *Mol Imaging*, 3, 324–32.

Beilhack, A., Schulz, S., Baker, J., Beilhack, G. F., Wieland, C. B., Herman, E. I., Baker, E. M., Cao, Y. A., Contag, C. H. & Negrin, R. S. 2005. *In vivo* analyses of early events in acute graft-versus-host disease reveal sequential infiltration of T-cell subsets. *Blood*, 106, 1113–22.

Bhaumik, S. & Gambhir, S. S. 2002. Optical imaging of Renilla luciferase reporter gene expression in living mice. *Proc Natl Acad Sci U S A*, 99, 377–82.

Branchini, B. R., Southworth, T. L., Deangelis, J. P., Roda, A. & Michelini, E. 2006. Luciferase from the Italian firefly Luciola italica: molecular cloning and expression. *Comp Biochem Physiol B Biochem Mol Biol*, 145, 159–67.

Branchini, B. R., Ablamsky, D. M., Murtiashaw, M. H., Uzasci, L., Fraga, H. & Southworth, T. L. 2007. Thermostable red and green light-producing firefly luciferase mutants for bioluminescent reporter applications. *Anal Biochem*, 361, 253–62.

Branchini, B. R., Ablamsky, D. M., Davis, A. L., Southworth, T. L., Butler, B., Fan, F., Jathoul, A. P. & Pule, M. A. 2010. Red-emitting luciferases for bioluminescence reporter and imaging applications. *Anal Biochem*, 396, 290–7.

Campbell, A. K., Patel, A., Houston, W. A., Scolding, N. J., Frith, S., Morgan, B. P. & Compston, D. A. 1989. Photoproteins as indicators of intracellular free Ca^{2+}. *J Biolumin Chemilumin*, 4, 463–74.

Cao, Y. A., Bachmann, M. H., Beilhack, A., Yang, Y., Tanaka, M., Swijnenburg, R. J., Reeves, R., Taylor-Edwards, C., Schulz, S., Doyle, T. C., Fathman, C. G., Robbins, R. C., Herzenberg, L. A., Negrin, R. S. & Contag, C. H. 2005. Molecular imaging using labeled donor tissues reveals patterns of engraftment, rejection, and survival in transplantation. *Transplantation*, 80, 134–9.

Cao, F., Wagner, R. A., Wilson, K. D., Xie, X., Fu, J. D., Drukker, M., Lee, A., Li, R. A., Gambhir, S. S., Weissman, I. L., Robbins, R. C. & Wu, J. C. 2008. Transcriptional and functional profiling of human embryonic stem cell-derived cardiomyocytes. *PLoS One*, 3, e3474.

Caspi, O., Huber, I., Kehat, I., Habib, M., Arbel, G., Gepstein, A., Yankelson, L., Aronson, D., Beyar, R. & Gepstein, L. 2007. Transplantation of human embryonic stem cell-derived cardiomyocytes improves myocardial performance in infarcted rat hearts. *J Am Coll Cardiol*, 50, 1884–93.

Chen, I. Y. & Wu, J. C. 2011. Cardiovascular molecular imaging: focus on clinical translation. *Circulation*, 123, 425–43.

Close, D. M., Patterson, S. S., Ripp, S., Baek, S. J., Sanseverino, J. & Sayler, G. S. 2010. Autonomous bioluminescent expression of the bacterial luciferase gene cassette (lux) in a mammalian cell line. *PLoS One*, 5, e12441.

Contag, C. H., Contag, P. R., Mullins, J. I., Spilman, S. D., Stevenson, D. K. & Benaron, D. A. 1995. Photonic detection of bacterial pathogens in living hosts. *Mol Microbiol*, 18, 593–603.

de Almeida, P. E., van Rappard, J. R. & Wu, J. C. 2011. *In vivo* bioluminescence for tracking cell fate and function. *Am J Physiol Heart Circ Physiol*, 301(3), H663–71.

de Wet, J. R., Wood, K. V., Helinski, D. R. & Deluca, M. 1985. Cloning of firefly luciferase cDNA and the expression of active luciferase in Escherichia coli. *Proc Natl Acad Sci U S A*, 82, 7870–3.

de Wet, J. R., Wood, K. V., Deluca, M., Helinski, D. R. & Subramani, S. 1987. Firefly luciferase gene: structure and expression in mammalian cells. *Mol Cell Biol*, 7, 725–37.

Edinger, M., Cao, Y. A., Hornig, Y. S., Jenkins, D. E., Verneris, M. R., Bachmann, M. H., Negrin, R. S. & Contag, C. H. 2002. Advancing animal models of neoplasia through *in vivo* bioluminescence imaging. *Eur J Cancer*, 38, 2128–36.

Fowler, M., Virostko, J., Chen, Z., Poffenberger, G., Radhika, A., Brissova, M., Shiota, M., Nicholson, W. E., Shi, Y., Hirshberg, B., Harlan, D. M., Jansen, E. D. & Powers, A. C. 2005. Assessment of pancreatic islet mass after islet transplantation using *in vivo* bioluminescence imaging. *Transplantation*, 79, 768–76.

Chapter 15

Gould, S. J. & Subramani, S. 1988. Firefly luciferase as a tool in molecular and cell biology. *Anal Biochem*, 175, 5–13.

Hansson, E. M., Lindsay, M. E. & Chien, K. R. 2009. Regeneration next: toward heart stem cell therapeutics. *Cell Stem Cell*, 5, 364–77.

Hiona, A. & Wu, J. C. 2008. Noninvasive radionuclide imaging of cardiac gene therapy: progress and potential. *Nat Clin Pract Cardiovasc Med*, 5 Suppl 2, S87–95.

Huang, M., Chen, Z., Hu, S., Jia, F., Li, Z., Hoyt, G., Robbins, R. C., Kay, M. A. & Wu, J. C. 2009. Novel minicircle vector for gene therapy in murine myocardial infarction. *Circulation*, 120, S230–7.

Huang, M., Chan, D. A., Jia, F., Xie, X., Li, Z., Hoyt, G., Robbins, R. C., Chen, X., Giaccia, A. J. & Wu, J. C. 2008. Short hairpin RNA interference therapy for ischemic heart disease. *Circulation*, 118, S226–33.

Jenkins, D. E., Oei, Y., Hornig, Y. S., Yu, S. F., Dusich, J., Purchio, T. & Contag, P. R. 2003. Bioluminescent imaging (BLI) to improve and refine traditional murine models of tumor growth and metastasis. *Clin Exp Metastasis*, 20, 733–44.

Kimura, T., Hiraoka, K., Kasahara, N. & Logg, C. R. 2010. Optimization of enzyme-substrate pairing for bioluminescence imaging of gene transfer using Renilla and Gaussia luciferases. *J Gene Med*, 12, 528–37.

Kutschka, I., Kofidis, T., Chen, I. Y., von Degenfeld, G., Zwierzchoniewska, M., Hoyt, G., Arai, T., Lebl, D. R., Hendry, S. L., Sheikh, A. Y., Cooke, D. T., Connolly, A., Blau, H. M., Gambhir, S. S. & Robbins, R. C. 2006. Adenoviral human BCL-2 transgene expression attenuates early donor cell death after cardiomyoblast transplantation into ischemic rat hearts. *Circulation*, 114, I174–80.

Laflamme, M. A., Chen, K. Y., Naumova, A. V., Muskheli, V., Fugate, J. A., Dupras, S. K., Reinecke, H., Xu, C., Hassanipour, M., Police, S., O'Sullivan, C., Collins, L., Chen, Y., Minami, E., Gill, E. A., Ueno, S., Yuan, C., Gold, J. & Murry, C. E. 2007. Cardiomyocytes derived from human embryonic stem cells in pro-survival factors enhance function of infarcted rat hearts. *Nat Biotechnol*, 25, 1015–24.

Lee, A. S., Tang, C., Cao, F., Xie, X., van der Bogt, K., Hwang, A., Connolly, A. J., Robbins, R. C. & Wu, J. C. 2009. Effects of cell number on teratoma formation by human embryonic stem cells. *Cell Cycle*, 8, 2608–12.

Li, Z., Wu, J. C., Sheikh, A. Y., Kraft, D., Cao, F., Xie, X., Patel, M., Gambhir, S. S., Robbins, R. C. & Cooke, J. P. 2007. Differentiation, survival, and function of embryonic stem cell derived endothelial cells for ischemic heart disease. *Circulation*, 116, 146–54.

Li, Z., Lee, A., Huang, M., Chun, H., Chung, J., Chu, P., Hoyt, G., Yang, P., Rosenberg, J., Robbins, R. C. & Wu, J. C. 2009. Imaging survival and function of transplanted cardiac resident stem cells. *J Am Coll Cardiol*, 53, 1229–40.

Loebinger, M. R., Eddaoudi, A., Davies, D. & Janes, S. M. 2009. Mesenchymal stem cell delivery of TRAIL can eliminate metastatic cancer. *Cancer research*, 69, 4134–42.

Loening, A. M., Fenn, T. D., Wu, A. M. & Gambhir, S. S. 2006. Consensus guided mutagenesis of Renilla luciferase yields enhanced stability and light output. *Protein Eng Des Sel*, 19, 391–400.

Loening, A. M., Wu, A. M. & Gambhir, S. S. 2007. Red-shifted Renilla reniformis luciferase variants for imaging in living subjects. *Nat Methods*, 4, 641–3.

Luker, K. E. & Luker, G. D. 2008. Applications of bioluminescence imaging to antiviral research and therapy: multiple luciferase enzymes and quantitation. *Antiviral Res*, 78, 179–87.

Lyons, S. K., Lim, E., Clermont, A. O., Dusich, J., Zhu, L., Campbell, K. D., Coffee, R. J., Grass, D. S., Hunter, J., Purchio, T. & Jenkins, D. 2006. Noninvasive bioluminescence imaging of normal and spontaneously transformed prostate tissue in mice. *Cancer Res*, 66, 4701–7.

Paroo, Z., Bollinger, R. A., Braasch, D. A., Richer, E., Corey, D. R., Antich, P. P. & Mason, R. P. 2004. Validating bioluminescence imaging as a high-throughput, quantitative modality for assessing tumor burden. *Mol Imaging*, 3, 117–24.

Pike, L., Petravicz, J. & Wang, S. 2006. Bioluminescence imaging after HSV amplicon vector delivery into brain. *J Gene Med*, 8, 804–13.

Pineda, J. R., Rubio, N., Akerud, P., Urban, N., Badimon, L., Arenas, E., Alberch, J., Blanco, J. & Canals, J. M. 2007. Neuroprotection by GDNF-secreting stem cells in a Huntington's disease model: optical neuroimage tracking of brain-grafted cells. *Gene Ther*, 14, 118–28.

Raty, J. K., Liimatainen, T., Unelma Kaikkonen, M., Grohn, O., Airenne, K. J. & Yla-Herttuala, S. 2007. Noninvasive imaging in gene therapy. *Mol Ther*, 15, 1579–86.

Roth, D. J., Jansen, E. D., Powers, A. C. & Wang, T. G. 2006. A novel method of monitoring response to islet transplantation: bioluminescent imaging of an NF-kB transgenic mouse model. *Transplantation*, 81, 1185–90.

Runnels, J. M., Carlson, A. L., Pitsillides, C., Thompson, B., Wu, J., Spencer, J. A., Kohler, J. M., Azab, A., Moreau, A. S., Rodig, S. J., Kung, A. L., Anderson, K. C., Ghobrial, I. M. & Lin, C. P. 2011. Optical techniques for tracking multiple myeloma engraftment, growth, and response to therapy. *J Biomed Opt*, 16, 011006.

Santos, E. B., Yeh, R., Lee, J., Nikhamin, Y., Punzalan, B., La Perle, K., Larson, S. M., Sadelain, M. & Brentjens, R. J. 2009. Sensitive *in vivo* imaging of T cells using a membrane-bound Gaussia princeps luciferase. *Nat Med*, 15, 338–44.

Sanz, L., Santos-Valle, P., Alonso-Camino, V., Salas, C., Serrano, A., Vicario, J. L., Cuesta, A. M., Compte, M., Sanchez-Martin, D. & Alvarez-Vallina, L. 2008. Long-term *in vivo* imaging of human angiogenesis: critical role of bone marrow-derived mesenchymal stem cells for the generation of durable blood vessels. *Microvasc Res*, 75, 308–14.

Sasportas, L. S., Kasmieh, R., Wakimoto, H., Hingtgen, S., van de Water, J. A., Mohapatra, G., Figueiredo, J. L., Martuza, R. L., Weissleder, R. & Shah, K. 2009. Assessment of therapeutic efficacy and fate of engineered human mesenchymal stem cells for cancer therapy. *Proceedings of the National Academy of Sciences of the United States of America*, 106, 4822–7.

Sato, T., Ramsubir, S., Higuchi, K., Yanagisawa, T. & Medin, J. A. 2009. Vascular endothelial growth factor broadens lentivector distribution in the heart after neonatal injection. *J Cardiol*, 54, 245–54.

Segers, V. F. & Lee, R. T. 2008. Stem-cell therapy for cardiac disease. *Nature*, 451, 937–42.

Shah, K., Bureau, E., Kim, D. E., Yang, K., Tang, Y., Weissleder, R. & Breakefield, X. O. 2005. Glioma therapy and real-time imaging of neural precursor cell migration and tumor regression. *Annals of Neurology*, 57, 34–41.

Sheikh, A. Y., van der Bogt, K. E., Doyle, T. C., Sheikh, M. K., Ransohoff, K. J., Ali, Z. A., Palmer, O. P., Robbins, R. C., Fischbein, M. P. & Wu, J. C. 2010. Micro-CT for characterization of murine CV disease models. *JACC Cardiovasc Imaging*, 3, 783–5.

Stell, A., Belcredito, S., Ciana, P. & Maggi, A. 2008. Molecular imaging provides novel insights on estrogen receptor activity in mouse brain. *Mol Imaging*, 7, 283–92.

Studeny, M., Marini, F. C., Champlin, R. E., Zompetta, C., Fidler, I. J. & Andreeff, M. 2002. Bone marrow-derived mesenchymal stem cells as vehicles for interferon-beta delivery into tumors. *Cancer Res*, 62, 3603–8.

Sun, A., Hou, L., Prugpichailers, T., Dunkel, J., Kalani, M. A., Chen, X., Kalani, M. Y. & Tse, V. 2010. Firefly luciferase-based dynamic bioluminescence imaging: a noninvasive technique to assess tumor angiogenesis. *Neurosurgery*, 66, 751–7; discussion 757.

Tannous, B. A., Kim, D. E., Fernandez, J. L., Weissleder, R. & Breakefield, X. O. 2005. Codon-optimized Gaussia luciferase cDNA for mammalian gene expression in culture and *in vivo*. *Mol Ther*, 11, 435–43.

van der Bogt, K. E., Sheikh, A. Y., Schrepfer, S., Hoyt, G., Cao, F., Ransohoff, K. J., Swijnenburg, R. J., Pearl, J., Lee, A., Fischbein, M., Contag, C. H., Robbins, R. C. & Wu, J. C. 2008. Comparison of different adult stem cell types for treatment of myocardial ischemia. *Circulation*, 118, S121–9.

van Laake, L. W., Passier, R., den Ouden, K., Schreurs, C., Monshouwer-Kloots, J., Ward-Van Oostwaard, D., van Echteld, C. J., Doevendans, P. A. & Mummery, C. L. 2009. Improvement of mouse cardiac function by hESC-derived cardiomyocytes correlates with vascularity but not graft size. *Stem Cell Res*, 3, 106–12.

Villalobos, V., Naik, S., Bruinsma, M., Dothager, R. S., Pan, M. H., Samrakandi, M., Moss, B., Elhammali, A. & Piwnica-Worms, D. 2010. Dual-color click beetle luciferase heteroprotein fragment complementation assays. *Chem Biol*, 17, 1018–29.

Viviani, V. R., Arnoldi, F. G., Brochetto-Braga, M. & Ohmiya, Y. 2004. Cloning and characterization of the cDNA for the Brazilian Cratomorphus distinctus larval firefly luciferase: similarities with European Lampyris noctiluca and Asiatic Pyrocoelia luciferases. *Comp Biochem Physiol B Biochem Mol Biol*, 139, 151–6.

Vooijs, M., Jonkers, J., Lyons, S. & Berns, A. 2002. Noninvasive imaging of spontaneous retinoblastoma pathway-dependent tumors in mice. *Cancer Res*, 62, 1862–7.

Wang, X., Rosol, M., Ge, S., Peterson, D., Mcnamara, G., Pollack, H., Kohn, D. B., Nelson, M. D. & Crooks, G. M. 2003. Dynamic tracking of human hematopoietic stem cell engraftment using *in vivo* bioluminescence imaging. *Blood*, 102, 3478–82.

Weissleder, R. 2001. A clearer vision for *in vivo* imaging. *Nat Biotechnol*, 19, 316–7.

Widder, E. A. 2010. Bioluminescence in the ocean: origins of biological, chemical, and ecological diversity. *Science*, 328, 704–8.

Wu, J. C., Sundaresan, G., Iyer, M. & Gambhir, S. S. 2001. Noninvasive optical imaging of firefly luciferase reporter gene expression in skeletal muscles of living mice. *Mol Ther*, 4, 297–306.

Chapter 15

Zhao, H., Doyle, T. C., Wong, R. J., Cao, Y., Stevenson, D. K., Piwnica-Worms, D. & Contag, C. H. 2004. Characterization of coelenterazine analogs for measurements of Renilla luciferase activity in live cells and living animals. *Mol Imaging*, 3, 43–54.

Zhao, H., Doyle, T. C., Coquoz, O., Kalish, F., Rice, B. W. & Contag, C. H. 2005. Emission spectra of bioluminescent reporters and interaction with mammalian tissue determine the sensitivity of detection *in vivo*. *J Biomed Opt*, 10, 41210.

Zhou, R., Acton, P. D. & Ferrari, V. A. 2006. Imaging stem cells implanted in infarcted myocardium. *J Am Coll Cardiol*, 48, 2094–106.

Zinn, K. R., Chaudhuri, T. R., Szafran, A. A., O'Quinn, D., Weaver, C., Dugger, K., Lamar, D., Kesterson, R. A., Wang, X. & Frank, S. J. 2008. Noninvasive bioluminescence imaging in small animals. *ILAR J*, 49, 103–15.

Index

Page numbers followed by f and t indicate figures and tables, respectively.

Milton Keynes UK
Ingram Content Group UK Ltd.
UKHW052029141024
449569UK00017B/753